ELECTRICITY, ELECTRONICS, AND CONTROL SYSTEMS FOR HVAC

Fourth Edition

Thomas E. Kissell
Terra Community College

PEARSON
Prentice
Hall

Upper Saddle River, New Jersey
Columbus, Ohio

Library of Congress Cataloging-in-Publication Data

Kissell, Thomas E.
 Electricity, electronics, and control systems for HVAC/Thomas E. Kissell.—4th ed.
 p. cm.
 ISBN-13: 978-0-13-199568-0
 ISBN-10: 0-13-199568-5
 1. Buildings—Electric equipment. 2. Heating—Control. 3. Ventilation—Control.
4. Air conditoning—Control. 5. Electric controllers. I. Title.
 ATK4035, A35K568 2008
 621. 319'24—dc22

 2007036864

Editor in Chief: Vernon Anthony
Editor: Eric Krassow
Associate Managing Editor: Christine Buckendahl
Editorial Assistant: Nancy Kesterson
Production Coordination: Rebecca K. Giusti, GGS Book Services
Project Manager: Holly Shufeldt
Design Coordinator: Diane Ernsberger
Senior Operations Specialist: Deidra Schwartz
Text Designer: Kristina Holmes
Director of Marketing: David Gesell
Executive Marketing Manager: Derril Trakalo
Marketing Assistant: Les Roberts

This book was set in Times by GGS Book Services. It was printed and bound by LSC Communications. The cover was printed by LSC Communications.

Pearson Education Ltd.
Pearson Education Singapore Pte. Ltd.
Pearson Education Canada, Ltd.
Pearson Education—Japan

Pearson Education Australia Pty. Limited
Pearson Education North Asia Ltd.
Pearson Educación de Mexico, S.A. de C.V.
Pearson Education Malaysia Pte. Ltd.

ISBN-13: 978-0-13-199568-0
ISBN-10: 0-13-199568-5

13 2020

To Kathy,
for all her support

Preface

Electricity, Electronics, and Control Systems for HVAC was written to help students understand how to install, troubleshoot, and repair electrical parts of air-conditioning, heating, and refrigeration systems. Electricity appears to be a difficult subject to understand for a beginning student, but this book has a long history of making the subject much easier to comprehend. The author uses over 20 years of experience to simplify electrical theory and practical applications so that vocational/technical students can quickly and easily learn how to install, troubleshoot, and repair HVAC and refrigeration systems. The overall objective of this book is to give students the necessary electrical skills to obtain a job and be successful on the job.

The author understands that a majority of students are visual learners, so this book uses an extensive number of pictures and cut-away diagrams of electrical components as well as electrical diagrams to explain the theory of operation for all components and show test points for troubleshooting. This text is designed to be simple enough for the beginner yet comprehensive enough to be a complete reference when the student finishes school and needs to review a troubleshooting technique or theory of operation for a control when on the job.

How This Book Is Organized

The book is organized so it can easily be used in heating, air-conditioning, and refrigeration programs in a wide variety of vocational/technical educational institutions. The content begins with the basics of electricity and how individual components operate. Each section increases in complexity until students have mastered installation, troubleshooting, and repair of systems. The book goes into great detail to explain the theory of operation of all electrical components as well as present detailed pictures and diagrams to help students see what they are learning. Students are shown how to use electrical schematics and wiring diagrams for installation, troubleshooting, and repair.

Important Features of This Edition

Many new features have been added to this edition to make it a stronger text for both students and instructors. Each chapter begins with a **Job Assignment**. The Job Assignment gives students a task that they would be expected to do on the job that relates directly to the subject for each chapter. The **Solution to the Job Assignment** is provided at the end of each chapter so that students can see what they are expected to do. **Objectives** are also provided at the beginning of each chapter so students know exactly what they will be learning. **Terms You Will Learn in This Chapter** are provided at the beginning of each chapter. The definition of each term is provided in the **Glossary** at the

end of the text so students can quickly look up any term. The material in each chapter is introduced in simplest terms in the beginning of the chapter and becomes more complex throughout the chapter as students build on simple things they have learned. Important safety notices are highlighted throughout each chapter to ensure that students learn to work safely around electrical circuits.

At the end of each chapter is a review that helps summarize all the important topics in the chapter. A complete set of questions is provided at the end of each chapter and includes a variety of question types, including multiple choice, true/false, open-ended questions, and problems, to solve. All of the questions are also available in an electronic test bank so students can take the test by paper and pencil or electronically. Additional job assignments are provided throughout the chapters to focus students on learning job skills.

Each chapter has a large number of electrical schematic and wiring diagrams to explain the operation of individual components as well as wide variety of HVAC and refrigeration systems. Chapter 6 provides a detailed explanation of how to read and write schematic (ladder) diagrams and wiring diagrams, which is one of the most important skills a student must have to be successful in this field.

THINGS THAT ARE NEW IN THIS EDITION

The chapters in this book have all been revised and new photographs and diagrams have been added to each. Chapter 1 covers an introduction to safety and provides an understanding of how to work safely around electricity, which is one of the greatest concerns for new students. Students can carry the important lessons learned in this chapter into their labs and onto the job. Chapters 2 through 4 provide a comprehensive introduction to basic electricity and reading meters. Chapters 5 and 6 have been strengthened to provide students with the important skills of reading and writing electrical schematic and wiring diagrams. Many texts fail to show students how to read diagrams, which is the prerequisite to learning how to effectively troubleshoot.

Chapters 7 and 8 provide all the information students need to understand magnetics and installing electrical systems. Chapters 9, 10, and 11 provide an in-depth study of the operation of motors and how to troubleshoot them as well as installation practices. Many students end up on the job with just minimal understanding of motors, yet the majority of their time on the job includes installing and troubleshooting motors and compressors.

Chapters 12 through 17 provide an in-depth presentation of the controls used in all types of HVAC and refrigeration systems. These chapters have been organized so that students can learn the controls as they apply to each system. This material can be quickly found for review for later use. Chapter 18 provides a comprehensive overview of how to troubleshoot all of these systems, which allows students to "put it all together" and learn the important skills needed for success on the job.

Chapters 19 and 20 provide a look into the future. Today modern HVAC and refrigeration systems are designed to have the highest efficiency ratings, which means they are using more electronic and computer control devices and systems. These chapters were specifically written to help students entering the HVAC field understand that the electronic devices are similar to the ones they have just learned, and that the computer controls are just doing the work of older electrical controls. These chapters have been written so that they can be included in the curriculum where needed, or they can be omitted and students can study them later when they are on the job.

The final chapter is specifically written to help students understand the steps required to get their first job and how to keep that job. This is the first HVAC book to

provide a detailed procedure on how to search for a job, write a résumé, and prepare for the job interview. This is perhaps the most important chapter in the book, as the most important goal for every student is to gain the skills to obtain a job.

SUPPLEMENTS

Over the past few years it has become vitally important for a textbook to become an integral part of a complete learning system. Just a few years ago a textbook was expected to be the only thing a student needed to learn a subject. This book has been designed to be an important part of a complete learning system that makes the difficult job of an instructor much easier. An Instructor's Manual and Lab Manual are available for this text. Other supplements include a complete set of PowerPoint slides of the figures and diagrams of each chapter in color in electronic form for instructors to use for presentation in the classroom. These can also be printed out for student notes. A comprehensive set of Test Bank questions are available for each chapter, so that instructors can test the knowledge of students electronically and limit the amount of time required to grade the tests.

To access supplementary materials online, instructors need to request on instructor access code. Go to **www.prenhall.com,** click the Instructor Resource Center link, and then click Register Today for an instructor access code. Within 48 hours after registering you will receive a comfirming e-mail including an instructor access code. Once you have received your code, go to the site and log on for full instructions on downloading the materials you wish to use.

ACKNOWLEDGMENTS

I would like to thank my wife, Kathleen Kissell, for her help in keeping all the figures straight for this book. I would also like to thank her for the support she has provided every time we write a book. I would also like to thank my daughter Kelly Starkey, who provided so much help in creating the Test Bank, reviewing this book, and helping with the questions and answers in the Instructor's Manual. I would like to thank Christine Buckendahl and Eric Krassow for being very supportive and providing help at all stages.

I would like to thank the following people at Terra Community College in Fremont, Ohio, for their help in taking pictures and providing material for the book: Steve Bender, Jayne Bowersox, John Carpenter, Tim Roth, Brian Skeel, and Kevin Walker.

I would also like to thank the following people for reviewing the first edition of this book at various stages:

Michael Dean Alderman, Central Virginia Community College;

William H. Burklo, Indian River Community College;

Hugh Cole
John A. Chemsak, Technical College of the Low Country;

Jeffery Cox, Central Community College;

Henry Gaus, Lamar Institute of Technology;

Clay Harbour, Lawson State Community College;

Charles Herman, Gateway Technical College;

Lawrence Hobbs, Gateway Technical College, Kenosha, WI;

Lynn Hudson, Richland Community College;

James Hutchinson, Roanoke Chowan Community College;

Daniel Leathers, College of DuPage;

Antonio Lozano, Laredo Community College;

William Matthews, Linn State Technical College;

Ron Meyer, Central Community College;

Michacll Mutarelli, Lehigh Carbon Community College;

Les Pullins, Washtenaw Community College;

Chris Sterrett, Fort Scott Community College; and

Freddie Williams, Lanier Technical College.

I would also like to thank the following people and companies for their help with this book. This book would not have been possible without their help and support in providing pictures, diagrams, and information about their products.

Mr. Tom Falconi
Acme Electric
Acme Transformer Division

Mr. Joe Daleo
Alco Controls Division, Emerson Electric Co.

Ms. Linda Sorenson and Ms. Kate Wagner
Allen-Bradley, a Rockwell Automation Business

Ms. Debbie Beihl
Armstrong Manufacturing

Mr. Bill Steel
Bard Manufacturing Co.

Ms. Patricia Pruis
GE Motors and Industrial Systems,
Fort Wayne, Indiana

Mr. John Rhoads, Jr.
Heat Controller, Inc.

Ms. Chari Paul
Honeywell

Mr. John Bernaden and Ms. Donyel Smith
Johnson Controls

Ms. Catherine Burkhardt
Magnetek, Century Motors

Mr. Larry Hagman
Manitowoc Equipment Works

Ms. Linda Capcara
Motorola Semiconductor

Ms. Terri Johnston
North American Capacitor Co.

Mr. Bob Rank
Paragon Electric Company, Inc.

Mr. Donald Faulhaber
R. W. Beckett Corporation

Mr. Ron Smith
Reliance Electric

Mr. Joe Clark
Scotsman

Ms. Susan Carlin
Carlyle Compressor Company, Division of Carrier Corp.

Mr. Gary Forcey
Challenger Electrical Equipment Corp.

Ms. Lil Nichols
Copeland Corporation

Mr. Bob Roderique
Copper Industries, Bussmann Division

Mr. Max Robinson
Danfoss Automatic Controls

Ms. Jennifer Toffler
Square D Groupe Schneider

Mr. John Mullen
Supco Sealed Unit Parts Co., Inc.

Mr. Charlie Dunham
SymCom, Inc.

Mr. Kenneth Goldman
Tecumseh Products Company

Mr. Jim Holbrook
THERM-O-DISC, Incorporated
Subsidiary of Emerson Electric

Mr. Ted Suever
Tripplet Corporation

Mr. Wayne Gray
Tyler Refrigeration Corp.

Ms. Lisa Abbott and Ms. Christine Domorat
United Technologies Carrier Corporation and Gibbs & Soell, Inc.

Ms. Yissell Fernandez
Watsco Components, Inc.

Mr. Elwood Spangler
York International Corporation

Ms. Sandra Darling and Ms. Karen Rollo
Fluke Corporation

Mr. Ron Powell
Klein Tools

Mr. Harold Ray
Leonard Safety Equipment

Mr. Peter Haywood
Ridge Tool Company

Mr. Richard Kirwin
Snap-on Tools

Mr. Martin Leslie
Therm-O-Disc

Mr. Gary Guthrie and Mr. Onnie Clem
Megger

Mr. Ronn Cronston
CamStat

Mr. Mark Kelly and Mr. John Connell
Crown Battery

Ms. Kathy Maynard
Invensys Controls Paragon and Ranco

Ms. Eileen Oleary
Sperry Instruments

Mr. Ralph Sheer
Nordyne

Mr. David Hoekje
Hot Surface Igniters

Mr. Bill Alderson
Publications Manager
Rheem Manufacturing Company

Mr. Dean D. Croucher
General Manager
Saint-Gobain Ceramics Igniter Products

Mr. Scott Koch
Marketing Manager
Carrier Corporation

Mr. John Black
Director of Marketing Services
White-Rodgers/Emerson Climate Technologies

Ms. Cindy Nuspl
White-Rodgers Division, Emerson Electric Co.

Ms. Pam Enstad
Marketing Communications Director
Honeywell

Mr. Daniel Wesner
Ranco Controls/Paragon Controls
Invensys

Mr. Keith Lester
Public Relations Manager
Rockwell Automation

Brief Contents

Contents

Chapter 3 Electrical Circuits 46

Chapter 16 **Electrical Control of Heat Pump Systems** 470

Chapter 17 **Electrical Control of Refrigeration Systems** 498

Chapter 1

Safety Practices for HVAC

A Word About Job Assignments

A Job Assignment will be presented at the beginning of each chapter to provide an assignment that you would actually use if you were working on the job for an employer. The Job Assignment is provided to help you see the link between what you are learning in the class and what your employer will expect you to know when you reach the job. You should look over the Job Assignment before you read the chapter to see what information in the chapter you will need to complete the task. Your instructor may use the Job Assignment to start a class discussion, or you might be asked to complete it as a written assignment.

Job Assignment for This Chapter

You have been assigned a service vehicle to use on service calls. Identify the safety gear that you should have on the truck at all times to use on your service calls. Also create a list of electrical safety tips you should review weekly or prior to starting a service call that will help you prevent electrical shocks. You should also create an inspection procedure to check all of your electrical hand tools, extension cords, and any GFI equipment in a timely fashion.

Terms You Will Learn in This Chapter

Cardiopulmonary resuscitation (CPR)
Circuit breaker
Current
Electrical shock
Fuse
Ground
Ground fault interrupter (GFI)

Grounded
Grounding
Grounding adapter
Lockout/tagout
National Electrical Code ® (NEC®)
Short circuit
Tagout procedure

After reading this chapter, you will be able to:

1. List typical safety gear and equipment you should use on the job.
2. Identify electric shock levels and their effect on the body.
3. Understand the proper response when someone has received an electrical shock.
4. Identify ways to prevent electrical shock.
5. Explain why the grounding of HVAC equipment and power tools is important.
6. Understand how to work safely around HVAC equipment that has electrical power applied.
7. Understand the function and operation of fuses and circuit breakers.
8. Explain lockout and tagout procedures.
9. Safely use electrical meters and electrical tools.
10. Identify the proper fire extinguishers you should carry on your service truck.

1.0 OVERVIEW OF ELECTRICAL SAFETY

The most important part of working on the job is personal safety. You are responsible for your own safety and the safety of others working with you. In this chapter you will learn about protecting yourself, protecting others, and implementing basic tool safety that will ensure a safe workplace. You can work safely around HVAC systems your entire career; the trick is to understand the possible problems that can occur and to create safe working habits to keep yourself as safe as possible on the job. Your instructor will use many of the same safety rules and practices in your lab at school to ensure your safety in that work environment as well as to help you learn about what your employer will expect. As an HVAC technician, you will need to understand all parts of safety while you are working in the HVAC shop or when you are on a service call. Another safety concern is that you will often be working by yourself when you are making service calls. You must understand how to provide first aid for yourself in the event you have an accident or how to summon help quickly if your accident is severe. One last thing you need to remember when you are on a service call at a residence is that when you have parked your vehicle in a driveway and are working in the backyard of the residence, young children may play behind your vehicle or leave toys. You need to walk around the back of the vehicle before you start up and leave to ensure that the way is clear.

1.1 ELECTRICAL SHOCKS AND HOW THEY AFFECT YOUR BODY

One of the most important parts about working with electrical circuits around HVAC systems is to avoid coming in contact with a live circuit and receiving an **electrical shock**. In this text you will learn how to work around electrical circuits safely to avoid electrical shocks and how to minimize the effects of an electrical shock. You will learn that the more you understand electricity, the safer you will become.

3

When a person comes into contact with a live electrical circuit, electric current will flow from the circuit into the skin. If the voltage is strong enough or if the person is grounded, the current will flow through the body and has the ability to become lethal and stop vital organs such as your heart from working correctly. Voltage is the electrical force that causes electrical current to flow in any circuit. It is also called electromotive force or EMF. The higher the voltage, the more force is applied to make the **current** flow. For example, the amount of voltage at the thermostat for an HVAC system is 24 volts AC (alternating current). This voltage is set as low as possible to limit the potential of an electrical shock when people commonly touch the thermostat to set it. The voltage in a furnace fan may be 120 volts AC or 240 volts AC, and this higher voltage has the potential to cause a severe electrical shock if you would come into contact with a live wire or terminal.

When electrical voltage comes into contact with a person's skin, electrical current will begin to flow into the body. The amount of current that flows into the body will depend on how well the rest of the body is insulated or **grounded**. If the person has one hand on a metal cabinet and the other hand comes into contact with a live circuit, the body is considered to be grounded and current will flow from the point on the skin to the point where the other hand is touching the metal cabinet. This creates a serious problem because your heart depends on a small amount of electrical current to beat. If you receive an electrical shock and a small amount of electrical current enters the body through one arm and out through the other, the path for the electrical current is directly through your heart and the extra current may cause your heart to stop beating. If you do not receive immediate care in the form of **cardiopulmonary resuscitation (CPR)**, your heart may not get started again. Figure 1–1 shows how the level of current affects the body. You can see the lowest level a body can perceive is as small as 1 mA (milliamp) of current. A milliamp is one thousandth of one amp. At this small level the body is able to perceive that current is present, and you can pull away. Current levels between 1 mA and 5 mA may cause a slight shock; your reflex action may make you pull away quickly, which can cause more of a problem than the shock. If your body receives electrical current that ranges from 6 mA to 30 mA, you can reach a point where your muscles will contract and you will not be able to let go of the point where the shock is occurring. This

Amount of Current	Probable Effect of Electrical Current on the Human Body
1 mA	Perception level; slight tingling sensation. Still dangerous under certain conditions.
5 mA	Slight shock felt; not painful but disturbing. Average individual can let go. However, strong involuntary reactions to shocks in this range may lead to injuries.
6–30 mA	Painful shock; muscular control is lost. This is called the freezing current or "let-go" range.
50–150 mA	Extreme pain, respiratory arrest, severe muscular contractions. Individual cannot let go. Death is possible.
1000–4300 mA (1–4.3 amps)	Ventricular fibrillation (the rhythmic pumping action of the heart ceases). Muscular contraction and nerve damage occur. Death is most likely.
10,000 mA (10 amps)	Cardiac arrest, severe burns, and probable death.

Source: United States Department of Labor, Occupation Safety & Health Administration (OSHA).

Figure 1–1 The effects of electrical current on the human body.

is called the *let-go threshold*. Current in the range of 50 mA to 150 mA can cause extreme pain and severe muscle contraction and death is possible. Current levels between 1 amp and 4.3 amps can cause irregular heart beat called ventricular fibrillation (the rhythmic pumping action of the heart ceases), and muscular contraction, nerve damage, and even death may occur. If your heart stops beating and you go into cardiac arrest, someone must start CPR on you immediately and call 911. Emergency Medical Service (EMS) should transfer you to a hospital emergency room as soon as possible.

Figure 1–2 shows different points in an open electrical panel where you can receive an electrical shock while working with hand tools on the job. Figure 1–3 shows another way you might receive an electrical shock when you are working on a condensing unit that sits outside the house. This picture shows the wires going to the unit where the protective covering has become broken and frayed to the point where it may cut the insulation of the wire. If the condensing unit is not properly grounded and one of the wires providing voltage to the system develops a bare spot and the bare wire comes into contact with the metal on the condenser, a shock hazard is present. If the technician touches the metal case on the unit with damp shoes, the technician could receive a shock between his or her hand and feet. The more wet the shoes are, the more severe the electrical shock can be. If the technician is wearing leather boots with thick rubber soles, the chance for electrical shock is greatly reduced.

Figure 1–2 Possible shock hazards when the electrical panel of an HVAC unit is open. You must be careful not to touch any electrical terminals where you would receive an electrical shock.

Figure 1–3 Shock potential exists when the technician is touching the air-conditioning cabinet and the wiring to the condenser has a bare, exposed wire touching the metal cabinet and the system is not grounded. If the system were grounded, a fuse or circuit breaker would open the circuit as soon as the bare wire touched the metal.

Figure 1–4 shows how you can receive an electrical shock when you are working in a circuit breaker panel or other power source panel. When you have the cover off of the electrical circuit breaker panel, the wires and support metal where the circuit breakers are connected are exposed and can be dangerous. If you leave the main circuit breaker turned on, these bare and exposed metal points pose a severe electrical shock hazard. If you touch them with a metal tool and your hand is touching the metal tool, you can receive a serious shock. The best prevention for this type of shock is to ensure that the main circuit breaker has been turned off and you have tested the panel with a voltmeter to ensure that no power is present. Another way to limit the chance for electrical shock is to use tools that are properly insulated.

Figure 1–4 A technician can receive a shock by reaching into the circuit breaker panel if the screwdriver touches L1 and his or her hand then touches the metal part of the panel or screwdriver.

Figure 1–5 shows how you can receive an electrical shock from the main electrical panel that provides power to the entire home and the outdoor air conditioning unit. If you are installing a new air conditioning unit, you may need to install a new circuit breaker into the main electrical panel and you must take the panel cover off to do

Figure 1–5 A technician can receive a shock if both hands are in the circuit breaker panel and one hand touches L1 or L2 while working.

this. The safest way to work in this panel when the cover is off is to turn off the main disconnect switch (circuit breaker). When the disconnect switch is turned to the off position, electrical voltage will be turned off to all of the panel except for the very top of the panel where voltage will be present at the incoming terminals marked line 1 (L1) and line 2 (L2). If you put both hands into this panel while you are adding the circuit breaker or connecting the wire, you can receive a serious electrical shock between your left hand and right hand if either hand comes into contact with the live electrical parts at the very top of the panel. This is the most severe shock you can receive, because electrical current will flow from one hand to the other and your heart is directly in line with this path. You can receive an electrical shock so severe that you will not be able to "let go" of the metal in the disconnect, and the shock can continue for several seconds. The safety procedure that you should use to ensure that it is safe to work in a main electrical panel is to turn the circuit breaker that feeds the main disconnect to the off position and then test the line 1 and line 2 terminals in the disconnect where you are going to connect the circuit breaker to ensure that no power is present. When no power is present, it is safe to install the new circuit breaker and the wires to it. You should be aware that if you turn the main circuit breaker off, you will turn off power to all the appliances and circuits in the residence, and you will need to reset the clocks and other electronic devices in the home. However, this is a small problem to ensure that you can work safely in the main electrical panel.

If you are aware of electrical shock hazards, you can make the system relatively safe by making sure all power to the area you are working on is turned off. You should make sure that power is turned off anytime you are working on the system and removing and

replacing parts. You should also make it a rule that you always test for power anytime you're working on a system, so that you know what parts of the system are electrically energized.

Safety Notice!
It is very important that you test your voltmeter when you first use it to see if it is operating correctly. This means that you should connect the probes to the meter, set it for a voltage reading, find a known voltage source (such as a 120 volt AC receptacle), and plug the probes into it to verify that your meter is reading 120 volts AC. This test allows you to verify that your meter and its leads are working correctly.
One problem that new technicians often make is that they use their voltmeter without testing it to ensure that it can read voltage. They may put the meter probes on terminals to test for voltage and the meter indicates no voltage is present. If the voltmeter has a blown fuse or has a bad probe, it will indicate that no voltage is present and the technician will assume that it is safe to work in that area. If power is actually there, and the meter is broken and does not indicate it, the technician will assume voltage is not present and can receive a severe electrical shock. When you make a test to a known good voltage source and your meter indicates the proper amount of voltage, you can be sure your meter and its leads are working correctly. Then when you test for voltage and the reading is zero, you can be sure that no power is present.

1.2 FIRST AID AND CPR

As many people as possible at every work site should have training for first aid and **cardiopulmonary resuscitation (CPR)**. Local Red Cross agencies usually provide first aid and CPR training courses for workers. If someone receives a severe electrical shock or collapses and stops breathing, he or she will need the aid of someone who is certified in CPR. The same is true for individuals who stop breathing because of heart problems. Someone who is trained in CPR can begin helping such a person.

First-aid safety includes fast response to problems such as severe cuts and other lacerations or broken bones. If people are injured by moving parts such as a condenser fan blade, it is important to turn off power to the system as quickly as possible and determine whether the injured person can be safely moved. You must call for help immediately. First-aid courses will show you how to apply pressure to deep wounds to stop bleeding. These courses will also show when it is unsafe to move someone with a head or neck injury.

1.3 PROTECTING ELECTRICAL CIRCUITS AND EQUIPMENT

Another part of safety is providing protection for the equipment such as the compressor and fan motors. These HVAC loads can be protected by either circuit breakers or fuses. Figure 1–6 shows a circuit breaker panel that is found in a residence. The **circuit breakers** for electrical circuits are mounted in this panel, and they protect the electrical circuit against excessive current; they can be reset if they trip when current flow exceeds their rating. Figure 1–7 shows a fuse disconnect that may be used

Figure 1–6 Typical residential electrical circuit breaker panel with its cover removed.

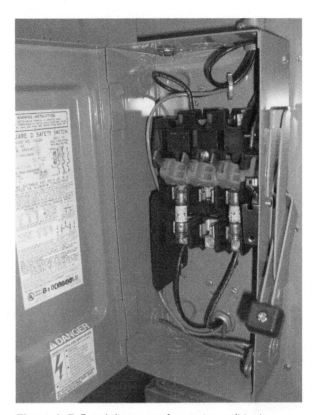

Figure 1–7 Fused disconnect for an air-conditioning system.

to protect the outdoor air conditioning unit. The **fuse** provides protection against excessive current like the circuit breaker, but the fuse must be replaced when it has been blown by excessive current. If a fuse has blown or circuit breaker has tripped, you will learn in future chapters how to test the system to ensure that the problem does not continue to occur. If you do not find the problem and it is not corrected before you change the fuses or reset the circuit breaker, they will trip again until the problem is corrected.

1.4 GROUNDING

All HVAC units must be grounded to protect people from potential electrical shocks if they touch the equipment. Grounding connects any metal in the cabinet to the common point in the electrical disconnect box or service panel. The **grounding** wire received its name because it is connected to a copper rod that is placed into the earth "**ground**" near the point where electricity is brought into the building. Grounding rods are typically 8 feet long and provide a means of making the electrical cabinets the same potential as the earth. In simple terms this means that anytime an electrical wire from line 1 or line 2 comes into contact with the metal cabinet, it will become grounded and a fuse will blow. If the metal cabinet were not grounded, it would have the same potential of the line 1 voltage if a wire

Figure 1–8 Typical residential grounding system, including grounding cable and grounding rods.

touched it, and any person who touched the cabinet would receive a severe electrical shock. Figure 1–8 shows an example of a system that is correctly grounded.

1.4.1 Ground Wires and Grounding Conductors

All electrical circuits and metal case power tools require an electrical ground conductor. The grounding wire can be a green wire or bare conductor, and it is used to provide a direct path to ground. If current flows through the ground wire, it will cause excessive electrical current to flow through fuses or circuit breakers, thus tripping them. If the grounding wire is not provided, a power cord could have a bare or exposed hot wire that could come into contact with the metal case of the power tool. When you use the power tool, you will come into contact with the metal case. If you are standing in water or moist grass or are touching the metal of an HVAC cabinet, the electrical current can pass from the bare wire, through the metal in the power-tool case, to you, and, finally, to the grounded surface. This will cause you to receive a severe electrical shock, which can be fatal. If a grounding wire is provided and the metal case of the power tool is properly grounded, the current from the bare wire will immediately be routed to ground and cause a circuit breaker or ground fault interrupter (GFI) to trip instead of posing a threat to you. If you are using a metal-case power tool and the circuit breaker continues to trip, you should suspect a problem with a bare wire somewhere. Another way to protect yourself against electrical shock when you are using power tools is to use double-insulated or plastic-case tools. If the tool has a plastic case or is double insulated, it

provides another layer of safety against short circuits, because plastic or double-insulated tools cannot conduct electricity.

The grounding terminal on the outlet receptacle is the center terminal and is rounded. The grounding terminal on any power cord that is plugged into a receptacle is slightly longer than the other two terminals, so the grounding terminal is the first to make contact when you are plugging something into the receptacle. This ensures that the power cord is safely grounded when the power terminals make contact. If a power tool requires a grounding circuit, it will have a third prong on its cord cap. You should use these tools only in an electrical outlet receptacle that also has a ground terminal. If you have a power tool with a three-prong cord cap, you should never cut the ground prong off the plug so that it will fit a two-prong (nongrounded) receptacle unless the receptacle is properly grounded and the adapter is properly grounded. You should use a GFI when possible. You should also never use a two-prong to three-prong adapter with a power tool that requires a three-prong cord. You will find some power tools that have a plastic case, and these tools have a two-prong plug that does not need to be grounded when it is used. You will learn more about short circuits, grounding circuits, fuses, and circuit breakers in a later chapter in this text. You should inspect the power cords of any electrical power tools that have a metal case to ensure that a grounding wire is provided.

The ground wire is usually a copper wire and should be connected in the electrical panel and on the unit to ensure that the unit is properly grounded. If an air conditioning system is not properly grounded, you can receive severe electrical shock if you come into contact with metal parts of the cabinet that has come into contact with a bare part of an electrical wire. If you have the system properly grounded, a fuse or circuit breaker would blow when the bare wire comes into contact with the metal of the case.

1.5 GROUND FAULT INTERRUPTION CIRCUIT BREAKERS

If you are using electrical hand tools or if you must work in areas that are damp or have standing water, it is important that all circuits for these locations have **ground fault interrupters**, called GFIs. The GFI-type breaker measures all the electrical current that is supplied to a circuit or power tool and compares it to the amount of current returning from the tool. If there is no electrical short circuit to ground (ground fault) and the power tool is working correctly, the amount of current returning from the circuit will be the same as the current supplied. If less current is returning, the GFI assumes a ground fault is occurring and a dangerous condition will result, so it trips to the open position. For example, if you are using a power tool that has a faulty power cord, as soon as a ground fault occurs and you start to get an electrical shock, the GFI will interrupt all current in the circuit and prevent you from becoming shocked. GFI protection is also available as a built-in feature for some extension cords or power cords that you use to provide power for your portable electrical power tools. You should always use a GFI with your extension cord.

Another type of GFI has an adjustable detection setting. The important point to remember when you are working with GFIs is that you should never increase their trip point or tamper with their mechanism if they are tripping. The GFI will trip when it detects a ground fault. If you tamper with the GFI, it may allow someone to receive a severe electrical shock. When GFIs are properly installed and maintained, they will prevent injury and death due to electrical shocks. You should test all GFI circuit breakers at their predetermined test intervals. Figure 1–9 shows a two-pole GFI, and Figure 1–10 shows a single-pole GFI circuit breaker.

Figure 1–9 Two-pole ground fault interrupter (GFI) circuit breaker.

Figure 1–10 Single-pole ground fault interrupter (GFI) circuit breaker.

1.6 **DUPLEX GROUNDED RECEPTACLES**

A duplex grounded receptacle is shown in Figure 1–11. The ground terminal is the round hole below the vertical blades. The ground terminal on the male plug is slightly longer than the two blades that carry line 1 and neutral voltage. This ensures that the ground terminal makes connection before the power-carrying terminals and will cause a circuit breaker to trip if there is a problem. Figure 1–12 shows the older style two-prong receptacle that does not provide a ground. The **grounding adapter** shown in Figure 1–13

Figure 1–11 A duplex receptacle that provides 120 V AC for tools and equipment.

Figure 1–12 Two-prong ungrounded receptacle.

Figure 1–13 A two-prong to three-prong adapter.

Figure 1–14 Three-prong plug on an extension cord.

Figure 1–15 Double-insulated drill has a plastic case and a two-prong plug. This drill is designed to limit electrical shock due to loss of ground or shorts, because the case is plastic.

shows a two-prong to three-prong adapter. If a residence is older and does not have grounded receptacles, you should not use an adapter to make the three-prong extension cord shown in Figure 1–14 able to plug into it. Instead you should have double-insulated tools (power drill, saws, etc.) to plug into these circuits. Figure 1–15 shows a double-insulated drill. The double-insulated tools have plastic cases and do not have metal parts that allow electrical shock to occur, so they have a two-prong plug and do not need a ground terminal. If your power tools have metal cases and have a three-prong plug, you must use a grounded receptacle or one with a GFI capability. Some power cords have a GFI-type receptacle included to provide an extra level of protection against electrical shock.

1.7 PORTABLE POWER TOOLS

Another way to protect yourself from electrical shock when using power tools is to use portable power tools that use rechargeable batteries as their power source rather than 120 volt AC power. A full range of portable power tools including saws, drills, and power shears are available with 12-, 14-, and 18-volt DC batteries. Figure 1–16 shows a picture of a battery-powered drill. Because the power tools use a portable rechargeable

Figure 1–16 Portable power drill uses a rechargeable battery as the power source, which totally eliminates the chances of electrical shock due to short or grounding problems.

battery as their power source, they do not pose any shock hazard. The portable battery source can be plugged into a range of equipment such as drills and saws, so you can continue working all day long using two or more batteries. You can remove the dead battery from the power tool and recharge one while you are using the second one. The batteries are charged through a 120 volt AC source.

1.8 TESTING THE GROUND WIRE

A ground wire is the bare wire or a wire that has a green coating on it. The ground wire should be connected to the ground terminal in the electrical circuit breaker box and at the end where it is connected in the air-conditioning condenser or the furnace you are installing. In some systems, electrical conduit (metal tubing) is used to protect the wiring, and this metal conduit can be used to provide the ground circuit for the system. If the metal conduit is used to provide the grounding system, you must be sure that all of the connections in the tubing are tight and provide a low resistance path for the ground circuit.

You can test the ground circuit with your voltmeter to ensure that it is working correctly. To perform this test, set your voltmeter to the proper voltage range of the system voltage such as 120 volt AC. If you are testing the ground of an electrical receptacle, you should put one probe in the L1 blade of the receptacle and the other one in the N blade of the receptacle. Next you can take the meter probe from the N terminal and touch it to the ground terminal. If the ground wire is connected correctly, you should measure the same voltage between L1 and the ground terminal as you would when you had the meter probes between L1 and N. If you read voltage between L1 and N but not between L1 and ground, you should suspect that the ground wire is not connected correctly or has been broken. If you are testing the grounding circuit at an air-conditioning condenser, you can touch the meter leads to the L1 and N terminals in the electrical cabinet or you can touch L1 and L2. Your next test is between L1 and the metal part of the cabinet. If the system is properly grounded, you should measure the same amount of voltage between L1 and the ground as you did between L1 and N. You need to be aware that if you are using an extension cord, it should be a three-wire cord that has a ground terminal. Figure 1–17 shows two problems with the grounding terminal that can occur—it is broken off or comes loose. If the grounding terminal comes loose or is missing, you should not use the extension cord and replace it immediately.

Figure 1–17 The ground terminal is broken off on one of these three-prong electrical cords, and the ground terminal is coming loose on the other. If this occurs, discard the cord and purchase a new one.

Missing Ground Terminal

Loose Ground Terminal

1.9 SHORT CIRCUITS

An example of the damage caused by a short circuit is shown in Figure 1–18. A **short circuit** occurs when a wire that is connected to line 1 (L1) touches a wire that is connected to line 2 (L2), neutral or ground. An L1 wire that touches any part of the metal cabinet that is grounded will cause an extreme amount of heat and energy. A short circuit is a condition where the equipment is responding to a problem whereby a live wire has the ability to touch the metal part of the cabinet that is grounded or a wire from neutral or line 2 (L2) and generates an enormous amount of heat energy, which could cause a fire that could burn down a house or building. If the system is properly protected with ground wires, fuses, and circuit breakers, a short circuit should not cause any damage to persons or equipment. The short circuit should cause a fuse to blow or circuit breaker to trip, which protects the equipment from severe damage. For this reason, when you find a blown fuse or tripped circuit breaker, you should inspect the system before you reset it. If you reset the circuit and it trips immediately, you can suspect that there is a problem with a short circuit. You will learn how to locate the short circuit and correct it in future chapters. If a circuit is not properly protected, the energy from the short circuit can cause the conductor and other metal to turn to a molten state, which will cause severe burns if it comes into contact with skin

1.10 ELECTRICAL SAFETY

Whether you are working in a shop or on an HVAC system at someone's home, it is important to completely understand how to work safely around electrical cabinets and circuits. This includes inspecting electrical safety grounds to ensure that they are connected correctly and are operational and working safely around circuits that are under power. The safest way to work around electrical circuits is to turn power off to the electrical circuit. This can be done by turning off the circuit breaker for the circuit you are

Figure 1–18 Results of a short circuit in an electrical receptacle that is not properly protected.

working on. If you are working on a larger system in a factory, you must observe lockout/tagout procedures that are required by OSHA.

In some cases you may be required to make a test of voltage and current in circuits that have electrical power applied. When possible, you should wear heavy rubber-soled shoes and work with only one hand in an electrical panel that is powered. If you work with both hands in an electrical panel, you may conduct electric current in one arm, through your body (near your heart), and out the other arm. This is the most dangerous type of shock, and it could cause severe damage to you or stop your heart. If you are working with only one hand at a time in the electrical panel, the chances of conducting electrical current directly through your heart during an electrical shock is minimal. It is important to understand that currents as small as 50 mA may cause heart or respiratory failure.

1.11 WORKING SAFELY AROUND EQUIPMENT

At times, you will be expected to work on HVAC equipment that must have electrical power applied. For example, if you're trying to troubleshoot the problem, you must have the voltage applied to the unit. When you must work on equipment that has the voltage applied, you must be very careful that you do not touch any live parts of the circuit with your hand or any other part of your body, so you do not receive an electrical shock. Figure 1–19 shows an example of how a technician can receive an electrical shock from touching the live electrical terminals in a fused disconnect. When you work around these circuits, you must be sure that your tools and electrical meters have enough insulation. Insulation on your tools will allow you to touch live electrical parts of the circuit

Figure 1–19 A technician can receive a shock if a ground wire is loose or broken and a hot wire makes contact with the metal case while the technician touches the metal box.

without receiving an electrical shock. If the insulation is worn, damaged, or missing (such as insulation on the end of a screwdriver or pliers) you may receive a shock if the metal comes into contact with a live electrical terminal. If the insulation is in good shape, you should not have any problems when you're touching test points that have voltage applied to them.

Another way to ensure your safety when working around equipment that has electrical voltage applied is to wear rubber-soled shoes. The rubber soles on your shoes will provide sufficient insulation to minimize any current flow through your body in the event that you touch a live electrical circuit with your hand or any other part of your body. You can also minimize the potential for severe shock by removing all watches and jewelry when you're working in the system and use one hand only if possible. This means you will have to set things up so that your meter is located where you can see it and connect one probe from your meter with an alligator clip to one side of the circuit so that you only need to use one voltmeter lead on the other side of the circuit. The advantage of using only one hand to make the test means that you're only exposing one part of your body to the possibility of electrical shock. You will remember that if you have both hands in the electrical box, you have the potential of receiving an electrical shock that would cause current to flow from one hand through your chest (near your heart) and out the other hand, which is the most dangerous situation you can encounter. When you have only one hand exposed to the cabinet, the electrical shock would be minimal if it occurred and would not jeopardize your heart.

You will begin to see that if you fully understand the serious problems that may occur from electrical shock and you fully understand how to prevent these problems, you will find that it is easy to work safely around HVAC equipment and conduct all of the electrical tests that you need.

1.12 ELECTRICAL LOCKOUT AND TAGOUT PROCEDURES

At times you may have to work on electrical systems where the electrical code requires that you lockout or tagout the equipment you are working on. A **lockout/tagout** procedure is used to turn off all electrical power to a larger system when it is too dangerous to work around the system with power lines. For example, you may test the system and find that the compressor needs to be changed out. When you have completed the electrical test and are ready to disconnect the compressor and remove it, you should turn all power off at the fused disconnect switch and place a padlock on the disconnect to ensure that it remains turned off while you are working on the system. The padlock for the lockout should include a tag with your name and photograph on it indicating that you're the person working on this system. Putting this tag with your padlock is called a **tagout procedure.** If you encounter a system that has been turned off and has a padlock keeping it off, you should understand that someone has put this padlock on the system to ensure that it is safe to work on. You should never remove a padlock used for lockout unless you are instructed to do so by the person who put the padlock in place. You should carry one or more sets of lockout/tagout equipment in your service truck at all times so that they are available to be used when you encounter the need. Figure 1–20 shows lockout devices for circuit breaker panels, and Figure 1–21 shows lockout devices for disconnect switches.

Figure 1–20 Circuit breaker lockouts. *(Used with permission from Leonard Safety Equipment)*

Figure 1–21 A typical lockout device that can accept multiple padlocks. If more than one technician is working on the system, each should place a padlock on the lockout device. *(Courtesy of Klein Tools)*

1.13 OVERVIEW OF LOCKOUT/TAGOUT SAFETY

You may also be requested to go to a factory to work on HVAC equipment, electric panels, coolers, or chillers used as part of the manufacturing processes. When you are working at a factory, you must abide by their safety procedures. A wide variety of safety procedures and precautions are used in factories today to ensure that workers are not endangered or injured while they are working, including the lockout/tagout program.

The lockout/tagout program is mandated for all factories by the Occupational Safety and Health Administration (OSHA). The exact standard is OSHA Standard 1910.147. This program explains techniques that are to be used to ensure that the

machine is locked in the off position prior to being worked on. The program is designed to prevent injury that may be caused by uncontrolled machine motion, electrical shock, electrical burns, or injuries from coming into contact with hazardous materials.

The National Institute for Occupational Safety and Health (NIOSH) has data that show a wide number of factory accidents have resulted in serious injury or death. These accidents include personnel being electrocuted by coming into contact with high voltages, being crushed between moving parts on an HVAC system, such as fans or pulleys, or having arms and legs caught in moving equipment or amputation of arms or legs from blades. These accidents occurred because coworkers did not pay attention when workers were inside equipment or did not understand the dangers of working around large HVAC equipment that still has power applied to it.

The lockout/tagout procedure is devised to allow maintenance and other personnel to safely work on and around HVAC equipment where unexpected start-up of fans or the unexpected release of pressurized energy may cause injury or death. The procedures should be used anytime equipment is being set up, inspected, or adjusted or anytime maintenance is being conducted.

These procedures are particularly important when guards must be removed to change fan belts. During this time workers are especially vulnerable to exposure of their arms and legs to moving parts. In these conditions, workers may sustain serious injury from a crushing accident or may have their clothing snagged, which can pull them deeper into moving parts of the HVAC system.

The procedure should also cover the disconnection and lockout of electrical power when necessary. When HVAC systems need to be cleaned by personnel, it is important that all power be turned off. This prevents personnel from being shocked if they come into contact with electrical terminals or if the cleaning fluids and equipment they are using come into contact with electrical terminals. You should be aware that you must pump down the refrigerant in any system that you need to open. This prevents accidents that can be caused by the sudden discharge of refrigerant when a refrigerant line is opened.

Exposure to dangerous fumes, such as those from fuels, must also be controlled. This includes exposure to exhaust gases and exposure to extreme heat, as from steam pipes or other exposed heating elements. You may also have to turn off the natural gas or other fuel supply when working on a furnace. The lockout/tagout procedure will ensure that you have the shutoff secured when it is in the off position and clearly identified so that it is not accidentally turned on while you have a line open to replace a part.

1.14 REMOVING THE LOCKOUT/TAGOUT DEVICES AND RESTORING POWER TO THE MACHINE

After all work has been completed on the HVAC system, you are ready to remove the lockout/tagout devices and return power to the HVAC equipment. You must make several checks prior to removing the lockout/tagout devices. The first step in this process is to ensure that all work is complete and that all tools and any test equipment have been removed. All guards and safety doors should be returned to the normal condition and tested to ensure that they operate correctly. Be sure that all parts of the HVAC system that have been disassembled are reassembled properly.

When you have completed these tasks, you must physically account for all personnel who are in the area to ensure that they know and understand that you are ready to turn on power to the machine. This is probably the most important part of the procedure,

especially if more than one person is working on the machine or if several employees routinely work on the machine when it is in normal operation.

When you are ready to apply power again, be sure you follow the operational checklist so that power is returned to the machine parts in the proper order. For example, some HVAC systems must have circulating pumps running before steam valves are returned to their open position.

1.15 REMOVING A COWORKER'S LOCKOUT/TAGOUT EQUIPMENT

The general rule is that the person who placed the lockout on the machine must be the one to remove it. But there will be times on larger jobs when work is started on the day shift and completed 14 hours later on another shift. In this case the person who placed the lockout/tagout equipment on the machine is not present to remove it. When this occurs, it is permissible for a supervisor or other person of responsibility to remove the lock. The personnel on each shift who may remove a lock should be identified when the lockout/tagout procedure is written. The supervisor may have a master key or know that it is permissible to cut off the lock with bolt cutters. The person that removes the lock should always file a report to indicate why the lock was removed and who performed the action.

1.16 SAFETY GLASSES, PROTECTIVE CLOTHING, AND EQUIPMENT

When you are working in the shop, on a job site, or around refrigerants, you will be required to wear some type of safety gear, such as safety glasses or safety face shields to protect yourself. Safety glasses with shatterproof glass lenses and side shields should also be worn in a lab and the shop at all times to protect your eyes from foreign debris. Problems can also occur with electrical equipment if there is an electrical short, when the metal from electrical contacts can melt and spray your face and eyes. Safety glasses with side shields protect your eyes against these problems. If you wear glasses, you can get safety glasses with prescription lenses to match your regular glasses so that you can feel comfortable wearing them continuously all day long. You need to make sure your safety glasses are comfortable to wear over extended periods.

Figure 1–22 shows examples of several types of eye-protection equipment. Safety glasses, safety goggles, and safety shields are shown in this figure. Safety goggles with colored lenses are required when you are using torches or welding equipment. Safety shields must be used when you mix or pour dangerous liquids that might splatter and get in your eyes or on your face.

(a) (b) (c)

Figure 1–22 a Safety glasses. **b** Safety goggles. **c** Face shield. *(Used with permission from Leonard Safety Equipment)*

1.17 FIRE SAFETY

Fire safety is important in all locations in which you work, including your shop or when you are on a service call. Fire safety consists of prevention and being able to fight fires with fire extinguishers. The most important part about fire safety is to understand the basic principles of fire. Fire requires fuel, oxygen, heat, and a chain reaction (source of ignition) to start and be sustained. When you are trying to prevent a fire or fight a fire, you must separate these ingredients. Fuels may include paper, oils, gases, or any other material that can burn. For example, you may need to use a torch when you are on a service call to solder copper lines. You must be aware that there is a potential to cause a fire to the HVAC equipment or the house itself.

1.18 SOURCES OF IGNITION

One way to prevent a fire is to remove the source of ignition from the fuel. Some sources of ignition that you should be aware of include open flames from torches and heaters, smoking and matches, electrical equipment and heating elements, friction, high-intensity lighting, combustion sparks from grinding and welding, hot surfaces, spontaneous ignition, and static electricity. You must also store flammable liquids and cleaning solvents in proper storage containers and keep them away from open flames and sources of ignition at all times.

Many HVAC companies have proactive programs that put specific policies into action when any of these ignition sources are used. For example, when you must use soldering equipment or cutting torches, safety covers and fire extinguishers should be moved into place and additional personnel should be used to watch for fire problems while the soldering is being completed. This is especially important when you are soldering a new compressor into a system in a residential or commercial location. You should always have a fire extinguisher with you and follow all rules. When you are working at commercial or industrial sites, you should always ask if they have additional rules that you will need to follow before you use an open flame or torch.

1.19 FIRE EXTINGUISHERS

If you work in a shop, a construction site, or from a truck, you should have a fire extinguisher available at all times. Fire extinguishers are rated by classification. A Class A fire extinguisher can be used on ordinary combustible material, such as paper, wood, or clothing. This type of fire can easily be extinguished using water. A Class B fire extinguisher is designed for fires in flammable liquids, grease, and other materials that can be extinguished by smothering or removing air (oxygen) from the chain reaction. Class C fire extinguishers are used on fires that involve live electrical equipment. The extinguishers for these types of fires must use material that is nonconducting, so that the person using the extinguisher is not exposed to additional hazard from electrical shock when fighting the fire. Class D fire extinguishers are used on fires in combustible metals, such as sodium, magnesium, or lithium. Special extinguishers must be used to lower the temperature and remove oxygen from these fires.

Color codes for these fire extinguishers have been developed, and it is important that you recognize the type of fire and use the proper extinguisher. Class A fire extinguishers are green, Class B fire extinguishers are red, Class C fire extinguishers are

blue, and Class D fire extinguishers are yellow. It is also important that you recognize the type of material in your shop area and have a specific plan for using fire extinguishers for small fires and escape plans for evacuation if the fire is too large to contain.

1.20 FIRE EXTINGUISHER SAFETY

Fire extinguishers must be inspected on an annual or semiannual basis. During the inspection each fire extinguisher should be checked to ensure that it is fully charged and ready for use. The material in the fire extinguisher should be loose when you shake the extinguisher. If the material is caked in the bottom of the extinguisher and it will not come out, it will render the fire extinguisher useless. All fire extinguishers must be clearly marked for the type of fire for which they are specified and clearly displayed where all employees can locate them when they are needed. Additional marking should be used to indicate the locations of all fire extinguishers, including those stored on your service truck.

1.21 NATIONAL ELECTRICAL CODE (NEC)

The **National Electrical Code (NEC)** is written by the National Fire Protection Agency, which ensures that wiring you are installing or troubleshooting will not cause a fire or pose a shock risk to those working on it. The NEC consists of nine chapters. Each chapter has a number of articles that spell out such things as wiring and protection, methods and materials, equipment for general use, special occupancies, special equipment, special conditions, and communication systems. The NEC also has a number of tables that spell out other conditions. You will learn more about the NEC in later chapters, and you will be provided specific examples and tables that you will be expected to look up and use.

1.22 NFPA 70E COMPLIANCE: PROTECTING TECHNICIANS AGAINST ELECTRICAL ARC FLASH AND ARC BLAST

The National Fire Protection Association has developed a comprehensive standard called NFPA 70E that establishes the best electrical safety practices on how to protect technicians who work with electrical circuits above 50 volts. This standard explains how to protect technicians who work with electrical circuits against electrical arc flash and arc blast exposure. The Occupational Safety and Health Administration (OSHA) has adopted regulations on safe electrical practices based on NFPA 70E. The goal of NFPA 70E is to protect technicians from the hazards of electrical shock, electrocution, arc flash, and arc blast. There are sources on the Internet that you can use to find out more about NFPA 70E.

1.23 REVIEW

In this chapter you learned about the importance of working safely. At times you will be working with others and you must be aware of their safety, and at other times you may be working alone and you are responsible for your own safety. You should be aware that working safely around electricity is important and it depends on you following safety

rules. You need to be aware anytime you are working in an open electrical panel that electrical shock hazard exists, and you need to use insulated tools and be careful not to touch exposed electrical connections. You also need to be aware that the safest way to work around electrical circuits is to turn off the circuit breaker providing power to the circuit you are working on. If you are working on an HVAC system in a factory, you will need to understand and implement lockout and tagout procedures. If you are working outdoors, you should make sure that all of your electrical tools that require grounding are properly grounded and that you are using a ground fault interrupter (GFI) circuit.

Another part of safety that you need to be aware of is fire safety. You will need to understand the color codes for fire extinguishers and use the proper fire extinguisher for the type of fire you encounter. It is also important to discuss safety issues on a regular basis so that you are aware of all safety problems on the jobs that you do.

Job Assignment for This Chapter

You have been assigned a service vehicle to use on service calls. Identify the safety gear that you should have on the truck at all times to use on your service calls. Also create a list of electrical safety tips you should review weekly or prior to starting a service call that will help you prevent electrical shocks. You should also create an inspection procedure to check all of your electrical hand tools, extension cords, and any GFI equipment in a timely fashion.

Solution to the Job Assignment

Your truck should have safety glasses for you to wear when working on systems to protect your eyes from damage if an electrical wire touches the metal case of a system and creates a spark that consists of molten metal. You should also have a variety of extension cords with three prongs and possibly have one with a GFI receptacle.

Your list for safety tips should include using your voltmeter to test to ensure that power is turned off if you are removing and replacing parts on a system. You should always test your voltmeter on a known voltage source before testing to see if power is off. This is an important step to ensure that the voltmeter is working properly and the probes are both good. Your list for inspecting tools should include electrical drills, saws, and other power tools that have an electrical cord. You do not need to worry about portable battery-powered tools.

You should also have a sign in the truck reminding you to look behind your truck before leaving so you do not run over young children who may be playing behind your vehicle or the toys they might leave behind your truck.

Assessment

QUESTIONS

1. Explain why you should wear safety glasses with side shields at all times while you are working on electrical circuits in HVAC.
2. Explain what you should be aware of to avoid electrical shock hazards while working on the job.

3. Explain how much electrical current can cause you to begin to have an irregular heartbeat.

4. Explain how a GFI circuit breaker may protect you from electrical shock.

5. Explain when you may need to use a lockout on the job.

TRUE OR FALSE

1. Class I fires include petroleum and solvent-type fires.

2. You can use a Class A fire extinguisher on an electrical fire.

3. A circuit breaker protects the HVAC system from drawing excessive current and damaging its wiring.

4. A GFI (ground fault interrupter) circuit breaker should be used with an extension cord in a damp or wet location.

5. If you are working at an older residence that does not have grounded receptacle outlets, it is safe to use a three-prong extension cord if you break off the ground terminal.

MULTIPLE CHOICE

1. The Class A type fire extinguisher is _____ and is used for paper or wood fires.
 a. red
 b. green
 c. yellow

2. A GFI (ground fault interrupter) receptacle is
 a. a special receptacle that has a third terminal for a ground wire.
 b. a special receptacle for power tools that allows more than one power tool to be plugged into it at the same time.
 c. a receptacle that has a special circuit breaker built into it that can determine whether a short circuit occurs and automatically opens to protect the circuit.

3. If a circuit breaker that protects an air conditioning system continually trips at a residence, you should
 a. put in a larger one.
 b. troubleshoot the system to determine what is causing the system to draw extra current.
 c. put a jumper wire around the circuit breaker so it is no longer in the circuit.

4. A three-prong plug
 a. is better than a two-prong plug because the extra prong help keeps the plug in the socket better.
 b. is needed if you have double-insulated tools.
 c. provides a third prong that is the grounding wire for the system and ensures that the ground wire is provided to all power tools that are plugged into it.

5. If an electrical power tool has a power cord with a three-prong plug and the electrical outlet has two prongs, you should
 a. cut the third prong (ground lug) off the power cord so that you can use the outlet.

b. find an adapter that allows you to plug a three-prong plug into a two-prong outlet.

c. not use this outlet because the power tool needs to be grounded for safe operation, and the two-prong outlet is not grounded. Then locate a GFI or a receptacle that is grounded.

PROBLEMS

1. Make a fire safety plan for the lab area of your classroom that includes an evacuation plan.
2. Make a list of all tools in your lab that should be inspected for damaged power cords.
3. Make a safety plan that lists all the potential fire safety problems in your lab.
4. Make an electrical safety plan that includes a weekly inspection of all the power cords used for power tools to ensure that they are properly grounded.
5. Make a safety plan that includes the proper periodic inspection of all safety guards for all equipment in your lab.

Websites

Websites change constantly. Use the key word search to locate sites that have similar content. The products that will be found at each site are indicated with the key word.

National Fire Equipment Products
Products: Fire extinguishers
Key word: Fire extinguishers

Occupational Health and Safety Organization (OSHA)
Products: Safety information and regulations
Key word: OSHA

Leonard Safety Equipment
Products: Safety equipment
Key word: Leonard Safety Equipment

Chapter 2

Fundamentals of Electricity

Job Assignment for This Chapter

You are on a service call to a customer who does not understand the basic theory of electricity and thinks you are trying to sell parts that are not needed. Your job is to use your knowledge of basic electricity and meters to explain how you can use your meter to test any of the switches in this system and how you can tell whether they are good or bad.

Terms You Will Learn in This Chapter

Alternating current (AC)

Ammeter

Ampere (amp)

Atom

Clamp-on ammeter

Conductor

Current

Direct current (DC)

Electrical control

Electrical load

Electrical potential

Electrical power

Electricity

Electromotive force (EMF)

Electron

Insulator

Neutron

Nucleus

Ohm

Ohmmeter

Ohm's law

Power

Proton

Resistance

Seasonal energy efficiency rate (SEER)

Static electricity

Volt

Voltage

Watts

After reading this chapter, you will be able to:

1. Explain the term *electricity*.
2. Describe the term *voltage*.
3. Explain the term *current*.
4. Describe the term *resistance*.
5. Understand the sources of electricity.
6. Understand Ohm's law.
7. Explain the difference between rechargeable and nonrechargeable batteries.
8. Explain the term *milli*.
9. Explain the term *Mega*.
10. Explain the term *micro*.
11. Explain the term *kilo*.

2.0 OVERVIEW OF ELECTRICITY

Understanding electricity will help you to troubleshoot the electrical components of systems such as compressors and controls. This chapter will provide you with the basics of electricity that are the building blocks for more complex circuits that you will encounter in the field. You do not need any previous knowledge of electricity to understand this material. The simplest form of electricity to understand is *direct current* (DC) electricity, so many of the examples in this chapter will use DC electricity. However, the majority of electrical components and circuits in field equipment use *alternating current* (AC) electricity, so some of the examples will use AC electricity. This chapter begins by defining basic terms.

2.1 EXAMPLE OF A SIMPLE ELECTRICAL CIRCUIT IN A HEATING SYSTEM

It is easier to understand the basic terms of electricity if they are described in terms of a working circuit that you can identify. We will use the fan that moves warm air in a furnace as an example. The fan must have a source of voltage and current. **Current** is the flow of electrons, and **voltage** is the force that makes the electrons flow. Since this is just an introduction, much more detail about the terms will be provided later in this chapter.

In the diagram in Figure 2–1 the source voltage is 120 V AC, which is identified by the labels L1 and N. The thermostat is a temperature-actuated switch mounted in the

Figure 2–1 Electrical diagram of a thermostat switch controlling a fan motor for a furnace.

furnance for this circuit that provides a way to turn the furnace fan motor on and off. When the temperature in the furnace rises, the thermostat switch closes. This allows voltage to push the current through the wires to the fan motor. When voltage and current reach the motor, its shaft begins to turn. The wires in the circuit are called **conductors**. They provide a path for voltage and current and a small amount of opposition to the current flow. This opposition is called **resistance**.

After the fan has run for a while, the temperature in the furnace cools down, and the thermostat switch opens. This causes the voltage and current to stop at the switch, and the motor stops. When the temperature in the furnace increases, the thermostat will close and allow voltage and current to reach the motor again. Now that you have an idea of the basic terms *voltage, current*, and *resistance*, you are ready to learn more about them.

The source of energy that makes the shaft in the motor turn is called **electricity**, which is defined as the flow of electrons. You should remember from the previous discussion that the definition of *current* is also the flow of electrons. In many cases the terms *electricity* and *current* will be used interchangeably. Electrons are the negative part of an atom, and **atoms** are the basic building blocks of all matter. All matter can be broken down into one or more of the 102 naturally occurring basic materials called elements. Two of the elements that are commonly used in electrical components and circuits are copper, which is used for wire, and iron, which is used to make many of the parts such as motors.

2.1.1 Atomic Theory

To understand electricity at this point, we need to understand only three parts of the atom: **electrons**, which have a negative charge; **protons**, which have a positive charge; and **neutrons**, which have no charge (also called a neutral charge). Figure 2–2 shows a diagram of a simple atom that has one electron, one proton, and one neutron. From this diagram you can see that the proton and neutron are located in the center of the atom, which is called the **nucleus**. The electron moves around the nucleus in a specific volume of space called an orbital. For the sake of simplicity we say that the electron orbits the nucleus in much the same way that the earth revolves around the sun, but electrons do not follow such precise paths.

Since the electron is negatively charged and the proton is positively charged, an attraction between these two charges holds the electron in orbit around the nucleus until sufficient energy, such as heat, light, or magnetism, is used to break it loose so that it can become free to flow. Anytime an electron breaks out of its orbit around the proton and becomes free to flow, an electrical current is produced. The energy to create electricity can come from burning coal, nuclear power reactions, or hydroelectric dams.

The atom shown in Figure 2–2 has only one electron, one proton, and one neutron. In the circuit presented in Figure 2–1, copper wire is used to provide a path for the electricity to reach the fan motor. Because copper is the element most often used for the conductors in circuits, it is important to study the copper atom. A copper atom is different from a hydrogen atom in that a copper atom has 29 electrons and a hydrogen atom has only one. The copper wire that provides a path for the electricity for the fan motor contains millions of copper atoms.

Figure 2–3 shows a simplified diagram of a copper atom with its 29 electrons moving about the nucleus in a series of orbits. In reality, the 29 electrons move about in regions of different energy levels called *shells*, each composed of a certain number of orbits that accommodate a specific number of electrons. Since there are also 29 protons in the nucleus of a copper atom, the positive charges of the protons attract the negative

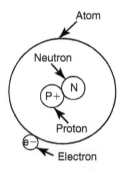

Figure 2–2 An atom with its electron, proton, and neutron identified.

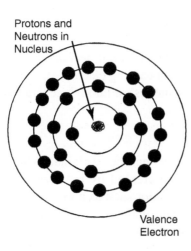

Figure 2–3 A copper atom has 29 electrons. This atom has one electron in its valence shell that can break free to create current flow.

Protons and Neutrons in Nucleus

Valence Electron

charges of the electrons and keep the electrons orbiting in their proper shells. The attraction for the electrons in the three shells that are closest to the nucleus is so strong that these electrons cannot break free to become current flow. The only electrons that can break free from the atom are the electrons in the outermost shell. The outermost shell is called the *valence shell*, and the electrons in the valence shell are called *valence electrons*. You should notice that a copper atom has only one electron in its valence shell. All good conductors have a low number of valence electrons. For example, gold is one of the best conductors and has only one valence electron; silver, which is a good conductor, has three valence electrons.

In the fan circuit we studied earlier, a large amount of electrical energy called *voltage* is applied to the copper atoms in the copper wire. The voltage provides a force to cause a valence electron in each copper atom to break loose and begin to move freely. The free electrons actually move from one atom to the next. This movement is called *electron flow* or, more precisely, *electrical current*. Since the copper wire has millions of copper atoms, it will be easier to understand if we study the electron flow between three of them. Figure 2–4 shows a diagram of the electrons breaking free from three separate copper atoms.

When the electron breaks loose from one atom, it leaves a space called a *hole* where the electron was located. This hole will attract a free electron from the next nearest atom. As voltage causes the electrons to move, each electron moves only to the *hole* that was created in an adjacent atom. This means that electrons move through the conductor

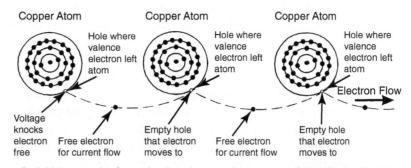

Figure 2–4 Voltage is the force that knocks one electron out of its orbit. In this diagram the valence electron in the atom on the far right breaks free from its orbit and leaves a hole. The electrons in the atoms to the left of the first atom move from hole to hole to create current flow.

by moving from hole to hole. The movement begins to resemble bumper-to-bumper cars on a freeway.

The flow of electrons can be accomplished in several ways. The simplest method is to create static electricity. Other means of causing electrons to flow include creating a potential voltage from chemical means (batteries) or through generation by electrical generators.

2.1.2 Static Electricity and Charges

Several important electrical concepts can be explained when static electricity moves electrons. **Static electricity** can be created by moving two nonmagnetic materials together such as a cloth over an insulator such as a glass rod. Even though most material is electrically neutral, which means that it has the same number of electrons and protons, some material has an excess of negative or positive charges. In our example the cloth has an excess of negative charges and deposits them onto the neutral glass rod when the cloth is rubbed across the glass. The electron charge will remain on the glass rod until it is brought into contact with something that will accept the electron charge. For example, if the glass rod is brought into contact with a material that has fewer electrons, such as a piece of metal, a "spark" of static electricity will be created, which is actually a short burst of electrons flowing from the glass rod to the metal. This static spark cannot sustain current, but it has the characteristics of current flow. The electrons are moving from a place they are stored in excess to the material that has fewer electrons. When the number of electrons moving to the new material equals the number of electrons on the rod, the difference of potential no longer exists and the flow stops. This is why the static exchange appears as a short spark. A battery or generator that can produce a continuous source of electrons is needed to create a sustained flow of electrons in circuits that are used in air-conditioning systems.

The number of electrons and protons in an element is usually the same. If an atom loses an electron, it will take on a positive charge; if it gains an electron, it becomes negatively charged. There is a set of laws that apply to charges. Like charges will repel and unlike charges attract.

2.1.3 Sources of DC Voltage

One source of electricity is DC voltage. DC stands for **direct current** and it means that the electrons flow in only one direction from the negative potential to the positive potential. The most common source of DC voltage is a battery. A battery creates electrons by a process called an electrochemical reaction. The battery consists of one or more voltaic cells that have electrolytes containing ions that react with chemicals in the electrode. The material in the battery can be lead, lithium, zinc, or other material that is made of a chemical paste from two dissimilar metals as a negative terminal and a positive terminal.

One of the most common types of batteries that you may be familiar with is the lead-acid battery in your car. Figure 2–5 shows a diagram of the lead plates in this type of battery. The lead-acid battery in your car is rechargeable, and each individual cell in the battery has sets of negative and positive plates. The negative plate consists of a spongy lead-acid material, and the positive plate consists of lead-dioxide active material. The plates are covered with an electrolyte, which is a liquid solution of 35% sulfuric acid and 65% water that is called a weak solution of sulfuric acid. When the electrolyte solution is covering the plates, a chemical reaction occurs that causes a buildup of excess electrons on the plates, creating voltage. Each cell typically produces 2 volts, so three sets of cells are connected in series for a 6-volt battery, and six sets of cells are connected in series for a 12-volt battery. The lead-acid battery can be recharged by

Figure 2–5 A diagram of a lead-acid battery showing lead plates inside. *(Courtesy of Crown Battery Mfg Company)*

putting electrical current from an alternator in the car, or a battery charger in the garage, back into the battery terminals, which reverses the chemical process between the plates and the electrolyte and makes the battery ready to produce electrons again.

Another type of rechargeable battery you will become familiar with are the rechargeable batteries used in power tools such as portable drills or portable screw drivers that you will use on the job. The battery used in a portable power tool is a nickel cadmium (NiCd) battery, which is also called the "Ni CAD" battery. Figure 2–6 shows a typical battery for a portable tool. Newer rechargeable batteries that you might find in portable power tools or portable test equipment include Lithium-ion (Li-ion) or Nickel-Metal Hydride (NiMH).

Figure 2–6 An example of a nickel cadmium (NiCd) battery used in portable power tools.

Figure 2–7 Typical 1.5- and 9-volt dry cell batteries, such as AAA, AA, 9 volt, B, and D.

During the recharging process the batteries allow electrons to become available for use again. The rechargeable battery allows the electrons in the battery to be used, and then be replaced during a recharging process, so the battery can be used over and over.

Storage batteries are batteries that can be recharged, whereas dry cell batteries are not designed to be recharged. Dry cell batteries are a low-cost way to produce 1.5 or 9 volts of electricity from a chemical reaction. Figure 2–7 shows examples of 1.5- and 9-volt dry cell batteries. The most common type of dry cell battery is called an alkaline battery, and it has two electrodes called cathode and anode. The anode is made of powdered zinc, and the cathode is made of manganese dioxide. The cathode and anode are placed into a metal container and a fibrous fabric separates them. The electrolyte for the battery is potassium hydroxide and water or manganese dioxide. When the electrolyte comes into contact with the anode, the anode is oxidized and electrons are conducted to a brass conductor pin that is connected to the negative terminal of the battery. A dry cell battery can only be used once and then it must be disposed of. The advantage of a dry cell battery is that it is inexpensive and it provides relatively long life in test meters and flashlights where the number of electrons needed is rather small. Rechargeable batteries such as those used in power tools use a large number of electrons in a short period of time.

2.1.4 Using Magnetism to Create Voltage

When current flows through a coil of wire, a magnetic field is created. The coil will have north and south poles that have flux lines. The more current that flows through the coil, the more flux lines that are produced. When a wire moves through the magnetic flux lines, a current flow and voltage potential is created in the wire. The simplest machine that creates voltage is a generator. The coils of wire in a generator are connected to a shaft and when the shaft is rotated, the wire in the coils moves past the magnetic flux lines, causing voltage and current to be created. The generator uses brushes to turn the voltage into DC voltage. An alternator works similar to the generator, but the alternator creates AC voltage. AC stands for alternating current and is the current used in households. In **alternating current**, the electrons in the current flow one way and then reverse. The frequency of the AC current will determine how fast it reverses. In the United States the frequency of AC is 60 cycles per second, which is also called 60 hertz. This means that the current flows in one direction for 1/60th of a second, then reverses for 1/60th of a second, which allows it to be transmitted much farther than DC voltage. You will learn more about the advantages of AC in later chapters.

A transformer uses two coils of wire that are located side by side to create a magnetic field that builds and collapses as AC current flows through the coils. The transformer changes the value of voltage, such as from 120 volts AC that comes into an HVAC system to 24 volts AC that is used in the control circuit and thermostat to turn the unit on and off. The HVAC system also uses a variety of coils of wire to create magnetic fields for relay coils, solenoid valve coils, and motors. In some applications, the coil of wire is specifically manufactured to change the current flowing through the coil into heat. An example of this is the electrical heating coils used in electric heat furnaces. When current flows through the heating coil, the resistance in the coil is designed to cause the coil to heat up quickly and give off a large amount of heat so that the furnace can heat a home.

2.2 **EXAMPLE OF VOLTAGE IN A CIRCUIT**

Voltage is the force that moves electrons through a circuit. Voltage is also called **electromotive force (EMF)** or *potential difference*. Figure 2–8 shows the effects of voltage in a circuit where two identical light bulbs are connected to separate voltage supplies. You should notice that the bulb connected to the 12-V battery is much brighter than the bulb connected to the 6-V battery. The bulb connected to the 12-V battery is brighter because 12 V can exert more force on the electrons in the circuit than 6 V, and higher voltage will also cause more electrons to flow. Voltage is also called electrical pressure, and it is the difference in **electrical potential** between two points. By definition 1 V is the amount of potential required to move one ampere of current through 1 ohm of resistance. Voltage is measured in volts and its abbreviation is V. Current is measured in amperes and its abbreviation is A. Resistance is measured in ohms and its abbreviation is Ω (omega), which is the Greek alphabet letter for "O." The letter O is not used as the abbreviation of ohms because it will get confused with the number 0. For example, if you wanted to write ten ohms and use the letter O it would look like 10 O, which might be confused for one hundred. If we use the Greek letter Ω you will not confuse 10 Ω for any other value.

You will need to become familiar with several other terms when you are measuring electricity with an electrical meter. At times the amount of voltage will be very small, such as when you measure the voltage in a thermocouple in the millivolt range. A millivolt is 1/1000 of a volt and can also be shown as 0.001 V. If you are trying to measure very small currents, the amount may be 1/1,000,000 (1 millionth) of an amp. The term for 1 millionth is *micro*, and it uses the Greek letter for *m*, which is μ. If you were measuring 4 micro amps you would write it as 4 μA. If you are working with electrical resistors, the amount of resistance may be very high such as 3000 ohms, which could also be

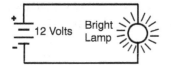

Figure 2–8 Two identical light bulbs with different voltages applied to them. The light bulb with 12 V applied to it is brighter than the one with 6 V applied to it.

Amount	Prefix Name	Value with Prefix Symbol
1 million (1,000,000)	Mega	1 M
1 thousand (1,000)	kilo	1 k
1 thousandth (0.001)	milli	1 m
1 millionth (0.000001)	micro	1 μ

Figure 2–9 Table showing prefix for Mega, kilo, milli, and micro.

stated as 3 kilo ohms or 3 kΩ where the letter k is used for kilo. Larger resistors may have up to 5 million ohms. The term *Mega* is used to indicate a million and the capital letter M is used, so the resistance value of 5 million ohms could be shown as 5 MΩ. A table is provided in Figure 2–9 to show values for all of these prefixes you will find shown on your electrical meter. You can use each of these prefixes with units of volts, ohms, or amperes.

In the diagram of the fan motor in Figure 2–1, the amount of voltage is 120 V, which is the voltage used in people's homes to make the air conditioning run. The utility company produces this voltage from a generator whose shaft is turned by a source of energy such as steam created by heating water to its boiling point by burning coal, or by utilizing the heat produced in a nuclear reaction. The energy to turn the generator shaft can also come from water moving past a dam at a hydroelectric facility. In the example circuit the voltage comes from a battery. The voltage in a battery is stored in a chemical form when the battery is originally charged. The larger the amount of voltage in a circuit, the larger the force that can be exerted on the electrons, which makes more of them flow.

2.3 EXAMPLE OF CURRENT IN A CIRCUIT

The definition of *current* is the flow of electrons. The actual number of electrons flowing at any particular time in a circuit can be counted, and their number will be very large. The unit of current flow is the *ampere* (A), which is usually shortened to *amp*. One amp is equal to 6.25×10^{18} electrons flowing past a point in 1 s. Figure 2–10 shows a meter measuring the number of electrons flowing. When 6,250,000,000,000,000,000 electrons have passed the meter in 1 s, an ampere will have been measured. This number of electrons is called a *coulomb* (C). The ampere is used in all references to heating, air-conditioning, and refrigeration equipment as the unit of electrical current.

Figure 2–10 Electrons passing a point where they are measured. When the number reaches 6.25×10^{18} in 1 s, 1 A has been measured.

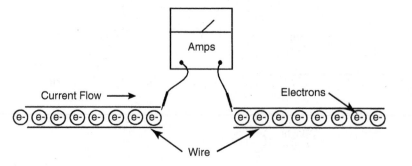

The amount of current flow in a circuit is determined by the current that each load draws. For example, if an air-conditioning system has a condenser fan motor that draws 2 amps and a compressor motor that draws 10 amps, the total current draw for both loads will be 12 amps. The amount of current needed at the motors will be supplied to them through wires, whose size must be large enough to safely carry the current. The larger the wire size, the more current it can carry. In future chapters you will work with a wire sizing table and the National Electric Code. The meter that is used to measure current is an **ammeter**. An ammeter that has a claw that opens to clamp around a wire to measurer its current is called a **clamp-on ammeter**.

2.4 ## EXAMPLE OF RESISTANCE IN A CIRCUIT

Resistance is defined as the opposition to current flow. The opposition is caused by friction in the conductor. Resistance is present in the wire that is used for conductors and in the insulation that covers the wires. The unit of resistance is the *ohm*, and its symbol is Ω. When the resistance of a material is very high, it is considered to be an **insulator**; if its resistance is very low, it is considered to be a *conductor*. The meter that measures resistance is called an **ohmmeter**.

Conductors must have very low resistance so that electrons do not require a lot of force to move through them. Also insulation that is used to cover wire must have very high resistance so that the electrons cannot move out of the wire into another wire that is nearby. The insulation also ensures that a wire on a hand tool such as a drill or saw can be touched without electrons traveling through it to shock the person using it. Examples of materials that have very high resistance and can be used as insulators are rubber, plastic, and air.

The amount of resistance in an electrical circuit can also be adjusted to provide a useful function. For example, the heating element for a furnace shown in Figure 2–11 is manufactured to have a specific amount of resistance so that the electrons will heat it up when they try to flow through it. The same concept is used in determining the amount of resistance in the electric heating element of an electric furnace. The heating element for the furnace has approximately $10\ \Omega$ of resistance, which is sufficient to cause the electrons to work harder as they move through the element. This causes it to heat up until it gives off heat. When the amount of resistance in a circuit such as a heating element or motor is designed to convert energy, it is referred to as a *load*.

Figure 2–11 Electric furnace heating element is an example of a resistance load.

2.5 ELECTRIC POWER IN A CIRCUIT

The definition of **electrical power** is the rate that work is done. In simpler terms electrical power is calculated by multiplying the amount of voltage times the amount of current that is used by a load such as a compressor motor (power = voltage × current). The unit of power is watts, and your utility uses a watt-hour meter to measure how many watts per hour that you use. Since the typical household uses thousands of watt-hours per month, the prefix kilo is used to indicate how many thousand watt-hours are used, so the bill will be shown in kilowatt-hours.

Other important units of measurement of work are the horsepower (hp) and the British Thermal Unit (BTU). One horsepower is equal to 746 watts and one watt is equal to 3.41 BTU/hour. The motors in an air-conditioning system will be rated in horsepower and the heat output of a furnace will be rated in BTU. Since current measurements indicate the amount of electrons flowing past a point in one second, utility companies use a larger period of time, one hour, for measuring power usage. For example, when a larger volume of watts is used when air-conditioning systems are running, it is easier to use a kilowatt-hour (kWh) meter to indicate how many thousands of amperes are used in one hour. The utility company then applies a rate such as 15 cents per kWh, which determines how much you must pay for electrical power usage. Since the electric utility company provides power to thousands of households, the output from the electric utility generator must be millions of watts per hour, which means their rating is in megawatts.

As an air-conditioning and heating technician, you will be expected to make calculations of the amount of energy a compressor uses when its rating is in hp or the heating output of an electrical furnace when its output is provided in BTU. The data plate for a furnace or air conditioner will show the amount of voltage, current, and kW that is used. The data plate shown in Figure 2–12 shows the amount of kW used, but it does not provide the amount of BTU output of a furnace, so you will need to complete a calculation to show the conversion from kW per hour to BTU per hour. The example will show you how to calculate the amount of BTU per hour for an electric furnace that draws 3000 watts (3 kilowatts).

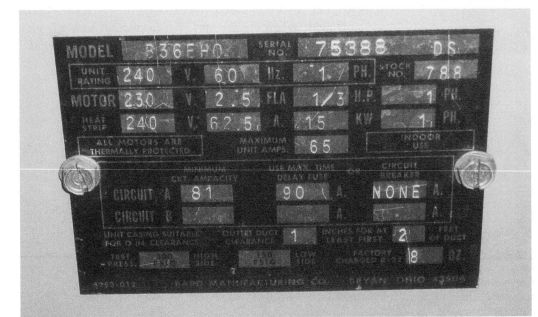

Figure 2–12 Data plate from an electric furnace.

Step 1 BTU/hour = kW × 3.41 BTU (1 kW = 3.41 BTU)

Step 2 BTU/hour = 3000 × 3.41 BTU

Step 3 BTU/hour = 10,230 BTU/h

Data plates also show the efficiency of the furnace and air conditioner so that home-owners can compare the efficiency of one system to another to determine which will be the cheapest to operate. All companies that make air-conditioning equipment have agreed on a formula, called the seasonal energy efficiency ratio (SEER), that is determined by the Air Conditioning and Refrigeration Institute. SEER ratings allow you to compare similar units to determine which is the most efficient or to determine the cost for running a unit you are purchasing. The formula for SEER is the total cooling output in BTU of the system divided by the power input, which is the total amount of energy that is consumed. For example, an air-conditioning unit with a SEER of 12 will produce 12 BTU of cooling per watt hour of electrical power consumed. The higher the SEER number, the more efficient the unit is. New federal energy regulations now mandate that all central air-conditioning units made in the United States after January 2006 must have an efficiency rating higher than 13.

2.6 IDENTIFYING THE BASIC PARTS OF A CIRCUIT

Each electrical circuit must have a *supply of voltage, conductors* to provide a path for the electrons to flow, and at least one *load*. The circuit may also have one or more controls. Figure 2–13 shows an example of a typical electrical circuit with the voltage supply, conductors, control, and load identified. In this circuit, the voltage source is supplied through terminals L1 and N, the load is a fan motor, and the two wires connecting the fan motor to the voltage source are the conductors. The thermostat switch is called an **electrical control**, since it controls the current flow to the motor. In a large system such as a residential air-conditioning system, it is possible to have more than one load. For example, the compressor motor, the condenser fan motor, and the evaporator fan motor are all loads in an air-conditioning system. You will learn more about these components later.

2.7 COMPARING ELECTRICITY WITH A WATER SYSTEM

It may be easier to understand electricity if you can compare it with a system that may be familiar to you, such as a water system. Water flowing through a hose, for example, would be equivalent to the flow of current (amps), and the water pressure would be

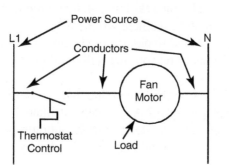

Figure 2–13 The basic parts of a circuit: the power source, conductors, control, and load.

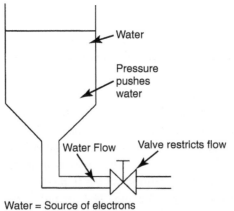

Figure 2–14 A diagram of similarities between a water system and an electrical system. The flow of water is similar to the flow of current (amps), the pressure on the water is similar to voltage, and the resistance from the valve causes the flow of water to slow just as resistance in an electrical system causes current flow to slow.

Water = Source of electrons
Water pressure = Voltage that pushes electrons
Water flow = Current flow
Valve restricts flow = Electrical resistance

equal to voltage. If you stepped on the hose or kinked the end, it would create a resistance that would slow the flow, just like resistance in an electrical circuit slows the flow of current. The diagram in Figure 2–14 shows the similarities between a water system and an electrical circuit.

By definition, 1 **volt** is the amount of force required to push the number of electrons equal to 1 ampere through a circuit that has 1 Ω of resistance. Likewise, 1 **amp** is the number of electrons that flow through a circuit that has 1 Ω of resistance when a force of 1 V is applied, and 1 **ohm** is the amount of resistance that causes 1 amp of current to flow when a force of 1 volt is applied. You can see that voltage, current, and resistance can all be defined in reference to one another.

2.8 USING OHM'S LAW TO CALCULATE VOLTS, AMPS, AND OHMS

There are times when you are working in a circuit that you need to calculate or estimate the amount of volts, amps, or ohms the circuit should have. In most cases when you are troubleshooting an electrical problem, you will also need to make measurements of the volts, amps, and ohms to determine if the circuit is operating correctly. The problem with taking a measurement is that you cannot be sure if the values are too high, too low, or just about right. A calculation can tell you if the measurements are within the expected values for the circuit.

A relationship called **Ohm's law** exists between the volts, amps, and ohms in every DC circuit. Ohm's law simply states that the amount of voltage in a DC circuit is always equal to the current multiplied by the resistance in that circuit. If a circuit has 2 A of current and 4 Ω of resistance, the total voltage will be 8 V.

The formula can be rearranged to determine the amount of amps or ohms. Since we know that 2 × 4 = 8, it stands to reason that 8 divided by 2 equals 4, and 8 divided by 4 equals 2.

If you measure the voltage in a circuit and find that you have 8 V, and measure the current in a circuit and find that you have 2 A, you can calculate that you have a resistance of 4 Ω by dividing 8 by 2. You can also determine the amount of current in the circuit by measuring the volts and ohms and calculating: 8 V divided by 4 Ω is 2 A.

2.9 OHM'S LAW FORMULAS

Scientists, engineers, and technicians use letters of the alphabet to represent units such as volts, amps, and ohms in calculations. The letters are commonly accepted abbreviations, so that when a calculation is passed from one person to another, it can be understood by everyone.

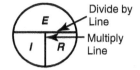

Figure 2–15 Ohm's law pie, a memory tool for remembering all the Ohm's law formulas.

In Ohm's law formulas, the letter E is used to represent voltage. The E derives from the first letter of the words *electromotive force*. The letter R is used to represent resistance. The letter I is used to represent current (amps). It derives from the first letter of the word *intensity*.

Ohm's law can be represented by the formula $E = IR$. When you use Ohm's law, the basic rule of thumb is that you should use E, I, and R when the values are *unknown*, and you should use V for volts, A for amps, and Ω for ohms when you have determined an answer or if the values are *known* or measured. Figure 2–15 shows a memory tool called an Ohm's law pie that can help you remember the formulas for voltage, current, and resistance.

2.10 USING OHM'S LAW TO CALCULATE VOLTAGE

When you are solving a calculation for voltage, you need to know the amount of current (amps) and the amount of resistance (ohms). Figure 2–16 shows the circuit with a 40 ohm resistor and 20 amps of current flow. If you know these two values, you can calculate the voltage. For example, if you are asked to calculate the amount of voltage when the current is 20 A and the resistance is 40 Ω, you can multiply 20 \times 40 to get an answer of 800 V. When the problem is stated, you should start with the formula and substitute the values for I and R. Then you can continue the calculation. Notice that the units A for amps and Ω for ohms are added when values are known.

Step 1 $\quad E = I \times R$

Step 2 $\quad E = 20\,\text{A} \times 40\,\Omega$

Step 3 $\quad E = 800\,\text{V}$

2.11 USING OHM'S LAW TO CALCULATE CURRENT

The formula for calculating an unknown amount of current in a circuit is $I = \frac{E}{R}$.

When you are calculating the current, you must divide the voltage by the resistance. For example, in Figure 2–17 if you have measured the circuit voltage and found that it is

Figure 2–16 A circuit with current and resistance provided. Calculate voltage using Ohm's law.

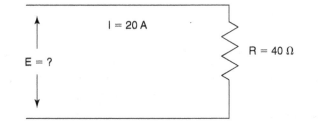

I = 20 A

R = 40 Ω

E = ?

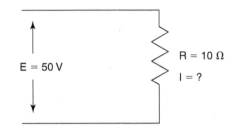

Figure 2–17 A circuit with voltage and resistance provided. Calculate current using Ohm's law.

50 V, and if you have measured the resistance and found that it is 10 Ω, you can find the current by dividing 50 V by 10 Ω, which equals 5 A.

Step 1 $I = \dfrac{E}{R}$

Step 2 $I = \dfrac{50\ V}{10\ \Omega}$

Step 3 $I = 5\ A$

2.12 USING OHM'S LAW TO CALCULATE RESISTANCE

The formula for calculating an unknown amount of resistance in a circuit is $R = \frac{E}{I}$.

When you are calculating the resistance of a circuit, you must divide the voltage by the current. In Figure 2–18 if you have measured the voltage to be 60 V and if you have measured the current and found that it is 12 A, you can find the resistance by dividing the voltage by the current. Again, you should start with the formula and substitute values to calculate an answer.

Step 1 $R = \dfrac{E}{I}$

Step 2 $R = \dfrac{60\ V}{12\ A}$

Step 3 $R = 5\ \Omega$

2.13 USING OHM'S LAW WHEEL TO REMEMBER OHM'S LAW FORMULAS

A learning aid has been developed to help you remember all the Ohm's law formulas. This aid, shown in Figure 2–19, is in the shape of a wheel or a pie. The top of the pie contains the letter E to represent voltage, the bottom left contains the letter I to represent current (amps), and the bottom right contains the letter R to represent resistance (ohms). If you want to remember the formula for voltage, put your finger over the E, and the letters I and R show up side by side. The vertical line separating the I and R is called the *multiply line* because that is the math function you are supposed to use with I and R. This use of the pie represents the formula $E = IR$.

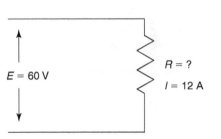

Figure 2–18 A circuit for calculating resistance using Ohm's law. Voltage and current are provided in this circuit.

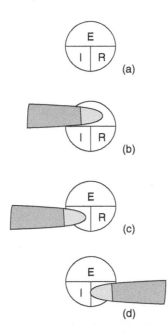

Figure 2–19 a Ohm's law pie is a memory tool for remembering all the Ohm's law formulas. **b** Put your finger over the *E* when you want the formula to find voltage. **c** Put your finger over the *I* when you want the formula to find current. **d** Put your finger over the *R* when you want the formula to find resistance.

If you want to know the formula for current *(I)*, you put your finger over the *I*, and the remaining letters are the *E* over the *R*. The horizontal line that separates the *E* and *R* is called the *divide by line* because you are supposed to divide the *E* by *R*. This use of the pie represents the formula $I = \frac{E}{R}$.

If you want to know the formula for resistance *(R)*, put your finger over the *R*, and the remaining letters will be the *E* over the *I*. This use of the pie represents the formula $R = \frac{E}{I}$.

2.14 CALCULATING ELECTRICAL POWER

Electrical power is the rate of work performed by electricity. The units for electrical power are **watts**. Watts are calculated by multiplying amps times volts *(P = IE)*. Most **electrical loads** that are resistive in nature, such as heating elements, will be rated in watts. For example, in Figure 2–20 a heating element is rated for 5 A and 120 V, and it will use 600 W of power. You can calculate the amount of power by using the formula $P = IE : 5\,A \times 120\,V = 600\,W$.

Step 1 $P = EI$

Step 2 $P = 120\,V \times 5\,A$

Step 3 $P = 600\,W$

Power is calculated by multiplying voltage times amperage and measured in watts. The power company charges for the amount of wattage that a household uses. Electrical systems such as air-conditioning and heating systems are the major users of power in a home.

Figure 2–20 A circuit with volts and amps provided. Use the formula to calculate power in watts.

$E = 120\,V$

$P = ?$

$I = 5\,A$

Figure 2–21 The formula pie for Watt's law formulas that are used to calculate electrical power.

The meter that measures electrical power usage at a home is called a watt/hour meter, which measures the amount of power (watts) over each hour. A thousand watts is called a kilowatt, and the bill you receive from the power company will be shown in kilowatt-hours.

The basic formula for power can be rearranged like the Ohm's law formula. You can calculate current by dividing the power (600 W) by the voltage (120 V), and you can calculate voltage by dividing the power (600 W) by the current (5 A). Figure 2–21 shows a pie that is a memory aid for remembering all the formulas for Watt's law. From this pie you can see that the original formula, $P = IE$, can be found by placing your finger over the P. The formula to solve for current $\left(I = \frac{P}{E}\right)$ can be found by placing your finger over the I, and the formula to solve for voltage $\left(E = \frac{P}{I}\right)$ can be found by placing your finger over the E.

2.15 PRESENTING ALL THE FORMULAS

When you are troubleshooting, you may be able to get a voltage reading easily. If you can find a data plate that gives the wattage for a heating element or other load, you will be able to use one of the formulas to calculate (estimate) the current or the resistance. This estimate will give you an idea of what the values should be so that you can determine whether the system is working correctly. When you are in the field making these measurements, it may be difficult to remember all the formulas, so a chart has been developed that provides all the formulas. (See Figure 2–22.) From the figure you can see that the chart is in the shape of a wheel with spokes emanating from an inner ring. A cross in the middle of the inner wheel separates the inner wheel into four sections. These sections are represented by the letters P (power or watts), I (amps), E (volts), and R (ohms). (Notice that power can be identified as either W for watts or P for power.)

The formulas in each section of the outer ring correspond to either $P, I, E,$ or R. The formulas for wattage are shown emanating from the inner section marked with the P. All the formulas for current are shown emanating from the inner section marked with an I. All the formulas for voltage are shown emanating from the inner section marked with an E, and all the formulas for resistance are shown emanating from the inner section marked with an R.

If you are trying to calculate the power in a circuit or the power used by any individual component, you can use any of the three formulas on the wheel emanating from the section identified by the letter P. This means that you can use the formula $P = EI$, $P = I^2 \times R$, or $P = E^2 / R$. Your choice of formula will depend on the two values that are given. *Note:* The function of multiplication can be shown using an asterisk (*) or by writing the two terms together, such as in $P = IE$.

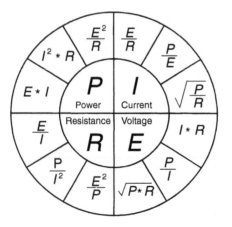

Figure 2–22 Formulas for calculating power, current, voltage, and resistance.

2.16 REVIEW

Electricity is defined as the flow of electrons. Direct current (DC) is electron flow in only one direction, whereas electron flow first in one direction and then the reverse for the other part of each cycle is called alternating current (AC). Electricity is also defined as the flow of current and is measured in amperes (amps). Voltage is the force that causes electrons to move. Voltage is also called electromotive force (EMF). The units for voltage are volts. Resistance is defined as the opposition to current flow. Its units are ohms. Current is the flow of electrons and the units for current are amps. Power is defined as the work performed by electricity. It is calculated by multiplying volts times amps. The units for power are watts.

Ohm's law defines the relationships among voltage, current, and resistance in a circuit. A set of formulas based on Ohm's law are used to calculate volts, ohms, or amps when any two of the values are known. The formulas are shown in Figure 2–19. Power can also be calculated by using voltage and current. Figure 2–22 shows all of the formulas for volts, ohms, amps, and power. There are four basic prefixes that you will find on a meter. The prefix for 1 million (1,000,000) is Mega, and for 1 thousand (1000) is kilo. The prefix for one thousandth (1/1000) is milli and for one millionth (1/1,000,000) is micro.

Job Assignment for This Chapter

You are on a service call to a customer who does not understand the basic theory of electricity and thinks you are trying to sell parts that are not needed. Your job is to use your knowledge of basic electricity and meters to explain how you can use your meter to test any of the switches in this system and how you can tell whether they are good or bad.

Solution to the Job Assignment

Your solution should include a basic statement about how current flows in a circuit. You should feel comfortable using terms such as *voltage, current,* and *resistance.* You should be able to explain that meters can measure voltage, current, and resistance and that if a circuit has a bad part, the current will not flow through it and the circuit will not operate correctly. You should also be able to explain that you could use a meter to test any switches that you thought were bad.

Assessment

QUESTIONS

1. Define the term *electricity.*
2. What is voltage?
3. What is current?
4. Explain what a valence electron is and why it is important to current flow.
5. What is resistance?
6. Define the term *power.*
7. Explain the term *infinite resistance.*
8. Explain how magnetism is created in a coil of wire.
9. What is EMF?

10. What is the unit for power?
11. What is static electricity?

TRUE OR FALSE

1. The insulation coating on wire has high resistance.
2. Voltage is the flow of electrons.
3. Resistance is the opposition to current flow.
4. Watts are the units for current.
5. A good conductor has low resistance.
6. The negative part of an atom is an electron.
7. Milli is the prefix for 1/1000.
8. A battery creates voltage from a chemical reaction.
9. Voltage is pressure that causes current to flow.
10. Kilo is the prefix for one thousand (1000).

MULTIPLE CHOICE

1. A valence electron is
 a. the electron closest to the nucleus.
 b. the electron in the outermost shell.
 c. an electron with a positive charge.

2. Wattage is
 a. the unit for power.
 b. the unit for current.
 c. the unit for voltage.

3. Resistance is
 a. the force that moves electrons.
 b. wattage.
 c. the opposition to current flow.

4. Current will _____ when voltage increases and resistance stays the same.
 a. increase
 b. decrease
 c. stay the same

5. Wattage will _____ when voltage increases and current stays the same.
 a. increase
 b. decrease
 c. stay the same

6. A milliamp is
 a. one-thousandth of an ampere.
 b. one-millionth of an ampere.
 c. one million amps.

7. The atom has which of the following parts?
 a. electron
 b. proton
 c. neutron
 d. nucleus
 e. all the above

8. A load is
 a. the part of a circuit that has resistance.
 b. a switch in a circuit that controls current flow.
 c. a power source for a circuit.

9. A conductor is
 a. a material that stops the flow of electrons.
 b. a material that allows electrons to flow easily.
 c. a source of voltage for a circuit.

10. A control is
 a. the part of a circuit that has resistance.
 b. a switch in a circuit that controls current flow
 c. a power source for a circuit.

PROBLEMS

1. Use Ohm's law to calculate the amount of voltage if current in a DC circuit is 5 A and resistance is 8 Ω.

2. Use Ohm's law to calculate the amount of current in a DC circuit if voltage is 25 V and resistance is 50 Ω.

3. Use Ohm's law to calculate the amount of resistance in a DC circuit if voltage is 40 V and current is 10 A.

4. Use the power formula to determine the wattage in Problem 1.

5. Use the power formula to determine the wattage in Problem 2.

6. Use the power formula to determine the wattage in Problem 3.

7. Calculate the BTU output for an electric heating element that is rated for 1000 kW.

8. Calculate the number of BTU in 10 hp.

9. Calculate the BTU output of the electric heating element that draws 10 amps and uses 240 VAC.

10. Calculate the total amperage draw for an air-conditioning condenser that has a fan motor that uses 3.5 amps and a compressor motor that uses 12 amps.

Websites

Websites change constantly. Use the key word search to locate sites that have similar content. The web sites can be used to learn more about the terms for this chapter.

Ammeter
Clamp-on ammeter
Dry cell batteries
Ohmmeter
Portable power tool batteries
Voltmeter

Chapter 3

Electrical Circuits

Job Assignment for This Chapter

A customer has a question about why a fuse continues to blow on the outdoor air conditioning unit that contains the compressor and condenser fan. Explain to this customer the function of a fuse and how it protects both motors when it blows. You should also explain to the customer how the fuse is connected in series with the fan motor and compressor and how they are connected in parallel with each other. You should also give the customer some ideas of how you could tell which component in the circuit is causing the circuit to draw more current than the fuse is rated for.

Terms You Will Learn in This Chapter

Closed switch

Control circuit

Good fuse

Open circuit

Open fuse

Open switch

Parallel circuit

Power circuit

Series circuit

Series-parallel circuit

Voltage drop

After reading this chapter, you will be able to:

1. Explain the operation of switches that are connected in series.
2. Explain the operation of electrical loads that are connected in series.
3. Explain the relationship among voltage, current, resistance, and power in a series circuit.
4. Utilize Ohm's law for series circuit calculations.
5. Explain the relationship among voltage, current, resistance, and power in a parallel circuit.
6. Use Ohm's law to calculate the voltage, current, resistance, and power in a parallel circuit.
7. Explain the operation of loads, such as motors, connected in parallel.
8. Explain the operation of switches connected in parallel.
9. Explain the relationship between switches and loads that are connected in series-parallel.
10. Calculate the voltage, current, resistance, and power in a series-parallel circuit.

3.0 INTRODUCTION

As a service technician who works on heating, air-conditioning, and refrigeration systems, you need to become familiar with electrical circuits. This chapter will help you understand how the electrical controls and switches combine to control fan motors, electric coils, and compressors. You will also learn how to read electrical diagrams and recognize the symbols of components that you will use to install and troubleshoot HVAC and refrigeration equipment.

This chapter will explain the fundamentals of series and parallel circuits, which are the building blocks of all electrical circuits. A **series circuit** has only one path for current flow; anytime a switch or control is opened, all current flow to the load or motor stops. In a **parallel circuit**, motors such as the compressor and condenser fan are connected so that current can flow to each independently, and each can be turned on or off by themselves. A **series-parallel circuit** allows switches and other controls to be connected in series and loads such as fan motors and compressor motors to be connected in parallel. Parallel circuits are designed to ensure that all loads that are connected in parallel receive the same amount of voltage.

3.1 FUNDAMENTALS OF ELECTRICAL CIRCUITS

A basic simple circuit consists of a power source, conductors (wires), one or more switches (controls) that can close to allow current to flow through them or open to stop current flow, and a load such as a motor. The basic diagram can be drawn in two ways. The diagram in Figure 3–1 shows these components drawn as a wiring diagram, and the

Figure 3–1 A power source, switch, and fan motor shown in a wiring diagram. The function of the wiring diagram is to show the location of components in a circuit.

Figure 3–2 A power
source, switch, and fan
motor shown in a
schematic (ladder)
diagram. Notice that the
overall appearance of the
ladder diagram resembles
a wooden ladder. This
diagram shows sequence
of operation.

diagram in Figure 3–2 shows the components in a ladder diagram. The wiring diagram is used to show the location of components in a circuit, whereas the schematic (ladder) diagram is used to show the sequence of operation. The reason the diagram in Figure 3–2 is called a ladder diagram is that as additional loads are added they are shown on additional "rungs," which makes the overall appearance of this diagram look like a wooden ladder. Basically the load in the first lines of the diagram will energize before the loads in the second or third lines (rungs). You will learn much more about ladder diagrams in future chapters. You will find that wiring diagrams and schematic (ladder) diagrams are used throughout this book. Equipment manufacturers will provide either the schematic (ladder) diagram or wiring diagram or both to help you install and troubleshoot the system.

Some basic terms are provided here to help you understand how electricity works. For example, a switch can have two basic conditions, open or closed. When a switch is open in a circuit, it is called an **open switch** and electrons cannot flow past the open contacts and all current in the circuit stops and the load is turned off. For example, if the switch is controlling a fan motor, when the contacts are opened, the current cannot flow through the switch and the motor is turned off. When the switch is closed, it is called a **closed switch** and the current is allowed to flow and the motor will run. (See Figure 3–3.)

3.1.1 Testing Wires, Switches, and Fuses to Determine Whether They Are Good or Bad

A good wire has low resistance, which allows current to flow freely. If the wire gets cut, it is said to have an open and the current cannot flow past the open part where the wire has been cut. If the wire is a part of a circuit when it has an open, the circuit is said to be an **open circuit**. Since the amount of resistance for a wire that has been cut is very high, an ohmmeter can measure the amount of resistance. Figure 3–3a shows an ohmmeter

(a) (b)

Figure 3–3 **a** An analog meter with leads not touching and the scale showing infinite resistance. **b** An analog meter with leads touching and the scale showing continuity (zero ohms).

that is measuring the resistance of a good wire that has low resistance, and Figure 3–3b shows the meter testing a bad wire that has been cut and has high resistance. When you use an ohmmeter to make this type of test, it is called a *continuity test*. If the resistance is very high, it is said to be *infinite* and the wire is *open*. The symbol for infinite is ∞, which looks like the number 8 laying on its side. If the resistance is very low or *zero ohms*, the wire is said to be good. A good wire is also said to have *continuity*. You might also refer to a closed switch as having continuity through it. Switches and wires can be tested with an ohmmeter if they are not connected to other components and if no power is applied to them.

Fuses act like wires or switches in that they pass current when they are good and stop current flow when they open or go bad. A **good fuse** is a fuse that has continuity and passes current. A blown fuse has an open circuit in its metal and stops all current flow. A bad fuse has an open in it and is referred to as an **open fuse**. You can test a fuse with an ohmmeter to determine if it is open or has continuity. You might think of the word *continuous* when you think of continuity test. If the wire is continuous, and passes current, it is said to have continuity and is considered to be good.

As a technician you must be able to test a wire, fuse, or switch to determine whether it can carry current or has an open. You can use an ohmmeter to test all of the components. You may have to show a homeowner that you have tested these components to determine whether they are good or bad; if they are bad, they will need to be replaced. If you are using an analog meter that has a needle to test a good wire, the needle will swing to the right side of the scale toward the zero ohms reading indicating low resistance. If the wire is open, it will have high resistance and the meter needle will swing to the left toward the infinite (∞) symbol indicating it has very high resistance. If you are using a digital meter, it will read zero when the resistance is very low indicating the wire is good. If the wire is open, the meter will show dashes or flash a symbol indicating it is reading infinite. Figure 3–4a shows the digital meter reading infinite, and Figure 3–4b shows a digital meter reading low resistance.

(a) (b)

Figure 3–4 a A digital meter showing high resistance (infinite resistance) in an open fuse. **b** A digital meter showing low resistance when fuse is good.

3.1.2 Relationship Between Voltage, Current, and Resistance in a Series Circuit Using Resistors

Simple circuits consisting of resistors will be used to explain the relationships among voltage, current, and resistance in a series circuit. The most common resistance loads in air-conditioning and refrigeration systems are crankcase heaters, defrost heaters, and resistive heating strips. Resistors are small components that are common in modern electronic circuit boards that are used to control furnaces, air conditioners, and refrigeration systems. The motors and other loads in air-conditioning and refrigeration systems will react in some part as resistances, so it is important for you to understand the relationships among multiple resistances in series circuits. Understanding resistances in series circuits and in parallel circuits is also essential for understanding the solid-state components that are commonly used in solid-state controls found in furnaces, air conditioners, and refrigeration systems.

As a technician you must understand the effects of switches in series circuits and electrical loads in series circuits. The first part of this chapter will explain the relationships among switches and loads that are connected in series. The second part of this chapter will explain the effects of electrical switches and loads that are connected in parallel.

The relationships among voltage, current, and resistance in series parallel circuits will be introduced in the last part of this chapter. It will be easier to understand these relationships when a known value of resistance is used. A component called a resistor is used to provide the resistance for these simple series circuits. Each resistor has a color code consisting of colored bands that are painted on the resistor to identify its resistance value. Ohm's law will also be used to help you understand how voltage and current should react in all types of series circuits. The basic rules discussed in this chapter will help you troubleshoot larger circuits that are commonly found in heating, air-conditioning, and refrigeration systems.

3.2 EXAMPLES OF SERIES CIRCUITS

All electrical circuits have at least a power source, one load, and conductors to provide a path for current. As circuits become more complex, switches are added for control. These switches can provide a variety of functions depending on the way they are connected in the circuit. In Figure 3–1, we saw a diagram of a simple circuit that included a power source, a switch for control, and a motor as the load. This circuit is called a *series* circuit because the current has only one path to travel to the load and back to the power source.

When additional switches are added to the circuit to control the load, they can be connected so that all current continues to have only one path by which to travel around the circuit. Figure 3–5 shows a diagram of a typical circuit that controls power to a compressor for a refrigeration case. This circuit has a thermostat that controls the temperature at which the compressor turns on and off, a high-pressure switch that protects the compressor against excessive pressure if the condenser coils should get dirty, and an oil pressure switch that protects the compressor against loss of oil pressure. This type of circuit is also called a **control circuit**.

This circuit is called a series circuit because the current has only one path by which to travel from L1 through the switches to the load and back to N. If any of the three switches is opened, current is interrupted and the motor will be turned off. When all the switches are closed, the compressor motor will run. When the refrigeration case becomes too cold, the thermostat (temperature switch) will open and turn the compressor motor off. If the case becomes warm again, the thermostat will close and reenergize the compressor. Thus the thermostat is called the *operational control*, since its job is to operate

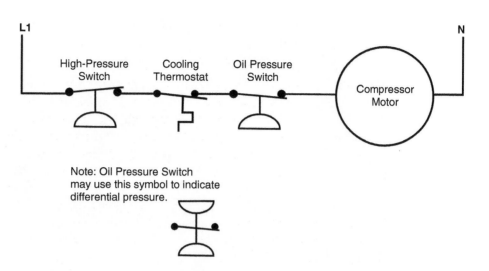

L1 N

High-Pressure Cooling Oil Pressure
Switch Thermostat Switch Compressor
Motor

Note: Oil Pressure Switch
may use this symbol to indicate
differential pressure.

Figure 3–5 Electrical diagram of a power source, high-pressure switch, thermostat, oil pressure switch, and compressor motor. All the switches are in series with the compressor.

the system at the setpoint temperature. The high-pressure switch and oil pressure switch are in the circuit for safety purposes. If the condenser becomes dirty and causes the high-side pressure (head pressure) to become excessive, the high-pressure switch will open and turn the compressor motor off. The same is true of the oil pressure switch. If the oil pressure becomes too low, the oil pressure switch will open and turn the motor off.

Since all three of these switches are connected in series, each is capable of opening and causing current to stop flowing to the compressor motor. If additional operational or safety switches need to be added to this circuit, they will be connected in series with the original switches. It is important to remember that when switches are connected in series, any one of them can open and stop current flow in the entire circuit.

3.3 CONNECTING SWITCHES IN SERIES

Switches are connected in series to control air-conditioning compressors and fans. The basic rule you will learn about series circuits is that any time a switch is open, all current stops and any loads that need the current will become de-energized. It is important to understand that all current in a series circuit stops whenever there is an open switch, open fuse, or open conductor.

3.4 OTHER COMPONENTS THAT ARE CONNECTED IN SERIES

A fuse is another electrical component that is connected in series with all wiring, motors, and controls. The fuses are designed to protect the circuit for the maximum amount of current the wire can handle. If for some reason the amount of current exceeds the normal amounts, the fuse will overheat and melt its fusible link and create an open. Since the fuse is in series with all loads and wires in the air-conditioning system, it will cause the power to the system to be turned off. Having the fuses wired in series makes them able to open and protect the entire circuit because an open anywhere in the series circuit will cause current to stop in all parts of the circuit.

A fuse disconnect is a switch that is combined with the fuses. The disconnect is connected in series with the fuses so it allows you to turn off all power to the unit when you need to change out an electrical component such as a fan motor or compressor. In this

Figure 3–6 A fused disconnect.

way, the disconnect switch is a safety device that is connected in series with every part of the electrical system, and since it is connected in series it will allow you to open it and stop current flow and voltage potential to every part of the system. Figure 3–6 shows a picture of a fused disconnect.

3.5 ADDING LOADS IN SERIES

In air-conditioning and refrigeration systems, loads such as motors are not connected in series with other loads; instead, they are connected in parallel with one another so that they all receive the same voltage. Some heating elements for electric furnaces are connected in series. You will also learn about how resistors are connected in series to create useful voltage drop circuits for electronic circuit boards and components used in HVAC electronic controls.

Loads are not generally connected in series because they would split the amount of applied voltage. For example, if you connected two light bulbs that have the same amount of resistance in their filaments in series with each other, and supplied the circuit with 120 V, each light bulb would receive 60 V. Each bulb would glow half as brightly as it would if it were to receive the entire 120 V.

In some cases loads are connected in series, such as in heating circuits. Figure 3–7 shows an example of two heating elements connected in series. If the amount of resistance

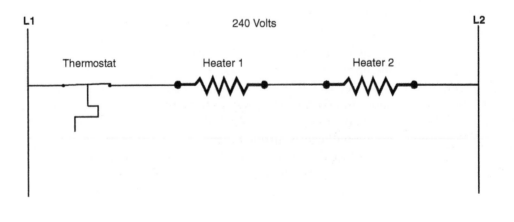

Figure 3–7 Electrical diagram of two 120-V heaters connected in series to a 240-V power supply.

for each heater is equal, the supply voltage will be split equally between them. In this example the supply voltage to the heaters is 240 V, and each heater will receive 120 V. This means that each heater is rated for 120 V. This type of circuit allows smaller heaters to be used on larger voltage sources.

Another problem with connecting loads in series is that if one of the loads has a defect and develops an open circuit, the current to all the loads in the circuit will be interrupted. An example of this type of circuit is a string of lights that is used to decorate a Christmas tree. If all the lights are connected in series, they will all go out if one of the lights burns out. The advantage of connecting 50 lights in series is that each of the light bulbs will receive approximately 2.4 V when the circuit is plugged into a 120-V power source.

3.6 **CALCULATING VOLTAGE, CURRENT, AND RESISTANCE IN SERIES CIRCUITS**

Some basic concepts about voltage, current, and resistance in series circuits can best be explained by simple circuits that have two or three resistors connected in series. This section will show you some basic series circuits that use resistors to change the amount of voltage and current in a circuit. These types of circuits are commonly found in the solid-state control boards that are now used in many heating, air-conditioning, and refrigeration systems. These simple circuits will help you understand how voltage, current, and resistance acts in a series circuit.

Figure 3–8 shows several resistors. The resistor is a component that has a specific amount of resistance. The electrical symbol for a resistor is the same as for the heaters shown in Figure 3–7. The amount of resistance will be identified on each resistor. The amount is printed on larger 5-watt and 25-watt resistors and with color stripes or bands on the smaller wattage resistors. Since the ½ watt and ¼ watt resistors are very small, it is easier to use a color code to indicate the amount of resistance, rather than print the amount on the side of the resistor in small print.

3.7 **RESISTOR COLOR CODES**

The resistor color bands are shown in their respective positions in Figure 3–9a. The resistor has four basic color bands that are grouped at one end of the resistor. (Note: A few resistors will have fifth and sixth bands that represent other manufacturing

Figure 3–8 Examples of resistors commonly found in electronic circuit boards used in heating, air-conditioning, and refrigeration systems. The resistor's physical size gets larger as the wattage rating increases.

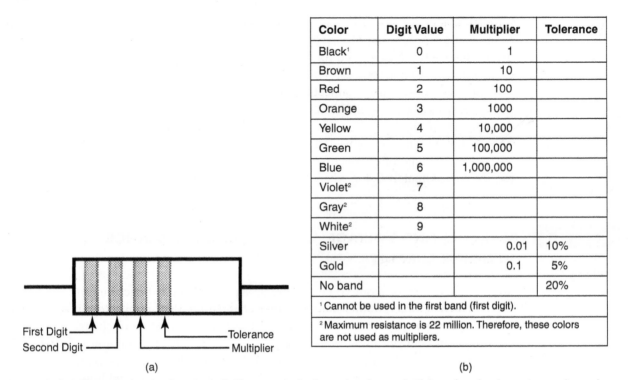

Color	Digit Value	Multiplier	Tolerance
Black[1]	0	1	
Brown	1	10	
Red	2	100	
Orange	3	1000	
Yellow	4	10,000	
Green	5	100,000	
Blue	6	1,000,000	
Violet[2]	7		
Gray[2]	8		
White[2]	9		
Silver		0.01	10%
Gold		0.1	5%
No band			20%

[1] Cannot be used in the first band (first digit).

[2] Maximum resistance is 22 million. Therefore, these colors are not used as multipliers.

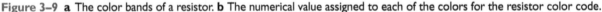

First Digit —
Second Digit —
— Tolerance
— Multiplier

(a) (b)

Figure 3–9 a The color bands of a resistor. **b** The numerical value assigned to each of the colors for the resistor color code.

information.) The color bands represent numbers that indicate the amount of resistance each resistor has. The first color band represents the first digit of the number, and the second color band represents the second digit of the number. Each digit can have a value from 0 through 9. The third band is a multiplier band that indicates the number of zeros to follow the two-digit numbers to show values above 99. The fourth band is called the *tolerance band*. The actual amount of resistance in the resistor will be close to the amount identified by the color bands, but it may be slightly more or less depending on the amount of tolerance. Typical tolerances are 5% and 10%. This means that the actual amount of resistance can be plus or minus the percentage identified by the tolerance band.

The color code is shown in Figure 3–9b, and you can see that each number 0–9 is represented by a color. The tolerances are represented by the colors silver and gold. A silver band means that the resistor has a tolerance of 10%, and a gold band means that the resistor has a tolerance of 5%.

3.7.1 Decoding a Resistor

A resistor with a brown first band, a blue second band, and a red third band is decoded as follows. The first two colors, brown and blue, mean that the first two numbers are 16. The third band, which is the multiplier band, is red, so you need to add two zeros (0s) to the first two digits, and the total resistance is 1600 Ω. If you want to use the multiplier for red, you multiply the first two digits by 100: ($16 \times 100 = 1600 \ \Omega$).

A second resistor has a violet first band, a white second band, and a green third band. This resistor is decoded as follows. The first two colors, violet and white, mean that the first two numbers are 79. The third band, which is the multiplier band, is green, so you need to add five zeros to the first two digits. This makes the total resistance 7,900,000 Ω. If you want to use the multiplier for green, you would multiply 79 by 100,000: ($79 \times 100,000 = 7,900,000 \ \Omega$).

Example 3-1

Determine the total resistance of a resistor that has a color code of yellow, brown, and orange.

Solution:

Since the first and second colors are yellow and brown, the first two digits are 41. The multiplier is orange so three zeros must be added to 41 to get a total of 41,000 Ω. If you use the multiplier for orange, you would multiply $41 \times 1000 = 41,000 \Omega$.

3.7.2 Calculating the Resistor Tolerance

The first resistor that was decoded had color bands of brown, blue, and red and a total resistance of 1600 Ω. If this resistor had a fourth band of silver, the resistor tolerance would be calculated as +10% and −10%. The first step in calculating this tolerance is to multiply the total resistance by 10%.

$$1600\ \Omega \times 0.10 = 160\ \Omega$$

The second step in calculating the tolerance is to separately add the 160 Ω and subtract the 160 Ω from 1600 Ω.

$$
\begin{array}{cc}
1600\ \Omega & 1600\ \Omega \\
+\ 160\ \Omega & -\ 160\ \Omega \\
\hline
1760\ \Omega & 1440\ \Omega \\
\end{array}
$$

These calculations show that the tolerance of the resistor with the color bands of brown, blue, red, and silver is 1760 Ω to 1440 Ω. If you were measuring this resistor with an ohmmeter and its total resistance fell between 1440 Ω and 1760 Ω, it would be considered a good resistor. If the total resistance was less than 1440 Ω or greater than 1760 Ω, the resistor would be considered out of tolerance and it would not be usable. If the fourth band was gold instead of silver, the multiplier would be 5% instead of 10%. The remainder of the calculation would be completed the same way in that you would separately add 5% and subtract 5% from the total resistance to get the upper and lower limits.

Example 3-2

Determine the total resistance with the tolerance of a resistor whose color code is yellow, red, orange, and gold.

Solution:

Since the first two colors are yellow and red, the first two digits are 42. The multiplier is orange, so you need to add three zeros to make the total resistance 42,000 Ω.

The tolerance band is gold, so the tolerance is +5% and −5%; 5% of 42,000 Ω is found by multiplying $42,000 \times 0.05$:

$$42,000 \times 0.05 = 2,100$$

The next step requires that you add 2,100 Ω to 42,000 Ω to get the upper tolerance, and subtract 2,100 Ω from 42,000 Ω to get the lower tolerance.

$$
\begin{array}{cc}
42,000\ \Omega & 42,000\ \Omega \\
+\ 2,100\ \Omega & -\ 2,100\ \Omega \\
\hline
44,100\ \Omega & 39,900\ \Omega \\
\end{array}
$$

The upper tolerance of this resistor is 44,100 Ω, and the lower tolerance is 39,900 Ω.

Figure 3–10 A series circuit with three resistors connected to create one path for current.

3.8

USING OHM'S LAW TO CALCULATE OHMS, VOLTS, AND AMPS FOR RESISTORS IN SERIES

In electronic circuits some of the loads are resistors. These resistors are sized so that the supply voltage will be dropped into smaller increments. These types of circuits are widely used in electronic circuits for air-conditioning, heating, and refrigeration equipment. Several such circuits will be used to explain how voltage and current are affected by changes in resistance. Figure 3–10 shows an example of three resistors connected in series with a voltage source (battery). Since the resistors are connected in series, there is only one path for current. The resistors are numbered R_1, R_2, and R_3 and the supply voltage is identified as E_T. The formula for calculating total resistance in a series circuit is $R_T = R_1 + R_2 + R_3$. If more than three resistors are used in the circuit, the additional resistors are added to the first three to obtain the total resistance. The arrows show conventional current flow. [Electron flow would show the flow of electrons (current) moving from the negative battery terminal to the positive terminal. This text will use conventional current flow in its explanations and diagrams.] If you would rather use electron-flow theory, the current would flow in the direction opposite to the arrows.

A series circuit has three resistors: R_1 is 30 Ω, R_2 is 20 Ω, and R_3 is 50 Ω. The total resistance for this circuit can be calculated by the formula.

$$R_T = R_1 + R_2 + R_3$$
$$R_T = 30\,\Omega + 20\,\Omega + 50\,\Omega$$
$$R_T = 100\,\Omega$$

3.9

SOLVING FOR CURRENT IN A SERIES CIRCUIT

The total current can be calculated by the Ohm's law formula $I = \frac{E_T}{R_T}$, $I = \frac{300\ V}{100\ \Omega}$, or $I = 3$ A. The total current in a series circuit is the same everywhere in the circuit because there is only one path for the current. This means that once you find the total resistance of a series circuit, you can divide total voltage by total resistance and find the total current, and you also have determined the current that flows through each resistor. The current (3 A) has been listed with each resistor in the diagram. Since the current is the same everywhere in the series circuit, it may be identified as I_T or I_1 where it is shown at resistor R_1.

3.10 **CALCULATING THE VOLTAGE DROP ACROSS EACH RESISTOR**

The voltage drop across each resistor can be calculated by the Ohm's law formula ($E = IR$). It is important to remember that the current in the series circuit is the same in all places, so if I_T is equal to 3 A, then I_1, I_2, and I_3 are also equal to 3 A. In this case the voltage drop across resistor R_1 is calculated by the formula

$$E_1 = I_1R_1 \qquad E_1 = 3\,\text{A} \times 30\,\Omega \qquad E_1 = 90\,\text{V}$$

If you placed the probes from a voltmeter on each side of resistor R_1, you would measure 90 V. This is the actual voltage that the resistor is causing to *drop* when current is flowing through it. The voltage drop across resistor R_2 is calculated by the formula

$$E_2 = I_2 \times R_2 \qquad E_2 = 3\,\text{A} \times 20\,\Omega \qquad E_2 = 60\,\text{V}$$

The voltage drop across resistor R_3 is calculated by the formula

$$E_3 = I_3 \times R_3 \qquad E_3 = 3\,\text{A} \times 50\,\Omega \qquad E_3 = 150\,\text{V}$$

From these calculations you can see that the voltage dropped across R_1 is 90 V, across R_2 is 60 V, and across R_3 is 150 V. If you add all these voltage drops, you will find that they are equal to the supply voltage. This means that $E_1 + E_2 + E_3 = E_T$ (90 V + 60 V + 150 V = 300 V).

Example 3-3

This circuit has three resistors connected in series and has a power source of 450 volts. Resistor R_1 is 40 Ω, R_2 is 30 Ω, and R_3 is 20 Ω. Calculate the total resistance for this circuit. After you have calculated the total resistance, divide the total voltage by the total resistance to calculate the total current. When you have determined the total current for this circuit, calculate the voltage drop that would be measured across each resistor.

Solution:

R_T is calculated by the formula $R_T = R_1 + R_2 + R_3$.

$$R_T = 40\,\Omega + 30\,\Omega + 20\,\Omega \quad R_T = 90\,\Omega$$

I_T is calculated by the formula $I_T = \dfrac{E_T}{R_T}$.

$$I_T = \frac{450\,\text{V}}{90\,\Omega} \quad I_T = 5\,\text{A}$$

Since current is the same in all parts of the circuit:

$$I_T = 5\,\text{A} \qquad I_1 = 5\,\text{A} \qquad I_2 = 5\,\text{A} \qquad I_3 = 5\,\text{A}$$

The voltage that is dropped across each resistor from the current flowing through it is calculated by the formulas

$$E_1 = R_1 \times I_1 \quad E_1 = 40\,\Omega \times 5\,\text{A} \quad E_1 = 200\,\text{V}$$
$$E_2 = R_2 \times I_2 \quad E_2 = 30\,\Omega \times 5\,\text{A} \quad E_2 = 150\,\text{V}$$
$$R_3 = R_3 \times I_3 \quad E_3 = 20\,\Omega \times 5\,\text{A} \quad E_3 = 100\,\text{V}$$

You can check your answers by adding all the drops to see that they equal the total supply voltage.

$$E_T = E_1 + E_2 + E_3 \qquad E_T = 200\,\text{V} + 150\,\text{V} + 100\,\text{V} \qquad E_T = 450\,\text{V}$$

3.11 **CALCULATING THE POWER CONSUMPTION OF EACH RESISTOR**

The power consumed by each resistor or the power consumed by the total circuit can be calculated by any of the three formulas for power (wattage): $P = EI$, $P = I^2R$, and $P = E^2/R$. We previously calculated the voltage, current, and resistance for each resistor in the circuit. This means that we can use any of the three formulas to calculate the power for any resistor. For this example we will use all three formulas to show that they all give the same result. We will use the values for R_1: $E_{R1} = 200$ V, $I_{R1} = 5$ A, and $R_1 = 40$ Ω.

$$P = EI \qquad P = 200 \text{ V} * 5 \text{ A} \qquad P = 1000 \text{ W}$$
$$P = I^2R \qquad P = (5 \text{ A})^2 * 40 \text{ }\Omega \qquad P = 1000 \text{ W}$$
$$P = E^2/R \qquad P = (200 \text{ V})^2/40 \text{ }\Omega \qquad P = 1000 \text{ W}$$

Since all three formulas are equivalent, you can use any of the three that you choose. The major factor in deciding which formula to use will be the values that you have been given or that you can determine by measuring.

3.12 **CALCULATING THE POWER CONSUMPTION OF AN ELECTRICAL HEATING ELEMENT**

The power formula can also be used to calculate the power consumption of any other type of resistance used in a circuit. The electric heating elements used in an electric furnace are *large resistance load*, so if you know the amount of resistance in the element and the amount of voltage applied to the circuit, you can calculate the current the element uses and the amount of power it consumes. For example, if the heating element has 10 ohms of resistance and it is connected to 240 V, it will draw 24 A, and it will consume 5760 W.

Example 3-4 A heating element in an electric furnace has a resistance of 2.5 Ω. Calculate the current and the power consumption of this heating element if the furnace is connected to 240 V AC.

Solution: The amount of current is calculated as 240 V/2.5 Ω = 96 A. The amount of power can be calculated by the formula $P = I^2R$ or $P = IE$. Using $P = I^2R$ we have $(96)^2 \times 2.5$ Ω = 23,040 W.

Using $P = IE$, we obtain $96 \times 240 = 23,040$ W.
(Note! This answer can be expressed as 23.040 kilowatts, or 23.040 kW.)

3.12.1 Troubleshooting Series Circuits

At times you will be requested to troubleshoot a series circuit. In most cases the problem will be an open in the circuit. When an open occurs at any point in a series circuit, current flow in all parts of the circuit will stop. The easiest way to locate an open in a series circuit is to start by measuring the voltage across the two terminals at the power supply of the circuit. For example, if you are troubleshooting the series circuit shown in Figure 3–5 because the compressor is not running, you would put the voltmeter leads on L1 and L2 and you should measure 240 V. If you do not measure the proper amount of voltage or no voltage, you have found the problem. If the amount of supply voltage is correct, you must continue to test the remainder of the circuit to locate the problem. To test the remainder of the circuit you must leave one voltmeter lead on the L2 terminal of the circuit. The meter lead on

terminal L2 is called the *point of reference*. Place the other meter lead on the terminal where voltage enters the high-pressure switch. *(Please notice that for the sake of troubleshooting we will say that voltage enters the terminal on the left side of each switch and leaves through the terminal on the right side of each switch.)* If you have voltage at the terminal on the left side of the switch, it means the wire that connects the high-pressure switch to L1 is good. If you measure 0 V, you have found the problem. The reason the voltage is 0 V is because there is an open in this wire.

If you have the proper amount of voltage at this terminal, you need to measure the amount of voltage at the output terminal of the high-pressure switch. If you measure 0 V at this terminal, you have found the problem—an open high-pressure switch. If the amount of voltage at this terminal is 240 V, the switch is good, and you need to test for voltage at the input terminal of the thermostat.

If you have 0 V at the input terminal on the thermostat, you have found the problem to be an open in the wire that connects the high-pressure switch to the thermostat. If you measure 240 V at this terminal, you have determined that the wire is good.

If you measure 240 V at the input of the oil pressure switch, your next test should be the output terminal of the oil pressure switch. If you measure 0 V at this terminal, the problem is an open in the oil pressure switch. If you measure 240 V at the output terminal of the oil pressure switch, you have found that the switch is good and the circuit is operational to this point. The next test is for voltage at the input terminal of the compressor motor. If you measure 0 V at this terminal, it means the wire between the oil pressure switch and the compressor is open. If you measure 240 V at this terminal, it means the wire between the oil pressure switch and the compressor is good, and power is being provided to the motor.

If you have determined that there is voltage to the left side terminal of the compressor, you can now move the voltmeter lead from L2 to the terminal on the right side of the compressor. If you measure 240 V when the meter leads are on the two motor terminals, you have found that the wire and switches are providing the correct amount of voltage to the compressor motor, and the problem is that the compressor has an open and is not operating correctly. If you measure 0 V when the meter leads are on the two motor terminals, the problem is an open in the wire that connects the compressor to L2.

Whenever you determine that a wire has an open, changing the wire should correct the problem. If you find an open in a switch, you can test the switch to ensure that it is set at the appropriate setpoint. If one of your tests indicates that a switch is open, you should try to determine what caused the switch to open before you change out the switch. You can also test any switch you suspect is open for continuity. When you make a continuity test on a switch, you should turn off power and isolate the wires to the switch.

3-13 ## PARALLEL CIRCUITS

Parallel circuits are used frequently in air-conditioning, heating, and refrigeration systems. Figure 3–11 shows an example parallel circuit, and you can see the difference between the series circuit and the parallel circuit is that all the loads in a parallel circuit have the same voltage supplied, whereas in a series circuit each load has a different amount of voltage depending on its resistance. The load components in air-conditioning, heating, and refrigeration systems, such as compressor motors and fan motors, must all be provided with the same voltage, so they must be connected in parallel when they are in the same circuit. Each point where a resistor is connected in parallel in this circuit

Figure 3–11 A parallel circuit with the voltage, resistance, and current shown at each load.

is called a *branch circuit*. The parallel circuit can have any number of branch circuits. A parallel circuit allows current to return to the power supply by more than one path. These paths are identified by the arrows in the diagram.

The current in a parallel circuit is additive, and the formula to calculate total resistance is $I_T = I_1 + I_2 + I_3 + \ldots$ (remember the ... means that additional currents would be added to the total). Another point to understand is that the current in a parallel circuit gets larger as more loads are added to the circuit. The parallel circuit also provides a means for disconnecting any load from the power source while still supplying voltage to the remaining loads. This is accomplished by placing a switch in each branch circuit just ahead of each resistor.

When a switch is placed in the branch circuit, it becomes a series-parallel circuit where some parts of the circuit are series in nature and other parts are parallel. For example, the fuse and disconnect for the circuit must be able to interrupt all power that is supplied to the loads in a circuit, so the fuse and disconnect must be connected in series with the power supply. The condenser fan motor and compressor must have the same amount of voltage supplied to each of them, so they must be connected in parallel with each other and with the power supply. The important point to remember when working with series-parallel circuits is that you use the formulas for series circuits for the part of the circuit that is in series, and the formulas for parallel circuits for the part of the circuit that is in parallel. It is also possible to combine parts of the circuit to make it a simplified series or simplified parallel circuit. You will learn more about series-parallel circuits at the end of this chapter after you calculate values in parallel circuits.

3-14 ## CALCULATING VOLTAGE, CURRENT, AND RESISTANCE IN A PARALLEL CIRCUIT

Voltage, current, and resistance can be calculated in a parallel circuit just as in a series circuit by using Ohm's law. Figure 3–11 shows an example circuit with the voltage, current, and resistance calculated at each point in the circuit. From this diagram you can see that the supply voltage is 300 V. Since the supply voltage is 300 V, you can determine that the voltage of each branch circuit across each resistor is also 300 V because the voltage at each branch circuit in a parallel circuit is the same as the supply voltage.

The current that is flowing through each resistor can be calculated by using Ohm's law formula for current: $I = E/R$. The current in each branch circuit is calculated from this formula and is placed in the diagram beside each resistor.

$$I_1 = E/R_1 \qquad I_1 = 300\text{ V}/30\ \Omega \qquad I_1 = 10\text{ A}$$
$$I_2 = E/R_2 \qquad I_2 = 300\text{ V}/20\ \Omega \qquad I_2 = 15\text{ A}$$
$$I_3 = E/R_3 \qquad I_3 = 300\text{ V}/60\ \Omega \qquad I_3 = 5\text{ A}$$
$$I_T = E_T/R_T \qquad I_T = 300\text{ V}/10\ \Omega \qquad I_T = 30\text{ A}$$
$$I_T = I_1 + I_2 + I_3 \quad I_T = 10\text{ A} + 15\text{ A} + 5\text{ A} \quad I_T = 30\text{ A}$$

If the voltage and current are given and the resistance needs to be calculated, the Ohm's law formula for resistance can be used: $R = E/I$.

$$R_1 = E/I_1 \quad R_1 = 300\text{ V}/10\text{ A} \quad R_1 = 30\ \Omega$$
$$R_2 = E/I_2 \quad R_2 = 300\text{ V}/15\text{ A} \quad R_2 = 20\ \Omega$$
$$R_3 = E/I_3 \quad R_3 = 300\text{ V}/5\text{ A} \quad R_3 = 60\ \Omega$$
$$R_T = E/I_T \quad R_T = 300\text{ V}/30\text{ A} \quad R_T = 10\ \Omega$$

If the total resistance and the total current in a circuit are known, you can calculate the voltage for the circuit. If you know the current and resistance at any branch circuit, you can also calculate the voltage using the Ohm's law formula: $E = IR$. The nice part about calculating voltage is that once you determine the voltage at any branch circuit, the same amount of voltage is at every other branch circuit and at the supply. The same is true if you calculate the voltage at the supply; you do not need to calculate the voltage at any other branch because it will be the same.

$$E_T = I_T R_T \quad E_T = 30\text{ A} \times 10\ \Omega \quad E_T = 300\text{ V}$$
$$E_1 = I_1 R_1 \quad E_1 = 10\text{ A} \times 30\ \Omega \quad E_1 = 300\text{ V}$$
$$E_2 = I_2 R_2 \quad E_2 = 15\text{ A} \times 20\ \Omega \quad E_2 = 300\text{ V}$$
$$E_3 = I_3 R_3 \quad E_3 = 5\text{ A} \times 60\ \Omega \quad E_3 = 300\text{ V}$$

Example 3-5

Use the circuit in Figure 3–12 to calculate the individual branch currents, total current, and total resistance.

Solution:

Use the following formulas to calculate the current for each branch:

$$I_1 = E/R_1 \quad I_1 = 240\text{ V}/60\ \Omega \quad I_1 = 4\text{ A}$$
$$I_2 = E/R_2 \quad I_2 = 240\text{ V}/30\ \Omega \quad I_2 = 8\text{ A}$$
$$I_3 = E/R_3 \quad I_3 = 240\text{ V}/15\ \Omega \quad I_3 = 16\text{ A}$$

Use the following formula to calculate the total current I_T:

$$I_T = I_1 + I_2 + I_3 \qquad I_T = 4\text{ A} + 8\text{ A} + 16\text{ A} \qquad I_T = 28\text{ A}$$

Use the following formula to calculate total resistance R_T:

$$R_T = E_T/I_T \qquad R_T = 240\text{ V}/28\text{ A} \qquad R_T = 8.57\ \Omega$$

Figure 3–12 Parallel circuit for Example 3-5.

3.15 CALCULATING RESISTANCE IN A PARALLEL CIRCUIT

At times you will need to calculate the total resistance of a parallel circuit when only the branch resistance and supply voltage are provided. You can calculate the individual currents at each branch circuit and then calculate the total with the formula $I_T = I_1 + I_2 + I_3$. After you have determined the total current, you can use the formula $R_T = E_T/I_T$.

Another method called the *product over the sum* method can be used, which requires you to use the formula for calculating total resistance in a parallel circuit. This method is called the product over the sum method because you multiply the two resistors to get the product, and you add the two resistors to get the sum. The third step in the calculation includes dividing the product by the sum (product over sum). The formula is written as follows:

$$R_T = \frac{R_1 \times R_2}{R_1 + R_2}$$

You should notice that with this formula you can calculate the total resistance of only two resistors at a time. Since this circuit has three resistors, you would need to find the total of the first two resistors in the branch and then use their total with the third resistor in the formula again to find the *grand* total resistance. We will use this method to calculate the total resistance for these resistors that are connected in parallel. $R_1 = 30\ \Omega$, $R_2 = 20\ \Omega$, $R_3 = 60\ \Omega$.

Since we can only use two resistances at a time in this method, in the first step we will find the resistance of R_1 and R_2. The amount of the resistance of R_1 and R_2 in parallel will be called equivalent resistance. Since this is the first equivalent resistance we will call this resistance R_{eq_1}. Next we will find the parallel resistance of R_{eq_1} and R_3. Since we only have three resistors, the amount of the resistance of R_{eq_1} and R_3 will be the total resistance (R_T) for the circuit.

$$R_{eq_1} = \frac{R_1 \times R_2}{R_1 + R_2} \qquad R_{eq_1} = \frac{30\ \Omega \times 20\ \Omega}{30\ \Omega + 20\ \Omega} \qquad R_{eq_1} = 12\ \Omega$$

$$R_T = \frac{R_{eq_1} \times R_3}{R_{eq_1} + R_3} \qquad R_T = \frac{12\ \Omega \times 60\ \Omega}{12\ \Omega + 60\ \Omega} \qquad R_T = 10\ \Omega$$

From these calculations you can see that the total parallel resistance is 10 Ω. *It is important to understand that in all parallel circuits, the total resistance will always be smaller than the smallest branch circuit resistance.* You can see that in this circuit, the smallest resistance in the branch circuits is 20 Ω, and the total resistance is 10 Ω.

The next calculation shows the total resistance calculated from the formula:

$$R_T = \frac{1}{\dfrac{1}{R_1} + \dfrac{1}{R_2} + \dfrac{1}{R_3}} \quad \text{or} \quad \frac{1}{R_T} = \frac{1}{R_1} + \frac{1}{R_2} + \frac{1}{R_3}$$

This formula is designed to be used with a calculator. If you do not have a calculator, it is recommended that you use the previous method. If you use a calculator, you should use the following keystrokes to get an answer using this formula:

$$R_T = \frac{1}{\dfrac{1}{R_1} + \dfrac{1}{R_2} + \dfrac{1}{R_3}} \qquad R_T = \frac{1}{\dfrac{1}{30\ \Omega} + \dfrac{1}{20\ \Omega} + \dfrac{1}{60\ \Omega}}$$

$R_T = 10\ \Omega$

Keystrokes (The boxes indicate keys on the calculator and numbers indicate number keys.) *Notice you might have a key identified as INVERSE, 1/X or a key might be identified as RECIPROCAL.*

Example 3-6

Use the parallel circuit in Figure 3–13 to calculate the current through each resistor, the total current, and the total resistance for this circuit. The supply voltage is 100 V.

Solution:

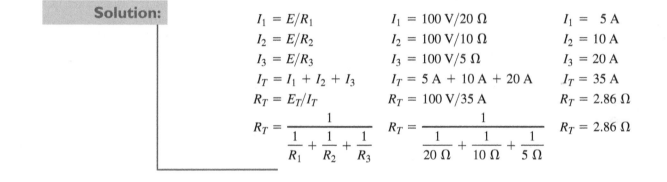

$$I_1 = E/R_1 \qquad\qquad I_1 = 100\ \text{V}/20\ \Omega \qquad\qquad I_1 = 5\ \text{A}$$
$$I_2 = E/R_2 \qquad\qquad I_2 = 100\ \text{V}/10\ \Omega \qquad\qquad I_2 = 10\ \text{A}$$
$$I_3 = E/R_3 \qquad\qquad I_3 = 100\ \text{V}/5\ \Omega \qquad\qquad I_3 = 20\ \text{A}$$
$$I_T = I_1 + I_2 + I_3 \qquad I_T = 5\ \text{A} + 10\ \text{A} + 20\ \text{A} \qquad I_T = 35\ \text{A}$$
$$R_T = E_T/I_T \qquad\qquad R_T = 100\ \text{V}/35\ \text{A} \qquad\qquad R_T = 2.86\ \Omega$$
$$R_T = \cfrac{1}{\cfrac{1}{R_1} + \cfrac{1}{R_2} + \cfrac{1}{R_3}} \qquad R_T = \cfrac{1}{\cfrac{1}{20\ \Omega} + \cfrac{1}{10\ \Omega} + \cfrac{1}{5\ \Omega}} \qquad R_T = 2.86\ \Omega$$

3.16 **CALCULATING POWER IN A PARALLEL CIRCUIT**

The formula for calculating power in a parallel circuit is the same as the formula for a series circuit: $P_T = P_1 + P_2 + P_3$. The formula for power at each individual resistor is found from the original Watt's law formula $P = IE$. This means that you must calculate the power consumed by each branch resistor and then add them all together to get the total power consumed. For example, in the parallel circuit in Figure 3–11 the voltage at R_1 is 300 V and the current is 10 A, the voltage at R_2 is 300 V and the current is 15 A, and the voltage at R_3 is 300 V and the current is 5 A. The following calculations are used

Figure 3–13 Parallel circuit for Example 3-6.

to determine the power consumed by each individual resistor and the total power used by the whole circuit.

$$
\begin{array}{lll}
P_1 = I_1E_1 & P_1 = 10\,\text{A} \times 300\,\text{V} & P_1 = 3000\,\text{W} \\
P_2 = I_2E_2 & P_2 = 15\,\text{A} \times 300\,\text{V} & P_2 = 4500\,\text{W} \\
P_3 = I_3E_3 & P_3 = 5\,\text{A} \times 300\,\text{V} & P_3 = 1500\,\text{W} \\
P_T = P_1 + P_2 + P_3 & P_T = 3000\,\text{W} + 4500\,\text{W} + 1500\,\text{W} & P_T = 9000\,\text{W}
\end{array}
$$

You could also calculate the total power consumed by this circuit by using the formula

$$
P_T = I_TE_T \qquad P_T = 30\,\text{A} \times 300\,\text{V} \qquad P_T = 9000\,\text{W}
$$

3.17 SERIES-PARALLEL CIRCUITS

As a technician you will work on circuits for residential and commercial heating, air-conditioning, and refrigeration systems. These systems will have a number of switches and loads that are connected in series and parallel. A typical circuit is shown in Figure 3–14. In this diagram you can see that a compressor motor and a condenser fan motor are connected in parallel. These two motors are connected in parallel so that each will receive the same amount of voltage. When loads such as motors are connected in parallel in a circuit, it is called a **power circuit**.

These loads are also connected in series with three control switches: a high-pressure switch, a thermostat, and an oil pressure switch. If any of these three switches is open, the current to both loads will be stopped. The switches are connected in series for this reason. Since this circuit has some switches connected in series, and two loads connected in parallel, it is called a series-parallel circuit.

The parts of the circuit that are connected in series will follow all the rules for series circuits, and the parts of the circuit that are connected in parallel will follow all the rules for parallel circuits. This series-parallel circuit will provide the best of series circuits and the best of parallel circuits to supply voltage to the loads and control their

Figure 3–14 An electrical diagram of a compressor motor and condenser fan motor connected in parallel, and a high-pressure switch, thermostat, and oil pressure switch connected in series.

operation. This type of series-parallel circuit is very common in field equipment that you will encounter.

3.18 SERIES-PARALLEL CIRCUITS IN AIR-CONDITIONING AND REFRIGERATION SYSTEMS

Series-parallel circuits exist in a variety of air-conditioning and refrigeration equipment. These types of circuits are also used in heating equipment. You can use the basic information that you learned about series-parallel circuits to make some simple observations about the size of wires and switches based on the voltage and current for which they will need to be rated. Figure 3–15 shows an example of a series-parallel circuit for an air-conditioning system. From this diagram you can see that the circuit has some switches in series with motors, and the motors are connected in parallel with one another. For example, if a disconnect switch was connected at the top of the circuit at L1, it would be connected in series with the entire circuit. This means that if you open the disconnect switch, the entire circuit is disconnected.

The thermostat is connected in series with the evaporator motor and with the parallel circuit that has the compressor motor connected to the condenser fan motor. The evaporator motor draws 5 A, the compressor motor draws 20 A, and the condenser motor draws 7 A; therefore, the thermostat must be rated for at least 32 A because all of this current will flow through it. If you were going to select a fuse to put in L1 to protect everything in the circuit, you would need to select a fuse that is larger than 32 amps.

The high-pressure switch and oil pressure switch are connected in series with the compressor motor and the condenser motor, so they must be rated to carry the 27 amps used by these two motors. This also means the current to these motors will be interrupted anytime either of these switches is opened. All three motors are connected in parallel with one another, so they will all receive the same amount of voltage, but the amount of current each will draw will be different unless their load characteristics are identical. The amount of total current that is needed to supply the three motors can be calculated by the formula $I_T = I_1 + I_2 + I_3$.

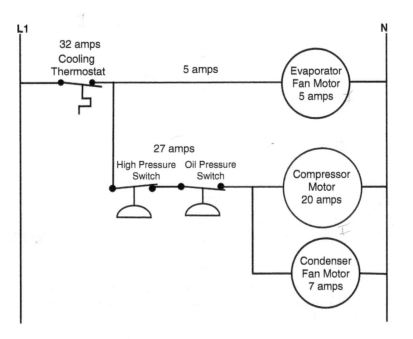

Figure 3–15 Electrical diagram of a typical series-parallel circuit for an air-conditioning system.

3.19 TROUBLESHOOTING SERIES-PARALLEL CIRCUITS

Troubleshooting a series-parallel circuit such as the one shown in Figure 3–15 is slightly different from troubleshooting a series circuit. In the series-parallel circuit shown in this figure you can do some preliminary troubleshooting by simply observing the circuit. For example, if the evaporator fan is running, but the compressor and condenser fan are not running, you can make some basic statements about this circuit. Since the evaporator is running, it means that the supply voltage at L1–N is correct. You can also predict that the thermostat is operating correctly, since it also supplies voltage to the evaporator fan. If both the compressor fan and the condenser fan are not running, you can state that the problem is more than likely in the wires and switches that supply voltage to both of these components. It is not likely that both the compressor motor and the condenser fan motor failed simultaneously.

Since you know the evaporator fan is running correctly, you can skip that part of the circuit and begin your troubleshooting process in the second and third rungs of the circuit, where the compressor and condenser fan receive power. When you are troubleshooting this type of circuit, you should use the same procedure that you used with the series circuit. This means that you should leave one voltmeter lead on the N terminal of the circuit as a point of reference. Since you know the evaporator fan is working, you can skip down to the second rung of the circuit. Your first test should be at the input terminal of the high-pressure switch. If the voltage reading at the input terminal is 0 V, the wire that connects the thermostat to the high-pressure switch is open, and you have found the problem. If the voltage is 120 V, the wire is good and you can continue your test.

For the next part of the test you should place the voltmeter lead on the output terminal of the high-pressure switch. If you measure 0 V, the contacts of the high-pressure switch are open, and you have found the problem.

Say you have determined that the input terminal of the oil pressure switch has 120 V and the output terminal has 0 V. This means that the problem is open contacts in the oil pressure switch.

You have found the location of the problem with this process but you have not found the actual source of the problem. For example, your first inclination is to replace the oil pressure switch, since voltage is not passing through its contacts. What you will learn in later chapters is that the oil pressure switch is doing its job, and it opened its contacts because it sensed a problem such as the compressor's not building oil pressure or a leak in the system and thus no refrigerant pressure. In later chapters you will learn how electrical controls work with refrigeration systems, and you will learn to troubleshoot the entire system for both electrical and refrigeration problems.

3.20 REVIEW OF SERIES CIRCUITS

In a series circuit all the loads will be connected in such a way that there is only one path for current to flow. If one or more switches are used to control current flow to the loads in the circuit, they must also be connected so that only one path exists. The total resistance in a circuit can be calculated by adding together all the resistor values ($R_T = R_1 + R_2 + R_3 + \ldots$). (Notice that the ... means that any number of resistors can be used in this formula, since they are all added.)

The total current of a series circuit can be calculated by the formula $I = E/R$, and it is important to remember that the current will be the same in all parts of the circuit. This

means that once you measure or calculate current at any point or through any component, it will be the same at all other components in the circuit.

The voltage that is measured across each component in the series circuit can be different. If the resistance values of any two resistors in a circuit are the same, the amount of voltage measured across them will be the same. The amount of voltage that is measured across each resistor is the product of the resistance in the resistor and the amount of current flowing through it. The formula for calculating voltage is $E = IR$. The amount of voltage measured across each resistor in a series circuit is also called the **voltage drop** for that resistor. The total of all the voltage drops in a series circuit will always be equal to the supply voltage for that circuit.

Typically, loads are not connected in series. The exceptions are heating elements in furnaces or in some industrial applications, where two 120-V heaters are connected in series and they are powered with 240 V. Resistors are small components that provide specific amounts of resistance. The amount of resistance a resistor has is identified by its color code, which consists of four color bands. The first band identifies the first digit of the value. The second band identifies the second digit of the value, and the third digit is the multiplier value. The fourth band is the tolerance band. Figure 3–9b lists the colors and values for each band. Tolerance is the acceptable range of variance (plus and minus) of the actual resistance of the resistor from the rated value identified by the color code. The physical size of the resistor indicates the power rating for the resistor, which is listed in watts.

The power consumed by each component in a series circuit can be calculated by the formula $P = IE$. The total power consumed by the components in a series circuit can be calculated by the same formula or by adding the amount of power consumed by each component. For example, $P_T = P_1 + P_2 + P_3$.

3.21 REVIEW OF PARALLEL CIRCUITS

Parallel circuits are used in air-conditioning, heating, and refrigeration systems because this type of circuit provides the same amount of voltage to each motor in the circuit. For example, if an air-conditioning system has a compressor that requires 240 V, and a condenser fan that requires 240 V, a circuit will supply this voltage to both motors if they are connected in parallel. A parallel circuit provides more than one path by which current can return to the power source. Each path is called a branch circuit. The amount of current in each branch of the parallel circuit can be determined by dividing the amount of voltage supplied to that branch by the amount of resistance for the load in that branch. The amount of total resistance in a parallel circuit will always be smaller than the smallest branch resistance. This also means that as additional loads (resistances) are added to a parallel circuit, the total amount of resistance will become smaller. Since the amount of resistance is reduced as loads are added, the total amount of current will increase.

Job Assignment for This Chapter

A customer has a question about why a fuse continues to blow on the outdoor air conditioning that contains the compressor and condenser fan. Explain to this customer the function of a fuse and how it protects both motors when it blows. You should also explain

to the customer how the fuse is connected in series with the fan motor and compressor and how they are connected in parallel with each other. You should also give the customer some ideas of how you could tell which component in the circuit is causing the circuit to draw more current than the fuse is rated for.

Solution to the Job Assignment

Your solution should include information about the difference between a series and parallel circuit. You should explain that the fuse is in series with all of the loads in the circuit. You should also explain how the compressor and the condenser fan both have current ratings posted on them. You can explain to the customer that you can put an ammeter on each device and see which one has exceeded its rating. Since you have not had information about meters yet, you are not expected to know how to take a current reading, just that a current reading should be made. Since the fuse is in series with both motors, if either motor is going bad and draws more current than it should, it will cause the main fuse to blow.

Assessment

QUESTIONS

1. Discuss the advantage of connecting switches in series with a load.
2. Explain why loads are typically not connected in series with one another.
3. In some cases loads may be connected in series with one another. Explain when this arrangement would be advantageous and what type of loads would be used.
4. Explain what a resistor is and where you would find it in HVAC circuits.
5. Explain what each of the four color bands on a resistor indicates.
6. Explain why the voltage at each branch of a parallel circuit is the same as the supply voltage.
7. What is the advantage of using exponents when you solve Ohm's law calculations?
8. Explain what happens to the voltage at the condenser motor if the compressor motor has an open in the diagram in Figure 3–14.
9. Determine the total current that flows through the thermostat in the diagram in Figure 3–15.
10. Explain why total power in a parallel circuit increases when additional loads are connected in parallel.

TRUE OR FALSE

1. When a series circuit has an open, current flow stops in all parts of the circuit.
2. The main reason the switches are connected in series is so that if any one of the switches opens, current to the load will be interrupted.
3. It is a good practice to connect the fan motor and compressor motor for an air conditioner in series so that when one quits the other will stop.

4. Two heating elements should be connected in series if you want them to split the amount of voltage applied to them.

5. All resistors have the same amount of resistance if they are the same physical size.

6. Current in each part of a parallel circuit will always be the same.

7. Voltage in the branches of parallel circuits will always be the same.

8. Total resistance in a parallel circuit becomes smaller as more resistors are added in parallel.

9. Total current in a parallel circuit becomes smaller as more resistors are added in parallel.

10. Total power in a parallel circuit will increase as more loads are added in parallel.

MULTIPLE CHOICE

1. In a series circuit the current in each part of the circuit is
 a. always zero.
 b. the same.
 c. equal to the voltage.

2. The amount of resistance in a resistor
 a. is the same if the resistors are the same size.
 b. can be determined by color codes.
 c. changes as the amount of current that flows through it changes.

3. When three resistors are connected in series, the amount of voltage that is measured across each one is
 a. determined by the amount of current flowing through it and the amount of resistance it has.
 b. determined by the wattage rating of each resistor.
 c. impossible to determine by calculations.

4. If a thermostat, high-pressure switch, and oil pressure switch are connected in series with a compressor motor, and the oil pressure switch is opened because of low oil pressure, the compressor motor will
 a. still run because two of the other switches are still closed.
 b. will not be affected because no other loads are connected to it in series.
 c. will stop running because current flow will be zero.

5. If the resistance of a heating element increases from 2 Ω to 5 Ω, the current it draws will
 a. increase if the voltage stays the same.
 b. decrease if the voltage stays the same.
 c. not change if the voltage does not change.

6. Current in a parallel circuit
 a. increases as additional resistors are added in parallel.
 b. decreases as additional resistors are added in parallel.
 c. may increase or decrease when resistors are added in parallel depending on their size.

7. Voltage in a parallel circuit
 a. increases as additional resistors are added in parallel.
 b. decreases as additional resistors are added in parallel.
 c. stays the same across parallel branches as additional resistors are added in parallel.

8. Resistance in a parallel circuit
 a. increases as additional resistors are added in parallel.
 b. decreases as additional resistors are added in parallel.
 c. may increase or decrease when resistors are added in parallel depending on their size.

9. When one branch circuit of a multiple branch parallel circuit develops an open, voltage in other branch circuits
 a. decreases to zero.
 b. increases because fewer resistors are using up the voltage.
 c. stays the same as the supply voltage.

10. When loads are added in a parallel circuit
 a. the total power increases.
 b. the total current increases.
 c. the voltage stays the same to all loads.
 d. all of the above

PROBLEMS

1. Decode the following resistors and indicate their tolerances.

First Band	Second Band	Third Band	Fourth Band	Answer
Red	Blue	Brown	Gold	
Green	Gray	Red	Silver	
Orange	Black	Blue	Gold	
Brown	Yellow	Violet	Gold	

Note: Please show the formula and all work for each problem.

2. Calculate the amount of voltage for a circuit that has 6 A and 20 Ω.
3. Calculate the amount of current for a circuit that has 100 V and 25 Ω.
4. Calculate the wattage for a circuit that has 120 V and 8 Ω.
5. Calculate the wattage for a circuit that has 15 Ω total and draws 25 A.
6. Draw a circuit that has three resistors (20 Ω, 40 Ω, and 80 Ω) connected in parallel, and calculate the total resistance for this circuit.
7. Calculate the current for each branch of the circuit that you drew for Problem 6.
8. Calculate the total current for the circuit that you drew for Problem 6.
9. Calculate the power consumed by each resistor in the circuit you drew for Problem 6.
10. Determine the current capacity for each switch in the circuit for Figure 3–15 if the evaporator motor uses 8 A, the compressor motor uses 15 A, and the condenser motor uses 5 A.

Websites

Websites change constantly. Use the key word search to locate sites that have similar content. The products that will be found at each site are indicated with the key word.

Online Electricity for Beginners
Products: Online electricity course
Key words: Series circuits, parallel circuits, series-parallel circuits

Chapter 4

Meters and Tools for HVAC

Job Assignment for This Chapter

You are on a service call working on an air conditioner and have found a bad fuse. The customer wants to know how you can tell whether a fuse is good or bad. The customer thinks you are trying to sell some parts that are not needed. Your job is to use the meters in your truck to convince the customer that you can tell whether a fuse is good or bad and to show that the fuse you found in the system is open and must be replaced.

Terms You Will Learn in This Chapter

Ammeter

Analog meter

Clamp-on ammeter

Digital meter

Magnet

Magnetic field

Meter

Meter movement

Milliammeter

Ohmmeter

Voltmeter

After reading this chapter, you will be able to:

1. Describe the operation of a meter movement.
2. Explain the difference between an analog meter and digital meter.
3. Explain the operation of an ohmmeter and why it needs a battery.
4. Use a VOM meter to measure voltage.
5. Use a VOM meter to measure resistance.
6. Use a VOM meter to measure milliamps.
7. Understand safety warnings for using ohmmeters and ammeters.
8. Explain the operation of a digital and analog clamp-on ammeter.
9. Describe a magnet and explain how it works.
10. Explain the difference between a permanent magnet and an electromagnet.
11. Identify ways to increase the strength of an electromagnet.
12. Explain what flux lines are and where you would find them.
13. Select the proper pliers for cutting and stripping wire.

4.0 MEASURING VOLTAGE, CURRENT, AND RESISTANCE

Measuring the amount of voltage (volts), current (amps), and resistance (ohms) in a heating, air-conditioning, or refrigeration system helps to determine whether or not it is working correctly. If you are checking a tire on your automobile, you can see whether it is flat by simply looking at it. Since the flow of electrons through a wire is invisible, you cannot simply look at the wire to determine whether it has voltage applied to it or current flowing through it, so a meter must be used to make voltage, current, and resistance measurements.

As an HVAC service technician, you will be expected to use a meter to troubleshoot each system. You'll find that 75% to 80% of all calls involve an electrical problem. This means that you'll need to measure the amount of voltage the system has, the amount of current the system is pulling, and the resistance of individual components such as fuses, switches, and motor windings. The material in this chapter will begin by showing you how an analog meter works. An **analog meter** is a meter that uses a needle that is moved by a magnetic field to indicate the amount of volts, ohms, or amps, whereas a **digital meter** is a meter that has a digital display. Figure 4–1 shows and analog meter, and Figure 4–2 shows a digital type meter.

4.1 THE VOM METER

Most **meters** are able to measure voltage, resistance, and current in one unit. These types of meters are called VOM meters. The letters *VOM* stand for volts, ohms, and milliamps. And you should remember that a milliamp is 1/1000 of an amp or 0.001 amp. The VOM meter can be digital or analog. When you want to measure volts, you would switch the range selector in the middle of the meter to one of the voltage ranges.

Figure 4–1 An analog volt, ohm, milliammeter (VOM). *(Courtesy of Triplett Corporation)*

Figure 4–2 A digital meter has a digital display to show the amount of voltage, current, or resistance being measured. Some digital meters have a knob to select the range of voltage, current, and resistance whereas others are self-ranging. *(Reproduced with permission of Fluke Corporation)*

If you want to measure resistance, you would rotate the range selector to one of the resistance ranges, and you would rotate the range selector to one of the milliamp ranges if you want to measure milliamps. The **milliammeter** is used to measure current that is less than one amp. If you are measuring current that is larger than one amp, you should use an ammeter. The meter has several terminals that you can plug the meter probes into.

If you're using an automatic or autoranging meter like the one shown in Figure 4–3, all you need to do is rotate the selector switch in the middle of the meter to AC or DC volts, AC or DC milliamps, or ohms. When you touch the probes to the circuit, the meter will automatically adjust to the amount you are reading to give you the most accurate

Figure 4–3 A self-ranging digital voltmeter. Rotate the selector switch to volts, ohms, or milliamps and the meter will automatically adjust to the amount that is being measured.
(Reproduced with permission of Fluke Corporation)

reading. If you are using a meter that has multiple ranges, you will need to start at the highest range and move the range selector to a smaller value if the range is too high. In this chapter you will learn to select the correct meter for each measurement, where to place the probes in a circuit, and how to read the meter.

4.1.1 Analog vs. Digital Meters

There has been much debate about whether analog meters are being used any longer, and whether students should only learn to use digital meters. It is also pointed out that digital meters are easier to read and lead to fewer errors when making measurements. This text will not debate this issue; rather it will point out that in special cases, analog meters may provide accuracy when variable frequencies and harmonics are present in a circuit, or when very small currents are being measured. For this reason, the next sections of this chapter will explain how to read the analog meter so that technicians can refer to these sections if they ever need to use them.

4.2 HOW DOES THE METER WORK?

This section will explain how the analog meter that has a needle and a face works. The analog meter shown in Figure 4–1 has a needle that is moved from left to right as it measures voltage, resistance, or current. The part of the meter that causes the needle to move is called the **meter movement**. The meter movement is made of a coil of wire that has current flowing through it. The stronger the current flow, the more magnetized the coil becomes. To fully understand this concept we will study basic magnetic theory.

4.3 MAGNETIC THEORY

Magnetism can be explained with several simple magnetic theories. **Magnet** is the name given to material that has an attraction to iron or steel. This material was first found naturally about 4000 years ago as a rock in a city called Magnesia. The rock was called magnetite and was not usable at the time it was discovered. Later it was found that when pieces of this material were suspended from a wire, the material would always orient itself so that the same ends always pointed in the same direction, aligned with the earth's north and south poles. Scientists soon learned from this phenomenon that the earth itself is magnetic. At first the only use for magnetic material was in compasses.

As scientists gained more knowledge, and equipment became available to study magnets more closely, a set of principles and laws evolved. The first of these discoveries showed that every magnet has two poles, called the north pole and the south pole. When two magnets are placed end to end so that similar poles are near each other, the magnets repel each other. It does not matter if the poles are both north or both south, the result is the same. When the two magnets are placed end to end so that the south pole of one magnet is near the north pole of the other magnet, the two magnets attract each other. These concepts are called the first and second laws of magnets.

A *permanent magnet* is a magnet whose dipoles remain aligned for a very long time, which means that the magnet will retain its magnetic properties. Figure 4–4 shows a diagram of nonmagnetic metal with its dipoles randomly placed, and Figure 4–5 shows a magnetic piece of metal with its dipoles aligned to make a strong magnet.

Figure 4–4 Example of a metal with its dipoles randomly placed, which makes it a very weak magnet or not magnetic at all.

Figure 4–5 Example of a metal with its dipoles aligned to make a very strong magnet.

4.4 A TYPICAL BAR MAGNET AND FLUX LINES

Figure 4–6 shows a bar magnet that is made of soft iron that has been magnetized. The magnet is in the shape of a bar, and its north and south poles are identified. Since the bar remains magnetized for a long time, it is called a permanent magnet. The magnet produces a strong magnetic field because all its dipoles are aligned. The **magnetic field** produces invisible *flux lines* that move from the north pole to the south pole along the outside of the bar magnet. The diagram in Figure 4–6 shows that these flux lines are lines of force that move from the north pole to the south pole in a slight arc. Any ferrous metal such as iron will become attracted to the ends of a magnet.

Since the flux lines are invisible, you can perform a simple experiment to allow you to see that flux lines do exist and what they look like as they surround the bar magnet. For this experiment you will need a piece of clear plastic film such as the plastic sheets used for overhead transparencies, and some iron filings. Place the plastic sheet over a bar magnet. Make the plastic as flat as possible and sprinkle iron filings over it. The filings will be attracted by the invisible flux lines as they extend in an arc from the north pole to the south pole along the outside of the magnet. Since the flux lines begin at one pole and stretch to the other, the highest concentration of flux lines will be near the poles. The iron filings will also concentrate around the poles, but a definite pattern of flux lines can be seen along each side of the bar magnet. If an overhead projector is available, the image of the flux lines can be projected onto a projector screen or blackboard so that they can be seen more easily. The pattern of these filings will look similar to the diagram in Figure 4–6. The number of flux lines around a magnet is directly related to the strength of the magnet. A stronger magnet will have more flux lines than a weaker magnet. The strength of a magnet's field can be measured by the number of flux lines per area. Since the strength of a magnet's field is based on the alignment of the magnetic dipoles, the number of flux lines will increase as the alignment of the magnetic dipoles increases.

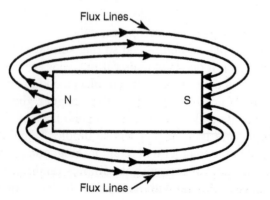

Figure 4–6 Example of a bar magnet. Notice that the poles are identified as north (N) and south (S). Flux lines move in an arc from north pole to south pole.

4.5 | **ELECTROMAGNETS**

An electromagnet is made by passing current through a coil of wire. The electromagnet has properties that are similar to a permanent magnet. When a wire conductor is connected to the terminals of a source of voltage, a current will begin to flow through the wire and create magnetic flux lines around it like concentric circles. If the wire is placed near a pile of iron filings while current is flowing through it, the filings will be attracted to the wire just as if the coil were a permanent magnet. Figure 4–7 shows a diagram indicating the location of magnetic flux lines around a conductor when current is flowing through it. As current flow through the wire increases, the flux lines will also concentrate closer and closer, making the magnetic field stronger until the current reaches a point of *saturation*. When the flux lines reach the saturation point, any additional increase of current in the wire will not produce any more flux lines and the magnet does not become any stronger.

You can increase the strength of the magnetic field by taking the straight wire and coiling it up so that the flux lines will concentrate and become stronger. Figure 4–8 shows an example of flux lines around a coil of wire that has current flowing through it. Since the flux lines are much stronger in a coil of wire, most of the electromagnets that you will encounter will be in the form of coils. For example, the part of the meter that measures current is a coil that is called the meter movement. In later chapters you will learn how coils are used in transformers, relays, solenoids, and motors.

Figure 4–9 shows how to set up an experiment so you can demonstrate how the magnetic field around a coil of wire can be increased by increasing the amount of current flowing through it. For this experiment, wrap a length wire into a coil, connect the ends to a dry cell battery, and place it near a pile of iron filings. Add a variable resistor to the

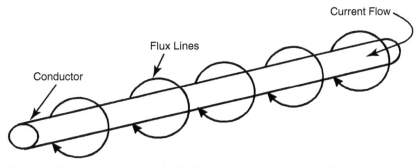

Figure 4–7 Example of magnetic flux lines that are created around a straight wire that is carrying current.

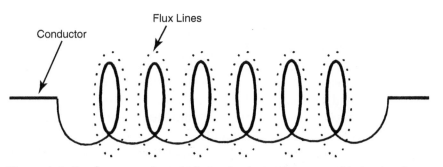

Figure 4–8 Flux lines around a coil of wire that is carrying current. Notice that the number of flux lines increases when the wire is coiled.

mA Meter

mA Meter

Figure 4–9 A small amount of current flowing in a coil of wire creates a small number of flux lines that attract iron filings.

Figure 4–10 A large amount of current flowing in a coil of wire creates a large number of flux lines that attract iron filings.

circuit to increase or decrease the current. When the current is set to a minimum, as in Figure 4–9, the magnetic field around the coil of wire will attract only a few filings. As the current is increased, as in Figure 4–10, the number of filings the magnetic field will attract also increases. If your battery is strong enough, you can create enough current to cause the coil of wire to reach *saturation*. When the saturation point is reached, any additional current flowing in the wire will not produce additional flux lines.

If a switch is added to this circuit, the magnetic field can be turned on and off by turning the switch on and off to interrupt the flow of current in the coil. When the switch is opened, current is interrupted and no flux lines are produced, so the magnetic field will not exist.

4.5.1 Advantages of an Electromagnet

One advantage an electromagnet has over a permanent magnet is that the magnetic field can be energized and deenergized by interrupting the current flow through the wire and the strength of the magnetic field can be increased by increasing the amount of current flowing through it. Another advantage of an electromagnet is that its field can be reversed by reversing the direction of current flow through the coil of wire. This will be very useful when the coil is used in the meter movement since it will be able to measure positive or negative current.

4.6 HOW THE METER MOVEMENT WORKS

The meter movement used in an analog meter consists of a permanent magnet with a north and south pole and an electromagnet that consists of a coil of wire wrapped around a needle that is a pointer. As current flows through the coil of the electromagnet, it will be attracted to one of the poles of the permanent magnet. When the current flowing through the electromagnet, it becomes stronger, and the electromagnet and pointer is pulled closer to the permanent magnet. Basically the more current that flows through the electromagnet the more the needle moves to indicate the strength of the current.

Figure 4–11 shows an example of the permanent magnet and electromagnet in the meter movement. Figure 4–12 shows a picture of a meter movement with a meter needle used as a pointer. In this figure the needle moves to the left when no current is flowing and the pointer indicates zero current. When the current is increased to half strength, the electromagnet is pulled toward the permanent magnet on the right, which makes the needle show up in the middle of the scale and point to 50. When maximum current is measured, the electromagnetic field becomes stronger and the meter movement is pulled farther to the right to the permanent magnet and the needle points to 100 on the face. A small spring mechanism provides a small amount of pressure or tension to keep the

Figure 4–11 A meter movement consisting of a stationary north and south pole of a permanent magnet and the movable electromagnet connected to a meter needle.

Permanent Magnet Housing

Pointer

Upper Spring

Lower Spring

N S

Movable Magnetic Core

Pointer

Permanent Magnet Housing

Movable Magnet

Figure 4–12 A meter movement with pointer, permanent magnet housing, and movable magnet identified.

needle pulling back to the left, so that if current is decreased, the movement is pulled back to the left and the needle will point to zero when no current is being measured. The amount of current that causes the meter to move to full scale is called *full-scale deflection*, and most meter movements are rated for 1 mA to 10 mA.

If current were put through the meter in the reverse direction, the needle would be pulled to the left toward the zero and the reading would not be accurate. This is called *reversing the polarity* of the current through the meter. If the meter is measuring AC (alternating current), the current must be rectified (changed to DC) so the meter movement can measure it correctly.

It is important to understand that all meter movements can only measure DC current in the milliamp range. If you are measuring larger currents than the full-scale rating of the movement, resistors must be used to ensure that the current flowing through the meter movement never exceeds the full-scale rating. If you are trying to measure voltage, a circuit in the meter will convert the voltage to current so the meter movement can measure it. The internal part of an analog meter consists of resistors that are used to

convert voltage to current and to ensure that the maximum amount of current flowing through the meter movement is not exceeded.

If the meter is used to measure resistance, the internal battery is required to provide a current source to make the meter movement move the needle to the appropriate reading. For this reason, you must make sure that there is no voltage applied when you are making any resistance measurement. When the meter is set to read resistance, a 1.5-, 9-, or 30-volt battery inside the meter is used as a power source; any outside voltage will damage the meter movement.

4.7 HOW THE DIGITAL VOLTMETER MEASURES VOLTAGE, CURRENT, AND RESISTANCE

The digital voltmeter uses an integrated circuit, which is also called a "chip," to sample voltage and determine whether it is AC or DC. Another integrated circuit takes the raw voltage or current signal that is in an analog value and coverts it to a digital format. This chip is called an A/D converter, which means it converts analog voltage or current to a digital value. The final integrated circuit in the digital meter is a digital display driver. This chip takes the digital value that represents the measured voltage, current, or resistance and sends it to the LED (light-emitting diode) display. Since the digital meter uses integrated circuits to sample the voltage, current, or resistance, it provides a very high internal impedance (up to 20 million ohms), which makes the digital meter very safe to use in any type of circuit without posing a danger to the person using it or the circuit it is testing.

Digital meters, like the analog meters, require in internal battery to provide the current source for making resistance measurements. One important point to remember is that you do not have to zero a digital meter when it is set to measure resistance. It has an internal circuit that samples the internal battery voltage prior to each measurement and compensates for differences in voltage levels in the battery due to it slowly discharging over time.

Some companies make a digital multimeter (DMM) that measures volts, ohms, and amps. DMMs may also have additional features that allow them to measure temperature and capacitance, which makes them valuable to HVAC technicians. Figure 4–13 shows an example of a DMM that can measure voltage, current, resistance, and capacitance; temperature can be measured by adding a thermocouple module, which is shown in Figure 4–14.

Figure 4–13 A Fluke 117 Multimeter that can accept a thermocouple module to allow it to measure temperature.
(Reproduced with permission of Fluke Corporation)

Figure 4–14 A thermocouple module that plugs into the multimeter and allows it to measure temperature.
(Reproduced with permission of Fluke Corporation)

4.8 MEASURING VOLTAGE

The meter used to measure voltage is called a **voltmeter**. The voltmeter can be part of a digital or analog VOM. The meter in Figure 4–1 is called an *analog* voltmeter when it uses a needle and a scale to indicate the amount of voltage being measured. The meter in Figure 4–2 is called a *digital* voltmeter when the display uses numbers to indicate the amount of voltage that is being measured.

The voltmeter has very high internal resistance, which is approximately 20,000 Ω per volt, so its probes can be safely placed directly across the terminals of the power source without damaging the meter. The amount of voltage that is measured by the voltmeter is shown on the display. The range selector switch adjusts the internal resistance of the meter to set the maximum amount of voltage the meter can measure. The voltmeter can be used to make many different voltage measurements. For example, it is used to measure the amount of supply voltage in a system as well as to measure the amount of voltage that is provided to each load. Figure 4–15 shows three examples where the probes of a voltmeter should be placed to safely make voltage measurements in a circuit. In the diagram in Figure 4–15a, the voltmeter probes are shown across the terminals of a battery to measure the amount of battery voltage supplied. Figure 4–15b shows the proper location to place the voltmeter probes to measure the voltage available to lamp 1, and Figure 4–15c shows the proper location to place the voltmeter probes to measure the voltage available to lamp 2. In each case the voltmeter probes are placed *across* the terminals where the voltage is being measured in parallel to the load.

Safety Notice!
It is important to remember that the voltage to a circuit must be turned on to make a voltage measurement. This presents an electrical shock hazard. You must use extreme caution when you are working around a circuit that has voltage applied so that you do not come into contact with exposed terminals or wires when you are making voltage measurements.

Typical meter ranges for digital voltmeters are 0–3 V, 0–30 V, 0–300 V, and 0–600 V. Special high-voltage probes must be used to measure voltages above 600 V. Typical voltages found in HVAC systems are 24 V, 120 V, 208 V, 240 V, and 480 V. The

Figure 4–15 **a** The proper location to place the probes of a voltmeter to measure the amount of voltage available at the battery. **b** The proper location to place the voltmeter probes to measure the voltage available at lamp 1. **c** The proper location to place the voltmeter probes to measure the voltage available at lamp 2.

actual amount of voltage in your area may be slightly different; for example, the 120 V may actually measure 110 V, and the 480 V may be closer to 440 V.

The ranges for analog voltmeters are typically 0–3 V, 0–12 V, 0–60 V, 0–300 V, and 0–600 V. If you try to measure 120 V when the selector switch is set to 60 V, the needle will try to go past the maximum reading, and the meter will be damaged. If you are using a digital voltmeter and you try to read a voltage that exceeds the range selector switch setting, the display will indicate that you are over the range, but the meter will not be damaged.

4.9 ## SOLENOID TYPE VOLTAGE TESTER

Another type of voltmeter used by service technicians is called a solenoid voltmeter. Figure 4–16 shows a Sperry V-11A Voltector, which is a solenoid voltmeter. The solenoid meter requires more current to make the voltage meter make a measurement. One of the problems that a digital voltmeter has is that it may display a small voltage measurement when a small voltage is present without any current. This occurs frequently in electronic circuits such as variable frequency drives (VFDs) or electronic thermostats. The solenoid meter needs current as well as voltage to make a voltage measurement. If you suspect the voltage in an electronic circuit is phantom and does not seem to be measurable with an analog meter, you can use a solenoid meter to verify that there is sufficient current in the line and that the voltage is not a phantom voltage.

Figure 4–16 A Sperry Voltector solenoid type voltage meter. *(Courtesy of Sperry Instruments)*

Figure 4–17 **a** A digital VOM (volt-ohm-milliamp meter) that can read currents from milliamps through 10 amps. *(Courtesy of Triplett Corporation)* **b** A digital VOM. *(Courtesy of Sperry Instruments)*

4.10 MEASURING ELECTRICAL CURRENT

An **ammeter** can be analog or digital and is used to measure electrical currents. Figure 4–17 shows two examples of digital ammeters used to measure current that is less than 10 amps. This type of ammeter is built inside the VOM, which allows you to measure voltage, ohms, or milliamps through 10 amps. Since the meter can measure volts, ohms, or milliamps, it is called a *VOM meter* (volt-ohm-milliamp meter). The ammeter is actually designed to read very small amounts of current, such as 1/1000 A, which is called a *milliamp* and is also written as 0.001 A. The prefix *milli-* means one thousandth. The analog VOM meter shown in the beginning of this chapter in Figure 4–1 has a single meter movement that causes a needle to deflect when it senses current flow. This means that when a voltage is applied to a VOM meter, it must be routed through resistors that convert it to current that actually flows through the meter movement. When the VOM meter is set to measure current or resistance, the same meter movement is used. The only part of the meter that changes as it is switched from a voltmeter, ammeter, or ohmmeter is the arrangement of resistors to limit the amount of current flow through the meter movement.

The meter shown in Figure 4–18 is called a **clamp-on ammeter**, and it is used to measure current in larger AC circuits (some clamp-on ammeters can measure DC current, too). From the figure you can see this type of meter has two *claws*, or jaws, at the top that form a circle when they are closed. A button on the side of the meter is depressed to cause the jaws to open so that they can fit around a wire without having to disconnect the wire at one end. The jaws of the clamp-on ammeter are actually a transformer that measures the amount of current flowing through a wire by *induction*. When current flows through a wire, it creates a magnetic field that consists of flux lines. Since

Figure 4–18 A clamp-on ammeter that is used to read high currents.

(Courtesy of Triplett Corporation)

AC voltage reverses polarity once every 1/60 of a second, the flux lines collapse and cause a small current to flow through the transformer in the claw of the ammeter. This small current is measured by the meter movement, and the display indicates the actual current flowing in the conductor clamped by the meter. Induction will be explained in more detail in Chapter 7.

4.11 MEASURING MILLIAMPS

At times you will need to measure the current in a circuit in milliamps, since the amount of current is less than 1 amp. For example, you will need to measure the amount of current used by a gas valve so that you can set the heat anticipator on a heating thermostat. You will also need to make milliamp measurements on the solid-state control boards found in many newer furnaces and air conditioners. The most important part about reading milliamps is that the ammeter must be placed in series with the current it is reading. For the meter to be put in series with the current it is reading, an open must be created in the circuit. Figure 4–19 shows that an open is made in the circuit when a milliamp measurement is made, and the terminals of the ammeter should be placed on the wires where the hole has been made in the circuit. The open is created by removing one of the wires from the battery terminal. One of the meter probes is placed on the battery terminal, and the other is placed on the end of the wire that was removed from the battery. This places the meter in series with the circuit, and all the electrons flowing in the circuit must go through the meter because the ammeter has become part of the circuit.

Since the ammeter must become part of the circuit to measure the current, it must have very low internal resistance so that it does not change the total resistance of the circuit. When the VOM meter is set to read milliamps, it is vulnerable to damage because it has very small internal resistance.

Figure 4–19 The proper way to place the meter probes to make a current measurement.

Safety Notice!
If you mistakenly place the meter leads across a power source as you would to take a voltage reading when the meter is set to read current, the meter will be severely damaged. You can also be severely burned because the meter can explode, since it is a very low resistance component when it is set to read current. This would be similar to taking a piece of bare wire and bending it so it could be inserted into the two holes of an electrical outlet. You must always be sure to check how you have the meter set when you are making voltage, current, or resistance measurements.

As a technician you will be expected to use both the milliamp meter and the clamp-on ammeter. Milliamp meters are used to measure small amounts of current and clamp-on meters are used to measure larger currents such as the amount of current a motor is using. If a motor is using (pulling) too much current, you will be able to determine that it is overloaded and that it will soon fail. Later chapters in this book will explain how to use the clamp-on ammeter to measure motor current.

Digital VOM meters and some analog VOM meters have circuits to measure both DC and AC current. These types of meters have several ranges for measuring very small current in the range of milliamps up to 10 amps.

4.12

READING CURRENT WITH A CLAMP-ON AMMETER

The second type of ammeter that you will use on the job is called a clamp-on ammeter. An analog clamp-on meter is shown in Figure 4–20 and a digital clamp-on is shown in Figure 4–21. The clamp-on ammeter allows you to open its claw, close it around a wire, and then measure the current flowing through the wire. Figure 4–22 shows the clamp-on ammeter measuring current in the wire to an air-conditioning compressor. The clamp-on ammeter operates on the principal of magnetic induction. The claws of the ammeter are connected to the meter movement, and when the claw is closed around any wire that has current flowing through it, the current will be induced through the claw and converted in the meter as a current measurement. The clamp-on ammeter will only measure AC current unless it is specifically designed to measure DC. It can only read AC current because the alternating current builds and collapses 60 times a second, which causes inductive current to be produced in the clamp-on ammeter.

The clamp-on ammeter allows you to check the amount of current a compressor is pulling when it is running so you can check the measurement against the rating supplied on the equipment data sheet. You can also use the clamp-on meter to check the total amount of current flowing to a piece of equipment during start up and verify whether the supply wire is the proper size.

You can encounter several problems when using a clamp-on ammeter. If you wrap the claws around the L1 and the L2 wires feeding voltage to a compressor, the

Figure 4–20 A Sperry SPR-300 analog clamp-on ammeter. *(Courtesy of Sperry Instruments)*

Figure 4–21 A Sperry DSA-4500 digital clamp-on ammeter. *(Courtesy of Sperry Instruments)*

Figure 4–22 A clamp-on ammeter used to measure current flowing to an air-conditioning compressor.

current in L1 is out of phase with the current in L2, so the two currents will cancel each other. As a rule, you should only place the clamp-on ammeter around one wire at a time. The second problem occurs if the claws do not go completely closed. If they do not completely close, the magnetic pickup in the ammeter claws will not have a compete circuit and no current will be read. Some ammeters come with a feature that locks in the highest reading that the ammeter measures. Sometimes you will need to measure the current on a fan motor that will not function properly if the door to the fan cabinet is open. The lock feature allows you to set the ammeter to save the highest reading and put the ammeter around the wire to the fan you want to measure and close the cabinet door. When you start the motor, the meter will take the measurement and save it. When you shut off the motor and open the door, you can look at the ammeter and see the amount of current that has been saved. Be sure to have the lock in the off position when you are taking a normal reading or the meter may not take a new reading.

A newer type of current-measuring meter is called a digital multimeter with an open jaw current sensor, and it allows you to measure AC or DC current up to 100 amps. Figure 4–23 shows an example of this type of meter.

4.13 VOM IN A CLAMP-ON AMMETER

Another type of clamp-on ammeter has a VOM built into it. Figure 4–24 shows this type of clamp-on ammeter. The VOM allows this meter to measure voltage, resistance, and milliamps as well as higher currents. This type of meter allows you to carry only one meter when you are making service calls or doing installation work. You can make all of the measurements with one meter.

Figure 4–23 A Sperry DM-2001 digital multimeter with open jaw current sensor. This meter can measure AC and DC current up to 100 amps. *(Courtesy of Sperry Instruments)*

Figure 4–24 A Sperry TD-2608 clamp-on ammeter with built-in VOM. *(Courtesy of Sperry Instruments)*

4.14 CREATING A MULTIPLIER WITH A CLAMP-ON AMMETER

Sometimes the amount of current that is measured with the clamp-on ammeter is very small, and the needle will not deflect sufficiently to create an accurate measurement. Since the jaws of the clamp-on ammeter are actually a transformer, the amount of current in the reading can be multiplied by wrapping the wire in which you are measuring the current around the claw several times to cause the reading to be multiplied. For example, if you wrapped the wire around the jaw two times and the wire has .5 A flowing through it, the meter reading would indicate 1 A, which is double the amount. You would need to take the meter reading and divide by 2 to determine the exact current flowing in the wire. If you wrapped the wire around the claw four times, the meter reading would be 2 A, and you would need to divide the meter reading by 4 to get the true current flow in the wire. Creating a multiplier makes very small current measurements easier to read and more accurate. Figure 4–25 shows an example of a wire twisted around the jaws of a clamp-on ammeter to create a multiplier to read the amount of current flowing to the circuit.

4.15 MEASURING ELECTRICAL RESISTANCE

The analog VOM meter is also able to measure resistance when the selector switch is moved to one of the ohms positions, such as $R \times 1$, $R \times 10$, or $R \times 1000$. When the VOM meter is set in the ohm configuration, it uses an *internal battery* to provide voltage to make the meter movement deflect the needle. When this voltage is applied to a resistance, a small amount of current will flow, which the meter detects.

Figure 4–25 A clamp-on ammeter with wires twisted around its claws to create a multiplier.

Safety Notice!

Since the battery in the meter provides the power source for the current that the meter movement will read, it is important that the components being tested not be connected to any other source of voltage. It is also important to remove the component from the circuit when resistance measurements are made. When a component is disconnected from a circuit and all sources of power, it is said to be *isolated*.

The **ohmmeter** can be used to measure the resistance of a wire or fuse to ensure that the resistance is near zero ohms (0 Ω). A wire is said to have *continuity* and is considered to be a good wire when it has near 0 Ω. If the wire has infinite resistance, it is considered to be bad since it must have an *open* that is causing the high resistance. The same terms apply to fuses. If a fuse is good, it will be described as having continuity, and if it is bad, it will be described as an *open fuse*. When a wire or fuse is open, it is also said to have *infinite resistance*. The word *infinite* indicates that the amount of resistance is very high because the wire or fuse has an open. Since a typical VOM meter cannot accurately measure resistance more than 2 million ohms, its scale will be marked with the symbol for infinity (∞) at the highest point on its scale. Digital VOM meters flash an over-range display when an infinite resistance reading is made.

Figure 4–26a shows an example of an ohmmeter indicating 0 Ω when a good wire is tested, and Figure 4–26b shows the ohmmeter indicating ∞ Ω when the meter is testing a wire with an open in it.

The electrical loads in air-conditioning, heating, and refrigeration systems, such as motors or heating elements, may have some amount of resistance that falls between zero ohms and infinite ohms. For example, a fan motor may have 12 Ω of resistance in its run winding and 30 Ω of resistance in its start winding. The ohmmeter can be used to accurately

(a) (b)

Figure 4–26 a An ohmmeter indicating 0 ohms when a good wire is tested. **b** An ohmmeter indicating infinite ohms when a wire with an open is tested.

measure the resistance of each winding and determine whether it is 12 Ω or 30 Ω. The ohmmeter can be used to test motor leads and determine whether they are open. If they are not open, the run and start windings can be distinguished by the amount of resistance in each. You will see in later chapters that the start winding of most AC induction motors will always have more resistance than its run winding, and the ohmmeter allows you to easily tell the run winding from the start winding.

4.16 ## ZEROING THE ANALOG OHMMETER BEFORE MAKING A RESISTANCE READING

The ohmmeter uses an internal battery to create the current that makes the meter movement move. Since a battery slowly goes dead over time, and because each range in the meter uses different circuits of internal resistance, you must always zero an analog meter with a needle before you make a precision resistance reading. Zeroing an ohmmeter consists of moving the range selector to the range you intend to use and then putting the two leads together. The needle on the meter should swing towards zero. You will need to adjust the "zero adjust" knob, which is part of a variable resistor that is connected to the resistor circuit and battery inside the meter. You can adjust the knob until

the needle reads exactly zero. If the meter needle will no longer reach zero, the battery is too low and must be replaced. It is important to remember you need to zero the ohmmeter if you are trying to measure the resistance accurately, such as measuring the resistance of a motor winding.

4.17 SIMPLE REPAIR OF YOUR VOM

At times your meter may not work correctly. The two basic problems are a low battery in the ohmmeter or a meter that will not work on any setting. You can determine this when you set the meter on any ohms range and put the two leads together as though you were trying to zero the meter. If the meter does not work, you should suspect that the internal fuse for the meter is open. Most meters have a spare fuse located inside the meter. You will need to remove the back of the meter and locate the primary fuse. Since your meter is inoperable, you will need to remove and replace the fuse, or you can check the bad fuse with another meter.

If the battery is low, you can remove and replace it with a battery that has the same voltage. The meter may have one or two batteries. Some analog meters use a 30-volt battery for the highest resistance ranges, so you may need to locate a 30-volt meter battery. Other meters use a 9-volt battery or a D-cell 1.5-volt battery, both of which can be purchased in any convenience store. Most digital meters use a 9-volt battery.

Notice!

You must be aware that some electronic circuit boards found in modern furnaces and air conditioners may have components that are damaged by the 30-volt battery found in some analog VOMs. For this reason, you should use a digital meter, which has a lower voltage battery, or an analog meter that has a range specifically designed for testing electronic components or circuit boards. The range on these meters is identified with the diode symbol ($\rightarrow\!\!\vdash$) that looks like an arrow against a bar. This symbol indicates that the meter has a safe range to use on solid-state components or integrated circuits.

4.18 READING THE SCALES OF THE ANALOG VOM METER

The scale of a typical analog VOM meter is shown in Figure 4–27. From this figure you can see that the face has three separate scales (lines that are shown in the shape of an arc). The bottom scale is used to measure AC amps, and the second scale from the bottom, which is identified as AC DC, is used to measure AC and DC voltages and DC milliamps. The top scale, identified by the word *Ohms*, is used to measure resistance. Notice that the needle that is attached to the meter movement should rest at the far left corner of the meter face over the zeros when voltage, current, or resistance is not being measured.

A VOM meter is able to measure volts, ohms, and milliamps with the same meter movement by switching a number of resistors into and out of the circuit that connects the two probes to the meter movement. The switch is moved by a large knob on the front of the meter. Figure 4–28 shows the knob and the range selections on the face of the meter. If we use the hands of a clock to explain the range selections, you can see that the AC voltage circuit is located between 1 and 3 o'clock. The AC voltage circuit has five

Figure 4–27 The face of a typical VOM meter showing the voltage, current (amps), and resistance (ohms) scales.

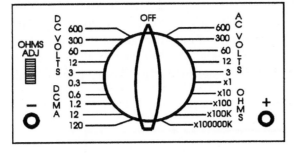

Figure 4–28 A VOM meter showing the ranges and the knob that is attached to the switch that connects the meter movement to different circuits of resistors inside the meter.

ranges: 0–3 V, 0–12 V, 0–60 V, 0–300 V, and 0–600 V. It is important to understand that some meters may use other ranges.

The ohms circuits used to measure resistance are located between the 3 and 6 o'clock positions. Notice that an × precedes each of the ranges because you will need to multiply the reading on the scale of the meter times the multiplier factor that is identified by the switch. You should get into the habit of saying the word *times* when you encounter the ×. For example, the first range of the ohms scales is identified as × 1. You will call this range the *times 1 range*. The other ohms ranges are × 10 (times 10), × 100 (times 100), × 1000 or × 1k (times 1000), and × 100000 or 100k (times 100,000).

Notice that when the range reaches 1000 and 100,000, the letter *k* is substituted. The letter *k* stands for *kilo*, which represents 1000. The *k* is often used where there is not enough space to write in all the zeros.

The DC milliamp circuits are located between the 6 and 9 o'clock positions. The ranges for milliamps are 0–0.06, 0–1.2, 0–12, and 0–120. The DC voltage ranges are located between the 9 and 12 o'clock positions. The ranges for DC voltage are 0–3 V, 0–12 V, 0–60 V, 0–300 V, and 0–600 V.

4.19 MEASURING DC VOLTS AND READING THE VOM SCALES

When you are making a DC voltage measurement, you will place the meter leads (probes) in the sockets identified as + and −, and you will set the selector switch on the face of the meter to the highest range. For this meter the selector switch would be set to 600 V. Setting the selector switch to 600 V will protect the meter, since you are not sure of how large a voltage you will be measuring. When you touch the tips of the probes to a voltage source, the needle that is attached to the meter movement will begin to move

from the position at the far left where it rests when no voltage is being sensed toward the middle. If you have the positive meter probe touching the negative voltage source, the meter movement will try to deflect farther to the left. If this occurs, simply switch the two probes so that the positive meter probe is touching the positive voltage source. This is called *changing the polarity* of the meter.

If the voltage that is being measured is small, the needle will not move very far, and you can change the switch on the meter face to the next lower scale. If the needle still does not move sufficiently, you can continue to change the switch to successively lower scales. Figure 4–29 shows a meter being used to measure the amount of DC voltage in a circuit that has approximately 4 V. Since the needle would not move sufficiently when the switch was set on the 0–600-V scale, the switch was changed to successively lower ranges until it was switched to the 0–12-V range and the needle moved to the position shown in the figure.

When you are ready to read the position of the needle on the face of the meter, you start by looking at the position of the switch. Since the switch in this example is at the 12-V DC setting, you must use the scale marked AC DC for this reading. Figure 4–30 shows an enlarged diagram of only the face of the meter. The AC DC scales are located directly below the ohms scale, which is the top scale. The voltage scales are highlighted in this figure so they are easier to identify. Notice that the scale marked AC DC has three

Figure 4–29 Voltmeter with its selector switch set on the 0–12-V range and its needle indicating that 4 V are being measured.

Figure 4–30 The face of a VOM meter with the 0–12-V DC scale shown in bold.

numbers shown at the far right of the scale: 300, 60, 12. Since the selector switch is set on 12 V, we will use the scale that has the numbers 0–12 marked on it. These numbers are also highlighted in the figure.

When you look closely at the meter face, you can see that the needle is pointing to the 4 on this scale. This means that the meter is measuring 4 V DC. You may be confused at first because the numbers 100 and 20 are also marked at this same position. You can be sure that the reading is 4 V by starting at the far right side of the scale and locating the number 12. Since the number 12 is the bottom number, you should use the bottom set of numbers for this scale. (A good habit to get into is to start with the number on the right side of the scale that matches the number the selector switch is pointing to and count backward as you locate the exact position to which the needle is pointing. For this example you would start at 12 and count backward 10, 8, 6, 4 as you moved to the left where the needle is pointing.)

4.20 MEASURING AND READING THE MILLIAMP SCALES ON THE VOM

When you are making a current reading with a milliamp meter, you must open the circuit and connect the meter probes so that the meter becomes part of the circuit. You should set the selector switch to the highest milliamp setting (120 DC mA range). If the needle does not deflect enough so that you can read it easily, you can switch the meter to the next lower range. When the selector switch is set on this setting, you will need to read the amount of current from the same AC DC scale that you used for the DC voltage reading.

Notice!

The scale marked AC Amps is used only when measuring AC current, and it is not usable for DC current readings

In Figure 4–31 the selector switch on the meter is set on the 120 DC mA setting. Since there is not a scale that ends with 120, we will use the 0–12 scale, and move the decimal point or multiply the reading by 10 to determine the exact amount of current flowing through the wire.

For example, in the figure you can see that the needle is pointing to the number 2. You can determine this by finding the 0–12 scale and begin counting backward (12, 10, 8, 6, 4, 2) until you reach the place where the needle is resting. Since the selector switch is set to the 120-mA setting, you must treat the 12 as 120. You can do this by multiplying the original reading by 10 or by counting backward from 120 to 20.

If the needle was pointing to the same position and the selector switch was pointing to 1.2 mA, the value the meter was measuring would be 0.2 mA. The major point to remember when reading the milliamp scales is that all three ranges use the numbers "one-two" (12), and the difference between each range is the placement of the decimal point.

4.21 MEASURING AND READING RESISTANCE SCALES ON THE VOM METER

When you are making a resistance measurement with the VOM meter, you must ensure that the device or component you are measuring does not have power connected to it and that it is isolated from other components. You should remember that this is a necessary

Figure 4–31 A meter reading 20 DC mA.

step because the VOM meter uses an internal battery as the power source when measuring resistance. When the meter is set to read resistance, the voltage from the battery is rather small: 3 V, 9 V, or 30 V, depending on the brand name of meter. The meter would be seriously damaged if you touched its probes to a power source such as 120 VAC by mistake when the meter was set for resistance.

Another point to remember when using the analog VOM to read resistance is to zero the meter. Since the meter uses a battery for the power source, the voltage in the battery must be able to move the needle to the zero mark on the scale when the two probes are touched together. This process is called *zeroing the meter* and it should be done prior to all resistance measurements. If the battery is weak, the needle will not be able to move to the zero point and all resistance readings will be faulty. This step is necessary because the actual voltage in the battery will constantly change as the battery is used and it starts to drain down.

The first step in the process of zeroing the meter is to place the selector knob on the ohms range you want to use. Next, you need to touch the two meter probes together, and the meter needle should swing toward the right side of the scale. Since the VOM meter is being used to measure resistance, the top scale, which is identified with the word *Ohms*, will be used. If the meter is zeroed correctly, the needle will rest directly on the 0 ohm setting. If the needle does not move to this position, you can adjust the needle position by changing the OHMS ADJ knob that is located to the left of the selector switch, just above the terminal for the negative probe. The OHMS ADJ knob is actually a potentiometer, and when you adjust it, you are changing the amount of resistance in the meter movement circuit, which will change the amount of current the battery is sending to the meter movement. If you change the OHMS ADJ knob and the meter needle does not reach all the way to the 0 ohm position, the meter has a weak battery that should be replaced.

Figure 4–32 VOM used to measure resistance. The scale indicates that the needle is resting on the 10 Ω value, and the selector is set to the × 1 (times 1) range, so the meter is measuring 10 ohms.

Notice!
You must zero the VOM meter whenever you switch resistance scales or change from one resistance scale to another because a different amount of internal resistance is used for each circuit. You do not need to zero the VOM meter when it is used to make voltage or current measurements because these types of readings do not use the meter's internal battery.

Figure 4–32 shows a VOM meter measuring resistance. In this example, the meter selector switch is set to the × 1 (times 1) setting. Because the meter is being used to measure resistance, the top scale is used. The needle is resting on the value 10 on the top scale. Since the selector knob is set on the × 1 (times 1) scale, you will need to multiply the reading by 1. This means that the VOM meter is measuring 10 Ω.

If the selector switch was set on the × 10 (times 10) range and the needle was resting in the same position, the meter would be measuring 100 Ω because you would need to multiply the scale reading by 10.

4.22 MAKING MEASUREMENTS WITH DIGITAL VOM METERS

When you are making a measurement with a digital VOM, you should follow the same rules as when you use an analog VOM. The major difference is that the measurement is presented as a digital number on the meter display. All you need to do is check the range selector switch to determine the units that the meter is measuring, and use the value that is displayed as the measurement. Be sure to observe polarity (+/−) if the voltage or current

is DC. Some digital meters also provide an audible signal for resistance measurements, which is useful when you are locating wires in multiconductor cables.

The digital voltmeter (DVM) may also be called a digital multimeter if it has additional functions. Some digital meters provide a function called autoranging or autoselection. This function provides two separate functions. First, if you inadvertently set the meter for a resistance test and then place the leads across terminals that have more than 4.5 volts applied, the meter automatically changes the meter circuit to read voltage instead of resistance. This protects the internal meter circuit from a short circuit. The second feature automatically changes the internal meter setting to the highest voltage if the original meter setting is too low. For example, if you have the meter set for 120 V and the circuit actually has 240, the internal circuit switches the meter to a range that can safely measure the 240 V. In most DVMs, overload protection is provided in all measurement modes. Figure 4–33 shows an example of a digital multimeter.

On some digital meters you must change the voltage range selector manually. To do this, you press the button that is located on the range-selector switch. For example, if you have the selector switch set on AC volts, you can set the voltage range to a range that is high enough to safely measure the higher voltage. Another way to approach this problem is to set the meter on its highest range and then set it lower as you need more accuracy. Some meters provide accuracy down to 10 µV. This range allows you to measure volts down to ten-millionths (10/1,000,000) of a volt.

Most digital multimeters provide an analog bar graph that provides a function similar to the needle movement on an analog meter. The meter indicates the amount of voltage on the bar graph as well as numerically on a screen.

If you are working in the field and need to measure the capacitance of a capacitor or if you need to measure temperatures, you can select a digital multimeter that can measure capacitance and temperatures. You can measure capacitance with the same probes you use to measure voltage. If you want to measure temperature, you must purchase the temperature probes to match the temperature range you want to measure.

Some digital multimeters, such as the Fluke 180 meter, provide data logging. This feature allows you to leave the DVM attached to equipment or a circuit over a period of time, and the meter will record the measured value and store the reading in its memory. Later you can review the recorded values to determine what has happened in the circuit while the meter was connected. This meter also allows you to connect a laptop computer

Figure 4–33 A Fluke 189 digital multimeter (DMM). This meter can measure true rms AC voltage, DC voltage, AC and DC current, resistance, capacitance, and temperature. It also has the capability of data logging. (Reproduced with permission of Fluke Corporation)

and record more data over a longer time. Some meters have a graphical display to show the values that you have measured and stored in the meter.

4.23 MEGOHMMETERS (MEGGERS)

A megohmmeter is an electrical meter that a technician can use to measure resistance leakage in motor windings and transformer windings. It is important to understand that you must ensure that all power is turned off to the motor or compressor you are testing with a megger. The megger is a resistance tester, so it would allow excessive current to flow through the meter like a short circuit, which could cause danger to the technician if the meter is connected to a circuit that has voltage applied. The problem occurs because the megger is a very low resistance meter and would act like a short circuit if connected to a circuit that has voltage applied to it.

Resistance leakage is an indication of problems in compressor motors, transformers, and other motors that can cause complete winding failure. When a compressor motor has excessive moisture in its windings, the moisture will cause the resistance leakage that will allow excessive current draw in the compressor motor, which eventually causes the compressor to fail. If you can measure the motor windings with a megger and determine that the windings are beginning to fail due to excessive moisture, you can draw the system into a vacuum and heat the compressor to draw the excess moisture out before the motor fails. This practice is usually done where the compressor is over 1 hp or when the compressor motor or other motor is expensive. Figure 4–34 shows a typical megohmmeter. This meter is a battery-powered digital insulation resistance tester. The megger can also make other measurements such as insulation resistance, current, capacitance, cable length, voltage, and resistance. The testers can make tests at 50 V, 100 V, or 600 V, and a voltmeter in the tester automatically detects and accurately displays AC/DC voltages in the test circuit on each of the insulation test voltage ranges. Since it is important that you have all voltage disconnected from the winding you are testing, some meggers provide an audible and visual signal when any voltage is present. Some meters have a circuit that checks for voltage and inhibits further tests if any voltage is present, thus protecting the user and the meter.

Figure 4–34 An example megger that is used to measure insulation resistance. This meter is used to measure insulation resistance from 5 kV to 10 kV in motors and transformers.
(Courtesy of Megger)

4.24 OVERVIEW OF ELECTRICAL TOOLS USED FOR HVAC

An HVAC technician must be able to properly use a wide variety of wire cutters, screwdrivers, wrenches, and other tools during the installation and troubleshooting of air-conditioning and refrigeration systems. As an HVAC technician, you will be responsible for a large number of tools. You may own your tools or the company you work for may provide tools for you. In either case, you will need to keep the tools clean, organized, and well cared for so you will have them every time you need them.

The following sections are provided to allow you to become familiar with the names and pictures of tools. Many of these tools are very simple to operate whereas others are more complex, so you may not fully understand how they work until you use them in the shop or on the job.

4.25 ELECTRICAL PLIERS, WIRE CUTTERS, AND WIRE STRIPPERS

HVAC technicians and electricians must use a wide variety of pliers, wire cutters, and wire strippers specifically designed for the HVAC trade. Figure 4–35 shows an insulated fuse puller. You should always use an insulated fuse puller to remove and replace fuses in a fuse disconnect rather than use metal tools. You can keep several sizes of insulated fuse pullers in your tool box for all sizes of fuses. Figure 4–36 shows a variety of wire cutters and strippers you may use on the job. The wire stripper is designed so that only the insulation on the wire is cut when the jaws are closed tightly, leaving the wire undamaged. The stripper has a variety of holes that fit wire gages from size 8 to size 22. When you are stripping the insulation from wires, be sure that you do not nick the wire or cut any strands, because this will reduce the ampacity of the wire.

Figure 4–37 shows a variety of slip joint, side cutting, needle nose, and clamping pliers that electricians use on the job. The wide variety of pliers are provided to give you access to hard-to-reach places.

Figure 4–35 A fuse puller is the correct tool to use to remove fuses safely.

Figure 4–36 a Klein six-in-one tool that cuts bolts, cuts and strips wire, measures stud sizes, and crimps insulated terminals. *(Courtesy of Klein Tools)* **b** Klein combination wire stripping and cutting tool. *(Courtesy of Klein Tools)* **c** Klein modular plug crimping tool. *(Courtesy of Klein Tools)* **d** Klein wire stripper and cutter. *(Courtesy of Klein Tools)* **e** Klein ratcheting cable cutter. *(Courtesy of Klein Tools)* **f** Klein electrical maintenance multitool. *(Courtesy of Klein Tools)* **g** Rigid cable cutter. *(Courtesy of the Ridge Tool Company)* **h** Klein bolt cutter. *(Courtesy of Klein Tools)*

4.26 ## SCREWDRIVERS FOR ELECTRICAL INSTALLATION AND MAINTENANCE

You will be required to use a wide variety of screwdrivers to install, troubleshoot, and maintain HVAC equipment. Figure 4–38 shows a variety of regular and Philips screwdrivers. Some of the screwdrivers in this figure are called self-starting screwdrivers because they hold a screw securely so that you can start the screw into a threaded hole with one hand. Other screwdrivers are specifically offset to provide access to very tight spaces so that you can turn screws slightly to tighten or loosen them. You should inspect the tips of your screwdrivers often and discard the ones that are worn. Do not try to

Figure 4–37 a Klein slip joint pliers. *(Courtesy of Klein Tools)* **b** Klein side-cutting pliers with smooth, streamlined nose. *(Courtesy of Klein Tools)* **c** Klein side-cutting pliers with crimping die. *(Courtesy of Klein Tools)* **d** Klein electronic diagonal cutting pliers. *(Courtesy of Klein Tools)* **e** Klein pump pliers. *(Courtesy of Klein Tools)* **f** Long-nose side-cutting pliers. *(Courtesy of Klein Tools)* **g** Klein locking pliers with curved jaw and cutting knives. *(Courtesy of Klein Tools)* **h** Klein curved-nose pliers. *(Courtesy of Klein Tools)*

grind the tip of your screwdriver when it becomes worn, because grinding will remove the temper in the steel and make it dangerous to use.

4.27 WRENCHES FOR HVAC MAINTENANCE

Figure 4–39 shows a variety of wrenches electricians may use on the job. Figure 4–39a shows combination box and open-end wrenches that are used to tighten or loosen nuts and bolts. These types of wrenches are available in metric and standard sizes. Figure 4–39b shows a set of socket wrenches. The sockets snap onto a ratchet drive, which allows you to tighten or loosen nuts and bolts quickly. The ratchet is also very useful in tight spaces.

Figure 4–39c shows a box-end wrench that has a ratchet feature. This allows you to use it to tighten and loosen nuts and bolts in tight spaces with a minimum of travel. Figure 4–39d shows a set of nut drivers. Nut drivers are special wrenches that have an end that looks like a socket wrench and a handle that is similar to a screwdriver. The nut driver is very useful for removing and replacing small sizes of screws or bolts that have

Figure 4–38 a Vaco 6-in. flat-blade screwdriver. *(Courtesy of Klein Tools)* **b** Vaco #2 Philips screwdriver. *(Courtesy of Klein Tools)* **c** Vaco six-piece screwdriver set. *(Courtesy of Klein Tools)* **d** Vaco four-in-one screwdriver. *(Courtesy of Klein Tools)* **e** Klein starting screwdriver. *(Courtesy of Klein Tools)* **f** Klein starting screwdriver. *(Courtesy of Klein Tools)* **g** Snap-on offset screwdrivers. *(Courtesy of Snap-on Tools)*

Figure 4–39 a Klein combination wrench set. *(Courtesy of Klein Tools)* **b** Klein socket wrench set. *(Courtesy of Klein Tools)*

Figure 4–39 (continued) c Klein box reversing ratchet set. *(Courtesy of Klein Tools)* **d** Klein nut-driver set. *(Courtesy of Klein Tools)* **e** Klein hex key Allen wrench set. *(Courtesy of Klein Tools)* **f** Klein T-handle hex key set. *(Courtesy of Klein Tools)* **g** Klein adjustable wrench. *(Courtesy of Klein Tools)* **h** Rigid pipe wrench. *(Courtesy of the Ridge Tool Company)* **i** Rigid chain wrench. *(Courtesy of the Ridge Tool Company)* **j** Rigid strap wrench. *(Courtesy of the Ridge Tool Company)*

hex heads. Figure 4–39e shows a set of hex key wrenches. These wrenches are sometimes called *Allen wrenches*. Figure 4–39f shows hex key wrenches with T-handles. The T-handle allows more torque to be placed on the wrench so that larger hex head bolts can be tightened or loosened.

Figure 4–39g shows an adjustable wrench. One brand of adjustable wrench is called a *crescent wrench*. This type of wrench is extremely useful when you can carry only one

wrench instead of a complete set of wrenches. The adjustable wrench can be used for a wide variety of bolt and nut sizes by simply changing the opening of its jaws.

Figure 4–39h shows a pipe wrench. The pipe wrench has an adjustable jaw that grips the smooth surface of a pipe or conduit so that it can be rotated to tighten or loosen its threads. Figure 4–39i shows a chain wrench, and Fig. 4–39j shows a strap wrench; both are used like a pipe wrench. These wrenches are adjustable-type wrenches used to loosen or tighten pipe and conduit.

4.28 ELECTRICAL CONDUIT BENDING, CUTTING, AND THREADING TOOLS

At times you may be requested to install conduit on larger commercial or industrial installations. You will need to be able to cut, thread, and bend conduit to complete the installation process. The objective for this section is to introduce you to the equipment that is designed for this purpose. You will not learn to operate this equipment until you get in the lab or in the field. Figure 4–40a shows a conduit bender that is used to make measured bends in metal conduit. You use this bender by placing your foot on the base plate of the bender and pulling on the handle. The bender has a set of indicators that show the amount of the bend in degrees. A hydraulic bender is available for bending larger conduits.

Figure 4–40b shows a flexible metal fish tape that is used to pull wire through a conduit. The fish tape is stored in a plastic case when it is wound up, and it is extended

(a) (b)

(c)

(d)

Figure 4–40 a Rigid tubing bender. *(Courtesy of the Ridge Tool Company)* **b** Klein flexible fish tape with grip handle. *(Courtesy of Klein Tools)* **c** Rigid pipe conduit cutter. *(Courtesy of the Ridge Tool Company)* **d** Rigid pipe reamer. *(Courtesy of the Ridge Tool Company)*

Figure 4–40 (continued) **e** Rigid portable pipe vise. *(Courtesy of the Ridge Tool Company)* **f** Rigid pipe (conduit) threader. *(Courtesy of the Ridge Tool Company)* **g** Rigid pipe tripod stand with vise. *(Courtesy of the Ridge Tool Company)* **h** Rigid tripod stand with electric pipe threader and pipe cutter. *(Courtesy of the Ridge Tool Company)*

into the empty conduit until the hook on its end extends out of the conduit. You can attach the wire that you are pulling into the conduit to the hook of the fish tape. The fish tape is strong enough to pull multiple wires into the conduit. You must be very careful not to allow the metal fish tape to come into contact with exposed electrical wires or terminals, because the metal tape will conduct current that can cause severe electrical shock. Figure 4–40c shows a cutting tool that is used to cut metal conduit. Figure 4–40d shows a tool that is used to ream pipe and remove rough edges after it is cut by the pipe cutter. This is an important step, because the rough edges of the pipe can cut or damage the insulation on wire that is pulled into the pipe or conduit.

Figure 4–40e shows a portable pipe vise, and Figures 4–40g and 4–40h show models of vice clamps that are mounted on a tripod. The tripod provides a stable work surface to hold conduit or pipe while you are cutting or threading it. Figure 4–40f shows a pipe-threading tool. This tool is used to put pipe threads on metal conduit so that it can be joined to threaded fixtures.

4.29 TOOL POUCHES AND TOOL BELTS

Figure 4–41a shows a general tool set that has wrenches and screwdrivers in it. The tool set has the most common tools that an electrician needs for maintenance. Figure 4–41b shows a tap-and-die set. You may need to use a tap to make interior threads inside a hole or a die to make exterior threads on a bolt. The die is used to "dress," or clean up,

(a)

(b)

(c)

(d)

(e)

(f)

Figure 4–41 **a** Klein general tool set. *(Courtesy of Klein Tools)* **b** Snap-on tap-and-die set. *(Courtesy of Snap-on Tools)* **c** Klein tool pouch. *(Courtesy of Klein Tools)* **d** Klein electrician tool pouch. *(Courtesy of Klein Tools)* **e** Klein electrician pocketknife. *(Courtesy of Klein Tools)* **f** Klein hammer. *(Courtesy of Klein Tools)*

threads on a bolt, and the tap is used to make initial threads or to dress up threads that are worn or damaged. You may need to drill a hole and tap threads into it to allow you to mount electrical components in an electrical panel with a screw or small bolt. The tapped hole takes the place of a nut, so you do not need to have a wrench on the back side of the panel to hold the nut.

Figures 4–41c and 4–41d show two different types of tool pouches. The tool pouch allows you to carry tools easily and safely so that they are always within reach. Figure 4–41e shows an electrician's pocket knife that is used to remove, or "skin," the insulation from a wire. Figure 4–41f shows a hammer that is used for maintenance. The hammer has a claw on one end to help remove nails or other fasteners.

4.30 REVIEW

When you are learning about components that utilize electromagnetic principles, you can return to this section to review the basic concepts about them. The important electromagnetic principles follow:

1. The dipoles of a magnet are aligned.
2. Every magnet has a north pole and a south pole.
3. Like magnetic poles repel, and unlike magnetic poles attract.
4. A permanent magnet is useful because its magnetic field is residual and can remain strong over a number of years.
5. When current passes through a conductor, magnetic flux lines will form around the conductor in concentric circles.
6. The strength of the magnetic field around a conductor will be proportional to the amount of current flowing through it until the amount of current reaches the saturation point of the electromagnet.
7. The magnetic field of an electromagnet can be turned on and off by interrupting the current flow through the coil of wire.
8. The polarity of an electromagnetic coil is determined by the direction of current flow through the conductor.
9. If the conductor is wrapped in coils, the strength of the magnetic field is increased.
10. When more coils of wire are added to an electromagnet, the magnetic field strength is increased.
11. A core will cause the field of an electromagnet to be strengthened.
12. The core must be made from laminated steel when AC voltage is used to power the coil so that the magnetic field can quickly build and collapse in the core as the AC voltage changes polarity every 1/60 of a second.

A volt-ohm-milliammeter (VOM) is an electrical meter that can measure volts, ohms, or milliamps. A milliamp is one-thousandth of an amp. A digital VOM meter displays measured values as numbers on the meter face. An analog VOM has a needle that moves when measurements are made. An analog ohmmeter must be zeroed to compensate for loss of battery power. A digital meter does not need to be adjusted or zeroed for ohms.

It is important to remember to set the range for a voltmeter higher than the amount of voltage you expect to measure and place the leads across the two terminals where the measurement is to be made. When you make a resistance measurement, you must be sure the power is turned off and the wire or component you are testing is isolated. When you are making an amperage measurement, you must provide an open in the circuit and place the ammeter in series with the open. If you place the meter leads across any component or power source when the meter is set to read milliamps, severe damage to the meter and to the circuit will result.

A clamp-on ammeter has claws that open so they can be clamped around a wire. The clamp-on ammeter allows a technician to make an AC current measurement without disconnecting any wires.

Job Assignment for This Chapter

You are on a service call working on an air conditioner and have found a bad fuse. The customer wants to know how you can tell whether a fuse is good or bad. The customer thinks you are trying to sell some parts that are not needed. Your job is to use the

meters in your truck to convince the customer that you can tell whether a fuse is good or bad and to show that the fuse you found in the system is open and must be replaced.

Solution to the Job Assignment

Your solution to the job assignment should include using an ohmmeter to show the customer that the fuse is open. A simple way to prove this is to show the two leads of the ohmmeter touching each other and the meter indicating 0 ohms. Next you would move the leads apart and show the customer that the meter now reads infinite resistance. Then you would place the leads on the fuse that you are testing and show the customer that the reading on the meter is infinite, which proves that the fuse has an open.

Assessment

QUESTIONS

1. Explain the term *infinite resistance*.
2. Identify four types of pliers.
3. Name two different types of screwdrivers.
4. Use the diagrams in Figure 4–15 to explain where you would place the probes of a voltmeter when making voltage measurements.
5. Use the diagram in Figure 4–19 to explain where you would place the probes of an ammeter when making a current measurement.
6. List two things that you should be aware of when making resistance measurements.
7. Explain why you should zero an analog ohmmeter every time you change scales or make a measurement.
8. Explain why you should set a voltmeter and ammeter to the highest setting when you first make a measurement of an unknown voltage or current.
9. Explain the difference between a digital and an analog meter.
10. Explain what you would check on an ohmmeter if you did not get a reading when you tried to zero the meter.

TRUE OR FALSE

1. An insulated fuse puller rather than a metal pliers should always be used to remove and replace a fuse in a fused disconnect.
2. All voltage should be turned off when making a resistance measurement.
3. Infinity is the highest resistance reading for an ohmmeter.
4. A VOM meter must be zeroed prior to making voltage, current, or resistance measurements.
5. The voltage should be turned off for safety reasons when making a voltage measurement.
6. The ohmmeter uses an internal battery to supply the power for all resistance readings.

MULTIPLE CHOICE

1. A milliamp is
 a. one-thousandth of an ampere.
 b. one-millionth of an ampere.
 c. one million amps.

2. When making a current measurement with a VOM meter, you should
 a. place the meter probes across the load.
 b. turn off all power to the circuit during the current reading so you don't get shocked.
 c. create an open in the circuit and place the meter probes so that the meter is in series with the load.

3. When making a voltage measurement, you should
 a. place the meter probes across the load.
 b. turn off all power to the circuit during the voltage reading so you don't get shocked.
 c. create an open in the circuit and place the meter probes so that the meter is in series with the load.

4. When making a resistance measurement, you should
 a. keep all voltage applied to the component you are measuring so you can determine the true resistance.
 b. turn off all power to the circuit and isolate the component you want to measure.
 c. create an open in the circuit and place the meter probes so that the meter is in series with the load.

5. When you are measuring current with a clamp-on ammeter, you should
 a. open the jaws of the meter and place them around the wire where you want to measure the current.
 b. create an open in the circuit and place the meter probes so that the meter is in series with the load.
 c. turn off all power to the circuit so that you don't get shocked.

6. You would use _____ to strip a wire to remove its insulation.
 a. wire cutters
 b. common pliers
 c. wire strippers
 d. side-cut pliers

7. Offset screwdrivers are used to
 a. provide more torque than a regular screwdriver.
 b. tighten or loosen hard-to-reach screws that have small clearance or are in tight spaces.
 c. reach screws in deep recesses.
 d. do all the above.

8. A nut driver is
 a. a type of pliers.
 b. a type of wrench.
 c. a tool used to strip wires.
 d. a special offset screwdriver.

PROBLEMS

1. Use the diagram in Figure 4–29 to determine the voltage reading for the meter if the selector switch is set on 60 V.

2. Use the diagram in Figure 4–30 to determine the current reading for the meter if the selector switch is set on 12 mA.

3. Use the diagram in Figure 4–31 to determine the resistance reading for the meter if the selector switch is set for × 1K.

4. Determine the amount of current flowing in a wire if a clamp-on ammeter has four turns of the wire twisted around its claw, and the reading on the scale is 2 A.

5. List the kind of problems that could be present in a circuit if a milliamp meter reading indicates the circuit has 0 mA.

Websites

Websites change constantly. Use the key word search to locate sites that have similar content. The products that will be found at each site are indicated with the key word.

Fluke Inc.
Products: Voltmeter, ammeter, VOM
Key words: Voltmeter, ammeter, VOM, DVM

A. W. Sperry Instruments
Products: Voltmeter, ammeter, VOM
Key words: Voltmeter, ammeter, VOM, DVM

Triplett Corporation
Products: Voltmeter, ammeter, VOM
Key words: Voltmeter, ammeter, VOM, DVM

Chapter 5

Symbols and Diagrams for HVAC and Refrigeration Systems

Job Assignment for This Chapter

You are on a service call and have removed the access door for the outside air-conditioning unit. The door has a wiring diagram on it that shows where the components in the system are located. An electrical symbol is used in the diagram for each component in the system. Your job is to match the component symbol with the actual component and explain to the customer the name of each component in the electrical panel and what its function is in the system. You should also explain the difference between the wiring schematic diagram and the ladder diagram.

Terms You Will Learn in This Chapter

Capacitor	Evaporator fan motor	Normally open
Coil	Fan	Overload
Compressor	Fuse	Pressure switch
Compressor contactor	Fuse disconnect	PSC motor
Condenser	Gas valve	Relay
Condenser fan motor	Heater	Schematic (ladder) diagram
Contactor	Ladder diagram (see *schematic diagram*)	Single-pole, single-throw switch
Control	Load	Solenoid
Crankcase heater	Motor	Switch
Cutaway diagram	Motor starter	Thermostat
De-energized	Normally closed	Transformer
Energized		Wiring diagram

After reading this chapter, you will be able to:

1. Identify the electrical symbols for electrical components in HVAC systems.
2. Explain the difference between an electrical load and an electrical control.
3. Explain the difference between a schematic (ladder) diagram and a wiring diagram.
4. Identify the components in the electrical system of an air-conditioning, heating, and refrigeration system.

5.0 OVERVIEW OF SYMBOLS AND DIAGRAMS FOR ELECTRICAL SYSTEMS

All electrical systems for air-conditioning, heating, and refrigeration equipment are represented in electrical diagrams by electrical symbols. This chapter will introduce you to all of the electrical components that are found in air-conditioning, heating, and refrigeration systems. You will learn the electrical symbol that is used to represent each component. You will also learn the details of the electrical components and how they are represented in an electrical diagram. You will also begin to see the slight differences in symbols for each component that is shown in a wiring diagram and how it is different in a schematic (ladder) diagram. The wiring diagram shows the location of all the components, and the schematic (ladder) diagram shows the sequence of operation.

You will find that once you can identify the components individually, you will begin to see how they work together in systems. You will also begin to see that air-conditioning, heating, and refrigeration systems all have similar components; once you understand the component in one system, you will be able to identify it in other systems. This makes it easier to understand all the electrical components in these systems because the same components are used in different equipment such as gas furnaces, oil furnaces, electric furnaces, refrigerators, ice makers, chillers, large commercial air conditioners, and even window-unit air conditioners. This chapter will help you begin to see how the same components and circuits are used in all the equipment you will be expected to work on. This chapter will also introduce several types of electrical wiring diagrams and the symbols that are used in them.

5.1 THE ELECTRICAL SYSTEM FOR A SIMPLE AIR-CONDITIONING UNIT

The air-conditioning system that you work on in the field consists of a compressor motor, a condenser fan motor, an evaporator fan motor, and the controls that turn it on and off. The compressor motor, evaporator fan motor, and condenser fan motor are all examples of **loads** in the electrical system. The definition of an *electrical load* is a component that has electrical resistance and uses (consumes) electricity to convert it to another type of energy such as motion or heat. This electrical system also has components called **controls**, which are devices or components that can switch electrical voltage and current on or off to energize or de-energize a load such as a motor. Controls for the air-conditioning system include the thermostat and the relays that **energize** and **de-energize** these motors.

Figure 5–1 A typical residential air-conditioning system. The condenser section is located outside the house, and the furnace part of the system is located inside the house. *(Courtesy of York International Corporation)*

Figure 5–1 shows a **cutaway diagram** of a typical residential air-conditioning system with all the components identified. This figure will help you understand which components of the system are inside the house and which are located outside. In this diagram you can see that the compressor and condenser fan are located in the condenser unit that is found outside the house, and the furnace fan, which is also called the evaporator fan in the air-conditioning system, is located in the furnace inside the house. A **fan** consists of a motor and a fan blade that moves air. Controls for this system include the **thermostat** that controls the temperature of the system. The thermostat is located inside the house and is usually mounted on a centrally located wall. The thermostat controls the compressor relay, also called the **compressor contactor**, which is located in the electrical panel of the condenser and energizes the compressor and condenser motors. The fan relay is located inside the furnace and energizes the furnace fan. From this point on, the furnace fan will be called the *evaporator fan* for the air-conditioning system because that is the function it performs.

Figure 5–2a shows a cutaway picture of the condenser so that you can see where the compressor motor and **condenser fan motor** are located and where the electrical panel is located. Figure 5–2b shows a cutaway diagram of a furnace. If the system is designed for air conditioning, it will have an evaporator coil on top of it. The **evaporator fan** (also called the furnace fan) blows indoor air over the evaporator where the air is cooled and then distributed throughout the house. The remainder of this chapter will show you the electrical symbol for each of the components in the system, and you will see some simple electrical diagrams for these systems.

Figure 5–2 a A cutaway
picture of the
outdoor part of the
air-conditioning system
shown in Figure 5–1.
b A cutaway picture of a
furnace, which is the
indoor part of the
air-conditioning system.
*(Courtesy of Armstrong Air
Conditioning, Inc.)*

(a) (b)

5.2 ELECTRICAL SYMBOLS

Each electrical control and load has a unique symbol that represents that component in an electrical diagram. The electrical symbols are standardized and all manufacturers use similar symbols for each component, although some manufacturers alter the symbols slightly. The symbols may not be used in **wiring diagrams** where the physical location and the outline of each part is being presented, but the symbols are generally used in electrical **schematic (ladder) diagrams**. Figure 5–3 shows a complete list of the electrical symbols, and you should notice that all the switches are shown with **normally open** or **normally closed** contacts. The symbol for each switch also shows the *operator* for the switch, which is the symbol for pressure, temperature, flow, and other actions and forces that cause the switch contacts to move open or closed. The symbol also indicates whether the force or action will cause the contacts to open when the force or action increases or when it decreases. If the switch contact is located on the top of the output terminal of the symbol, and the contacts are shown as normally closed, the symbol shows that an increase of action such as temperature will cause the switch contacts to open, and a decrease in temperature will cause the switch contacts to close. If the contact is located on the bottom of the output terminal, and the contacts are shown as normally closed, the symbol shows that a decrease in temperature causes the contacts to open, and an increase in temperature causes the contacts to close.

5.2.1 Symbols for Common Components in Air-Conditioning, Heating, and Refrigeration Systems

This section of the chapter will show the electrical symbol with the picture of the component it represents. At first it is difficult for technicians to make the connection between the actual electrical device and its electrical symbol. One of the symbols you will find on many electrical diagrams is an electrical **switch**. The simplest switch is the **single-pole, single-throw (SPST) switch**. More complex switches include the single-pole, double-throw (SPDT) switch; the double-pole, single-throw (DPST) switch; and the double-pole, double-throw (DPDT) switch. Figure 5–4 shows the electrical component for each of the switches, and Figure 5–5 shows the electrical symbol for each of these switches.

STANDARD ELEMENTARY DIAGRAM SYMBOLS

The diagram symbols shown below have been adopted by the Square D Company and conform where applicable to standards established by the National Electrical Manufacturers Association (NEMA).

Figure 5-3 Electrical symbols for motors and controls used in air-conditioning, heating, and refrigeration systems. (*Courtesy of Square D/Groupe Schneider. Square D/Groupe Schneider assumes no liability for accuracy of information.*)

SPST SPDT DPST DPDT

Figure 5–4 Electrical components for single-pole, single-throw (SPST) switch; single-pole, double-throw (SPDT) switch; double-pole, single-throw (DPST) switch; and double-pole, double-throw (DPDT) switch.

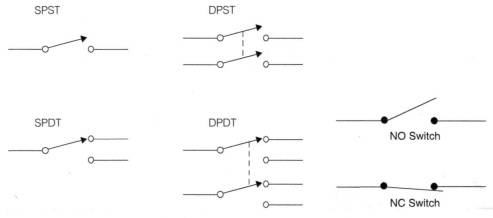

Figure 5–5 Electrical symbols for single-pole, single-throw (SPST); single-pole, double-throw (SPDT); double-pole, single-throw (DPST); and double-pole, double-throw (DPDT).

Figure 5–6 Symbols for normally open (NO) and normally closed (NC) switch.

ALS	Auxiliary Limit Switch	IDM	Induced Draft Motor
BM	Blower Motor	IDR	Induced Draft Relay
BMR	Blower Motor Relay	LED	Light-Emitting Diode
CAP	Capacitor	LEVS	Level Switch
COM	Common	LGPS	Low Gas Pressure Switch
CPU	Central Processing Unit	LS	Limit Switch
DPDT	Double-Pole, Double-Throw	NC	Normally Closed
DPST	Double-Pole, Single-Throw	NO	Normally Open
DSS	Draft Safeguard Switch	OL	Overload
EAC	Electronic Air Cleaner	PCB	Printed Circuit Board
FRS	Flame Rollout Switch	PRI	Primary
FS	Flow Switch	PS	Pressure Switch
FU	Fuse	SEC	Secondary
GV	Gas Valve	SPDT	Single-Pole, Double-Throw
GVR	Gas Valve Relay	SPST	Single-Pole, Single-Throw
HSI	Hot Surface Igniter	TRAN	Transformer
HUM	Humidifier		

Figure 5–7 Abbreviations for common HVAC electrical symbols.

Figure 5–6 shows the symbol for a normally open (NO) switch and a normally closed (NC) switch. You will see the term NO or NC used with switch contacts to indicate whether they are normally open or normally closed. The term *normal* indicates the condition of the switch when no one is touching it. The normal condition of a switch is also referred to as the "out of the box" condition when no one is touching the switch.

Abbreviations are sometimes used in diagrams along with the symbols. Figure 5–7 shows examples of common abbreviations.

Another type of symbol that you will encounter in electrical diagrams is the symbol for an indicator lamp. Figure 5–8a shows the symbol for a red indicator lamp, Figure 5–8b

(a) (b) (c) (d)

shows the symbol for an amber indicator lamp, and Figure 5–8c shows the symbol for a green indicator lamp. Indicator lamps are used on commercial system control panels to indicate when the system is operating correctly. The letter inside the lamp indicates the color of the lens. For example, R stands for red, A stands for amber, and G stands for green. Figure 5–8d shows typical indicator lamps you would find on a system.

5.2.2 Relay Coil and Contacts

One of the components that you will encounter in air-conditioning systems and furnaces is called a relay. The **relay** consists of a **coil** and contacts and is used to turn the compressor motor and condenser fan motor on and off in a system. The relay uses a magnetic coil to open or close one or more sets of contacts that control current flow to the motors. When the contacts are closed, current flows through them to the motors, and when the contacts are open, current is stopped. The voltage that controls the coil in an HVAC system is typically 24 volts. This low voltage flows through the thermostat that is mounted on the wall in the home or business and on to the coil to make it a magnet. Since people put their hands on the thermostat, the low voltage (24 volts) does not pose an electrical shock hazard. The voltage that flows through the contacts to the fan motors is typically 240 volts in a residence and may be 480 volts in a commercial building. The relay allows the coil to be controlled by low voltage, and the contact can control the larger voltage and current to the compressor and fan motor. Figure 5–9a shows the coil and contacts of a typical relay. This picture shows the parts exposed so you can easily see the coil and contacts. Figure 5–9b shows a relay that is use in air-conditioning systems. The coil and contacts are not as visible in the relay shown in Figure 5–9b, and the terminals for connections are brought out the top of the relay. The relay consists of a coil and one or more sets of contacts. Figure 5–10 shows the

Figure 5–9 a Example of a relay with its coil and contacts exposed. **b** A different type of relay used in an HVAC system. Notice that the contacts and coil are enclosed in plastic, but they are identified on the relay with symbols and terminal numbers.

(a) (b)

NO Contacts

NC Contacts

Figure 5–10 Symbols for normally open (NO) and normally closed (NC) contacts.

(a)

(b)

Figure 5–11 a A symbol for a relay coil or solenoid coil. **b** A second symbol for the relay coil.

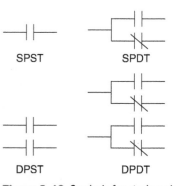

SPST SPDT

DPST DPDT

Figure 5–12 Symbols for single-pole, single-throw (SPST) relay contacts; single-pole, double-throw (SPDT) relay contacts; double-pole, single-throw (DPST) relay contacts; and double-pole, double-throw (DPDT) relay contacts.

symbols for normally open (NO) and normally closed (NC) relay contacts. The normally open contacts are held open by a spring, and when power is applied to the relay coil, the contacts move to the closed position. The normally closed contacts are held closed by a spring, and when power is applied to the coil, they are moved to the open position.

Figure 5–11 shows the electrical symbol for the coil of the relay. Notice it can be shown several different ways. You will learn more about the operation of the relay in Chapter 12. The coil has two leads, and when power is applied to it, current flows through the coil and causes it to create a magnetic field that changes the position of the contacts.

Figure 5–12 shows the electrical symbols for single-pole, single-throw (SPST) relay contacts; single-pole, double-throw (SPDT) relay contacts; double-pole, single-throw (DPST) relay contacts; and double-pole, double-throw (DPDT) relay contacts. These contacts act the same way as the switch-type contacts. The only difference is that the relay coil pulls the contacts in the relay open or closed, whereas the manual switch contacts must be switched by hand.

The ratings for the relay contacts include a voltage rating and current rating that should not be exceeded. The voltage rating on the coil should be matched exactly. You will learn more about voltage and current ratings in later chapters.

5.2.3 Solenoid Coils

A **solenoid** is a coil that uses its magnetic energy to open or close a valve. A typical solenoid is shown in Figure 5–13a and its symbol is shown in Figure 5–13b. A solenoid valve is used to open and close valves in a refrigeration system such as a hot gas solenoid and in a heat pump system to control the flow of refrigerant in the reversing valve. Solenoid valves are also used in liquid line applications to control the flow of liquid refrigerant. When the coil of the solenoid valve receives current, a magnetic field is created that pulls the inner workings of the valve to an open or closed position, depending on the type of valve. When power is turned off to the solenoid coil, the valve position is returned to its normal condition by a spring. Sometimes the coil of a relay is referred to as a solenoid, but typically a solenoid is used to describe a magnetically controlled valve.

5.2.4 Contactors

A **contactor** is a larger version of a relay. It has a coil and can have one or more sets of contacts that can be normally open or normally closed. When the contact rating exceeds 20 amps, the control is called a contactor instead of a relay. The operation of the contactor is

(a) (b)

Figure 5–13 a Example of a solenoid valve. **b** Electrical symbol for a solenoid valve.

exactly like the relay, in that when current flows through the coil a magnetic field is produced that changes the position of the contacts. Figure 5–14a shows an example of a typical two-pole contactor, and Figure 5–14b shows an example of a three-pole contactor used to switch three phase voltage to a compressor. You will normally find contactors in larger HVAC and refrigeration systems with compressors that pull more than 20 amps. Figure 5–15a shows the electrical symbol for a single-pole contactor, Figure 5–15b shows a two-pole contactor, and Figure 5–15c shows a three-pole contactor. You can see that the symbols for the contactors look very similar to the symbols for relays.

5.2.5 Motor Starters

A **motor starter** is similar to a contactor in that it has a coil and contacts, but the motor starter has additional components built into it that provide overcurrent protection for motors. The motor starter is shown in Figure 5–16a and its wiring diagram is shown in Figure 5–16b. Motor starters are typically used to protect larger compressors, larger fan motors, or larger pump motors. They normally are not used on residential systems. The motor starter has a heating element, shown in Figure 5–17b, that is connected in series with the motor load and the main contacts so that it samples the current going to the motor. When current flows through the contacts and the heating element to the motor, the heating element produces heat that is sent to the overload mechanism. The **overload** mechanism is designed so that when current exceeds the predetermined setting, the heat from the heating

(a) (b)

Figure 5–14 a Two-pole contactor. **b** Three-pole contactor.
(Courtesy of Honeywell)

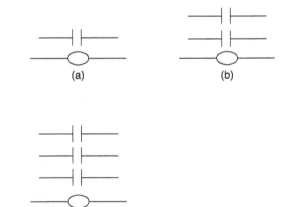

(a) (b)

(c)

Figure 5–15 a Symbol for a single-pole contactor. **b** Symbol for two-pole contactor. **c** Symbol for a three-pole contactor.

Keyhole Mounting Slot — Easy to reach with large screwdriver or power driver; permits the mounting screw to be in place before installing the starter.

Lineside Power Terminals — Up-front for accessability; self-lifting saddle clamps for ease of wiring; clearly marked in contrasting white for quick identification. Optional top wiring kit for easy connection of power factor correction capacitor ahead of the overload relay.

Arc Hood Cover Screws — Up-front for easy accessability; allows removal of the arc hood cover for inspection of the front movable and stationary contacts, and contact springs.

Coil Cover Screws — Up-front for easy accessability allows removal of the coil cover for coil and contact change and access to all internal components.

Auxiliary Contact Terminals — Angled and up-front for accessability; self-lifting pressure plates for ease of writing; clearly marked in contrasting white for quick identification.

Coil Terminals — Up-front for accessability; self-lifting saddle slamps for ease of wiring.

N.C. Overload Relay Contact Terminal — Self-lifting pressure plates for ease of wiring; clearly marked in contrasting white for easy identification.

Tie Point Terminal — Convenient access point for control circuit wiring; accessability; self-lifting pressure plates for ease of wiring.

Heater Element Screws — Up-front for easy installation; allows interchangeability of heater elements for Class 10, 20 and 30 operation.

Loadside Power Terminals — Up-front for accessability; self-lifting saddle clamps for ease of wiring; clearly marked in contrasting white for quick identification.

(b)

Figure 5–16 a A typical motor starter. *(Courtesy of Rockwell Automation)* **b** Wiring diagram (schematic) of a motor starter with overloads and heaters.

Figure 5–17 a Symbol for overload heater. **b** Symbol for overload contacts.

element will cause a set of small normally closed overload contacts that are shown in Figure 5–16b to trip. You can see in the electrical diagram of the motor starter shown in Figure 5–16b that the overload contacts are connected in series with the motor starter coil. When the overload contacts trip to the open position, current to the motor starter coil is interrupted and the main motor starter contacts are opened, causing voltage and current to stop flowing to the motor. The symbol for the overload heater is shown in Figure 5–17a and the symbol for the overload contact is shown in Figure 5–17b. The motor starter has a reset button that must be manually depressed to reset the motor starter to the operating condition. In this way the motor starter protects the compressor motor, fan motor, or pump motor from drawing too much current, which will damage the motor windings. It also causes a technician to manually reset it, which means the technician would inspect the

system and measure the current to determine what is causing the overloads to trip. You will learn more about motor starters and the way they protect systems in Chapter 12.

5.2.6 Heaters

A **heater** is an electrical component that has resistance and creates heat when power is applied. An example of a **crankcase heater** used in an HVAC system is shown in Figure 5–18a and 18b and its symbol is shown in Figure 5–18c. One type of heater used in an HVAC system is a crankcase heater that is connected around the base of a compressor to provide heat that keeps the oil and the refrigerant separated, as shown in Figure 5–18b. Another example of a heater used in an HVAC system is the electric heating element that is used as the main heating element in an electric furnace. In refrigeration systems heaters are used as defrost heaters that cause excess frost that accumulates on the evaporator coils to defrost. You will also find heating strips in the doors of refrigerators and refrigeration and freezer cases. You should notice that the symbol for the electrical heater looks like the symbol for a resistor. The reason for this is that an electrical heater is actually a resistance source. The amount of resistance in the heater coil will determine how much heat it will produce when voltage is applied to it.

5.2.7 Transformers

Transformers are used in air-conditioning and refrigeration systems to step voltage down from one level to another. For example, in a residential air-conditioning system the transformer primary winding is connected to 120 V and the transformer steps the voltage down to 24 V. The 24 V is used in the control circuit, which allows voltage to pass through the room thermostat to the coil of the relay that controls voltage to the compressor. The 24 volts is considered a low-voltage power source and does not pose an electrical shock threat as much as 120 V would. This means that residential customers can feel safe when they put their hand on the thermostat to make a change in the setting. Figure 5–19a shows a picture of a typical transformer, and Figure 5–19b shows the electrical diagram for the transformer. This type of transformer is called a control transformer because it provides voltage to the control circuit. In the electrical diagram you can see that the transformer has two separate windings. One winding is called the primary winding and is connected to 120 V; the other is called the secondary winding, where it provides 24 V.

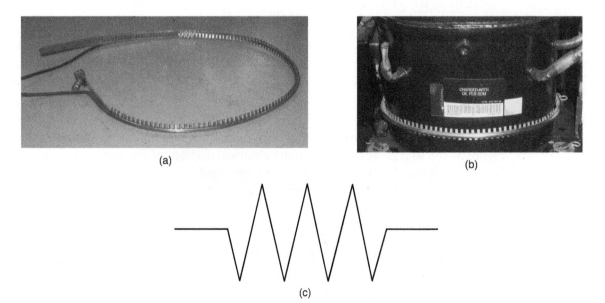

(a)

(b)

(c)

Figure 5–18 **a-b** Electric crankcase heater band for compressor. **c** Symbol for electric heater.

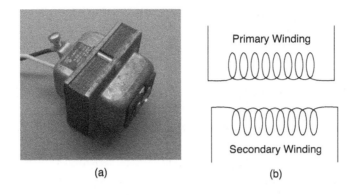

Primary Winding

Secondary Winding

(a) (b)

Figure 5–19 **a** Picture of a 24-volt transformer. **b** Electrical symbol for a transformer.

5.2.8 Fuses and Protective Devices

A **fuse** is a protective device that has an element inside that will melt when the amount of current flowing through it increases its temperature beyond its setting. Figure 5–20a shows the wiring diagram symbol for the fuse, Figure 5–20b shows the schematic (ladder) diagrams symbol for the fuse, and Figure 5–20c shows pictures of several types of fuses. The fuse is wired in series with the entire circuit or individual components, and it is used to monitor the amount of current flowing to the components. In the HVAC system, the fuse is designed to protect the compressor motor and condenser fan against overcurrent. If a compressor starts to have a problem, it will begin to pull excessive current beyond its rating. If the compressor is allowed to continually pull excessive current, it will damage the windings and the compressor will have to be replaced, which is very expensive. The fuse, on the other hand, usually costs a few dollars, and it will blow when the overcurrent condition begins, thereby protecting the motor. When the fuse blows, the compressor motor no longer receives voltage and current and turns off, and the customer will notice that the temperature in the home will begin to rise and call for a service call. When you arrive for the service call and find a fuse has blown, you should replace the fuse with a new one and put a clamp-on ammeter on the line that feeds current to the compressor to test for the exact amount of current. If the overcurrent condition continues, you will have to determine whether the condenser coil is dirty or low voltage has caused the overcurrent condition. The condition causing the overcurrent must be repaired before you complete the service call. If you do not repair the condition causing the overcurrent, the fuses will continue to blow.

The **fuse disconnect** provides a location to house or mount the fuses in a way that they are connected in series with the system. The fuse disconnect also has a switch that allows you to disconnect all power to the system and fuses when you are doing maintenance. The switch handle for the fuse disconnect is located on the side of the disconnect.

(a)

(b)

(c)

Figure 5–20 **a** Fuse symbol for wiring diagram. **b** Fuse symbol for schematic (ladder) diagram. **c** Examples of fuses.

Figure 5–21 a A fuse disconnect that provides a location to connect the fuses in the circuit and a means to disconnect all power to the system. **b** Symbol for a fuse disconnect.

When the handle is in the up position, the switch is closed; when the handle is in the down position, the switch is opened, and the door of the disconnect can be opened. You will need to open the door to test for voltage or to remove and replace a blown fuse. The fuse disconnect is shown in Figure 5–21a and its symbol is shown in Figure 5–21b.

Safety Notice!

You should always treat the fuse disconnect switch as though it is powered. The line side terminals at the top of the disconnect switch will still have voltage applied even when the switch is in the open position. The only time that power is not applied to the line side terminals is if the circuit breaker that supplies power to the disconnect is turned off. If you are working on the system, you must turn the switch to the off position, open the door, and immediately test for the presence of voltage at the bottom terminals. You should never assume the power is turned off until you actually test for voltage even though the handle is in the down position. You should remember that you need to remove the fuse to test it for continuity with your ohmmeter. If you are removing and replacing fuses to test them or replace them, you should always use an insulated fuse puller rather than a metal pair of pliers. You should also close the door prior to turning power back on, because the door will protect you from any electrical problems that may occur because of a short circuit or other condition that would cause the fuses to blow or explode.

At times you may need to test for the loss of voltage in the electrical system of the air conditioner, and it is permissible to make these tests with voltage applied. However, you need to be aware that power is turned on and make the test safely.

5.2.9 Overload Protective Devices

The overload protective device is used to protect individual motors such as the compressor motor or pump motors. The overload protective device is sized for the maximum amount of current the compressor or motor will draw. The overload device works similarly to a fuse in that it protects against overcurrent, but it is different in that it also is able to sense an overtemperature condition and is able to be reset manually. Hermetically sealed compressors depend on refrigerant to provide cooling for the motor winding. If the air-conditioning system begins to leak and the refrigerant gets low, not enough refrigerant will return to the compressor to cool it, so it will overheat and become damaged. It is important to understand that fuses are usually connected in the circuit at the disconnect switch that supplies voltage and current to the wires and loads in the circuit, and they cannot detect an

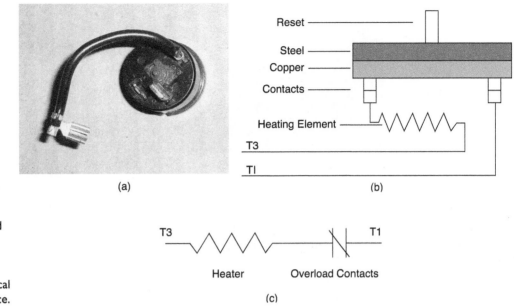

(a)

Reset

Steel

Copper

Contacts

Heating Element

T3

TI

(b)

T3 — Heater — T1

Heater Overload Contacts

(c)

Figure 5–22 a Overload device that protects compressor motor. **b** Electrical diagram of overload device. **c** Electrical symbol for overload device.

overheating condition. If you have a compressor motor that draws 15 amps and a condenser fan that draws 5 amps, the fuse would need to be sized to protect the system wiring for 20 amps. In some systems an additional overload device is connected in series with the compressor to protect it against exceeding 20 amps and to test to see if its temperature becomes too hot. When a compressor is working too hard, it will draw excessive current, but it may also be damaged by not receiving enough refrigerant back to cool its windings. Figure 5–22a shows a picture of an overload device; it is mounted directly on the outside case of the compressor so that it can also test the compressor to see if its temperature increases. In some compressors, the thermal overload is mounted inside the compressor and it will reset automatically if it trips. All you need to do is provide sufficient time, usually several hours, for the compressor to cool down so the overload contacts will reset.

Operational Tip

Some compressors will have internal thermal overload devices that will open and stop voltage and current from flowing to the motor winding if the compressor becomes too warm or draws excessive current. For this reason, it is important to allow a compressor to cool down for several hours if you suspect it has an open winding and will not run. If you allow the compressor sufficient time to cool down, the overload contacts will automatically reset. If you have allowed the compressor to cool down and your tests indicate there is still an open in the windings, you must presume the motor windings have opened and are permanently damaged.

Most technicians will allow a compressor that they suspect has open windings to cool down overnight to provide additional time for the overload contacts to reset. Another trick technicians may use is to hit the side of the compressor with a rubber mallet to jar the contacts in case they are stuck in the open position after it has cooled down. If after leaving the compressor motor to cool for an extended time, the resistance test still indicates an open in the compressor motor windings, you can feel assured the compressor is bad and must be removed and replaced.

It is important to remember you must fix the problem that is causing the compressor to overload, such as low refrigerant, or you'll have the same condition occur over and over.

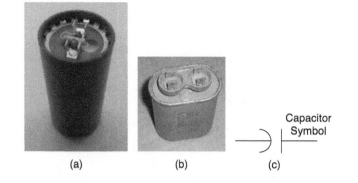

Capacitor Symbol

Figure 5–23 a A start capacitor used to start compressors and other motors. **b** Typical run capacitor. **c** Symbol for the capacitor.

Figure 5–22b shows a diagram of an external overload device, and Figure 5–22c shows the electrical symbol for the overload. The overload has a heater coil that converts the current that flows through it to heat. When the temperature increases past the setting for the overload, it will trip and cause the normally closed contacts to open. The thermal overload is connected in series with the motor windings.

You may find bimetal overloads, thermal overloads, or magnetic overloads in a system. You will learn more about these types of overloads in future chapters.

5.2.10 Capacitors

You will find two types of capacitors in HVAC and refrigeration systems: the start capacitor and the run capacitor. Figure 5–23a shows a picture of a start capacitor, Figure 5–23b shows a run capacitor, and Figure 5–23c shows the symbol for both types of capacitors. **Capacitors** are devices that are connected in series with motor windings to provide additional torque to allow the motor to start. For example, a start capacitor may be used with a compressor to help it get started when pressure inside the refrigeration system is placing resistance against a compressor piston to keep it from moving. You will learn more about capacitors in the chapters that explain the operation of compressors and other motors.

5.3 ELECTRICAL DIAGRAM OF AN AIR-CONDITIONING SYSTEM

As you know, the electrical components of the air-conditioning system consist of three main loads: the compressor motor, the condenser fan motor, and the evaporator fan motor. This system has a thermostat, a compressor relay, and an evaporator fan relay that acts as controls. The primary side of the control transformer will be treated as a load because it consumes power, and the secondary side will be treated as part of the control circuit because it provides voltage for the control circuit.

Figure 5–24 shows an electrical diagram of these components. This diagram shows the electrical symbols for each component and their physical location. It also presents the physical outline of each component; if electrical terminals or wires come into or out of the component, this diagram will indicate their exact location and the color of wire that is used. This diagram is called a *wiring diagram* and is generally used to help locate components and the electrical terminals on those components.

Figure 5–25 shows an example of a schematic (ladder) diagram of the electrical system for the air conditioner. This diagram shows the same components as the wiring diagram, but it shows them in the sequence the events must occur for the system to operate correctly. Note that it is easier to understand the operation of the system when

Figure 5–24 An electrical wiring diagram of the air-conditioning system.

Figure 5–25 A schematic (ladder) diagram of the air conditioner's electrical system. Notice that the overall appearance of this diagram looks like a wooden ladder, which gives the diagram its name. This diagram shows the sequence of operation for the electrical system.

using the schematic (ladder) diagram. The schematic (ladder) diagram derives its name because the overall appearance is that of a wooden ladder. The schematic (ladder) diagram is divided into two sections. The top section shows the high voltage that is used to energize the compressor, evaporator fan, and condenser fan, which all use 240 V. The evaporator fan is powered with 120 V AC, and it is supplied voltage by L1 and neutral. The compressor and condenser fans can be operated on a different voltage from the evaporator fan voltage because they are located in the outdoor part of the system called the **condenser**, which is powered by 240 V, and the evaporator fan is located in the furnace, which is powered by 120 V. The condenser unit is supplied higher voltage because this is the part of the system that will consume the most power when the system is operating.

The bottom part of the diagram shows the control circuit. The control circuit is powered by 24 V AC, coming from a step-down transformer called the *control transformer*. The transformer is physically located in the furnace and it steps 120 V down to 24 V. The thermostat is a switch that is located inside the house on a wall and is activated by a change in room temperature. This switch energizes or de-energizes power to the relay coils for the compressor relay and evaporator fan relay. For this introduction of the electrical control system, you need only to understand that the relay coil becomes magnetized and the magnetic force causes its contacts to move from their normally open position to the closed position. The contacts of the compressor contactor are wired in series with the compressor and condenser fan motor, and when they close they provide power to these motors. The contacts of the evaporator fan relay are connected in series

with the evaporator fan, and when they close they provide power to the evaporator fan. Additional information about schematic (ladder) diagrams and wiring diagrams is provided in Chapter 6.

LOADS FOR THE AIR-CONDITIONING SYSTEM

The main load for the air-conditioning system is the compressor motor. In a residential system the compressor motor is supplied with single-phase 240 V. In a commercial system the compressor is supplied by either 240 V three phase, or 480 V three phase. A **motor** is a device that has coils and a rotating shaft that is part of an armature. When electricity is applied to the motor coils, the shaft will rotate. The function of the compressor motor is to pump refrigerant and increase its pressure. The compressor is the only load required to cause the refrigerant to move through the system. Figure 5–26a shows a picture of a compressor. The electrical diagram of the compressor is shown in Figure 5–26b. The diagram in this figure shows a single-phase compressor.

The second major load in the air-conditioning system is the condenser fan. The condenser fan is generally powered by the same voltage as the compressor because they are mounted in the same cabinet. The condenser fan moves outside air over the fins of the condenser coil to provide sufficient cooling to cause the refrigerant to change from a vapor to a liquid. From the electrical diagrams in Figures 5–24 and 5–25, you should notice that the wires for the condenser fan are connected to the same terminals on the relay contacts as the compressor. This means that the condenser fan will run anytime the compressor is running and that the compressor relay also controls the condenser fan even though the name implies that this relay controls only the compressor motor. Figure 5–27a shows a picture of a condenser fan motor that is a permanent split capacitor (PSC) motor that has a run capacitor mounted directly on its frame, and Figure 5–27b shows an electrical diagram of a typical condenser fan motor that is wired as a **PSC motor**.

The third load for the air-conditioning system is the evaporator fan. The evaporator fan is mounted near the evaporator in the furnace. As you know, the evaporator fan is also used as the furnace fan, but it is referred to as the evaporator fan when you are working on the air-conditioning system. Figure 5–28a shows a picture and Figure 5–28b

(a)

(b)

Figure 5–26 a A typical compressor motor. **b** Electrical diagram of a compressor motor.
(Courtesy of Copeland Corporation)

(a)

(b)

Figure 5–27 a A typical condenser fan motor. **b** Electrical diagram of a typical condenser fan motor.

(a)

(b)

(c)

Figure 5–28 a A typical evaporator fan. Since this fan is housed in the furnace, it is sometimes called the furnace fan. *(Courtesy of GE Motors & Industrial Systems, Fort Wayne, Indiana)* **b** Electrical diagram of an evaporator fan. **c** Cutaway diagram of a furnace. Notice the location of the electrical loads and controls for this furnace. *(Courtesy of United Technologies Carrier Corporation)*

shows an electrical diagram of a typical evaporator fan motor. This fan is wired as a PSC motor. Figure 5–28c shows the typical location of a furnace fan in the furnace. If the furnace has an evaporator coil on it, the fan will be used as the evaporator fan motor when the system is operating in the air-conditioning mode and operate as the furnace fan when the system operates in the heating mode.

From the diagrams in Figures 5–24 and 5–25, you should notice that the evaporator fan is **energized** by the contacts of the evaporator fan relay. The thermostat in this system supplies voltage to the coil of the evaporator fan relay whenever the temperature becomes warmer than the setpoint on the thermostat. The setpoint is the temperature setting on the thermostat. In the diagrams in Figures 5–24 and 5–25 notice that the terminals in the thermostat that provide this circuit are identified as terminal G. Since the "hot" side of a 24-V transformer is connected to terminal R of the thermostat, the circuit for the fan relay is between R and G. Whenever the thermostat switches power to terminal G, voltage will be supplied to the coil of the fan relay.

You should also notice from these diagrams that the contacts of the evaporator fan relay are powered by 120 V AC (L1–N). The fan is powered by 120 V because it is a part of the furnace. In cases where a furnace is not part of a system, the evaporator fan will be housed in a unit called an air handler, which is basically an evaporator fan and evaporator coil. Since the furnace is located inside the house and does not consume a large amount of power, it generally is powered by 120 V.

5.5 THERMOSTAT CONTROLS FOR THE AIR-CONDITIONING SYSTEM

The thermostat is the main control in the air-conditioning system that turns it on when the space needs cooling and turns it off when the temperature is lowered enough. The control circuit for the air-conditioning system provides the means to energize and de-energize the compressor, the condenser fan, and the evaporator fan. The control circuit is identified in the diagrams in Figures 5–24 and 5–25 as the 24-V circuit that comes from the secondary winding of the control transformer. The control circuit consists of the coil for the compressor relay, the coil for the evaporator fan relay, and the thermostat. You should notice in the diagrams that the thermostat circuit that energizes the compressor relay coil is located between terminals R and Y in the thermostat. An in-depth presentation of information about the thermostat will be provided in a later chapter. Figure 5–29a shows a picture of a typical air-conditioning thermostat. Figure 5–29b shows the symbol for the cooling thermostat. You can tell it is a cooling thermostat because the switch contact is

(a) (b) (c)

Figure 5–29 **a** Thermostat for air-conditioning and heating system. **b** Electrical symbol for a cooling thermostat. Notice that the cooling thermostat will close when temperature increases. **c** Electrical symbol for a heating thermostat. Notice that the heating thermostat will close when temperature decreases.

below the terminal. When temperature increases, the thermostat switch raises up and closes and turns on the air-conditioning equipment. The symbol in Figure 5–29c shows a heating thermostat, which is identified by the switch contact being above the contact terminal. When temperature falls, the switch closes and turns on the heating system.

The cooling thermostat closes whenever the temperature inside the house becomes warmer than the setpoint on the thermostat. This action makes the circuit between R and Y complete. The letters R and Y are a standard identification for thermostat terminals and in the furnace. These letters stand for the colors of the wires that are generally connected to these terminals: R for red, and Y for yellow. The red wire is connected to the thermostat terminal R and at the furnace terminal R. The yellow wire is connected to terminal Y at the thermostat and at the Y terminal of the furnace.

The thermostat circuit that controls the evaporator fan is located between terminals R and G. G represents the color green. The coil of the evaporator fan relay can be energized in two different ways. First, it will be energized whenever the compressor is energized. This is accomplished through the fan switch in the thermostat which is also called the *auto/on switch*. When the auto/on switch is in the *auto* position, it connects the G terminal of the thermostat directly to the Y terminal in the thermostat. This enables power to be sent to the G terminal and the Y terminal anytime the system calls for cooling, which will energize the compressor relay and the evaporator fan relay.

The second way the evaporator fan relay coil is energized is when the auto/on switch is moved to the *on* position. When the switch is moved to the on position, a circuit is provided between the R terminal in the thermostat and the G terminal. This provides power to the coil of the evaporator fan relay from terminal G. Anytime the fan switch is in the on position, the fan will be energized 100% of the time. Anytime the fan switch is in the auto position, it will be energized only when the system is calling for cooling and the compressor relay is energized.

5.6 ELECTRICAL DIAGRAMS OF A SIMPLE HEATING SYSTEM

Figure 5–30 shows a picture of a typical residential gas furnace. This type of furnace is called an up-flow furnace because air is pulled in at the bottom of the furnace and is forced over the heat exchanger and out the top of the furnace into the ductwork, where it is sent to all parts of the house. The major electrical load for the gas furnace is the furnace fan. You should remember that this same fan is called the evaporator fan if the system is also used for air conditioning. This means that sometimes the fan will be referred to as the furnace fan, and sometimes it will be referred to as the evaporator fan. In this section it will be called the furnace fan. Another load in this system is the **gas valve**. The gas valve is generally powered with 24 V AC, which comes from the transformer. The furnace fan is located near the bottom of this gas furnace, and the gas valve is located near the burners in the middle of the furnace.

The electrical wiring diagram for a simple gas furnace is shown in Figure 5–31. Figure 5–32 shows a schematic (ladder) diagram of the same heating system. Recall that the wiring diagram shows the components in the approximate position that you will physically find them when you inspect the furnace. You can refer to the cutaway picture of the furnace shown in Figure 5–30 to help you determine the location of the major electrical components in the furnace. The schematic (ladder) diagram presents the components in the sequence in which they operate. Remember that the high-voltage circuit (120 V AC) is generally shown at the top part of the schematic (ladder) diagram, and the control voltage circuit (24 V AC) is generally shown at the bottom part of the diagram.

Figure 5–30 A typical upflow furnace. *(Courtesy of Bryant)*

Figure 5–31 Wiring diagram of a typical gas furnace.

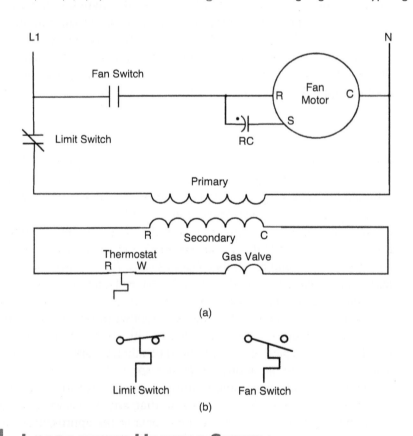

Figure 5–32 a Ladder diagram of a typical gas furnace. **b** Alternate symbols for limit switch and fan switch.

5.7 LOADS IN THE HEATING SYSTEM

The loads in the heating system consist of the furnace fan motor and the gas valve. From the diagram you can see that the furnace fan is energized and de-energized by the fan relay contacts. Anytime the contacts are closed the fan will be energized, and anytime the contacts are opened the fan will be de-energized. The furnace fan can be

(a)

or

(b)

Figure 5–33 **a** A typical gas valve. *(Courtesy of Honeywell)* **b** Two ways to show the electrical symbol for a gas valve.

energized by the fan relay or by a fan switch. The fan switch is wired in series with line 1 and the fan motor. The switch is physically located in the furnace at a point where it will be activated when the temperature in the furnace increases to approximately 110°F to 120°F. This means that the fan can be energized to run continuously by the fan relay or it can be run as needed when the temperature in the furnace is above 110°F. The temperature in the furnace should be above only 115°F if the burners are turned on.

The furnace fan is an open-type, fractional horsepower motor. It can be a direct-drive or belt-driven motor. Most newer furnaces use multispeed, direct-drive, permanent split-capacitor (PSC) motors. Chapter 9 provides more in-depth information about the theory of operation and troubleshooting for these types of motors. The electrical diagrams in Figures 5–31 and 5–32 identify the motor terminals and their locations. Figure 5–33a shows a picture and Figure 5–33b shows an electrical symbol for a typical gas valve. You should notice that there are two acceptable symbols for the gas valve solenoid.

5.8 CONTROLS FOR THE HEATING SYSTEM

The controls for the heating system consist of the thermostat, control transformer, and temperature switches. The gas valve is considered an electrical load for the furnace. In some cases it may be called a furnace control because it provides the gas that is burned to create the heat energy for the furnace. It is an electrical load by definition because it has resistance and consumes power. It can be classified as a control for the heating system because it energizes and de-energizes a solenoid valve, which allows gas to flow to the burners. The gas valve is explained in more depth in Chapter 14.

Other controls in the heating system include the thermostat and the overtemperature safety switch (also called the limit switch). The thermostat provides a circuit to energize the gas valve when the temperature falls below the thermostat setpoint. This means that if the thermostat is set for 70°, and the temperature goes below 70°, the circuit between R and W in the thermostat will be completed, and voltage will be passed to the gas valve. When the room heats up, and the temperature increases above the setpoint, the circuit between R and W will open in the thermostat, and the gas valve will become de-energized.

A typical furnace may include one or more overtemperature safety switches. In the electrical diagrams of the furnace in Figures 5–31 and 5–32, you should notice that an overtemperature (limit) switch is wired in series with the gas valve. If the temperature inside the furnace exceeds 170°F, the overtemperature switch will open and de-energize power to the gas valve. This provides a degree of safety for the furnace so it cannot cause a fire or other unsafe condition that could possibly melt wires on the furnace. When the furnace experiences an overtemperature condition, the overtemperature switch will open

(a) (b)

Figure 5–34 a A typical overtemperature switch. *(Courtesy of Honeywell)* **b** The electrical diagram of a furnace overtemperature switch can be shown with either symbol.

and the gas valve will become de-energized so a hazard no longer exists. This overtemperature switch is sometimes called a *high-limit switch*. The high-limit and other safety switches will be discussed in more depth in Chapter 14. Figure 5–34a shows an example of a typical overtemperature switch, and Figure 5–34b shows its electrical diagram.

5.9 ELECTRICAL DIAGRAM OF A SIMPLE REFRIGERATION SYSTEM

Figure 5–35 shows a picture of a typical refrigeration system, a refrigeration display case often found in a grocery store. The electrical wiring diagram for a refrigeration system is shown in Figure 5–36. From this diagram you can see that the main components of this system are similar to those of an air-conditioning system. The compressor and

Figure 5–35 A typical refrigeration system. This refrigeration system is part of a display case you would find in a grocery store. *(Courtesy of Tyler Refrigeration Corporation)*

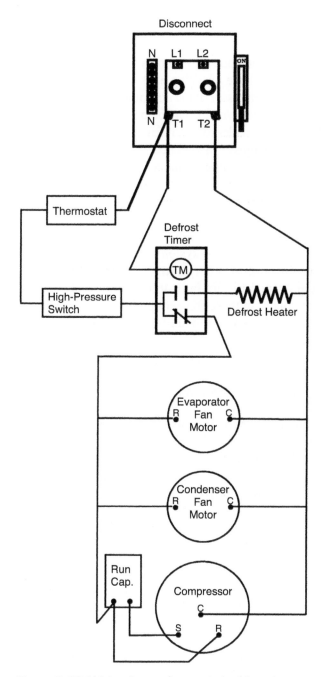

Figure 5–36 Wiring diagram for a typical refrigeration system.

Figure 5–37 Electrical schematic (ladder) diagram for a refrigeration case.

condenser fan are powered by 240 V AC single phase. The evaporator fan is powered by 240 V AC or 120 V AC.

Figure 5–37 shows the schematic (ladder) diagram of the refrigeration system. You can see that it looks similar to the schematic (ladder) diagram of the air-conditioning system. The main difference between the refrigeration system and the residential air-conditioning system is that the control voltage for the thermostat is 120 V for the refrigeration system and 24 V AC for the residential air-conditioning system. The operation of the refrigeration system is similar to that of the air-conditioning system.

5.10 LOADS FOR THE REFRIGERATION SYSTEM

The electrical loads for the refrigeration system include the compressor, the condenser fan, and the evaporator fan. Some systems may also have a heating element to help defrost the system. The compressor looks similar to the compressor in the air-conditioning system and electrically operates exactly like it. The compressor motor may be a single-phase or three-phase motor. The condenser fan will be slightly smaller for the refrigeration system than the condenser fan for the air conditioner. The compressor and condenser fan may be located outside the store where it will have ample air for cooling available. The condenser fan can also be either a three-phase or a single-phase motor. It will normally be the same type of motor as the compressor because they operate from the same voltage.

The third load for the refrigeration system is the evaporator fan. The evaporator fan on smaller refrigeration systems is usually a blade-type fan that is powered by a small single-phase motor called a shaded-pole motor. In some systems the evaporator fans use squirrel-cage blowers to move more air, and in other systems a fan is not needed because the refrigeration coils for the evaporator are built into a large plate and the products to be frozen are stacked on the plate. In these freezer applications the evaporator fan may not be required because the products will remain frozen without circulating evaporator air.

5.11 CONTROLS FOR THE REFRIGERATION SYSTEM

The controls for the refrigeration system in Figure 5–35 consist of a temperature control (thermostat), high pressure control, and defrost control. On some refrigeration systems the temperature control may be a low-pressure switch that monitors the pressure of the refrigerant that is under pressure inside the tubing. It is important to understand that there is a direct correlation between the amount of refrigerant pressure and the temperature at the evaporator. When the temperature in the display case becomes warmer, the pressure will increase inside the tubing and the **pressure switch** will close and start the

(a) (b)

Figure 5–38 a Low-pressure control used for temperature control for a refrigeration system. **b** Electrical diagram of a low-pressure switch controlling a compressor motor. *(Courtesy of Johnson Controls)*

compressor. When the compressor runs long enough, the temperature in the display case becomes cooler and causes the refrigerant pressure to drop, which causes the pressure switch to open and de-energize the compressor. Figure 5–38a shows a picture of a low-pressure switch that is used as a temperature control, and Figure 5–38b shows a diagram of a low-pressure switch used as a temperature control for a refrigeration system.

The main difference between the temperature control in a refrigeration system and one used in a residential furnace is that the refrigeration temperature switch typically uses high voltage (120 V AC or 240 V AC) as the control voltage, and the residential furnace uses 24 V. This means that the contacts in the refrigeration control and its overall appearance will be slightly larger than a residential thermostat.

5.12 | REVIEW

Electrical symbols are used to represent all the electrical controls and loads in electrical circuits. The electrical symbols can be used in a wiring diagram to show the location of each component or in a schematic (ladder) diagram to show the sequence of operation. In the electrical symbol for a load such as a motor or compressor, the motor terminals and the internal windings will be shown. The most important point to remember about electrical symbols is that each of them represents an actual electrical component, and you should begin to relate them to one another.

The electrical symbol for switches such as thermostats and oil pressure switches consists of two parts. The first part shows the switch contacts, which can be shown in either their open or their closed position. The second part of the switch symbol shows the physical operator of the switch. In the case of a thermostat, temperature causes the contacts to open or close, and in the case of a pressure switch, pressure causes the contacts to open or close. If the switch symbol shows the contacts open above the terminal, as in Figure 5–29c, the symbol is indicating that the switch is a temperature-operated switch for heating applications. When the temperature lowers, the switch will go closed. This chapter also showed you how to combine a number of electrical symbols to represent an entire electrical system for an air-conditioning, furnace, or refrigeration system.

Job Assignment for This Chapter

You are on a service call and have removed the access door for the outside air-conditioning unit. The door has a wiring diagram on it that shows where the components in the system are located. An electrical symbol is used in the diagram for each component in the system. Your job is to match the component symbol with the actual component and explain to the customer the name of each component in the electrical panel and what its function is in the system. You should also explain the difference between the wiring schematic diagram and the ladder diagram.

Solution to the Job Assignment

Your solution should include the picture and diagram for the compressor, condenser fan, evaporator fan, compressor contactor, thermostat, transformer, fan relay, fused disconnect, fuses, capacitors, and any other controls in the furnace or condenser. You should be familiar enough with these terms and components so that you can find the pictures and symbols of each. You should be able to follow the wiring diagram and ladder diagram for each.

ASSESSMENT

QUESTIONS

1. Define the term *electrical load*.
2. Explain why the evaporator fan in an air-conditioning system may also be called the furnace fan.
3. Define the term *electrical control*.
4. Explain the difference between a wiring diagram and a schematic (ladder) diagram.
5. Explain where you would typically find the evaporator and condenser on a refrigeration display case.

TRUE OR FALSE

1. A wiring diagram shows the relative location of each component in a circuit.
2. An electrical load is a device that has resistance and uses current.
3. A schematic (ladder) diagram shows the relative location of each component in a circuit.
4. A furnace fan can also be the condenser fan in some systems.
5. The thermostat for a furnace is an electrical load.

MULTIPLE CHOICE

1. The control in the furnace that protects the furnace from overheating and causing a fire is the
 a. fan switch.
 b. high-limit switch.
 c. fan relay.

2. The "operator" of an electrical switch is the part of the switch that
 a. causes the action that makes the switch change from open to closed or closed to open.
 b. conducts current.
 c. is the same as the contacts.

3. The auto/on switch on the thermostat allows the fan to
 a. operate continuously or with the gas valve.
 b. operate continuously or with the transformer.
 c. operate continuously or with the compressor.

4. The thermostat terminals are identified with the letters R, Y, W, and G, which stand for
 a. the colors of the thermostat wires (red, yellow, white, and green).
 b. the operation of the thermostat (auto, manual, and semiautomatic).
 c. the primary and secondary windings on the control transformer.

5. The fan and limit switch in the gas furnace control
 a. the temperature at which the fan turns on and the temperature at which the gas is ignited.
 b. the temperature at which the fan turns on and the temperature at which the air conditioner turns off.
 c. the temperature at which the fan turns on and the maximum temperature the furnace can reach before power to the gas valve is turned off.

PROBLEMS

1. Sketch a wiring diagram or a schematic (ladder) diagram of a simple air-conditioning system and identify the loads and controls.

2. Sketch a wiring diagram or a schematic (ladder) diagram of a simple heating system and identify the loads and controls.

3. Sketch a wiring diagram or a schematic (ladder) diagram of a simple refrigeration system and identify the loads and controls.

4. Sketch the wiring diagram symbol and the schematic (ladder) diagram symbol of any load or control and explain why they are different.

5. Sketch a wiring diagram or schematic (ladder) diagram of the fan auto/on switch on a thermostat and explain what causes the fan to turn on and off in each mode.

Websites

Websites change constantly. Use the key word search to locate sites that have similar content. The products that will be found at each site are indicated with the key word.

Electrical symbols
Products: Electrical symbols on line
Key word: Electrical symbols

Products: Danfoss Controls
Key word: HVAC and refrigerator controls

Products: Johnson Controls
Key word: HVAC and refrigerator controls

Products: Air-conditioning compressors
Key word: Scroll compressors

Products: Evaporator fan motors
Key word: PSC fan motors

Products: Condensor fan motors
Key word: PSC fan motors

Chapter 6

Reading and Writing Schematic (Ladder) Diagrams and Wiring Diagrams

Job Assignment for This Chapter

You are on a job where there is only a wiring diagram that shows the location of electrical components and the color of wires connected to each part of an air-conditioning system. However, you need a schematic (ladder) diagram to help you troubleshoot the air conditioner. Your job assignment is to convert the wiring diagram to a schematic (ladder) diagram.

Terms You Will Learn in This Chapter

Capacitor
Compressor
Compressor contactor
Condenser fan
Controls
Cooling thermostat
Electrical control
Electrical load
Evaporator fan
Fan relay

Heating thermostat
HVAC electrical symbols
Legend
Loads
Schematic (ladder) diagram
Symbols
Thermostat
Transformer
Wiring diagram

After reading this chapter, you will be able to:

1. Explain the difference between a schematic (ladder) diagram and a wiring diagram.
2. Create a schematic (ladder) diagram from a wiring diagram.
3. Create a wiring diagram from a schematic (ladder) diagram.
4. Identify loads and controls in an air-conditioning diagram.
5. Identify loads and controls in a window air-conditioning diagram.
6. Identify loads and controls in a refrigeration diagram.

6.0 OVERVIEW OF DIAGRAMS AND SYMBOLS

When you are working on the job, you will need to be able to read schematic (ladder) diagrams as well as wiring diagrams. Some equipment manufactures provide a schematic, whereas others provide a wiring diagram. Some companies provide both, but sometimes you will need to convert a wiring diagram into a schematic (ladder) diagram or vice versa. You will also need to be able to identify the loads and the controls in each diagram and identify what controls provide power to each load in the system. This chapter will provide a variety of diagrams and teach you how to identify the **HVAC electrical symbols** and read the diagrams and determine their sequence of operation.

6.1 READING AND WRITING SCHEMATICS (LADDER) DIAGRAMS AND WIRING DIAGRAMS

Before you can begin to understand how to read a **schematic (ladder) diagram** or a wiring diagram, you must understand how they are created. In this chapter we will present the schematic diagram, which is also called a ladder diagram, because it shows the sequence of operation for the system and we will show the **wiring diagram** because it shows the location of all the components and wires in the systems. As an HVAC technician, you must be able to read wiring diagrams and ladder diagrams and convert back and forth between them. This chapter will show you how to create a ladder diagram from a wiring diagram and then how to create a wiring diagram from a ladder diagram. The chapter will then provide a number of diagrams from air-conditioning and heating systems that we will analyze circuit by circuit just as if you were in the field trying to analyze the diagram to be able to troubleshoot the system. You may need to refer to the **symbols** and **legends** from Chapter 5 for some of the components in the HVAC system diagrams.

Often you will be troubleshooting a system that provides a wiring diagram of the electrical circuit but not a ladder diagram. You can develop your own ladder diagram by tracing the circuits in the wiring diagram. The material in this section will provide an example of converting a wiring diagram for a packaged air-conditioning system to a ladder diagram. Figure 6–1 shows a wiring diagram of a typical packaged air-conditioning system that is connected to 240 V AC. You can see in this diagram that the major components in this system are the compressor motor, the condenser fan motor, the evaporator fan motor, the compressor contactor, the evaporator fan relay, the transformer, and the thermostat.

Figure 6–1 Wiring diagram of a packaged air conditioner.

Figure 6–2 shows the skeleton or the outline of the ladder diagram. Typically, the compressor and condenser fan loads will be shown at the top of this diagram, and the transformer and thermostat will be shown at the bottom of this diagram. You can start developing the ladder diagram by placing the compressor motor, the condenser fan motor, and the evaporator fan motor between the two rails of the ladder diagram, as shown in Figure 6–3.

The next step in the procedure is to find the source of power. In the wiring diagram, voltage comes into the electrical disconnect at terminals L1 and L2. Voltage flows through the disconnect to terminals T1 and T2. For this exercise, we will refer to the incoming voltage in the diagram as L1 and L2. You should begin the process by tracing the wire that connects L1 to the contacts of the compressor contactor. In Figure 6–4,

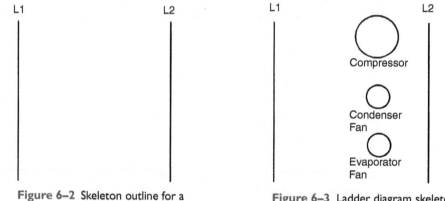

Figure 6–2 Skeleton outline for a ladder diagram.

Figure 6–3 Ladder diagram skeleton with compressor motor, condensor fan motor, and evaporator fan motor.

Figure 6–4 Wires A, B, C, and D are highlighted in the wiring diagram.

you can see that a wire identified as wire A runs terminal L1 of the disconnect to terminal L1 of the compressor contactor. A wire identified as wire B runs from terminal T1 of the **compressor contactor** to terminal R of the **compressor**. On the other side of the compressor a wire identified as wire C is connected to terminal C on the compressor and terminal T2 of the compressor contactor. You can follow a wire identified as wire D from terminal L2 of the compressor contactor to terminal T2 of the disconnect. These wires are highlighted in the wiring diagram shown in Figure 6–4, and they are placed in the ladder diagram you are developing in Figure 6–5. The next component you can add to the ladder diagram is the **condenser fan** motor. You should notice in the wiring diagram in Figure 6–6 that a wire identified as wire E is highlighted, and it runs from the compressor contactor terminal T1 to the R terminal on the condenser fan and the run capacitor. You should also notice a wire identified as wire F that runs from terminal T2 of the compressor contactor to the C terminal of the condenser fan. Since the

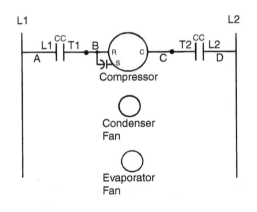

Figure 6–5 Ladder diagram with wires A, B, C, and D.

Figure 6–6 The wires connected to the condenser fan are shown highlighted in the wiring diagram.

condenser fan is wired as a PSC-type motor, the run **capacitor** is connected between the R and S terminals of the motor. Figure 6–7 shows the condenser fan motor added to the ladder diagram you are developing.

The next component you will add to your diagram is the **evaporator fan**. In Figure 6–8 you will notice that the evaporator fan is controlled by the fan relay. Wire G runs between terminal L1 of the compressor contactor and terminal L1 of the **fan relay**. Wire H runs between terminal T1 of the fan relay and terminal R of the fan and the run capacitor. Wire I runs between terminal C of the fan and terminal T2 of the fan relay. Wire J connects terminal L2 of the fan relay to terminal L2 of the compressor contactor, where it picks up supply voltage. Figure 6–9 shows the evaporator fan added to the ladder diagram.

Figure 6–7 Wires E and F that connect the condenser fan are shown in the ladder diagram.

Figure 6–8 The wires to the evaporator fan are highlighted in the wiring diagram.

Figure 6–9 Wires G and H that connect the evaporator fan are added to the ladder diagram.

Figure 6–10 shows the primary side of the **transformer** is connected with wire K at terminal L1 of the compressor contactor and to terminal T1 of the disconnect. The second wire of the primary side of the transformer is identified as wire L, and it is connected to terminal L2 of the compressor contactor and to terminal T2 of the disconnect. The primary side of the transformer is connected to a 240-V AC power source whenever the disconnect switch is turned on. The secondary side of the transformer is identified as terminals R and C. Figure 6–10 shows the transformer highlighted in the wiring diagram, and Figure 6–11 shows the transformer primary and secondary added to the ladder diagram.

When you are ready to draw the low-voltage control side of the circuit, you can place the secondary transformer winding, the coil of the compressor contactor, and the coil of the fan relay between the power wires of the ladder diagram skeleton, as shown in Figure 6–12.

Figure 6–10 The wires connected to the transformer are highlighted in the wiring diagram.

Figure 6–11 The transformer and the wires connected to it are shown in the ladder diagram.

Figure 6–12 The coil of the compressor contactor and the coil of the fan relay are added to the ladder diagram.

Figure 6–13 The wires connected to the thermostat and the compressor contactor coil are shown highlighted in the wiring diagram.

In Figure 6–13 you can see a wire identified as wire M connects the **cooling thermostat** terminal R to terminal R on the secondary of the transformer. Terminal Y on the thermostat is connected to the terminal on the right side of the compressor contactor coil with wire N. Wire O is connected to the terminal on the left side of the compressor contactor coil and the terminal of the fan relay coil. An internal connection exists between the fan switch and terminal R inside the thermostat. Wire P is connected between terminal G of the thermostat and the terminal on the right side of the fan relay coil, and wire Q is connected between the terminal on the left side of the fan relay coil and terminal C of the transformer. Figure 6–13 shows these wires in the wiring diagram, and Figure 6–14 shows these wires in the ladder diagram.

The connection between the cooling thermostat contact and the fan switch is inside the thermostat. The fan switch can be set to auto, where it receives power from the Y terminal of the thermostat whenever the thermostat is calling for cooling, or it can be set to on, where it receives power from the R terminal of the thermostat constantly. When the fan switch is in the on position, it energizes the fan relay coil constantly, which causes the fan to run continuously.

Figure 6–15 shows a completed ladder diagram constructed from the wiring diagram. You can see the high-voltage components at the top of the diagram, and the low-voltage components at the bottom of the diagram. You should also notice that it is much easier to see the switches that control each of the loads in the diagram on the ladder diagram than on the wiring diagram. You should notice that the wiring diagram shows some of the wires crossing each other, which makes it difficult to use during troubleshooting. With practice, you will find it fairly easy to develop the ladder diagram from a wiring diagram, and the time it takes to convert the diagram will be saved during troubleshooting.

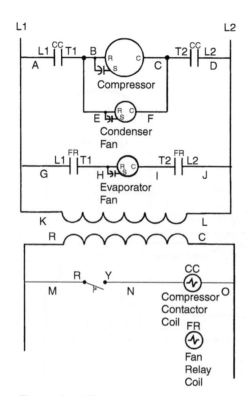

Figure 6–14 The skeleton for the low-voltage side of the ladder diagram.

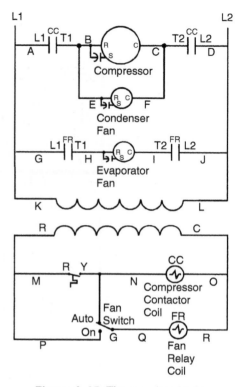

Figure 6–15 The completed ladder diagram.

6.2 CONVERTING A SCHEMATIC (LADDER) DIAGRAM INTO A WIRING DIAGRAM

At times you may have a schematic (ladder) diagram (also called an elementary diagram) that provides the sequence of operations for the air conditioner, but you will not have a wiring diagram that shows information such as the physical location of the terminals and the number of contacts on relays. This usually occurs when a new component that is not an exact replacement part is installed in an air conditioner or when the unit is rewired and a new circuit is added to the **electrical controls**. When you are wiring the new components into the circuit, you will use many of the same skills that you would use to convert a ladder diagram to a wiring diagram. With practice, you can create a wiring diagram from a ladder diagram in a very short time.

For this exercise we will use a package-type air-conditioning unit that you would normally find on a rooftop. The first step in this process is to make a visual inspection of the electrical cabinet and to make a sketch of the components that you see mounted inside the electrical box in the package unit. The sketch should be accurate with respect to the location and relative position of each component. It is also important to identify where terminal connections are on a component, such as the terminals where power comes into the device and goes out. The simplest way to identify these contacts is to use identifiers at the top and bottom of the component or at the left-side terminal and right-side terminal. We will use these identification terms as we develop the diagram.

This section provides an exercise that shows how to develop a wiring diagram from a ladder diagram. In this exercise we use a sketch of the components that shows their exact locations inside the electrical cabinet and the exact locations of components mounted elsewhere in the package system cabinet. From this sketch we develop the

Figure 6–16 Wiring diagram of components of the air-conditioning systems shown in the location you would find them in relation to each other. This first diagram does not show any wires connected between components.

skeleton for the wiring diagram. This sketch of the components is shown in Figure 6–16. The ladder diagram that we will use for this exercise is provided in Figure 6–17. All the terminals for the components are identified with component abbreviations such as CC for compressor contactor and FR for fan relay. If the component has line-side terminals, they are identified as L1 and L2, and load-side terminals are identified as T1 and T2. We will convert the entire ladder diagram for this exercise. In reality, you would need to convert only the part of the ladder diagram necessary to complete your job.

In Figure 6–18 the wires that we will use to start the conversion process to the wiring diagram are shown as bold lines in the wiring diagram, and they are identified by the letters A–Y. We will start the conversion process by copying the wires that are identified with A–Y in the ladder diagram onto the wiring diagram. These wires include the wires (A and B) that provide voltage to the L1 and L2 terminals of the compressor contactor (CC), the two wires (C and D) that provide wire to the L1 and L2 terminals of the fan relay (FR), and the two wires (E and F) that supply voltage to the primary winding of the transformer.

Figure 6–19 shows how these wires are placed in the wiring diagram. The first wire (A) is connected between T1 on the fuse disconnect to terminal L1 on the compressor contactor (CC). The second wire (B) is connected between T2 on the fuse disconnect to terminal L2 on the compressor contactor (CC). The third wire (C) is shown on the ladder diagram as connecting to the power rail for L1 on its left side and to the L1 terminal of the fan relay (FR) on its right side. In reality the wire does not have to go all the way

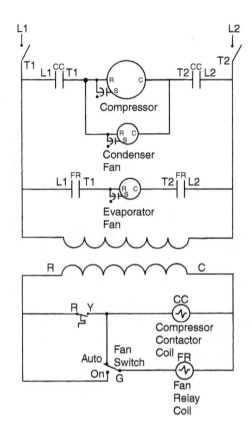

Figure 6–17 Ladder diagram of the same components shown in Figure 6–16. We will use this completed ladder diagram, which shows all of the wiring connections between components, to develop the wiring diagram in this exercise.

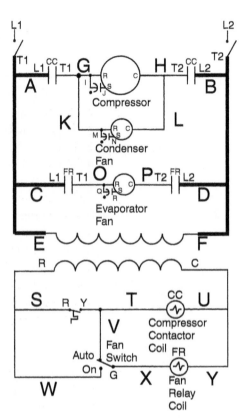

Figure 6–18 Ladder diagram of the air-conditioning system with letters A–Y placed on the wires to identify them.

Figure 6–19 Wiring diagram shown with wires A–F from the ladder diagram drawn in the wiring diagram. Refer to the ladder diagram in Figure 6–18 to see the exact connections of wires A–F between components.

back to the fuse disconnect to pick up power from L1; instead it can be connected to the L1 terminal on the compressor contactor. The reason we will connect the wire to the L1 terminal of the compressor contactor is that the compressor contactor is mounted within inches of the fan relay in the electrical panel of the package unit. We will draw it in the wiring diagram and show one end of it connected to the L1 terminal of the compressor contactor (CC) and the other end connected to the L1 terminal of the fan relay (FR). We will show the fourth wire (D) in the wiring diagram in the same manner. It will have one end connected to terminal L2 on the compressor contactor where it gets its power, and its other end will be connected to terminal L2 of the fan relay (FR).

The third set of wires, E and F, will be connected in the wiring diagram from the fused disconnect to the transformer. In this wiring diagram we will show one end of the E wire connected to terminal T1 of the fused disconnect and the other end connected to the L1 terminal of the primary side of the transformer. Depending on where the transformer is connected in the package air-conditioning unit, the wire could be connected to the L1 terminal of the compressor contactor (CC) instead of the L1 terminal of the fused disconnect. The other wire (F) has one end connected to the T2 terminal of the fused disconnect and the other end of the wire connected to terminal L2 of the transformer.

The next part of the diagram that we will convert is shown in Figure 6–20. In this diagram you can see that additional wires are identified by the letters G, H, I, J, K, L, M, and N, shown in bold so you can see where they are located. Figure 6–21 shows these wires added to the wiring diagram, boldfaced so you can easily see them. The first wire is identified by the letter G and it is used to provide L1 voltage to the compressor and the run capacitor. One end of wire I is connected to the T1 terminal of the compressor contactor (CC), and the other end is connected to the run capacitor for the compressor. The terminal on the run capacitor where this wire is connected is identified with a red dot. The red dot used as an identifying mark is physically placed on one

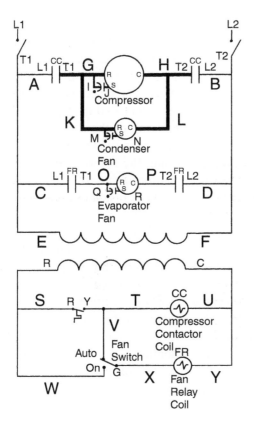

Figure 6–20 Ladder diagram of the air-conditioning system with letters G–N placed on the wires to identify them.

Figure 6–21 Wiring diagram shown with wires G–N from the ladder diagram drawn in the wiring diagram. Refer to the ladder diagram in Figure 6–20 to see the exact connections of wires G–N between components.

set of terminals in the capacitor, and it is always shown in the ladder and the wiring diagrams so that it can be used as a reference when you are connecting wires to the capacitor. It is important to remember that each terminal on the capacitor has a double-spade terminal connection, so two wires can be connected to each terminal point on the capacitor. One end of a second wire, identified as wire G, is connected to the red-dot terminal on the capacitor at one end and to the R terminal of the compressor on the other end. In the ladder diagram the wire is shown running directly between the T1 terminal of the compressor contactor and the R terminal of the compressor, and a small wire is used to connect the run capacitor to these wires. In reality, the compressor does not have two terminals for this extra wire; instead the capacitor has a double-spade terminal provided to make this extra connection. This means that the wire connected to the T1 terminal of the compressor contactor will go to the capacitor first and then to the compressor R terminal. A wire identified as wire J is used to connect the second terminal of the run capacitor to the S winding on the compressor. This is the wire that connects the capacitor in series with the start winding of the compressor. The reason the capacitor is called the run capacitor, even though it is connected to the start winding, is that it remains in the circuit, connected to the start winding at all times that the compressor is running. A run capacitor will provide voltage to the start winding with a phase shift any time there is a potential voltage difference between the start and run windings. This means that any time the compressor starts to slow down from its load, the run capacitor will provide voltage to the start winding with a phase shift that will

increase the torque on the compressor shaft and allow the motor to come back up to full speed. The next wire is identified as wire H and is used to provide L2 power to the compressor. One end of this wire is connected to terminal T2 on the compressor contactor and the other end is connected to terminal C on the compressor. This wire is connected directly between these two terminals.

The condenser fan also receives its power from the compressor contactor. This means that the condenser fan will be running any time the compressor is running. The wire that provides L1 power to the condenser fan is identified as wire K. One end of this wire is connected to terminal T1 on the compressor contactor, and the other end is connected to the run capacitor for the condenser motor. This circuit is similar to the compressor circuit in that wire K must be connected to the run capacitor first, and then another wire identified as wire M must be connected between the red-dot terminal on the run capacitor for the condenser fan and the R terminal on the condenser fan. The wire must be connected to the capacitor first since it has two terminals and the motor has only one terminal for the run winding. A third wire, wire N, is connected between the start winding terminal S on the condenser fan and the red-dot terminal on the run capacitor for the condenser fan.

Wire L is the next wire in the conversion of the diagram, and it is used to provide L2 power to the condenser fan. One end of wire L is connected to terminal T2 on the compressor contactor, and the other end is connected to the C terminal of the condenser fan.

The next component in the circuit to be converted in the diagram is the evaporator fan. Figure 6–22 shows the ladder diagram with the wires connected to the evaporator fan. Figure 6–23 shows the wiring diagram for the connecting wires to the evaporator fan. Wire O is the first wire, and it has one end connected to the T1 terminal on the

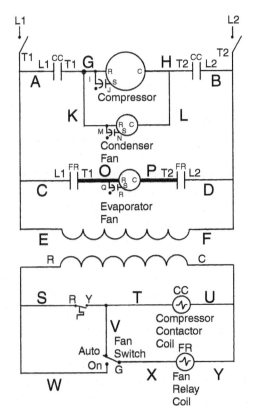

Figure 6–22 Ladder diagram of the air-conditioning system with letters O–R placed on the wires to identify them.

Figure 6–23 Wiring diagram shown with wires O–R from the ladder diagram drawn in the wiring diagram. Refer to the ladder diagram in Figure 6–22 to see the exact connections of wires O–R between components.

contacts of the fan relay (FR). The other end is connected to the red-dot terminal of the run capacitor for the evaporator fan. The second wire is wire Q, which is connected between the red-dot terminal on the run capacitor and the R terminal on the evaporator fan. The next wire is wire P. This wire provides L2 voltage to the evaporator fan. One end of the wire is connected to the T2 terminal of the fan relay contacts. The other end is connected to terminal C of the evaporator fan.

The next area of the ladder diagram that we will convert is the control circuit for the package unit. The control circuit includes the secondary side of the transformer, the cooling thermostat, the coil of the compressor contactor (CC), and the coil of the fan relay (FR). Figure 6–24 shows the ladder diagram for the wires that connect the compressor contactor coil to the thermostat in the control circuit. Figure 6–25 shows wires S, T, and U in the wiring diagram for this part of the circuit. The first wire, wire S, has one end connected to the R terminal on the secondary side of the transformer, and its other end is connected to terminal R on the thermostat. You should remember that terminal R–Y in the thermostat controls the cooling function. When the room becomes too hot, the contacts between R–Y go to their closed position and pass power from the R terminal to the Y terminal. In most systems you will find in the field, the S wire will be the red wire in the thermostat wire set. The next wire in this section of the diagram is wire T, and one end of it is connected to terminal Y on the thermostat, and the other end is connected to the compressor contactor (CC) coil. The Y wire is usually the yellow wire in the thermostat wire set. The final wire in this part of the circuit

Figure 6–24 Ladder diagram of the air-conditioning system with letters S–U placed on the wires to identify them.

Figure 6–25 Wiring diagram shown with wires S–U from the ladder diagram drawn in the wiring diagram. Refer to the ladder diagram in Figure 6–24 to see the exact connections of wires S–U between components.

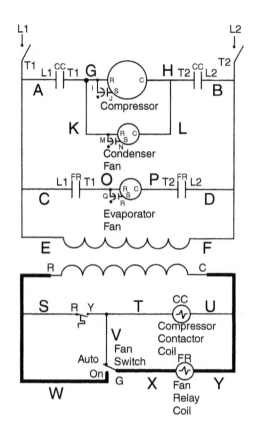

Figure 6–26 Ladder diagram of the air-conditioning system with letters V–Y place on the wires to identify them.

is wire U. One end of wire U is connected to the coil of the compressor contactor, and the other end is connected to the secondary terminal of the transformer that is identified as C.

The final section of the ladder diagram to be converted is the circuit that controls the fan relay (FR) coil. This part of the circuit is highlighted in the ladder diagram shown in Figure 6–26. Figure 6–27 is the wiring diagram that shows this section of the circuit. The wires that are included in this part of the diagram are identified as wires V, W, X, and Y. The actual wire that V represents in this circuit is an internal wire inside the thermostat. If you were wiring this circuit in the field, you would not need to add this wire, because it already exists in the thermostat. We have it in our wiring diagram because it shows how power gets between the Y terminal of the thermostat and the fan switch. This wire provides power to the coil of the evaporator fan relay when the fan switch on the thermostat is in the auto position. The next wire in the circuit is identified as the W wire, also an internal wire in the thermostat. It provides power from terminal R to the fan switch. When the fan switch is set to the on position, it will provide power from the R terminal to the G terminal to energize the fan relay coil.

The next wire in this circuit is identified as the X wire. One end of this wire is connected to the G terminal in the thermostat, and the other end is connected to the coil of the fan relay. The final wire is identified as wire Y. One end of this wire is connected to the terminal of the fan relay coil, and the other end is connected to the C terminal on the secondary of the transformer.

This completes the conversion of the ladder diagram to the wiring diagram. Figure 6–28 shows the completed wiring diagram. You can review the process used to convert this program any time you need to convert a ladder diagram to a wiring diagram.

Figure 6–27 Wiring diagram shown with wires V–Y from the ladder diagram drawn in the wiring diagram. Refer to the ladder diagram in Figure 6–26 to see the exact connections of wires V–Y between components.

Figure 6–28 The completed wiring diagram of the air-conditioning system converted from the ladder diagram. Refer to all the diagrams (6–16 through 6–27) to review the process of converting a ladder diagram to a wiring diagram.

6.3

ELECTRICAL DIAGRAM OF A RESIDENTIAL SPLIT AIR-CONDITIONING SYSTEM

As an HVAC technician you will need to read and understand an electrical diagram of a split residential air-conditioning system. Figure 6–29 shows a schematic (ladder diagram) of a typical split residential air-conditioning system. The schematic is broken into three parts. The **electrical loads** include the compressor motor, evaporator fan motor, and condenser fan motor and are shown in the top of the diagram. The compressor motor and condenser motor are powered by 240 volts and the indoor fan (evaporator fan) is powered by 120 volts. The controls are shown at the bottom of the diagram, and they include the thermostat, fan relay coil, and the compressor contactor coil. The contacts for the fan relay and compressor contactor are in the load part of the diagram as they control the voltage to the loads. The coils of the fan relay and compressor contactor are part of the control circuit. When you are reading the diagram, you will need to identify which load you are trying to energize and follow the control circuit components that must be turned on. As a rule when you are trying to energize a load, you must energize the part of the control circuit that controls it first. The next sections will explain how each load is controlled.

6.3.1 Turning on the Compressor and Condenser Fan Motor

The compressor motor is controlled by the contacts of the compressor contactor. You can see from the diagram that the normally open (NO) contacts are in series with the compressor and condenser motors. The coil of the compressor contactor is located in the control circuit. When the thermostat senses the room is getting too warm, the thermostat closes the circuit between R and Y. This sends 24 volts to the coil of the compressor contactor, which will create a magnetic field and cause its contacts to close and send voltage to the compressor. Since the condenser fan motor is connected in parallel with the compressor motor, it will also become energized and run at the same time as the compressor motor.

Figure 6–29 A schematic (ladder) diagram of a split system air conditioner.

The coil of the furnace fan relay (FR) will also become energized because the fan switch in the thermostat will be in either the auto or on mode. In either mode, 24-volt power is sent to the coil of the fan relay. When the fan relay coil receives voltage, it becomes magnetized and causes the fan relay contacts to close, which sends voltage to the evaporator fan motor. (It is important to understand that if the system is a package-type air-conditioning system, the evaporator fan will be powered with the same voltage as the compressor and condenser fan. If the system is a split system, the evaporator fan will be in the furnace inside the house and the voltage supplied to it will be 120 volts.)

When the room or conditioned space cools down, the thermostat will sense the change in temperature and open its circuit R–Y. When this occurs, the voltage to the coil of the compressor contactor (CC) is turned off and the contact of the compressor contactor opens and turns off the compressor and condenser fan. If the thermostat is set so the fan switch is in the auto position, voltage to the furnace fan relay (FR) coil will also be turned off, and the furnace fan will turn off when the thermostat senses the temperature change and turns off the compressor. If the thermostat fan switch is set in the on position, the circuit through the thermostat to the fan relay (FR) coil will remain energized and the indoor fan will stay running.

6.3.2 Turning on the Furnace Fan Only

It is possible to turn the furnace fan on and off without turning on the compressor and condenser fan. Some homeowners prefer to run the furnace fan all the time to ensure that the temperature in the home remains more constant. Running the furnace fan ensures that the air in the house is constantly moving and has a more even temperature. The fan switch can be set in the on position to turn the fan on, and it can be set back to the auto position to turn it off until the room temperature increases to the setpoint. When the room temperature reaches and exceeds the thermostat setpoint temperature, the circuit inside the thermostat between R and Y will turn on the compressor and the evaporator fan. When the temperature lowers, the thermostat will sense the temperature change and open the circuit between R and Y again and turn off the evaporator fan.

When the fan switch is on auto, the fan will turn on and off with the compressor, and when it is set in the on position, it will turn on and stay on until it is switched back to the auto position.

6.4 LOADS FOR THE AIR-CONDITIONING SYSTEM

When you are reading a diagram of a system, you can also identify the loads and the controls. The loads are the motors such as the compressor, condenser fan, and evaporator fan. By definition a **load** is an electrical component that consumes current and voltage and converts it to some form of energy such as mechanical energy. The loads in a system are controlled by controls. Typically the loads are in the top portion of the schematic (ladder) diagram, and they are connected to the higher voltage in the system. The contacts for the control relays are used to turn the loads on and off.

6.5 CONTROLS FOR THE AIR-CONDITIONING SYSTEM

The **controls** for the air-conditioning system include the thermostat, the compressor contractor coil, and the fan relay coil. The **thermostat** has a circuit connected between R and Y that close when the temperature inside the conditioned space gets too warm and goes above the setpoint setting of the thermostat. For example, if the thermostat is set at

74 degrees, the circuit between R and Y will close any time the temperature in the room gets above 74 degrees.

The thermostat has a second control circuit inside it, which controls the fan relay by sending voltage out terminal G. The fan circuit inside the thermostat is controlled by the setting of the fan switch. If the fan switch is set on the auto position, power will be sent to terminal G any time the circuit between R and Y is closed. This circuit is called the auto mode because the fan will automatically turn on any time the thermostat is "calling" for cooling.

When the fan switch is set to the on position, voltage is sent to terminal G from terminal R constantly or all the time. Some homeowners prefer this setting because it tends to keep the air temperature more constant inside each room. If the indoor furnace fan is running, it keeps the air in each room moving and mixed so that the temperature is even. If the air is not moving, the cooler air tends to drop near the floor and the warmer air rises, and homeowners will complain of cold feet and feeling warm in their face. When the indoor fan runs all the time, the homeowner may not be aware the outdoor unit is cycling on and off throughout the day.

The other controls in the air-conditioning system include the fan relay coil and the compressor contactor coil. Each relay and contactor controls one or more sets of normally open contacts. The coils become magnetic when voltage is applied to them, which causes current to flow through them. The magnetic energy from the coil causes the contacts to change from the open to the closed position, which sends voltage to the load that is connected to the contacts.

6.6 PACKAGE AIR CONDITIONERS

The schematic (ladder) diagram for the package air conditioner is shown in Figure 6–30. The main difference between the package air conditioner and the split system air conditioner is that the evaporator fan is mounted in the same unit with the compressor and

Figure 6–30 Schematic (ladder) diagram of a packaged heating and air-conditioning system.

condenser. Usually the compressor and condenser fan are located in one end of the unit, so that the condenser coil is exposed to the outside air. The evaporator fan is also called the indoor fan and is mounted in the opposite end of the unit, which is sealed so that the indoor air is blown over the evaporator. On the diagram for the package unit, the evaporator fan will be connected to the same power supply as the compressor and the condenser fan. In the split system, the evaporator fan is connected to the power supply for the furnace, which is usually 120 volts. The loads and controls for the package unit are the same as you would find in the split system.

6.7 WINDOW AIR CONDITIONERS

The window air conditioner is similar to the package unit in that the evaporator fan is in the same cabinet. In most window air conditioners the evaporator fan and condenser fan are a single motor with an extended shaft that comes out each end. A squirrel cage fan is mounted on the shaft end for the evaporator fan, and a blade-type fan is mounted on the shaft end for the condenser fan. Some window units put a piece of metal called a "slinger" on the condenser fan that catches moisture and throws it on the condenser fins to help cool the condenser faster. The moisture, called condensate, drips off the evaporator coil fins and flows to the window unit base pan and pools directly below the condenser fan. On some units this water is allowed to drip out of the base pan, and on other units the condenser fan slings the water on the condenser coils to help with cooling the condenser.

The loads and controls in the window air conditioner are similar to the package unit and the split system air conditioner. The main difference is that the thermostat for the unit is built into the front panel and the fan switch usually is more complex for the window unit in that it has several speeds for the evaporator fan, which include high, medium, and low speeds. Figure 6–31 shows the electrical wiring diagram for the

Figure 6–31 Window unit wiring diagram.

Figure 6–32 Schematic (ladder) diagram of window unit.

window air conditioner, and Figure 6–32 shows the schematic (ladder) diagram. In the wiring diagram you can see all the functions for the fan switch are shown as a unit, and in the ladder diagram each function of the fan switch is shown in the rung of the diagram where the control occurs.

If you are troubleshooting a window air conditioner, the wiring diagram will show you the terminals and the location on the switch exactly as you would see them if you are looking at the back of the switch. The wiring diagram will help you understand which circuit in the switch is connected and controls which load in the system.

6.8 ELECTRICAL DIAGRAMS OF A SIMPLE HEATING SYSTEM (GAS FURNACE)

Figure 6–33 shows an electrical diagram of a simple gas furnace. The loads in the gas furnace include the furnace fan and the gas valve. The controls include the **heating thermostat**, a fan and limit switch that contains a limit switch, and the fan switch. The heating thermostat provides voltage to the gas valve through the R–W circuit. When the temperature in the room gets too cool and drops below the setpoint on the thermostat, the thermostat closes the R–W circuit and passes 24 volts to the gas valve. If the fan switch on the thermostat is in the on position, 24 volts will be sent directly to the fan relay coil where it will become magnetic and cause its contacts to close, which will provide 120 volts to the furnace fan.

If the fan switch on the thermostat is in the auto position, the furnace fan will be energized by the increase in temperature in the furnace. When the gas valve opens and the burners are lit, the temperature in the furnace will begin increasing. When the temperature in the furnace reaches the setpoint of the fan switch part of the fan and the limit switch that is mounted in the furnace near the heat exchanger, the fan switch contacts will close and send 120 volts to the furnace fan. The temperature setpoint of the fan and limit fan switch is usually about 110 degrees. The fan switch, limit switch, and heating thermostat all use the temperature switch symbol. Notice in the heating thermostat symbol that the switch contact symbol is normally open and is shown above the terminal. This indicates that the thermostat will close when the temperature drops. The temperature symbol for the overtemperature "limit" part of the fan and limit switch shows the

Figure 6–33 Electrical diagram of a typical gas furnace.

contact symbol as normally closed and is shown above the terminal, indicating that when the temperature reaches the setpoint for this switch, the contacts will open. This means that the limit switch will protect the furnace against overheating if the temperature would get too hot. The last temperature symbol is in the fan switch part of the fan and limit switch. The switch symbol is normally open and is drawn below the terminal point on the switch. Since the symbol is drawn below the terminal point, it means that an increase in temperature will cause the switch to close when the temperature reaches the setpoint. When the room reaches its setpoint temperature, the thermostat switch will open and the furnace fan will continue blowing air across the heat exchanger, which ensures that all the remaining heat in the furnace is moved into the room before the furnace fan is turned off. When the temperature in the heat exchanger drops below the setpoint on the fan switch part of the fan and limit switch, the contacts open and turn off power to the furnace fan.

6.9 ## ELECTRICAL DIAGRAM OF A SIMPLE REFRIGERATION SYSTEM

Figure 6–34 shows an example of a simple refrigeration system. The schematic (ladder) diagram shows the sequence of operation of the system. You would use this diagram to troubleshoot the refrigeration system and identify which components should be energized.

Figure 6–34 Schematic (ladder) diagram for a freezer.

6.10 LOADS FOR THE REFRIGERATION SYSTEM

The loads in the refrigeration system may include the evaporator fan, the condenser fan, and the compressor. Some systems may have multiple evaporator fans or condenser fans if the coils are larger, while other systems do not have an evaporator fan, such as the one shown in the ladder diagram. You can see from the diagram that the compressor has a run capacitor, a start capacitor, and a potential relay, which makes it a capacitor-start, capacitor-run compressor. The extra capacitors are used to ensure that the compressor will start under all circumstances. The condenser fans are single-phase fans, and since they do not have any capacitors, they are probably shaded-pole motors.

6.11 CONTROLS FOR THE REFRIGERATION SYSTEM

The controls for the refrigeration system include the thermostat, which is in series with the compressor, condenser fan, and evaporator fan. The thermostat's contacts must be sized to carry the current of all three motors, and it will turn on and off to control the temperature. The other controls in the system include the high-pressure control that protects the system from high pressure and the defrost control that turns off the motors when the system is in defrost. The defrost control has a timer built into it and stops the system on a timed schedule to allow the evaporator coil to defrost periodically. The frost built up on the evaporator coils in the refrigeration system makes the system less efficient, so the coils must be defrosted periodically. The high-pressure switch will open and protect the system anytime system pressure gets too high. Some high-pressure controls have a reset function that causes the control to be manually reset if it opens on high pressure. This ensures that a technician checks out the system when the manual reset is depressed. All the controls in the refrigeration system must be sized for the full voltage of the motors, and they must be sized to handle the full load current of all three motors combined.

6.12 REVIEW

In this chapter you learned that a schematic (ladder) diagram shows the sequence of operation for a system and generally shows the loads in the top part of the diagram and the controls in the lower part of the diagram. Each line of the diagram is called a "rung," which is a reference to the term *ladder* which the diagram is named after. The companion to the schematic diagram is the wiring diagram. The wiring diagram shows each component and its terminals just as you would see them in their exact locations in the systems. The wiring diagram is useful in identifying the path of wires and terminals for each component. Some HVAC and refrigeration systems have both a schematic (ladder) diagram and a wiring diagram. If the system does not have both diagrams, you can use the material provided in this chapter to create a schematic (ladder) diagram from the wiring diagram or a wiring diagram from a schematic (ladder) diagram. The combination of the two diagrams will help you better understand the electrical systems and identify how to troubleshoot them.

Additional diagrams in this chapter show the split system air-conditioning unit and the package-type air-conditioning system. The main difference between the two systems is that the split system has a furnace or air handler inside the building that has the evaporator coil and evaporator fan in it. The condenser fan and compressor in the split system are located in the outdoor unit. The evaporator coil and evaporator fan in the package-type unit are located in the same unit that is usually mounted on the roof for commercial buildings or outdoors on the ground on a pad for residential systems. The ductwork for the package system connects the package unit from outdoors to the inside ductwork. The evaporator fan for the split system will operate from the same voltage supplied to the furnace or indoor unit, which is usually 120 volts. The evaporator fan for the package system is usually powered by the same voltage as the compressor and condenser fan, which is usually 240 volts. The loads for the package and split system air conditioners are usually controlled by relay contacts, and the coils for these relays are located in the control circuit.

The window unit air conditioner is similar to the packaged air conditioner in that the evaporator coil is mounted in the same unit as the condenser coil, compressor, and condenser fan. In fact most window air conditioners use a single motor with a double shaft that extends out both ends of the motor. The condenser fan is mounted on one end of the shaft and the evaporator fan is mounted on the other end. The window air conditioner has a thermostat and fan switch that controls the loads directly through line voltage and does not typically use relays.

The refrigeration system has a compressor and one or more evaporator fans and condenser fans. The controls for the refrigeration system include the thermostat, which is a line voltage thermostat that must have the same voltage and current ratings as the loads. The refrigeration system also has a high-pressure control and a defrost switch. The high-pressure switch protects the system against high pressure, and the defrost control periodically turns off all the loads for a timed period to allow the evaporator coils to defrost.

Job Assignment for This Chapter

You are on a job where there is only a wiring diagram that shows the location of electrical components and the color of wires connected to each part of an air-conditioning system. However, you need a schematic (ladder) diagram to help you troubleshoot the air conditioner. Your job assignment is to convert the wiring diagram to a schematic (ladder) diagram.

Solution to the Job Assignment

Your schematic (ladder) diagram of an air-conditioning system should look like the diagram in Figure 6–17. The loads should be shown in the top part of the diagram, and the controls should be in the lower part of the diagram. Each rung of the diagram should show the sequence of operation. The contacts for the relays should be shown in series with the motors, and the coils of these relays should be shown in the controls part of the diagram. Your teacher can check your diagram for accuracy.

Assessment

QUESTIONS

1. What is the function of the wiring diagram?
2. What is the function of the schematic (ladder) diagram?
3. Why is each line of the ladder diagram called a "rung"?
4. What is a load in a system or diagram?
5. What is a control in a system or diagram?

TRUE OR FALSE

1. The schematic (ladder) diagram shows the sequence of operation for a system.
2. A control is a device in an electrical system that uses current and produces work.
3. The loads in a schematic (ladder) diagram are usually located in the top part of the diagram.
4. The relay coil for the evaporator fan relay is located in the control part of the circuit, and the contacts for the relay are located in the load part of the circuit.
5. The main difference between a package unit and a split system air conditioner is that the evaporator fan of the split system is usually mounted in the indoor unit.

MULTIPLE CHOICE

1. The compressor motor in the air-conditioning system electrical diagram is located in
 a. the load circuit.
 b. the control circuit.
 c. both circuits, depending on the type of system.

2. The condenser fan motor
 a. is usually controlled by the same controls that control the compressor.
 b. is usually controlled by the same controls that control the evaporator fan.
 c. is usually controlled by the high-pressure switch.

3. In the refrigeration system, the compressor, condenser fan, and evaporator fan
 a. are each controlled by separate controls.
 b. are all turned on and off by the thermostat.
 c. can be turned off by the defrost switch or the high-pressure switch.
 d. all the above.
 e. only a and b.

4. In the package air-conditioning system, the
 a. evaporator fan is located in the indoor unit (furnace).
 b. evaporator fan is located in the same unit as the compressor.
 c. could be located in the indoor unit or the outdoor unit.
 d. all the above.
 e. only a and b.

5. In the window air conditioner, the
 a. evaporator fan and condenser fan are usually powered by the same motor that has a dual shaft that sticks out each end of the motor.
 b. evaporator fan is controlled by the multispeed switch on the front panel.
 c. evaporator fan is controlled by a separate switch than the condenser fan.
 d. all the above.
 e. only a and b.

PROBLEMS

1. Using the wiring diagram in Figure 6–1, draw the schematic (ladder) diagram of the part of the circuit that controls the compressor. Be sure to include the control part of the circuit and the load part of the circuit.

2. Using the wiring diagram in Figure 6–1, draw the schematic (ladder) diagram of the part of the circuit that controls the evaporator fan. Be sure to include the control part of the circuit and the load part of the circuit.

3. Using the schematic (ladder) diagram in Figure 6–15, create a wiring diagram of the parts of the system that control the compressor. Be sure to include the control part of the circuit and the load part of the circuit.

4. Using the schematic (ladder) diagram in Figure 6–15, create a wiring diagram of the parts of the system that control the evaporator fan. Be sure to include the control part of the circuit and the load part of the circuit.

5. Using the diagrams in Figures 6–29 and 6–30, explain the difference between the evaporator fan for the split system air conditioner and the package-type air conditioner.

Websites

Websites change constantly. Use the key word search to locate sites that have similar content. The products that will be found at each site are indicated with the key word.

Products: Split system air conditioner electrical diagrams
Key words: Wiring diagram, schematic diagram

Products: Package air conditioner electrical diagrams
Key words: Wiring diagram, schematic diagram

Products: Nordyne electrical diagrams
Key words: Wiring diagram, schematic diagram

Chapter 7

Magnetism, Alternating Current, and Power Distribution

Job Assignment for This Chapter

You are working on a job at a residence where the service call complaint is that the air conditioner will not run. After making several checks, you have determined that one of the wires in the fused disconnect has been severely burned and is not passing voltage to the air-conditioning unit. The homeowner is asking you questions about how power gets from the transformer on the power pole in front of his home to the point where it is supplied to the air conditioner. The homeowner also wants to know the reading of the voltage and its frequency that is supplied to his air conditioner. Your job is to draw a sketch of how power gets from the transformer on the power pole to the air conditioner. Be sure to include the meter set, the circuit breaker panel, the disconnect for the air conditioner, and the power terminals on the air conditioner.

Terms You Will Learn in This Chapter

AC (alternating current)
Apparent power
Capacitive reactance
Disconnect
Frequency
Hertz
Impedance
Inductive reactance
Magnet
Neutral
Peak-to-peak voltage

Peak voltage
Period
Power factor
Reactance
RMS voltage
Root mean square (rms)
Sine wave
Single-phase voltage
Three-phase voltage
True power

After reading this chapter, you will be able to:

1. Explain electromagnetic induction.
2. Explain the term *alternating current*.
3. Explain the terms *peak-to-peak* and *rms voltage*.
4. Calculate the frequency and period of an AC sine wave.
5. Explain the effects of a capacitor and an inductor in an AC circuit.
6. Explain how voltage is provided to an air conditioner in a disconnect.

7.0 MAGNETIC THEORY

The theory of operation for all types of transformers, motors, and relay coils can be explained with several simple magnetic theories. As a technician who works on air-conditioning, heating, and refrigeration systems, you will need to comprehend fully all magnetic theories so that you will understand how these components operate. You must understand how a component is supposed to operate before you can troubleshoot it and perform tests to determine if it has failed. Understanding magnetic theory will make this job easier. It is also important to understand that some of the magnetic theories rely on AC voltage. These theories will be introduced in this chapter, and more detail about AC voltage will follow in the next chapter. In this chapter you will learn about permanent magnets first, and then you will learn about electromagnets.

An overview of magnetic theory was introduced in Chapter 4 to help explain analog meter operation. In this chapter we will review basic magnetic theories again as they are used in creating electromagnetic induction and AC voltage, and go much deeper into magnetic theory as it is used in various HVAC electrical applications.

Magnet is the name given to material that has an attraction to iron or steel. This material was first found naturally about 4000 years ago as a rock in a city called Magnesia. The rock was called magnetite and was not usable at the time it was discovered. Later it was found that when pieces of this material were suspended from a wire the material would always orient itself so that the same ends always pointed in the same direction, aligned with the earth's north and south poles. Scientists soon learned from this phenomenon that the earth itself is magnetic. At first the only use for magnetic material was in compasses. It was many years later that the forces caused by two magnets attracting or repelling each other could be utilized as part of a control device or motor.

As scientists gained more knowledge, and equipment became available to study magnets more closely, a set of principles and laws evolved. The first of these discoveries showed that every magnet has two poles, called the north pole and the south pole. When two magnets are placed end to end so that similar poles are near each other, the magnets repel each other. It does not matter if the poles are both north or both south, the result is the same. When the two magnets are placed end to end so that the south pole of one magnet is near the north pole of the other magnet, the two magnets attract each other. These concepts are called the first and second laws of magnets.

(a)

Figure 7–1 a Example of a metal with its dipoles randomly placed, which makes it a very weak magnet or not magnetic at all. **b** Example of a metal with its dipoles aligned to make a very strong magnet.

(b)

When sophisticated laboratory equipment became available, it was found that this phenomenon is due to the basic atomic structure and electron alignments. When the atomic structure of a magnet was studied, it was found that its atoms are grouped in regions called *domains* or *dipoles*. In material that is not magnetic or that cannot be magnetized, the alignment of the electrons in the dipoles is random and usually follows the crystalline structure of the material. In material that is magnetic, the alignments in each dipole are along the lines of the magnetic field. Since each dipole is aligned exactly like the ones next to it, the magnetic forces are additive and are much stronger. In material where the magnetic forces are weak, the alignment of the dipoles is random and not along the magnetic field lines. The more closely the dipoles are aligned with the magnetic field lines, the stronger the magnet. Today we refer to a piece of soft iron with all its dipoles aligned as a *permanent magnet*. The name *permanent magnet* is used because the dipoles remain aligned for very long periods of time, which means that the magnet will retain its magnetic properties for long periods of time. Figure 7–1a shows a diagram of nonmagnetic metal with its dipoles randomly placed, and Figure 7–1b shows a magnetic piece of metal with its dipoles aligned to make a strong magnet.

7.1 A TYPICAL BAR MAGNET AND FLUX LINES

Figure 7–2 shows a bar magnet that is made of soft iron that has been magnetized. The magnet is in the shape of a bar, and its north and south poles are identified. Since the bar remains magnetized for a long period of time, it is called a permanent magnet. The magnet produces a strong magnetic field because all its dipoles are aligned. The magnetic field produces invisible *flux lines* that move from the north pole to the south pole along the outside of the bar magnet. The diagram in Figure 7–2 shows that these flux lines are lines of force that move from the north pole to the south pole in a slight arc.

Since the flux lines are invisible, you will need to perform a simple experiment to allow you to see that the flux lines do exist and what they look like as they surround the bar magnet. For this experiment you will need a piece of clear plastic film, such as the plastic sheets used for overhead transparencies, and some iron filings. Place the plastic sheet over a bar magnet. Make the plastic as flat as possible and sprinkle iron filings over it. The filings will be attracted by the invisible flux lines as they extend in an arc from the north pole to the south pole along the outside of the magnet. Since the flux lines begin at one pole and stretch to the other, the highest concentration of flux lines

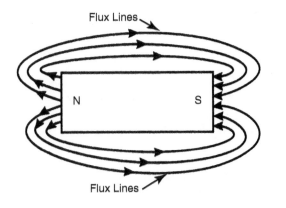

Figure 7–2 Example of a bar magnet. Notice that the poles are identified as north (N) and south (S). Flux lines move in an arc from north pole to south pole.

will be near the poles. The iron filings will also concentrate around the poles, but a definite pattern of flux lines can be seen along each side of the bar magnet. If an overhead projector is available, the image of the flux lines can be projected onto a projector screen or blackboard so that they can be seen more easily. The pattern of these filings will look similar to the diagram in Figure 7–2. The number of flux lines around a magnet is directly related to the strength of the magnet. A stronger magnet will have more flux lines than a weaker magnet. The strength of a magnet's field can be measured by the number of flux lines per area. Since the strength of a magnet's field is based on the alignment of the magnetic dipoles, the number of flux lines will increase as the alignment of the magnetic dipoles increases. Some materials, such as alnico and Permalloy, make better permanent magnets than iron, since the alignment of their magnetic domains (dipoles) remains consistent even after repeated use. You may find these materials used in some expensive controls and motors, but normally the permanent magnet will be made of soft iron. Permanent magnets are useful in many types of controls, especially in motors and generators, because the soft iron produces residual magnetism for long periods of time over many years. Permanent magnets have several drawbacks for use in some applications. One of these problems is that the magnetic force of a permanent magnet is constant and cannot be turned off if it is not needed. This means that if something is attracted to a magnet, it will remain attracted until it is physically removed from the force of the flux lines. Another problem with a permanent magnet's constant flux field is that it cannot easily be made stronger or weaker if circumstances so require.

7.2 ELECTROMAGNETS

An electromagnet is made by connecting a coil of wire to an electric cell (battery). The electromagnet has properties that are similar to a permanent magnet. When a wire conductor is connected to the terminals of the battery, current will begin to flow and magnetic flux lines will form around the wire like concentric circles. If the wire is placed near a pile of iron filings while current is flowing through it, the filings will be attracted to the wire just as if the coil were a permanent magnet. Figure 7–3 shows diagrams indicating the location of magnetic flux lines around conductors. Figure 7–3a shows that flux lines will occur around any wire when current is flowing through it. You can set up several simple experiments to demonstrate these principles. In one experiment you can insert a current-carrying conductor through a piece of cardboard and place iron filings around the conductor on the cardboard. When current is flowing in the wire, the filings will settle around the conductor in concentric circles showing where the flux

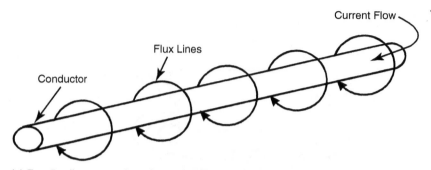

(a) Few flux lines around conductor that is not coiled.

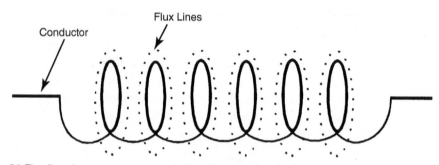

(b) Flux lines become more concentrated when wire is coiled.

Figure 7–3 a Example of flux lines around a straight wire that is carrying current. **b** Flux lines around a coil of wire that is carrying current. Notice that the number of flux lines increases when the wire is coiled.

lines are located. As the amount of current is increased, the number of flux lines will also increase. The flux lines will also concentrate closer and closer to the wire until the current reaches *saturation*. When the flux lines reach the saturation point, any additional increase of current in the wire will not produce any more flux lines.

When the straight wire is coiled, the flux lines will concentrate and become stronger. Figure 7–3b shows an example of flux lines around a coil of wire that has current flowing through it. Since the flux lines are much stronger in a coil of wire, most of the electromagnets that you will encounter will be in the form of coils. For example, coils are used in transformers, relays, solenoids, and motors.

One advantage an electromagnet has over a permanent magnet is that the magnetic field can be energized and de-energized by interrupting the current flow through the wire. The strength of the magnetic field can also be varied by varying the strength of the current flow through the conductor that is used for the electromagnet. This theory is perhaps the most important theory of magnetism, since it is used to change the strength of magnetic fields in motors, which causes the motor shaft to produce more torque so it can turn larger loads. Since the flux lines are invisible, you may need to perform an experiment to prove that the magnetic field becomes stronger as current flow through the coil is increased. Figure 7–4 shows how to set up this experiment so you can demonstrate it. Wrap a wire into a coil and connect the ends to a dry cell battery and place it near a pile of iron filings. Add a variable resistor to the circuit to increase or decrease the current. When the current is set to a minimum, as in Figure 7–4a, the magnetic field around the coil of wire will attract only a few filings. As the current is increased, as in Figure 7–4b, the number of filings the magnetic field will attract also increases until the current causes

(a) Few filings are attracted when a small amount of current passes through the conductor.

(b) Large numbers of filings are attracted as current is increased.

Figure 7–4 a A small amount of current flowing in a coil of wire creates a small number of flux lines that attract iron filings. **b** A large amount of current flowing in a coil of wire creates a large number of flux lines that attract iron filings.

the magnetic field in the wire to reach *saturation*. When the saturation point is reached, any additional current flowing in the wire will not produce additional flux lines. When a switch is added to this circuit, the magnetic field can be turned on and off by turning the switch on and off to interrupt the flow of current in the coil. When the switch is opened, current is interrupted and no flux lines are produced, so the magnetic field will not exist.

Components such as electromechanical relays and solenoids use the principle of switching a magnetic field on and off. When current flows in the coil, the magnetic field causes the contacts in the relay or the valve in the solenoid to close. When the current to the coil in these devices is interrupted and the magnetic field is turned off, springs cause the relay contacts or valves to open. More information about relays and solenoids is provided in later chapters.

This principle of magnetism is also used to turn motors on and off. When current flows through the coils of a motor, its shaft turns. When current is interrupted, the magnetic field diminishes and the motor stops rotating. This point is also important to remember if the wire that is used in the coil develops an open. When current stops flowing in the coil due to the open circuit, the shaft of the motor will stop turning.

7.3 ## ADDING COILS OF WIRE TO INCREASE THE STRENGTH OF AN ELECTROMAGNET

Another advantage of the electromagnet is that its magnetic strength can be increased by adding coils of wire to the original single coil of wire. The magnetic field is increased because the additional coils of wire require a longer length of wire to be used, which

provides additional flux lines. The magnetic field is stronger when the coil is more tightly wound because the flux lines are more concentrated. This means that very fine wire is used in some electromagnets to maximize the number of coils. You must be aware that as smaller wire is used the amount of current flowing through it must be reduced so that the wire is not burned open.

You will learn that some motors use this principle to increase their horsepower and torque rating. These motors have more than one coil that can be connected in various ways to affect the torque and speed of the motor's shaft. *Torque* is defined as the amount of rotating force available at the shaft of a motor. You will also learn that coils can be connected in series or in parallel to affect the torque and speed of a motor.

7.4 USING A CORE TO INCREASE THE STRENGTH OF THE MAGNETIC FIELD OF A COIL

The strength of a magnetic field can also be increased by placing material inside the helix of the coil to act as a core. The farther the core is inserted into the coil, the stronger the flux field becomes. When the core is removed completely from the coil, it is considered to be an *air-coil magnet* and the magnetic field is at its weakest point. If soft iron is used as the core, it will strengthen the magnetic field, but it also creates a problem because it has excessive residual magnetism, which is unwanted. Residual magnetism means that the core will retain magnetic properties when current is interrupted in the coil, which will make it like a permanent magnet. This problem can be corrected by using laminated steel for the core. The laminated steel core is made by pressing sheets of steel together to form a solid core. Figure 7–5 shows an example of layers of laminated steel pressed together to form a core. When current flows through the conductors in the coil, the laminated steel core enhances the magnetic field in much the same way as the soft iron does, and when current flow is interrupted, the magnetic field collapses rapidly because each piece of the laminated steel does not retain sufficient magnetic properties.

7.5 REVERSING THE POLARITY OF A MAGNETIC FIELD IN AN ELECTROMAGNET

When current flows through a coil of wire, the direction of the current flow through the coil determines the polarity of the magnetic field around the wire. The polarity of the magnetic field around the coil of wire is important because it determines the direction in which a motor shaft turns in an AC or DC motor. If the direction of current flow is reversed, the polarity of the magnetic field will be reversed, and the direction in which a motor shaft is turning will be reversed. In some motors used in heating, air-conditioning,

Figure 7–5 Example of pieces of steel pressed together to form a laminated steel core for use in electromagnets.

and refrigeration systems, such as fan motors and pump motors, the direction of rotation is very important. In these applications you will be requested to make changes to the connections for the windings in the motor or to the supply voltage for a three-phase motor to make the motor rotate in the opposite direction. You should understand that the changes you are making are taking advantage of changing the direction of current flow through a coil or changing the polarity of the supply voltage with respect to the other phases so that the motor will reverse its rotation.

As you learn advanced theories about motors and other electromagnetic components, diagrams will be presented to explain more complex concepts. When these diagrams are used to explain the direction of current flow through a wire, the direction of current flow will be identified by a universally accepted method. Figure 7–6a shows a diagram that indicates the location of the flux lines in a coil of wire and shows the direction the flux lines travel around a straight current-carrying conductor. A dot or a cross (X) is used to mark the conductor to indicate the direction in which current is flowing. In Figure 7–6a you should notice that a dot is used to indicate that current is flowing toward the observer. The flux lines in this diagram show that their flow is in a clockwise motion. In Figure 7–6b, a cross is used to indicate that current is flowing away from the observer, and the flux lines are shown moving in the counterclockwise direction. If you know the direction of current flow, the advanced theories will allow you to determine the polarity of the magnetic field and predict the direction in which a motor's shaft will rotate.

Figure 7–6c shows another way to determine the directions of the flux lines, the direction of current flow, and which pole is the north pole. This method is called the *left-hand rule*. From this diagram you can see that you need to know either the direction of the current flow or the direction of the flux lines to determine the polarity of the coil. Normally, the direction of the current flow is easy to determine with a voltmeter by determining which end of the wire is positive and which is negative. Current (electron flow) is from negative to positive.

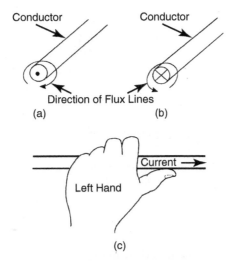

Figure 7–6 a The direction of flux lines around a wire when current is flowing in the wire toward the observer. The dot indicates that the direction of current flow is toward the observer. **b** The direction of flux lines around a wire when current is flowing in the wire away from the observer. The X indicates that current is flowing in the wire away from the observer. **c** An example of the left-hand rule. The thumb is pointing in the direction of the current flow in the wire, and the fingers are pointing in the direction of the flux lines as they move around the wire when current is flowing.

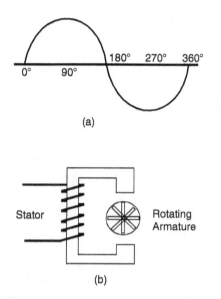

(a)

Figure 7-7 a A sine wave for AC voltage. **b** A simple alternator that has a moving coil and a stationary coil. The alternator produces the AC sine wave.

Stator Rotating Armature

(b)

7.6 AN OVERVIEW OF AC VOLTAGE

Before you can understand more about magnetic theories, you will need to know more about AC voltage and current. This section will provide an overview of AC voltage. The name **AC (alternating current)** is derived from the fact that AC voltage alternates positive for half of a cycle and then alternates negative for the other half of a cycle. Figure 7–7a shows the characteristic AC waveform, which is called a **sine wave**.

The AC waveform is created by an alternator that looks similar to a motor. Figure 7–7b shows a diagram of an alternator. The alternator has a rotating coil of wire that is mounted on a shaft, and a stationary coil that has current flowing through it, so that it creates a magnetic field. The shaft of the alternator is rotated by an energy source such as a steam turbine. When the rotating coil of wire passes through the magnetic field, an electron flow is created in the coil of wire. Since the coil of wire rotates, it will pass the positive magnetic field and then the negative magnetic field during each complete rotation of 360°. This action produces the voltage sine wave, and the sine wave is also identified as having 360°. In the United States, the speed of rotation of the alternator is maintained at a constant rate so that the sine wave will have a frequency of 60 cycles per second, which is also called 60 **hertz** (Hz). Hertz are the units for frequency.

7.7 ELECTROMAGNETIC INDUCTION

Electromagnetic induction is another magnetic principle that is used in transformer and motor theory. This theory states that if two individual coils of wire are placed in close proximity to each other, and a magnetic field is created in one of the coils so that flux lines are created around it, the flux lines will collapse across the second coil when the current is interrupted in the first coil. When these flux lines collapse across the coils of the second coil, they will cause a current to begin to flow in the second coil. This current flow will be 180° out of phase with the current flow that created the magnetic field in the first coil.

Since AC voltage periodically builds voltage to a positive peak and interrupts it when the sine wave reaches 180°, AC voltage is ideal for creating the magnetic field that builds and collapses. When the current flows, it causes the flux lines to build in the coil, and when the current flow is interrupted, it causes the flux lines to collapse. Since AC voltage in the United States is generated at 60 Hz, the magnetic field in the first coil will

build and collapse 60 times a second. The flux lines will also build and collapse across the windings of the second coil 60 times a second. This is called the frequency of the voltage and in measured in hertz (HZ), which are the number of cycles per second.

One component that uses two coils is called a *transformer*. The first coil in the transformer is called the *primary winding*, and the second coil in the transformer is called the *secondary winding*. The ratio of the number of turns of wire in the primary winding and in the secondary winding will determine the ratio of the voltage in the primary winding and secondary winding. Another important point to remember is that since the two coils are electrically isolated, the voltage produced in the second coil by induction is totally isolated from the voltage in the first coil. The isolation allows the number of turns of wire in the two coils to be different, so that a different amount of voltage can be created in the second coil. If the number of turns of wire in the first coil and second coil are identical, the amount of voltage induced in the second coil will be approximately the same as the voltage in the first coil.

In motors, electromagnetic induction is used to create a second magnetic field in a part of the motor that rotates. Since the shaft of the motor must become magnetic and rotate freely from the stationary part of the motor, it would not be practical to make a permanent connection with a wire to the rotating member so that it could receive current to create its magnetic field. Electromagnetic induction creates the second magnetic field without any connections. In future chapters you will learn more about transformer turns, ratios, and magnetic theories for motors.

7.8 WHERE DOES AC VOLTAGE ORIGINATE?

AC voltage is generated at a number of power stations around the country. The voltage that is used in your city is generated somewhere within a range of 600 miles. The energy source that turns the rotor of the alternator to produce this voltage can come from burning coal, burning fuel oil, or burning other fossil fuels to heat water to make steam. In some areas of the country the steam is generated from nuclear power stations. The steam is used to turn a turbine wheel that turns an alternator shaft. In other parts of the country the turbine wheel is turned by water that is stored behind hydroelectric dams. When the alternator shaft turns, AC voltage is produced.

The voltage that is produced at large utility power stations is generated from three equal fields in the alternator. This causes the generated voltage to be **three-phase voltage**. Three-phase voltage is transmitted through three separate wires that are identified as line 1 (L1), line 2 (L2), and line 3 (L3). The voltage is sent to transformers at the generating station, where it is stepped up to several hundred thousand volts so it can be transmitted over long distances. When the voltage arrives at a city, it is transformed down (stepped down) to a value of approximately 40,000 V. When the voltage reaches an industrial or commercial area, the voltage is stepped down again to 480 V or 240 V. The voltage that is used in residential areas is stepped down to 240 V.

Notice!

The single-phase line to neutral voltage in the United States may vary from location to location. These differences may be due to how much primary voltage is provided, how many customers are on a line, and how heavy the load usage is for a particular period of time. Typical line-to-neutral voltages are 110, 115, 120, and 125 V and typical line-to-line voltages are 220, 230, 240, and 250 V. This text uses 120 V as a standard for line-to-neutral voltages and 240 V as a standard for line-to-line voltages. Neutral is defined as the return line to a single-phase transformer where L1 or L2 is the supply line.

7.9 FREQUENCY OF AC VOLTAGE

The most important feature of AC voltage that is different from DC voltage is that AC voltage has a frequency. The typical frequency for AC voltage in the United States and much of North America is 60 Hz. The frequency of 60 Hz is determined by the speed at which the alternator rotates when the voltage is generated. The frequency of AC voltage that is used to supply voltage for heating, air-conditioning, and refrigeration equipment is constant. **Frequency** is defined as the number of cycles (sine waves) that occur in 1 second. Figure 7–8 shows four sine waves occurring in 1 second.

The **period** of a sine wave is the time it takes one sine wave to start from the zero point and pass through 360°, as seen in Figure 7–9. This represents one complete cycle and is one complete revolution of the alternator. Since the shaft of the alternator rotates through one full circle to produce the sine wave, we can equate one complete cycle to 360°. This means that the sine wave can be described in terms of 360°. In Figure 7–9 you can see the sine wave starts at 0°, reaches the positive peak at 90°, returns to zero at 180°, then reaches the negative peak at 270°, and finally returns to zero volts at 360°. We will use the number of degrees to identify points of the sine wave in future discussions. The period of a sine wave can also be described as $P = \frac{1}{freq}$, and frequency can be defined as $F = \frac{1}{period}$.

| **Example 7–1** | Determine the period of a 60-Hz sine wave. |

| **Solution:** | Use the formula $P = \frac{1}{F}$ (where F = frequency). |

$$P = \frac{1}{F} = \frac{1}{60} = 0.016 \text{ s}$$

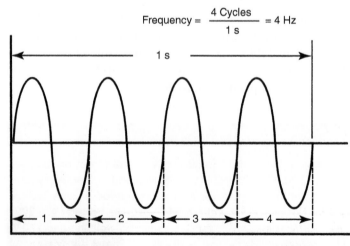

Figure 7–8 The frequency of AC voltage is calculated from the number of cycles that occur in 1 s.

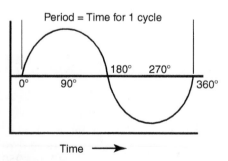

Figure 7–9 The period of AC voltage is calculated as the time it takes for one cycle to occur.

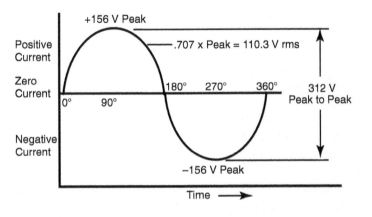

Figure 7–10 Sine wave with positive and negative peaks identified. The peak-to-peak (P-P) voltage is also identified.

7.10 PEAK VOLTAGE AND RMS VOLTAGE

Figure 7–10 shows a sine wave with the point where peak voltage occurs indicated. **Peak voltage** is the highest point of the sine wave. From this figure you can see that the peak voltage is 156 V for this example. The negative peak voltage is also identified as minus 156 (–156) V in this example. The total **peak-to-peak (P-P) voltage** is 312 V. The peak voltage can be measured only with a peak-reading voltmeter or an oscilloscope.

The voltage that you will read with a VOM meter is called **rms voltage**. The term *rms* stands for **root-mean-square**, which is the mathematical function used to calculate the voltage that a typical rms meter will read. The major difference between root-mean-square voltage and peak voltage is that the meter compensates for the difference between the rms voltage and the peak voltage at various times in any given cycle. The formula for calculating rms voltage from peak voltage (V_P) is

$$rms = 0.707 \times V_P$$

The formula for calculating peak voltage from rms voltage is Peak = 1.414 × rms. As you can see, the rms voltage is approximately 70% of the peak voltage. This means that if the peak voltage is 100 V, the rms voltage will be 70 V. In the example in Figure 7–10, the peak voltage is 156 V, so the rms voltage is 110.3 V.

Example 7–2 Calculate the rms voltage if the peak voltage is 156 V.

Solution:

rms volts = .707 × V_P
rms volts = .707 × 156 V
rms volts = 110 V

Example 7–3 Find the peak voltage if rms voltage is 120 V.

Solution:

peak volts = 1.414 × rms V
peak volts = 1.414 × 120 V
peak volts = 169.68 V

Figure 7–11 A typical load center.
(Courtesy of Cutler-Hammer)

Figure 7–12 A fused disconnect that is mounted near the air-conditioning or refrigeration equipment.

7.11 THE SOURCE OF AC VOLTAGE FOR HEATING, VENTILATING, AND AIR-CONDITIONING SYSTEMS

When you are working on a job in a residence or in a commercial building, you will need to locate the source of incoming voltage. The source of incoming voltage at these locations is typically a *circuit breaker panel*. The circuit breaker panel is also called a *load center*, and one is shown in Figure 7–11. The load center in a commercial building is typically three phase, and the load center in a residential building is typically single phase.

Voltage from the load center is sent to a disconnect box that will be mounted near the air-conditioning or refrigeration system on which you are working. Figure 7–12 shows a picture of a disconnect box. If the disconnect box has fuses, it is called a *fused disconnect*. The **disconnect** is used to turn power off to the unit when you are working on the equipment. The fuses in the disconnect also provide protection against overcurrent for the equipment connected to it.

7.12 MEASURING AC VOLTAGE IN A DISCONNECT

When you are installing or troubleshooting equipment in the field, you will need to measure the amount of AC voltage available at the load center or at the disconnect. You must be extremely careful when you take these readings because full voltage will be applied to the system at the time you are taking the measurement. Figure 7–13 shows the points in the disconnect where you should place the meter probes to measure high voltage and low voltage. Figure 7–13a shows the points to place the meter leads when

(a)

Figure 7–13 a Location to place meter probes for reading 240 V on the line side of a fusible disconnect. **b** Location to place meter probes for reading 120 V on the line side of a fusible disconnect.

(b)

you are reading voltage line to line (L1 to L2). From this figure you can see that if the meter probes are placed on line 1 (L1) and line 2 (L2) incoming terminals, the applied voltage will be 240 V if you are in a residence. If the fused disconnect is located in a commercial location, you may read 208 V when you place the meter on L1 and L2 or you may measure 480 V. These terminals are called the *line-side terminals*.

Figure 7–13b shows the meter probes on line 1 (L1) and neutral. The meter should measure 120 V if the disconnect is in a residence. This is the lower voltage that is used on smaller equipment. You may measure a higher voltage such as 208 V or 240 V in the disconnect in some cases.

The previous voltage measurements are made at the line-side terminals of the disconnect. You will need to make a second set of measurements at the *load-side terminals* that are located at the bottom of the disconnect. If you have voltage at the line-side but not at the load-side terminals, the disconnect may have a blown fuse. In some cases you can determine that the fuse is blown if you can measure full voltage across the top and bottom of a fuse. The reason you can read full voltage across the fuse is that it has an open, and you are reading voltage that feeds back from the bottom of the fuse through a load such as a transformer winding to the other supply voltage line. If you suspect you have a faulty fuse, you can remove one or both of them and check them for continuity.

7.13 VOLTAGE AND CURRENT IN AC CIRCUITS

If an AC circuit has only resistance in it, its voltage waveform and current waveform will be *in phase*. This means that both waveforms start at the same point in time and all the energy that the source delivers to the resistance is dissipated in heat. If the load in the

circuit is a resistance heating element, the voltage waveform and current waveform will be in phase with each other, and all the energy delivered by the source will be converted to heat by the heating element. (There will be some small losses due to the inefficiency of the heating element.) If a capacitor or inductor is used in an AC circuit such as a motor circuit, the voltage waveform and current waveform will not be in phase, and some of the energy that is delivered by the source is stored in the inductor or capacitor and then returned to the circuit during another part of the cycle. This will cause some additional amount of inefficiency in the circuit. A capacitor is a device that is made of two pieces of conducting foil that are separated by an insulator called a dielectric. Figures 9–19 and 9–20 show more information about capacitors used in HVAC systems. The unit of capacitance is called a farad, and the amount is so small in most capacitors used in HVAC systems that it is measured in millionths of a farad, or microfarads. The Greek letter mu (μ) is used to denote micro (millionths).

Figure 7–14 shows the voltage, identified by the letter *v*, and current waveform, identified by the letter *i*. You can see that these waveforms are in phase because they both start at the same point in time and end each of their cycles at the same time. The voltage in this circuit is larger than the current, so the voltage waveform is shown larger than the current waveform.

When voltage and current waveforms are in phase, there are no losses in the circuit due to inductive reactance or capacitive reactance. An AC circuit containing an inductor, such as a motor coil, has **inductive reactance**, and a circuit containing a capacitor such as a start capacitor for a motor, has **capacitive reactance**. **Reactance** is an opposition to the AC circuit that is similar to the opposition in a DC circuit caused by resistance. Inductive reactance is the opposition caused by the inductor, and capacitive reactance is the opposition caused by capacitors. If a circuit has inductors (transformer windings or motor windings) or capacitors, the voltage and current waveforms for that circuit have a phase shift. The amount of phase shift can be defined as a number of degrees or it can be defined as the total amount of reactance (opposition) in the circuit. As a technician you need to know that reactance exists in these types of circuits and that the effects of reactance can be calculated, predicted, and used to your advantage. For example, the more phase shift that can be created between the start winding and the run winding of an induction motor, the higher the reactance, which creates greater current and larger starting torque in the motor. This is useful for helping larger single-phase compressor motors to start.

The next sections discuss the theory of capacitive reactance and inductive reactance. Even though you may never do reactance calculations when you are troubleshooting a compressor motor, you must comprehend the theory about capacitive reactance and inductive reactance to understand why additional capacitance may be required to make a compressor motor start more easily. You will also need a basic understanding of reactance to learn about electronic control circuits that are designed to make compressor motors and fan motors run more efficiently.

Figure 7–14 Voltage waveform (*v*) and current waveform (*i*).

7.14 RESISTANCE AND CAPACITANCE IN AN AC CIRCUIT

In the previous section we learned that when capacitors or inductors are used in an AC circuit, they create an opposition that is similar to resistance. The opposition caused by a capacitor is called *capacitive reactance*, and the opposition caused by the resistor is called *resistance*. Both reactance and resistance are measured in ohms. The main difference between capacitive reactance and resistance is that a phase shift between the voltage and current waveform occurs. Figure 7–15a shows a diagram of a capacitor in an AC circuit with a resistor. Figure 7–15b shows the voltage and current waveforms for this circuit, and Figure 7–15c shows the vector diagram that is used to calculate the amount of phase shift for this diagram. Because the capacitive reactance and resistance have a phase shift, we cannot simply add the two values together. Instead, a vector diagram, which is a graph derived from a trigonometric calculation, is used to determine phase angle and the total amount of opposition the capacitor and resistor cause to the circuit.

When voltage encounters a capacitor in a circuit, it will take time to charge up the capacitor and then discharge it. This causes reactance, which becomes an opposition to the voltage waveform. The capacitor does not affect the current waveform. To put this in simpler terms, the energy in a circuit with only a capacitor in it will be delivered by the source during part of the voltage cycle and stored in the capacitor. Later, in another part of the cycle, the capacitor will release the stored energy back into the circuit without

(a)

(b)

Figure 7–15 a Capacitor and resistor in an AC circuit. **b** Waveform of voltage and current for the AC circuit. Notice that the current waveform leads the voltage waveform. **c** Vector diagram showing the relationship between voltage across the resistor and the voltage across the capacitor.

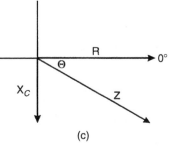

(c)

any energy losses. If the circuit has both a resistor and a capacitor in the circuit, some of the energy is stored and released by the capacitor, and the resistor dissipates some of the energy. In Figure 7–15b you can see that the voltage waveform *lags*, or starts later than, the current waveform for the circuit that has both a capacitor and resistor.

7.15 CALCULATING CAPACITIVE REACTANCE

The total amount of opposition caused by the capacitor is called *capacitive reactance* and is represented by the symbol X_c. Even though you may never need to calculate capacitive reactance, you should understand the effects of changes in capacitance and frequency on the amount of capacitive reactance. If frequency of the supply voltage increases or the size of the capacitor increases, the amount of capacitive reactance becomes smaller in a circuit. Because the frequency of the supply voltage to residential and commercial air-conditioning systems will remain at 60 Hz in the United States, the only thing that can change the amount of capacitive reactance is changing the amount of capacitance. This can be done by removing the capacitor that is in a system and replacing it with a different-value capacitor. For example, if you change a 20 μF capacitor with a 40 μF capacitor, the amount of capacitive reactance will become smaller, which will cause the amount of current in the circuit to become larger. This will cause the motor to have more starting current and starting torque. Capacitive reactance can be calculated by the following formula:

$$X_c = \frac{1}{2\pi FC}$$

where

$\pi = 3.14$
$F = $ frequency
$C = $ capacitance in microfarads

XC = Capacitive reactance

From this formula you can see that you must know the value of the capacitor and the frequency of the AC voltage. For example, if a 40-μF capacitor is used in a 60-Hz circuit, the amount of opposition (capacitive reactance) will be 66.35 Ω. Notice that since the capacitive reactance is an opposition, its units are ohms. The solution of this problem follows:

$$X_c = \frac{1}{2\pi FC} = \frac{1}{2 \times 3.14 \times 60 \times 0.000040} = 66.35 \ \Omega$$

Example 7–4 Calculate the capacitive reactance of a 60-Hz AC circuit that has a 90-μF capacitor.

Solution:

$$X_c = \frac{1}{2\pi FC} = \frac{1}{2 \times 3.14 \times 60 \times 0.000090} = 29.49 \ \Omega$$

7.16 CALCULATING THE TOTAL OPPOSITION FOR A CAPACITIVE AND RESISTIVE CIRCUIT

Impedance is the amount of total opposition caused by a capacitor and resistor in an AC circuit and it can be calculated. The units of measure for impedance are ohms, and impedance is represented by the letter Z in all formulas. Figure 7–16c shows the diagram

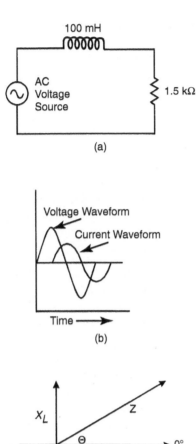

100 mH

AC
Voltage
Source

1.5 kΩ

(a)

Voltage Waveform

Current Waveform

Time ⟶

(b)

X_L

Z

Θ

R

0°

(c)

Figure 7–16 a An inductor and a resistor in an AC circuit. **b** Waveforms for the voltage and current in the inductive and resistive circuit. **c** Vector diagram showing voltage across the inductor leading the voltage across the resistor.

↓ example does not match paragraph

Capacitance and resistance

that is used to determine the impedance for this type of circuit. For example, if a circuit has 70 Ω of resistance and 50 Ω due to capacitive reactance, the total impedance must be calculated with a vector diagram because the voltage and current in this circuit are out of phase. The formula for calculating impedance of this circuit is

Z = impedance

$$Z = \sqrt{R^2 + X_c^2} \quad Z = \sqrt{70^2 + 40^2} \quad Z = 80.62 \ \Omega$$

| 7.17 | **RESISTANCE AND INDUCTANCE IN AN AC CIRCUIT** |

When inductors (coils of wire) are used in an AC circuit, they create an opposition that is similar to resistance. The opposition caused by an inductor is called *inductive reactance*. The main difference between capacitive reactance, inductive reactance, and resistance is that a phase shift between the voltage and current waveforms occurs. Figure 7–16a shows a diagram of an inductor in an AC circuit with a resistor. Figure 7–16b shows the voltage and current waveforms for this circuit, and Figure 7–16c shows the vector diagram that is used to calculate the amount of phase shift for this circuit.

When current encounters an inductor in a circuit, it will take time to charge up the inductor and then discharge it. This causes the reactance, which becomes an opposition

to the current waveform. The inductor does not affect the voltage waveform. In Figure 7–16b you can see that the current waveform *lags* or starts later than the voltage waveform for this circuit.

7.18 CALCULATING INDUCTIVE REACTANCE

$X_L =$ inductive reactance

The total amount of opposition caused by the inductor, inductive reactance, is represented by the symbol X_L. Like capacitive reactance, inductive reactance can be calculated, and you should remember that even though you will not calculate reactance in the field when you are troubleshooting, it is important that you understand the effects that changing the size of the inductor or changing frequency have on the total amount of inductive reactance. Inductive reactance can be calculated by the following formula:

$$X_L = 2\pi FL$$

where

$\pi = 3.14$
$F = $ frequency
$L = $ inductance in henries (H)

From this formula you can see that you must know the value of the inductor and the frequency of the AC voltage. For example, if an 80-H inductor is used in a 60-Hz circuit, the amount of opposition (inductive reactance) will be 30,144 Ω (30.144 kΩ). Notice that since the inductive reactance is an opposition, its units are ohms. The solution of this problem follows:

$$X_L = 2\pi FL = 2 \times 3.14 \times 60 \text{ Hz} \times 80 = 30{,}144 \ \Omega$$

Example 7–5 Calculate the inductive reactance of a 60-Hz AC circuit that has a 20-H inductor.

Solution: $$X_L = 2\pi FL = 2 \times 3.14 \times 60 \times 20 = 7536 \ \Omega$$

7.19 CALCULATING THE TOTAL OPPOSITION FOR AN INDUCTIVE AND RESISTIVE CIRCUIT

$Z =$ impedance

The amount of total opposition (impedance) caused by an inductor and resistor in an AC circuit can be calculated. This impedance is the same as the impedance caused by capacitive reactance and resistance in the previous circuit. This means that impedance is still represented by the letter Z in any formulas. Figure 7–17c shows the diagram that is used to determine the impedance for this type of circuit. For example, if a circuit has 70 Ω of resistance and 50 Ω due to inductive reactance, the total impedance must be calculated with a vector diagram because the voltage and current in this circuit are out of phase. The formula for calculating impedance of this circuit is

$$Z = \sqrt{R^2 + X_L^2} \quad Z = \sqrt{70^2 + 50^2} \quad Z = 86.02 \ \Omega$$

| 7.20 | ## TRUE POWER AND APPARENT POWER IN AN AC CIRCUIT |

In a DC circuit you can simply multiply voltage times current to determine the total power of the circuit. In an AC circuit you must use the vector diagram when you calculate the current caused by any resistors and inductors. The current that is caused by a resistor is called **true power** (TP), and current caused by capacitive reactance or inductive reactance is called *reactive power*. True power is measured in watts, and if the amount of power is more than 1000 W, the kilowatt (kW) is used, where k represents a thousand.

When true power from the resistor is added to the reactive power from the capacitor or inductor in the vector diagram, the resultant power is called **apparent power** (AP). The unit for apparent power is VAR, which stands for *volt ampere reactance*. If the amount of power is more than 1000 VAR, the unit kVAR is used, where k stands for thousand.

The main point to remember is that all the energy that the source supplies to the resistor is dissipated in the resistor and is called true power. The energy that the source supplies to the capacitor or inductor is stored in one part of the cycle and returned in another part of the cycle, so there is no net loss of energy. This means that if you have a 120-V AC supply providing 1000 W of power to an electric heating element, the heating potential for this circuit is 1000 W of true power.

| 7.21 | ## CALCULATING THE POWER FACTOR |

The ratio of true power to apparent power in an AC circuit is called the **power factor** (PF). The formula for the power factor is

$$PF = \frac{TP}{AP}$$

Example 7–6

Calculate the power factor for a circuit that has 600 VA (volt-amps) of apparent power and 500 W of true power.

Solution:

$$PF = \frac{TP}{AP} = \frac{500 \text{ W}}{600 \text{ VA}} = 0.830 \text{ or } 83\%$$

The true power in a circuit is always smaller than the apparent power, so the power factor is always less than 1. In a pure resistive circuit, the true power and the apparent power are the same value, so the power factor is equal to exactly 1. This is due to the way the vector diagram is calculated. When the circuit has several motors, such as compressor motors or pump and fan motors, the reactance of the inductors and capacitors in these devices causes the circuit to have different true and apparent power, which causes a power factor less than 1. If the power factor is substantially less than 1, the power company will assess a penalty, which can result in larger electric bills. This will become a major issue if a department or grocery store has a large number of compressors and fans running for long periods of time. One way to limit this penalty is to add capacitors to this circuit to correct the power factor so that it is closer to 1. The power-factor-correction

capacitors are different from the capacitors that are connected directly to the compressors, because they are connected at the power disconnect. These commercial power-factor-correction systems may automatically adjust the amount of capacitance that is added or subtracted to the system as compressors and other motors cycle on and off. You will not be expected to work on or adjust the power-factor-correction system; instead, you just need to understand why it is used in the circuit.

7.22 OVERVIEW OF POWER DISTRIBUTION FOR HVAC

You will encounter AC voltage and power distribution when you install a new furnace or air-conditioning system and supply new wiring. AC power is supplied to a residential site as 240 volts in the form of three wires called L1, L2, and neutral, which come from a transformer that may supply power to more than one home. The voltage available between L1 and neutral or between L2 and neutral is 120 V. Figure 7–17 shows the three wires as they come from a transformer through an electrical mast, which extends above the roof to ensure that wires do not touch the house, and a drip leg is provided to ensure that water does not get into the mast (notice that the wire enters the mast head from below, so that water does not run uphill into the mast). A plastic or metal conduit (pipe) is connected to the electric meter set at the bottom and has the mast head, which is also called the *service entrance cap*, at the top. The wires run from the transformer to the service entrance cap, through the pipe, and down to the meter set. The meter set may also be called a *meter base* and has line-side terminals at the top where the wires from the transformer are connected and load-side terminals at the bottom where the wires that lead into the main electrical circuit breaker panel are

Figure 7–17 The transformer supplies 240-volt power through the meter set, main circuit breaker panel, circuit breaker for the air conditioner, fused disconnect, and finally to the air-conditioning unit.

connected. The electric utility company will set an electric watt-hour meter in the meter socket, which connects the electric meter in series with all power that is consumed in the home. The meter measures the voltage and the current used in the home, which become watts when multiplied together. The unit of measure for the meter is called a watt-hour, so this electrical meter is sometimes called a watt-hour meter. The wires that are connected to the load side of the meter set run to the top of the main circuit breaker panel in the house and are connected to the line-side terminals of the main circuit breaker. When the home was built, the power lines called the entrance cable were run from the utility feed to the meter set and from the meter set to the main circuit breaker panel. If the home is in a newer neighborhood, the power may be supplied from underground service. In an underground service, the transformer is set in the backyard and may supply power to several homes. All wiring to and from the transformer is buried underground and the wiring must be rated for underground service. In an underground system, the conduit that is connected to the meter base will come up from underground into the bottom of the meter set. Generally all service entrance wire will be in place by the time you are requested to install an air conditioner, and you will only need to supply wire between the main circuit breaker panel to the fused disconnect.

If you are installing an air conditioner in a home, you will be required to add circuit breakers to the circuit breaker panel to provide power to the new system. The circuit breaker will be a two-pole breaker to supply 240 volts through L1 and L2 to the air conditioner. The equipment manufacturer will specify the current rating for the breaker. If you are working on an existing furnace or air conditioner, they will be connected to circuit breakers in the main electrical panel, and you may need to switch the breaker off to perform some maintenance.

The next part of the power distribution for the air conditioner is a fused disconnect that is mounted within several feet of the air conditioner. Since the fused disconnect sits outside and is exposed to rain or snow, it must be rated as weather tight and all wiring to and from it must be rated to withstand moisture from rain or snow. The disconnect provides a place to mount fuses for the system as well as a means to disconnect power from the system when you must work on it. The disconnect must be mounted close enough to the air-conditioning system so anyone working on it can switch it to the off position so they can safely work on the air-conditioning system. Wires are run between the disconnect and the air-conditioning system. They are connected to the line-side terminals in the disconnect and the load-side terminals in the air conditioner and are usually mounted on the main contactor.

A ground wire is supplied from the transformer at the street to the meter base and on to the main circuit breaker panel. A grounding rod, which is an 8-foot rod, is usually buried into the ground directly below the meter set. This provides additional grounding protection for the electrical system. All electrical components must have a continuous ground circuit to protect humans from electrical conditions that may create a shock hazard. If one of the power lines, L1 or L2, come in contact with any metal in the air-conditioner case or anywhere else in the circuit, a person could touch the metal case and receive a severe electrical shock. If the system is properly grounded, the circuit breaker and/or fuses for the system would interrupt power the instant the power line came into contact with anything that is metal or could conduct electricity. A properly grounded system will ensure that a shock hazard cannot exist, because it would be interrupted by a fuse or circuit breaker that opens when the hazard first exists. If the system is not properly grounded, any person touching the metal cabinet could receive an electrical shock if line 1 or line 2 power lines came into contact with the metal cabinet.

The power that is supplied to a residential or commercial site originates at a generating station in your area. The power is generated as three-phase voltage at the generating station and is transported across country at very high voltages up to 100,000 to 200,000 volts. The lines for this power system are mounted on large steel towers that you will see outside of towns. The three-phase power lines come into a town to a set of transformers called a substation where they are reduced to approximately 4800 volts to be distributed throughout the town or city as three-phase voltage. The three-phase lines are routed to industrial and commercial areas of the city, and two of the three lines are routed to residential areas. Some residential areas use L1 and L2, and others use L2 and L3, and still others use L1 and L3. The amount of power for each phase of the three phases is balanced so that none of the lines is overloaded. When only two lines are used, the voltage is called **single-phase voltage**.

Single-phase residential transformers are located in neighborhoods to further step the voltage down to 240 volts line to line and 120 volts line to neutral. One or more homes may be connected to transformers that are mounted on poles for overhead supplies or on the ground for underground service supplies. You will not work on the transformers, but you will be expected to put new circuit breakers into residential or commercial circuit breaker panels for new installations. The next chapter will explain how three-phase and single-phase distribution is wired into circuit breaker panels for installations of HVAC systems.

7.23 REVIEW

When you are learning about components that utilize electromagnetic principles, you can return to this section to review the basic concepts about them. The important electromagnetic principles follow:

1. The dipoles of a magnet are aligned.
2. Every magnet has a north pole and a south pole.
3. Like magnetic poles repel, and unlike magnetic poles attract.
4. A permanent magnet is useful because its magnetic field is residual and can remain strong over a number of years.
5. When current passes through a conductor, magnetic flux lines will form around the conductor in concentric circles.
6. The strength of the magnetic field around a conductor will be proportional to the amount of current flowing through it until the amount of current reaches the saturation point of the electromagnet.
7. The magnetic field of an electromagnet can be turned on and off by interrupting the current flow through the coil of wire.
8. The polarity of an electromagnetic coil is determined by the direction of current flow through the conductor.
9. If the conductor is wrapped in coils, the strength of the magnetic field is increased.
10. When more coils of wire are added to an electromagnet, the magnetic field strength is increased.
11. A core will cause the field of an electromagnet to be strengthened.
12. The core must be made from laminated steel when AC voltage is used to power the coil so that the magnetic field can quickly build and collapse in the core as the AC voltage changes polarity every 1/60 of a second.

Alternating current (AC) is identified as a sine wave that shows the electron flow in one direction for half of the cycle and in the reverse direction for the second half of the cycle. Since the sine wave is produced by a generator shaft that rotates in a circular motion, the points on the sine wave are identified in values from 0° to 360°. The sine wave peaks positive at 90° and peaks negative at 270°.

Frequency is defined as the number of cycles (sine waves) that occur in 1 second. Period is the inverse of frequency and is defined as the amount of time it takes for one cycle to occur.

Peak voltage is defined as the amount of voltage measured from the zero point of the sine wave to the peak of the sine wave. Peak-to-peak voltage is defined as two times the peak voltage. Generally, the peak voltage must be measured with an oscilloscope. Rms (root-mean-square) voltage is the working voltage that is measured with a voltmeter. The value of rms voltage can be calculated by the formula $0.707 \times V_P$. You can find the peak voltage by multiplying $1.414 \times$ rms value.

When you are on the job and need to install a new air conditioner or furnace, you will find the source of AC voltage in a residential or commercial location in a circuit breaker box called the entrance panel. If you are making a service call, you will probably find the source of voltage for the system at a disconnect switch box that may be a fused disconnect.

You can calculate voltage, current, and impedance in an AC circuit using Ohm's law. The difference between a DC circuit and an AC circuit is that an AC circuit contains elements in addition to resistance that create an opposition to current flow. For example, capacitors create capacitive reactance, and inductors create inductive reactance. Total opposition in an AC circuit is called impedance, which is a combination of resistance, capacitive reactance, and inductive reactance. Sections 7.15 through 7.19 give the formulas for these calculations.

Power in an AC circuit is measured in watts. The power that is produced from only resistance loads is called true power (TP). Power due to capacitive reactance and inductive reactance is called apparent power. The power factor is the ratio of true power to apparent power.

Power distribution it is a system of wires and components that supply power from an electrical power pole to a residence and can be used to power an air conditioner. Residential power is generally 240 volts line to line and 120 volts line to neutral. It is important to have a proper grounding system with a grounding rod installed at the base of the electrical meter where the entrance wires come into the house.

Job Assignment for This Chapter

You are working on a job at a residence where the service call complaint is that the air conditioner will not run. After making several checks, you have determined that one of the wires in the fused disconnect has been severely burned and is not passing voltage to the air-conditioning unit. The homeowner is asking you questions about how power gets from the transformer on the power pole in front of his home to the point where it is supplied to the air conditioner. The homeowner also wants to know the reading of the voltage and its frequency that is supplied to his air conditioner. Your job is to draw a sketch of how power gets from the transformer on the power pole to the air conditioner. Be sure to include the meter set, the circuit

breaker panel, the disconnect for the air conditioner, and the power terminals on the air conditioner.

Solution to the Job Assignment

Your solution should include a drawing of the voltage coming from the transformer at the power pole to the customer's service entrance. From the entrance your diagram should show the voltage going to the circuit breaker panel and on to the disconnect for the outdoor unit for the air conditioner. You should show that the voltage at the transformer at the power pole is stepped down to 230 or 240 single-phase voltage (two wires; L1 and L2 with a third wire for neutral). In the United States the frequency is 60 Hz.

Assessment

QUESTIONS

1. Draw a diagram of an AC sine wave and identify its peak voltage and peak-to-peak voltage levels.
2. Explain the term *frequency* as it refers to AC electricity.
3. Discuss the difference between a load center and a fused disconnect.
4. Explain how you would test for voltage in a fusible disconnect on the line-side and load-side terminals.
5. Trace the path of a generated AC voltage from the generating station where it is produced to the point where you would connect an air conditioner at a home in a residential area. Be sure to identify all the important parts of the system between the generation station and the residence.
6. Explain the difference between a permanent magnet and an electromagnet.
7. Define the term *saturation* as it applies to an electromagnet.
8. Explain the principle of electromagnetic induction.
9. Identify two ways to increase the strength of an electromagnet.
10. Explain why it is important that the magnetic field of an electromagnet can be turned on and off.

TRUE OR FALSE

1. Impedance is the total opposition to an AC circuit caused by inductive reactance, capacitive reactance, and resistance.
2. The units of impedance are ohms.
3. A main circuit breaker panel (load center) is the point where incoming voltage is connected in a residence and where circuit breakers are mounted.
4. Rms voltage is the voltage measured by a VOM meter.
5. If the rms voltage in an AC circuit is 110 V, the peak voltage will also be 110 V.
6. When dipoles are aligned in a material, it is a good magnet.

7. Whenever current flows through a conductor, magnetic flux lines are created around the conductor.

8. The polarity of the magnetic field of an electromagnet is determined by the direction of current flow through the conductor.

9. Like poles of magnets attract.

10. The strength of the magnetic field of a permanent magnet can be changed easily.

MULTIPLE CHOICE

1. If the size of a capacitor in an AC circuit increases from 10 μF to 20 μF and the frequency stays the same, the capacitive reactance will
 a. increase.
 b. decrease.
 c. remain the same.

2. When a capacitor is added to an AC circuit, the voltage waveform will _____ the current waveform, which provides additional torque to a motor if it is connected to the circuit.
 a. lag
 b. lead
 c. remain in phase with

3. If a circuit has two large refrigeration or air-conditioning compressor motors and has a low power factor, the power factor can be raised by _____ to make the true power equal to the apparent power, which will avoid a penalty on the electric bill.
 a. adding capacitance
 b. removing capacitance
 c. adding inductance

4. The frequency of AC voltage is defined as
 a. the period of the voltage.
 b. the number of cycles per second.
 c. the peak voltage of the voltage.

5. The number of cycles per second in an AC voltage is called the _____ of the voltage.
 a. frequency
 b. period
 c. impedance

6. The strength of the field of _____ can be changed easily.
 a. a permanent magnet
 b. an electromagnet
 c. a dipole

7. The polarity of an electromagnet can be changed by
 a. changing the direction of current flow through the coil of wire.
 b. changing the amount of current flow through the coil of wire.
 c. changing the frequency of the current flow through the coil of wire.

8. A laminated steel core is used in electromagnets that have AC voltage applied to them because
 a. laminated steel is more economical to use than soft iron.
 b. laminated steel is easily formed to any shape, which makes it more usable in complex components.
 c. laminated steel allows the magnetic field to build and collapse quickly.

9. A magnetic field is created in a coil of wire when
 a. current flows through the wire.
 b. current stops flowing through the wire.
 c. the coil has a core.

10. The left-hand rule is used to
 a. determine the amount of magnetic flux in a coil.
 b. determine the number of coils in an electromagnet.
 c. determine the polarity of the magnetic field in an electromagnet.

PROBLEMS

1. Calculate the rms voltage for an AC sine wave that has 80 V peak.
2. Calculate the peak voltage for an AC sine wave that has 208 V rms.
3. Calculate the period of a 60-cycle AC sine wave.
4. Calculate the frequency of an AC sine wave that has a period of 16 ms.
5. Calculate the impedance of an AC circuit that has 40 Ω of resistance and 30 Ω of capacitive reactance.
6. Draw a sketch of a permanent magnet and show where flux lines occur.
7. Draw a sketch of an electromagnetic coil and show where flux lines occur.
8. Draw a sketch of an electromagnetic coil with an open somewhere in the coil of wire. Explain what effect the open will have on the magnetic field.
9. Draw two permanent magnets with their poles placed so the two magnets will attract each other.
10. Perform the experiment presented in Section 7.2 and explain why the flux lines are concentrated around the ends of the permanent magnet.

Omit

Don't do

Websites

Websites change constantly. Use the key word search to locate sites that have similar content. The products that will be found at each site are indicated with the key word.

Keyword: capacitive reactance, inductive reactance

Disconnect switches
Products: Disconnect switches
Key word: Disconnect

Acme Electric Acme Transformer Division
Products: Transformers
Key word: Acme Electric

Three-phase and single-phase transformers
Products: Power transformers
Key word: Three-phase transformer

Transformer formulas
Products: Free transformer training course on line
Key word: Transformer formulas

Chapter 8

Installing HVAC and Refrigeration Systems, Transformers, and Power Distribution

Job Assignment for This Chapter

You are working on a residential air-conditioning system and you suspect the control transformer is bad. The homeowner is not sure that the transformer is bad. Your job assignment is to explain the basic parts of a transformer and explain to the homeowner what tests you will use to show whether the transformer is good or bad. You should also sketch a diagram to show the homeowner how the transformer is connected in the circuit and how it changes 240 volts to 24 volts. If you must replace the transformer, you should explain what ratings on the transformer should match exactly and which ones can be larger or smaller.

Terms You Will Learn in This Chapter

Control transformer
Delta-connected transformer
Fusible disconnect (fused disconnect)
Neutral
Primary winding
Secondary winding
Sine wave
Single phase

Step-down transformer
Step-up transformer
Three-phase voltage
Transformer
Turns ratio
VA rating for transformer
Wye-connected transformer

After reading this chapter, you will be able to:

1. Explain the operation of a transformer.
2. Discuss the difference between step-up and step-down transformers.
3. Explain how 480, 240, and 120 V AC are developed from a delta-connected transformer.
4. Explain how 440, 208, and 125 V AC are developed from a wye-connected transformer.

8.0 OVERVIEW OF INSTALLING HVAC AND REFRIGERATION SYSTEMS

One of the jobs you will be involved in is the installation of air-conditioning, heating, or refrigeration systems. You will be responsible for locating a source of power, which is generally the circuit breaker panel in the residence or commercial establishment. The HVAC technician is responsible for reading the data plate on the equipment and identifying the voltage and current requirement for the system. From the current rating on the data plate, you can determine the circuit breaker size, wire size, and fuse size for the system. You will be responsible for installing the electrical power, circuit breakers, and fused disconnect for the system. You will also need to run the 24-volt thermostat lines. The information in this chapter will help you learn how to install the electrical power for these systems.

8.1 WIRE SIZING AND VOLTAGE DROPS

One part of the power distribution system that is involved at all points in the system is wiring, which is also called the conductor. Conductors can be made of aluminum or copper wire. Aluminum is generally used for long-distance distribution because of its lighter weight. Once the power is inside the plant, copper conductors are generally used for distribution. Solid copper conductors are used in bus bar and busway applications, and stranded conductors are used for all other applications.

The conductors are sized by the amount of amperage they can carry. A table of typical conductor sizes and their ampacities is provided in Figure 8–1. The standard used to determine the size of each conductor is called the American Wire Gauge (AWG). These tables have been established by the National Electrical Code (NEC), and they are used to select the proper size of conductor to carry the load. Conductors are also classified by the type of insulation that is used as a cover. The type of cover also determines the voltage rating of the wire. Typical voltage ratings for conductors used in motor control applications include 300, 600, and 1000 V. The voltage and current ratings of the conductor should not be exceeded under any circumstances.

The type of covering for wiring insulation is listed in Figure 8–2. Abbreviations for each of these coverings are listed in the NEC tables and help determine the type of wire that should be selected for each application. The outer covering of the wire serves several purposes, including protecting the conductors from coming in contact with metal in the cabinet or other conductors. The cover also provides a location to stamp all specification data regarding the wire, including the voltage rating, AWG size, temperature specification, and type of covering.

Table 310.16 Allowable Ampacities of Insulated Conductors Rated 0 Through 2000 Volts, 60°C Through 90°C (140°F Through 194°F), Not More Than Three Current-Carrying Conductors in Raceway, Cable, or Earth (Directly Buried), Based on Ambient Temperature of 30°C (86°F)

Handwritten note: 14 AWG = 15A
12 AWG = 20A
10 AWG = 30A

Size AWG or kcmil	Temperature Rating of Conductor (See Table 310.13.)						Size AWG or kcmil
	60°C (140°F)	75°C (167°F)	90°C (194°F)	60°C (140°F)	75°C (167°F)	90°C (194°F)	
	Types TW, UF	Types RHW, THHW, THW, THWN, XHHW, USE, ZW	Types TBS, SA, SIS, FEP, FEPB, MI, RHH, RHW-2, THHN, THHW, THW-2, THWN-2, USE-2, XHH, XHHW, XHHW-2, ZW-2	Types TW, UF	Types RHW, THHW, THW, THWN, XHHW, USE	Types TBS, SA, SIS, THHN, THHW, THW-2, THWN-2, RHH, RHW-2, USE-2, XHH, XHHW, XHHW-2, ZW-2	
	COPPER			ALUMINUM OR COPPER-CLAD ALUMINUM			
18	—	—	14	—	—	—	—
16	—	—	18	—	—	—	—
14*	20	20	25	—	—	—	—
12*	25	25	30	20	20	25	12*
10*	30	35	40	25	30	35	10*
8	40	50	55	30	40	45	8
6	55	65	75	40	50	60	6
4	70	85	95	55	65	75	4
3	85	100	110	65	75	85	3
2	95	115	130	75	90	100	2
1	110	130	150	85	100	115	1
1/0	125	150	170	100	120	135	1/0
2/0	145	175	195	115	135	150	2/0
3/0	165	200	225	130	155	175	3/0
4/0	195	230	260	150	180	205	4/0
250	215	255	290	170	205	230	250
300	240	285	320	190	230	255	300
350	260	310	350	210	250	280	350
400	280	335	380	225	270	305	400
500	320	380	430	260	310	350	500
600	355	420	475	285	340	385	600
700	385	460	520	310	375	420	700
750	400	475	535	320	385	435	750
800	410	490	555	330	395	450	800
900	435	520	585	355	425	480	900
1000	455	545	615	375	445	500	1000
1250	495	590	665	405	485	545	1250
1500	520	625	705	435	520	585	1500
1750	545	650	735	455	545	615	1750
2000	560	665	750	470	560	630	2000

CORRECTION FACTORS

Ambient Temp. (°C)	For ambient temperatures other than 30°C (86°F), multiply the allowable ampacities shown above by the appropriate factor shown below.						Ambient Temp. (°F)
21–25	1.08	1.05	1.04	1.08	1.05	1.04	70–77
26–30	1.00	1.00	1.00	1.00	1.00	1.00	78–86
31–35	0.91	0.94	0.96	0.91	0.94	0.96	87–95
36–40	0.82	0.88	0.91	0.82	0.88	0.91	96–104
41–45	0.71	0.82	0.87	0.71	0.82	0.87	105–113
46–50	0.58	0.75	0.82	0.58	0.75	0.82	114–122
51–55	0.41	0.67	0.76	0.41	0.67	0.76	123–131
56–60	—	0.58	0.71	—	0.58	0.71	132–140
61–70	—	0.33	0.58	—	0.33	0.58	141–158
71–80	—	—	0.41	—	—	0.41	159–176

* See 240.4(D).

Figure 8–1 American Wire Gauge (AWG) wire sizing table provided by the National Electrical Code (NEC). (National Electrical Code® and NEC® are registered trademarks of the National Fire Protection Association, Quincy, MA 02269.) *Reprinted with Permission from NFPA 70–2005. National Electrical Code®, Copyright © 2004, National Fire Protection Association, Quincy, MA, 02169. This reprinted material is not the complete and official position of the NFPA on the referenced subject which is represented only by the standard in its entirely.*

8.1.1 Voltage Drops

When current flows through a wire, a small drop in voltage will occur for every foot of length in the wire. This means that if you use a 50-foot length of wire, the voltage drop will be larger than if you use a 10-foot length of the same wire. The NEC allows for no more

Insulation Type	Application	Type of Insulation
THHN	Installed in dry or wet locations	Thermoplastic
TW	Residential, commercial, and industrial buildings for general-purpose lighting, appliance, power, control, and relay panel applications	Thermoplastic
THWN	Installed in dry or wet locations	Thermoplastic
THW	General-purpose lighting, power, control, and relay panels, wet or dry locations	Thermoplastic
RH		
THHW		
UF		
MTW (Machine Tool Wire)	Machine tool, internal appliance wiring and control circuits	PVC
TFN		Thermoplastic
TFFN		Thermoplastic
XHHW		Thermoset
USE-2		Thermoset
RHH		Thermoset
RHW-2		Thermoset
CT use		PVC
Notes		
T = Thermoplastic N = nylon 2 = 90 degree C rating		
H = heat resistant F = Flexible		

Figure 8–2 Wiring coverings (insulation types) for common wires used in HVAC.

than 2% drop in voltage in any wiring application. When you are installing wiring for an HVAC application, you can use a table from the NEC to calculate the voltage drop for any given wire size and length of wire. One way to ensure that you do not have excessive voltage drop is to turn on the loads in your system, measure the voltage at the disconnect or circuit breaker where the power goes into the wire, measure the voltage at the end of the wire where it is attached to the load, such as the compressor contactor, and then compare your two readings. You can calculate the amount of voltage drop by dividing the voltage that you measured at the load by the voltage that you measured at the supply. The difference should not be more than 2%. If the voltage drop is more than 2%, you will need to increase the wire ampacity size (select a lower number wire gage), which will increase the diameter of the conductor and result in a smaller voltage drop. Typically you will not calculate voltage drops in the field; rather you will use a table that indicates that you should increase the wire ampacity one size for every 20 additional feet of wiring that is used. For example, if your HVAC unit indicates a 20-amp FLA rating on its name plate, you could use number 14 TW wire to supply this circuit because 14 TW is rated for 25 amps. If the distance between the service panel where the circuit breakers are located is more than 20 feet away from where you are mounting the outdoor unit, you would be safe to increase the wire size to a number 12 TW, which would provide a rating for up to 30 amps. The larger wire size will decrease the amount of voltage drop over the longer distance and the wire will not overheat.

Notice!
You should always measure the voltage at the unit and compare it to the voltage at the circuit breaker panel with all loads running to ensure that there is not an excessive voltage drop when you install a system for the first time, or if you are making a service call. You should also measure the full load amperage of all wires at the point where the wires are connected to the outdoor unit when all your loads are running to make sure the amperage is within the rating for the wires. You can also check for any wires that are overheating or any signs of overheating. Be sure that you do not touch or come into contact with any bare wires or components in the system when you are making this test and the measurements of voltage and current.

8.2 OVERVIEW OF TRANSFORMERS

Transformers are needed to step up or step down voltage levels. As you know, power losses are calculated with I^2R, which means the higher the current, the larger the line losses. This means that when voltage is generated, it needs to be stepped up to several hundred thousand volts so its current will be lower when it is transmitted. Since the current is lower, the size of wire used to transmit the power can be smaller, so the power can be transmitted over longer distances. When the voltage arrives at a city, it needs to be stepped down to approximately 40,000 V so that it is less dangerous. This level of voltage is sufficient to transmit power throughout a city. When the voltage arrives at a residence, it must be further stepped down to 240 V and 120 V for all the power circuits in the home. Transformers provide a means to step up or step down this voltage. **Step-up transformers** are used in oil furnaces to increase the line voltage (120 V AC) to several thousand volts to send to the igniter to cause the spark for ignition.

A **step-down transformer** is used in heating and air-conditioning equipment to step down 120 V AC or 240 V AC to 24 V. The 24 V AC is used as control voltage for a residential heating or air-conditioning system so that higher voltages are not used in thermostats or other controls that people touch when they are setting them. The lower voltage provides a degree of safety against electrical shock and also allows these controls to last longer. This type of transformer is called a **control transformer**. Examples of a foot-mounted transformer and a flange-mounted transformer are shown in Figures 8–3a and 8–3b.

As a technician you will encounter different voltages such as 480, 240, 208, 120, and 24 V AC while you are working on air-conditioning, heating, or refrigeration equipment. The equipment you are working with will require a specific amount of single-phase or three-phase voltage, and you will be responsible for making the proper connections to provide the correct voltage at the disconnect or load center where supply voltage is provided. When the proper amount of voltage is not present, you will need to know which transformers are required to step the voltage up or down to the proper level. This chapter will explain how transformers operate and how the various levels of voltage are derived from connecting transformers to provide the proper amount of three-phase and single-phase voltage. The chapter will also explain how to take voltage readings at the transformer and disconnect to troubleshoot the loss of a phase.

Figure 8–3 a A foot-mounted control transformer that is used in residential heating and air-conditioning equipment. **b** A flange-mounted control transformer. *(Courtesy of Honeywell)*

(a)

(b)

8.3 OPERATION OF A TRANSFORMER AND BASIC MAGNETIC THEORY

The transformer consists of two windings (coils of wire) that are wrapped around a laminated steel core. The winding where voltage is supplied to the transformer is called the **primary winding**. Line 1 (L1) and neutral are connected to the primary terminals. The winding where voltages come out of the transformer is called the **secondary winding**. The terminals for the secondary winding are identified by the letters R and C. The letter R stands for *red*, which is the color of wire that is traditionally connected to it, and the letter C stands for *common*. Figure 8–4 shows a diagram of a typical transformer with the primary coil, secondary coil, and the windings identified.

A transformer works on the principle called *induction*. Induction occurs when current flows through the primary winding and creates a magnetic field. The magnetic field produces flux lines that emanate from the wire in the coil as current is flowing through it. When this current is interrupted or stopped, the flux lines collapse, and the action of the collapsing flux lines causes them to pass through the winding of the secondary coil, which is placed adjacent to the primary coil. Figure 8–5 shows the transformer operation in four stages as voltage builds up in the transformer and the flux lines building and collapsing at each point as the sine wave flows through the primary coil. Recall that the AC sine wave continually starts at zero volts and increases to a peak value and then returns to 0 V and repeats the waveform in the negative direction. The process of increasing to peak and returning to zero creates flux lines in the wire as current passes through it and then allows the flux lines to collapse when the voltage returns to zero. Since the AC voltage follows this pattern naturally 60 times a second, it produces the perfect type of voltage for operating the transformer.

When the AC voltage returns to zero during each half-cycle, the flux lines that were created in the primary winding begin to collapse and start to cross the wire that forms the secondary coil of the transformer. When the flux lines from the primary winding collapse and move across the wire in the coil of the secondary winding, the electrons in the secondary winding begin to move, which creates a current. Since the current in the secondary winding begins to flow without any physical connection to the primary winding, the current in the secondary winding is called an *induced current*.

The following provides a more technical explanation of the relationship between the AC voltage and the windings of the transformer. The magnetic field in the primary winding of the transformer builds up when AC voltage is applied during the first half-cycle of the sine wave (0° to 180°). When the sine wave voltage reaches its peak at 90°, the voltage has peaked in the positive direction and it begins to return to 0 V by moving from the 90° point to the 180° point. When the voltage reaches the

Note: Primary wires are identified as L1 and N for 120 V AC.
Primary wires are identified as L1 and L2 for 240 or 208 V AC.

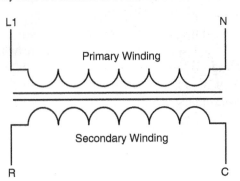

Figure 8–4 Diagram of a typical control transformer. The coil where incoming voltage is applied is called the primary, and the other coil is called the secondary.

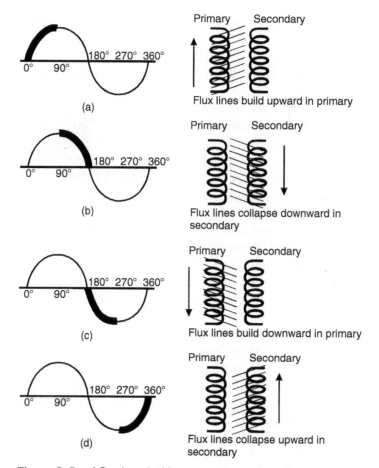

Figure 8–5 **a** AC voltage builds to a positive peak as the sine wave moves from 0° to 90°. Flux lines are shown building in the primary coil during this time. **b** AC voltage drops off from peak at 90° to 180°. During this time, the flux lines in the primary coil begin to collapse and cut across the coils of the secondary. **c** AC voltage builds to a negative peak as the sine wave moves from 180° to 270° and the flux lines build in the primary coil again. **d** When voltage decreases from the negative peak back to zero as the sine wave moves from 270° to 360°, the flux lines collapse.

180° point, the voltage is at 0 V and the current flow is interrupted. When the sine wave voltage is 0 V, the flux lines that have been built up in the primary winding collapse and cross the secondary coils, which creates a current flow in the secondary winding.

The sine wave continues from 180° to 360° and the transformer winding is energized with the negative half-cycle of the sine wave. When the sine wave is between 180° and 270°, the flux lines build again, and when the sine wave reaches the 360° point, the sine wave returns to 0 V, which again interrupts current and causes flux lines to cross the secondary winding. This means the transformer primary energizes a magnetic field and collapses it once in the positive direction and again in the negative direction during each sine wave. Since the sine wave in the secondary is not created until the voltage in the primary moves from 0° to 180°, the sine wave in the secondary is *out of phase by 180°* with the sine wave in the primary that created it. The voltage in the secondary winding is called *induced* voltage because it is created by induction.

The induced voltage is developed even though the primary coil of the transformer does not make electrical contact with the secondary coil in any way. This means that no electrons flow between the primary and secondary windings.

Figure 8–6 A control transformer connected to T1 and N in a disconnect to test the transformer. The primary voltage will be 120 V AC, and the secondary voltage will be 24 V AC.

Figure 8–7 A control transformer connected to T1 and T2 terminals in a fusible disconnect. The primary voltage is 240 V AC, and the secondary is 24 V AC.

8.4 CONNECTING A TRANSFORMER TO A DISCONNECT FOR TESTING

An easy way to test a control transformer is to connect it directly to a disconnect and apply power to the transformer primary circuit. When voltage is applied to the primary circuit, voltage should be available at the secondary. The amount of voltage at the secondary will depend on the rating of the transformer. Figure 8–6 shows a transformer connected to T1 and N, which is the source for 120 V at the bottom terminals of the disconnect. The secondary voltage available at this transformer at terminals R and C will be 24 V. Figure 8–7 shows a similar transformer connected to terminals T1 and T2 in a disconnect to provide 240 V to the primary circuit. Since this transformer is rated as 240/24, the secondary voltage will also be 24 V.

8.5 TRANSFORMER RATED FOR 240 V AND 208 V PRIMARY

If the air-conditioning equipment is used in a residential application, the supply voltage is normally 240 V, and if the equipment is used in a commercial application, the supply voltage may be 240 or 208 V depending on the transformer connections that are used by the utility company that supplies the power. Since the equipment manufacturer does not know where its equipment will be used, it may provide a control transformer that can be

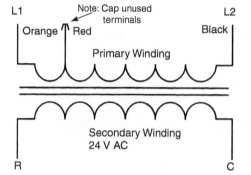

Figure 8–8 A control transformer that is made to operate with either 208 V AC or 240 V AC primary voltage and provide 24 V AC secondary volts.

wired to either 240 V or 208 V. This is accomplished by providing a second connection that is *tapped* in the primary winding. Figure 8–8 shows an example of this type of control transformer.

8.6 TRANSFORMER VOLTAGE, CURRENT, AND TURNS RATIOS

The amount of voltage a transformer will produce at its secondary winding for a given amount of voltage supplied to its primary is determined by the ratio of the number of turns in the primary winding compared with the number of turns in the secondary. This ratio is called the **turns ratio**. The amount of primary current and secondary current in a transformer is also dependent on the turns ratio.

Figure 8–9 shows a diagram that indicates the primary voltage (E_p), the primary current (I_p), the number of turns in the primary winding (T_p), the secondary voltage (E_s), the secondary current (I_s), and the number of turns in the secondary winding (T_s).

The primary and secondary voltage, primary and secondary current, and the turns ratio can all be calculated with the following formulas. The turns ratio for a transformer is calculated from the following formula:

$$\text{turns ratio} = \frac{T_p}{T_s} \text{ or } \frac{E_p}{E_s} \text{ or } \frac{I_s}{I_p}$$

The ratio of primary voltage, secondary voltage, primary turns, and secondary turns is

$$\frac{E_p}{E_s} = \frac{T_p}{T_s}$$

Figure 8–9 Diagram of a transformer that shows the primary voltage (E_p), primary current (I_p), and the number of turns in the primary winding (T_p). This diagram also shows the secondary voltage (E_s), secondary current (I_s), and the number of turns in the secondary winding (T_s).

E_p = 120 V
I_p = 2 A
T_p = 500
Primary

Secondary
E_s = 24 V
I_s = 10 A
T_s = 100

From this ratio you can calculate the secondary voltage with the formula

$$\frac{E_p}{E_s} = \frac{T_p}{T_s}$$

$$E_s = \frac{E_p \times T_s}{T_p}$$

The ratio of primary voltage, secondary voltage, primary current, and secondary current is

$$\frac{E_p}{E_s} = \frac{I_s}{I_p}$$

(Notice that the ratio of voltage to current is an inverse ratio.)
From this ratio you can calculate the secondary current with the formula

$$\frac{E_p}{E_s} = \frac{I_s}{I_p}$$

$$I_s = \frac{E_p \times I_p}{E_s}$$

For example, if you were provided the information that the primary voltage was 120 V, the primary current was 2 A, the primary turns were 500, and the secondary turns were 100, you could easily calculate the secondary voltage and the secondary current:

$$E_s = \frac{E_p \times T_s}{T_p} \qquad \frac{120 \text{ V} \times 100 \text{ turns}}{500 \text{ turns}} = 24 \text{ V}$$

$$I_s = \frac{E_p \times I_p}{E_s} \qquad \frac{120 \text{ V} \times 2 \text{ A}}{24 \text{ V}} = 10 \text{ A}$$

| **Example 8–1** | Calculate the secondary current and secondary turns of the transformer shown in Figure 8–10. |

| **Solution:** | Find the turns ratio: |

$$\frac{E_p}{E_s} = \frac{240 \text{ V}}{24 \text{ V}} = 10:1$$

Since the number of turns in the primary is 500, the number of turns in the secondary will be 50.
The secondary current can be calculated from the formula

$$I_s = \frac{E_p \times I_p}{E_s} = \frac{240 \text{ V} \times 2 \text{ A}}{24 \text{ V}} = 20 \text{ A}$$

8.7 STEP-UP AND STEP-DOWN TRANSFORMERS

If the secondary voltage is larger than the primary voltage, the transformer is known as a *step-up transformer*. If the secondary voltage is smaller than the primary voltage, the transformer is known as a *step-down transformer*. The control transformer that was presented in the previous sections is an example of a step-down transformer because the secondary voltage is smaller than the primary voltage. Step-up transformers are used to boost the secondary voltage, which also has the effect of lowering the secondary current so that more power can be transferred on smaller-size wire. An application in heating and air conditioning where a step-up transformer is used is in the ignition system of an

oil furnace. The oil furnace uses an ignition transformer that is a step-up transformer to increase the 120 V AC that is used to power the furnace to approximately 16,000 V. The high voltage is supplied to the igniter points so that the high voltage can jump the gap between the ignition points to create a spark that will ignite the fuel oil as it comes out of the nozzle.

Notice!

Do not attempt to read the voltage from an ignition transformer with a VOM because the voltage will be in excess of 10,000 V.

8.8 VA RATINGS FOR TRANSFORMERS

The **VA (volt-ampere) rating for a transformer** is calculated by ~~multiplying the primary voltage and the primary current, or~~ the secondary voltage and the secondary current. In the example in Figure 8–10 you can see that the primary voltage is 240 V and the primary current is 2 A. The VA rating for this transformer is 480 VA. The VA rating indicates how much power the transformer can provide. As a technician you can say that the primary VA is equal to the secondary VA. If you were a circuit designer, you would need to be more precise and you would find that the secondary VA is slightly less than the primary VA because of transformer losses. It is important that the VA rating of the transformer is large enough for the application. If the VA rating is too small, the transformer will be damaged and fail prematurely. Whenever you need to replace a transformer you must be sure that the voltage ratings match and that the VA size of the replacement transformer is equal to or larger than the original transformer. Do not replace a transformer with one that has a smaller VA rating than the original.

8.9 THE 120-V AC CONTROL TRANSFORMER

Another type of control transformer provides 120 V AC in the secondary for use in commercial heating, air-conditioning, and refrigeration systems. You should remember that utility companies in different areas of the country provide voltage that is between 110 and 120 V. This means that one part of the country may refer to their lower voltage as 110, whereas another part calls their voltage 115 or 120 V. For this section the voltage will be referred to as 120 V AC. The 120 V AC secondary voltage is necessary because some of the control devices require 120 V AC, and the larger loads in the system such as a compressor and fan motors require 240 V AC. The control transformer allows the compressor and fan motors to be supplied with 240 V while it provides the control circuit with 120 V AC.

$E_p = 240$ V
$I_p = 2$ A
$T_p = 500$
Primary

Secondary
$E_s = 24$ V
$I_s =$
$T_s =$

R C

Figure 8–10 Diagram for Example 8–1.

Figure 8–11 Diagram of a control transformer whose secondary voltage is 120 V. The primary winding of this transformer is identified as H1, H2, H3, and H4. The secondary winding of this transformer is identified as X1 and X2.

The operating principle of the 240/120 V AC type of control transformer is similar to that of the 120/24 V AC transformer. The diagram for the former is provided in Figure 8–11. The primary windings of this transformer are identified with the letter H, and the secondary windings are identified with the letter X. The primary winding for this transformer is constructed in two equal sections (windings). The terminals of the first winding are identified as H1 and H2. The terminals of the second winding are identified as H3 and H4. The primary winding is divided into two separate windings so that the transformer primary side can be powered with 480 V or 240 V. If these two sections are connected in series with each other, the primary side of the transformer will be powered with 480 V, and if the two sections are connected in parallel with each other, the transformer will be powered with 240 V.

8.10 WIRING THE CONTROL TRANSFORMER FOR 480 V AC PRIMARY VOLTS

The control transformer can be powered with 480 V AC by connecting its two primary windings in series. Figure 8–12 shows three diagrams of the control transformer. The first diagram, Figure 8–12a, shows the transformer as you would see it on a wiring diagram without any jumpers connected. You should notice that the H2 and H3 terminals are positioned so that they can be connected either in series or in parallel. In this application the jumper is connected between terminals H2 and H3 to connect them in series. Figure 8–12b shows the physical location of the jumper when it is connected across these terminals. Figure 8–12c shows the equivalent electrical diagram as you may see it in the electrical diagram of the air-conditioning unit. The reason the two windings are connected in series is because each winding is rated for 240 V, and the primary voltage is 480 V AC. Since the windings are connected in series, each of them will receive 240 V of the total 480 V.

8.11 WIRING THE CONTROL TRANSFORMER FOR 240 V AC PRIMARY VOLTS

The control transformer can also be connected to operate with 240 V AC primary volts and produce 120 V AC on the secondary terminals. Figure 8–13 shows a set of three diagrams for the same control transformer that is connected for 240 V AC primary. Figure 8–13a shows the electrical diagram of the control transformer as you would see it in a circuit diagram without any jumpers. The reason this diagram is shown is that some equipment manufacturers will show the control transformer in the electrical diagram for the air-conditioning system, since they are not sure whether the system voltage will be 480 or 240. This diagram indicates that the control transformer is to be connected by the technician when the unit is installed and the supply voltage can be verified. Figure 8–13b shows the diagram with two jumpers in place to connect the two

Figure 8–12 a Electrical diagram of the control transformer. Notice that the H2 and H3 terminals are located so that a jumper can be placed on them to connect them in series or in parallel. **b** Diagram showing the jumper in place connecting the two primary windings in series. **c** The equivalent electrical diagram showing the two windings connected in series.

Figure 8–13 a Electrical diagram of control transformer. **b** Diagram showing the jumpers in place for the transformer to have 240 V AC primary. **c** Equivalent diagram showing the two primary windings connected in parallel.

primary windings in parallel with each other. This diagram will help you to connect the jumpers to the correct terminals so that the windings are connected in parallel. Figure 8–13c shows the equivalent circuit of the two windings in parallel as you would see them in the system electrical diagram provided by the equipment manufacturer. Sometimes it is difficult to see that the connections of the two jumpers actually connect the two windings in parallel, so the diagram in Figure 8–13c shows how the windings look when they are in parallel.

As a technician you will be responsible for making the proper jumper connections based on the primary voltage that the system will have. If the jumpers on the transformer are connected for 480 V AC primary when the unit is shipped, and the unit is connected to 240 V, you will need to make the changes in the jumper so that the secondary winding of the transformer provides exactly 120 V AC.

8.12 TROUBLESHOOTING A TRANSFORMER

As a technician you will need to troubleshoot a variety of transformers. You must be sure that the transformer you are testing is not connected to power when you test each of the windings with an ohmmeter for continuity. This is called isolating the windings. Each of the coils should have some amount of resistance that reflects the resistance of the wire in each coil. If you measure infinite resistance (∞), it indicates that the winding has an open and the transformer must be replaced.

Another way that you can test a transformer is by applying power to the primary winding. It is important that you provide the continuity test first to detect any shorts before you apply power. If the transformer windings are not shorted, you can apply any amount of AC voltage that is equal to or less than the primary rating of the transformer. If the transformer is operational, a voltage will be present at the secondary terminals of the transformer. If no voltage is present at the secondary, check all the connections to ensure they are correct. If no voltage is present at the secondary, the transformer is defective and should be replaced. It is important to remember that the transformer consists of a primary winding and a secondary winding that are placed in close proximity to each other, and when AC voltage is applied to one winding, induction will cause voltage to be available at the other winding.

The transformer may have the problem that its secondary voltage is higher or lower than its rating. If this occurs, it is possible that the amount of primary voltage is incorrect or that the transformer jumpers are not connected properly. For example, if the primary voltage is 208 V instead of 240 V, the secondary voltage will be less than 120 V, which will cause a problem. Be sure that the secondary is the proper amount before you use the transformer in a circuit.

8.13 NATURE OF THREE-PHASE VOLTAGE

Air-conditioning and refrigeration equipment installed in commercial and industrial buildings may require three-phase AC voltage instead of **single-phase voltage**. **Three-phase voltage** is usually indicated on equipment data plates by the symbol 3ϕ voltage. Three-phase voltage is generated as three separate AC **sine waves**. Figure 8–14 shows an example of the three sine waves. From this figure you can see that the sine waves are identified as Phase A, Phase B, and Phase C. Each of the phases is separated by 120° from one another. The 120° phase separation is caused by the generator windings being 120° out of phase with one another. Three-phase voltage is generated and transmitted because it is more efficient to produce than single-phase, since the generator can have three separate windings.

Figure 8–14 Sine waves for three-phase electricity. Notice that Phase A is 120° out of phase with Phase B, and Phase B is 120° out of phase with Phase C.

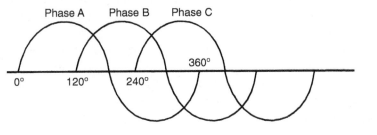

WHY THREE-PHASE VOLTAGE IS GENERATED

Three-phase voltage is used more often than single-phase voltage for larger commercial air-conditioning and refrigeration systems because it provides more power in smaller components than single-phase voltage for an equal-size system. For example, if a 10-hp (horsepower) motor is required as the air-conditioning compressor, a single-phase motor would need to be physically larger than the 10-hp three-phase motor because the latter has three sets of windings and receives three equal sources of voltage (L1, L2, and L3), whereas a single-phase motor receives only two sources of power. Since the power is shared by three circuits instead of two, the wire sizes, fuse sizes, and switch sizes are also smaller for a three-phase system.

For example, if a 10-hp motor is connected to a single-phase 240-V power source, each line (L1 and L2) will need to be capable of providing 50 A. A three-phase 10-hp motor connected to a three-phase power distribution will need only 28 A from each wire. This means that the three-phase system can use much smaller wire, which is lighter and less expensive.

THREE-PHASE TRANSFORMERS

Figure 8–15 shows examples of three-phase transformers you will encounter on the job. The transformer may be mounted near the air-conditioning and refrigeration equipment, or it may be located in a transformer vault (a special room where transformers are mounted). In some cases the transformers are mounted on a utility pole just outside the commercial site. Figure 8–16 shows a three-phase transformer with its cover removed so you can see three separate transformer windings. The three-phase transformer operates exactly like a single-phase transformer in that AC voltage is applied to the primary side of the transformer and induction causes voltage to be created in the secondary winding. In the case of the three-phase transformer, the 120° phase shift of the three-phase voltage applied to the primary windings will be maintained in the secondary side of the transformer.

Figure 8–15 Typical three-phase transformers with their covers in place.
(Courtesy of Acme Electric, Acme Transformer Division)

Figure 8–16 A typical three-phase transformer with its cover removed, showing the three independent transformers and connection terminals. *(Courtesy of Acme Electric, Acme Transformer Division)*

When you are working on a commercial system, you may need to make connections on the secondary side of the transformer to the disconnect box, or between the secondary side of the transformer and a load center. You may also need to make voltage tests at the terminals of the transformer. You can use the diagram that is provided on the side of the transformer to identify the correct terminals to use. In most cases, you will not be required to make any electrical connections at the primary side of the transformer, since the primary voltage may be rather high (1300–1700 V AC). The electrical technicians for the electric utility company or technicians from a high-voltage service company will make the primary voltage connections for the transformer. You will be expected to make the connections for the air-conditioning or refrigeration equipment at the disconnect or at the load center.

8.16 THE WYE-CONNECTED THREE-PHASE TRANSFORMER

Figure 8–17 shows a diagram of a **wye-connected three-phase transformer**. You should notice that the physical shape of the transformer windings looks like the letter Y, which gives this type of configuration its name. You should also notice that this diagram shows only the secondary coils of the transformer. It is traditional to show only the primary-side connections or only the secondary-side connections when discussing a transformer power distribution system, since showing both the primary and secondary connections in the same diagram tends to become confusing.

The amount of voltage measured at the L1–L2, the L2–L3, and the L3–L1 terminals on the secondary winding will be 208 V for the wye-connected transformer if its turns ratio is set for low voltage. If the turns ratio of the transformers is set for high voltage, the amount of voltage between each terminal will be 480 volts. The voltage indicated between the windings for the transformer shown in Figure 8–17 is 208 V, which indicates that the turns ratio for this transformer will provide the lower voltage. If 480 V

Figure 8–17 A diagram showing the secondary windings of a three-phase transformer connected in a wye configuration. The voltage available at L1–L2, L2–L3, and L3–L1 is 208 V for this transformer.

were needed, a transformer with a higher turns ratio would be used. The primary and secondary voltage for each transformer is provided on its data plate and can be specified when the transformer is purchased and installed.

8.17 THE DELTA-CONNECTED THREE-PHASE TRANSFORMER

Figure 8–18 shows a diagram of a **delta-connected three-phase transformer**. You should notice that the shape of the transformer windings in this diagram looks like a triangle (Δ). This shape is the Greek letter D, which is named *delta*. This diagram also shows only the secondary side of the three-phase transformer.

The amount of voltage measured at L1–L2, L2–L3, and L3–L1 is 240 V for the delta-connected transformer if it is wired for its lower voltage. If the delta-connected transformer is wired for its higher voltage, the voltage at L1–L2, L2–L3, and L3–L1 will be 480 V. If the delta-connected transformer is wired for its lower voltage, it is very easy to differentiate it from a wye-connected transformer. If the delta-connected transformer and wye-connected transformer are both connected for their higher voltage, you cannot tell them apart, since both will provide 480 V between their terminals.

It is important to understand at this point that you will purchase the air-conditioning and refrigeration equipment so that its voltage requirements match the voltage supplied by the transformers at the commercial or industrial site. This means that you do not need to determine if the transformer is connected in a delta configuration or a wye configuration. You will need only to measure the amount of voltage at the secondary of the transformer, and if it is 480 V, you will need to ensure that the air-conditioning system or refrigeration system is rated for 480 V. If the secondary voltage is 240 V, the air-conditioning system must be rated for 240, and if it is 208, the equipment must be rated for 208.

If you need to know whether the transformer windings are connected as delta or wye, you can check the physical connections or you can make an additional voltage

Figure 8–18 Electrical diagram of the secondary windings of a transformer connected in a delta configuration. Notice that the voltage at L1–L2, L2–L3, and L3–L1 is 240 V.

measurement between each line and the neutral terminal of the transformer if one is provided. The next section will explain these measurements.

8.18 DELTA- AND WYE-CONNECTED TRANSFORMERS WITH A NEUTRAL TERMINAL

Figure 8–19 shows the diagrams of a wye-connected transformer and delta-connected transformer, each with a neutral terminal. The neutral terminal on the wye-connected transformer is at the point where the three ends of the individual windings are connected together. This point is called the *wye point*.

Note that the neutral point for the delta-connected transformer is actually the midpoint of the secondary winding that is connected between L1 and L3. This point is essentially the center tap of one of the transformer windings. Traditionally, it is the winding that is connected between L1 and L3 if a neutral connection is used with the three-phase transformer.

Figure 8–20a shows the amount of voltage that you would measure between terminals L1 and L2 of a wye-connected transformer and between terminals L1 and N, and L2 and N. Notice that the voltage between L1 and L2 is 208 V, so this transformer is wired for the lower voltage. The voltage between L1 and N and between L2 and N is shown as 120 V. Figure 8–20b shows the amount of voltage that you would measure between terminals L1 and L3 and between L1 and N, and L3 and N for a delta-connected transformer. The voltage between L1 and L3 is 240 V, so this transformer is connected for its lower voltage. The voltage between L1 and N and between L3 and N is shown as 120 V. You should notice that since the neutral point for the delta-connected

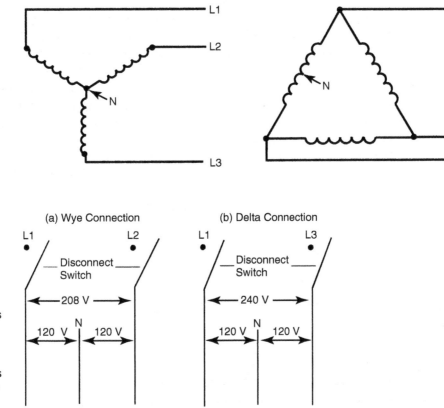

Figure 8–19 Electrical diagram of a wye-connected transformer with a neutral point and a delta-connected transformer with a neutral point.

Figure 8–20 a Voltage from L1–L2 for a wye-connected transformer is 208 V, and from L1–N or L2–N is 120 V. **b** Voltage from L1–L3 for a delta-connected transformer is 240 V, and from L1–N or L3–N is 120 V.

transformer is exactly halfway on the transformer winding, the voltage L1–N and L3–N will always be exactly half of the voltage between L1 and L3.

This is the main difference between a wye-connected transformer and a delta-connected transformer. The voltage at L1–N and at L3–N for any delta-connected transformer is always exactly half, and the L1–N or L2–N voltage for a wye-connected transformer will always be more than half the voltage at L1–L2. The exact amount of voltage L1–N can be calculated by dividing the voltage between L1 and L2 by 1.73, which is the square root of 3 ($\sqrt{3} = 1.73$). The square root of 3 is used because of the relationship among the phase shifts of the three phases. Note that 208 divided by 1.73 is 120 V.

The same relationship of voltage between L1 and L2 and L1 and N exists when the transformers have turns ratios for their higher voltage. Figure 8–21a shows a wye-connected voltage between L1 and L2 is 480 V, and between L1 and N is 277 V. The 277 V can be calculated by dividing 480 by 1.73. The 277 V that comes from L1–N or L2–N are generally used for fluorescent lighting systems in commercial buildings. Since the supply voltage originates from a three-phase transformer, the lighting system in a commercial or industrial building will also use L3–N, so that voltage is used from all three legs of the transformer. This voltage is referred to as single-phase voltage, since only one line of the three-phase transformer is used at each circuit. For example, L1–N is a single-phase circuit.

It is also important to understand at this time that L1–L2, L2–L3, or L3–L1 could each be used to supply complete power for an air-conditioning system or refrigeration system. Even though this type of power supply uses two legs of the transformer, it is still called a single-phase power supply because only one phase of voltage is used at any instant in time. For example, if the system is powered with voltage from L1–L2, during any given half-cycle of AC voltage, the power source for the system will come from L1, and then during the next half-cycle it will come from L2. The source of power will continue to oscillate between L1 and L2, but only one phase is in use during any instant in time.

Figure 8–21b shows that the higher voltage for the delta system between L1 and L3 is also 480 V, but this time the voltage between L1 and N or L3 and N is 240 V. Again the L1–N or L3–N voltage is exactly half of the supply voltage, since the neutral point on the transformer is the center tap of one of the transformers. It is important to understand that it is very easy to distinguish between a wye-connected power source and a delta-connected power source by measuring the voltages L1–L3 and L1–N. If the L1–N voltage is half of the L1–L3 voltage, the system is a delta-connected system. If the L1–N voltage is more than half, it is a wye-connected system. You should remember that the L1–N wye voltage can always be calculated by dividing the L1–L2 voltage by 1.73.

Figure 8–21 **a** Voltage from L1 to L2 for a wye-connected transformer is 480 V and from L1 to N or L2 to N is 277 V. **b** Voltage from L1 to L3 for a delta-connected transformer is 480 V and from L1 to N or L3 to N is 240 V.

8.19 THE HIGH-LEG DELTA SYSTEM

When a three-phase transformer system is used for the power source for an air-conditioning or refrigeration system, it may have the neutral tap. It is important to remember that a three-phase system does not need to have a neutral to operate correctly. The neutral is added only if the lower voltage (120 V) is needed for some part of the system. Generally, equipment manufacturers make all the components in a three-phase system a higher voltage, or they may supply a small transformer inside the equipment power panel to drop the higher voltage to the necessary voltage level. For example, if the equipment needs 208 V three phase for the compressor motor and fan motors, a control transformer can be provided in the power panel of the equipment to drop the 208 V L1–L2 to 24 V. The 24 V is used to provide power for the control circuit and thermostat. Since the primary side of the control transformer can be powered by L1–L2 (208 V), a neutral is not needed in this system. If the control transformer or fan motors require 120 V for power, a neutral tap, which is called **neutral,** will be needed, and these components will be connected between L1 and N, L2 and N, or L3 and N.

If the transformer is connected as a delta transformer, it is important to understand that a different voltage becomes available between L2 and N. Figure 8–22 shows the diagram for this voltage. From this diagram you can see that the secondary winding of this three-phase transformer is connected as a delta transformer. The voltage from L1–N and L3–N is 120 V, so the voltage between L1 and L2, L2 and L3, or L3 and L1 is 240 V.

The different voltage occurs between L2 and N. Since the winding between L1 and L3 has the center tap, it stands to reason that the voltage between L1 and N or between L3 and N will be exactly half of the voltage L1–L3. Since the center tap is not between L2 and L1, the voltage for L2–N must come from one complete phase (240 V) and half of the next phase (120 V). Since this voltage uses two phases, the 240 V and the 120 V are out of phase, and the resultant voltage from them is 208 V.

Since the 208 V between L2 and N comes from two phases, it will cause the transformer to overheat if it is used to power any components that require 208 V. For this reason the voltage is called the *high-leg delta voltage* to indicate that it is derived from L2–N of a delta-connected transformer and that it should not be used to power 208-V components. It is very important to understand that the L2 leg of the transformer is very usable when it is used with L1–L2 or L2–L3 as part of the three-phase system or 240-V single-phase system. The only problem occurs when the L2 terminal is used in conjunction with the neutral, which creates the L2–N voltage of 208 V.

The L2 terminal in any power distribution box for a delta-wired system should always be marked with an orange wire to identify it as the high-leg delta. In some areas of the country, the high-leg delta is also called the *wild leg, red leg,* or *stinger leg.*

Figure 8–22 A high-leg delta voltage occurs between L2 and N because the center tap to produce the neutral is the center tap of the L1–L3 winding.

8.20 THREE-PHASE VOLTAGE ON SITE

The on-site source for three-phase voltage will be a **fusible disconnect** or a load center. The fusible disconnect is also known as the fused disconnect. If it does not have fuses, it is just called the disconnect switch. As you know, the voltage actually comes from the three-phase transformers, but as stated before, as an air-conditioning or refrigeration technician, you will not be expected to make connections right at the transformers. Instead other technicians will connect the wires between the transformers and the load centers or main disconnects. The load center may also be called a circuit breaker panel. Figure 8–23 shows an example of a three-phase fusible disconnect, and Figure 8–24 shows an example of a three-phase load center. In the picture of the three-phase disconnect, you can see where the three fuses are mounted. Incoming voltage is connected to the line-side terminals at the top of the disconnect. The line-side terminals are identified as L1, L2, and L3. The disconnect has a handle on the right side that is used to turn its switch on and off. If the disconnect switch is depressed at the top, the switch is closed and the disconnect will have voltage at the load-side terminals at the bottom of the disconnect. The load-side terminals are identified as T1, T2, and T3. It is important to understand that the line-side terminals will always have power, even when the disconnect switch is open (in the off position).

The three-phase load center is similar to the circuit breaker panel for a single-phase system, except it will have three individual circuits. The load center is specifically designed to mount three-phase circuit breakers for each individual circuit that will be used. This means that if you have four individual air-conditioning systems connected to the load center, each one will be connected to its own individual circuit breaker. When you are installing an air-conditioning system or refrigeration system, you will need to locate the load center and circuit breakers for the circuit you intend to use, or you must locate the fusible disconnect if one is installed near the unit location.

8.21 INSTALLING WIRING IN A THREE-PHASE DISCONNECT

When you are ready to connect the wires between the disconnect and the equipment you are installing, you will need to ensure that the source of three-phase voltage is turned off and a padlock is used to lock out the system. The padlock should also have a tag attached to it that has your picture and name on it to identify you as the technician who has locked out the system. The lockout/tagout procedure ensures that no one will accidentally turn the power on when you are still working in the panel, so that you can safely work on the system.

Figure 8–23 A three-phase fusible disconnect that is used to provide three-phase voltage to air-conditioning and refrigeration equipment in commercial and industrial applications. *(Courtesy of Challenger Electrical Equipment Corp.)*

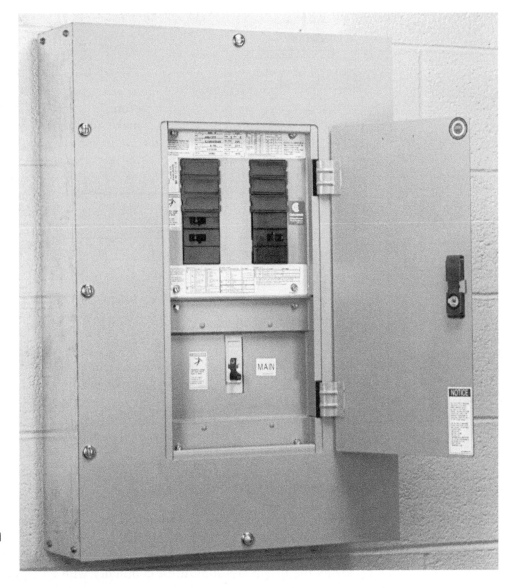

Figure 8–24 A three-phase load center that contains the circuit breakers for the electrical system. *(Courtesy of Challenger Electrical Equipment Corp.)*

Figure 8–25 shows a diagram of the locations in the disconnect where you will connect the equipment wires. This diagram shows how to provide single-phase voltage from a three-phase disconnect. The utility company will have the supply voltage wires previously installed at the three terminals at the top of the disconnect box. These terminals are called the *line-side terminals* and are identified as L1, L2, and L3. You will make your connections at the bottom terminals that are identified as T1, T2, and T3. These terminals are called the *load-side terminals*. The other end of the wires will be connected in the power panel for the air-conditioning or refrigeration equipment or to a single-phase disconnect. This process is called *connecting the field wiring*.

Sometimes you must get the supply voltage from a single-phase circuit breaker panel. The circuit breaker panel has single-phase voltage connected to its *main breaker* at the top of the panel as incoming voltage, and it has individual circuit breakers that receive power whenever the main breaker is in the on position. You will need to add a circuit breaker for each piece of equipment for which you are providing power. Wires must be run from the circuit breaker to the incoming power terminals in the air-conditioning or refrigeration equipment.

Figure 8–25 A single-phase disconnect is connected to the load-side terminals of a three-phase disconnect to provide a single-phase power source.

8.22 TESTING FOR A BAD FUSE IN A DISCONNECT

At times you will be working on a system that receives its power from a three-phase disconnect. The system will not run and you may suspect that one or more of the fuses are bad. You can check for a blown fuse by measuring the voltage that comes through it or by removing the fuse and testing it for continuity with an ohmmeter. If you select to check the fuses with a voltage test, you will need to test the incoming voltage at L1, L2, and L3 to determine that the supply voltage is good. Check the voltage between terminals L1 and L2, L2 and L3, and L3 and L1. The voltage at each of these sets of terminals should be the rated voltage for the system, such as 480 V. If you do not get the same voltage reading at each set of terminals, the incoming power has a problem and you will need to determine where the voltage has been interrupted.

The next test should be at the bottom terminals of the disconnect at T1–T2, T2–T3, and T3–T1, with the disconnect in the on position. Each of these readings should be the same as the voltage readings at the line side of the disconnect. If any of the voltage readings is lower than the line-side readings or 0 V, you have an open fuse. At this point the simplest way to find the bad fuse is to turn the disconnect off and remove all the fuses and test them for continuity. Be sure to use a fuse puller when you are removing or installing fuses in a disconnect. It is also important to remember that the line-side terminals of the disconnect are "hot" (powered) even though the disconnect switch is open.

The reason the load-side voltage may be lower than the line-side voltage rather than 0 V is because voltage may feed back through two of the windings in any three-phase motor or three-phase transformer that is connected to the load side of the disconnect.

It is difficult to determine the exact amount of voltage that is dropped when voltage feeds back through other windings, so you must test for an open fuse if the line-side voltage is not the same as the load-side voltage at each set of terminals where you take readings. If you find an open fuse, you will need to replace it with one that has the same current rating and same voltage rating. If a load is connected to the fuses, you can also measure for voltage across each fuse, and if you measure full voltage across any fuse, the fuse is open and the voltage is actually *back-feed voltage*.

8.23 SINGLE-PHASE VOLTAGE FROM A THREE-PHASE SUPPLY

At times you will need to provide a single-phase source of voltage for an air-conditioning or refrigeration system. The single-phase voltage can be 480 V, 208 V, or 240 V L1–L2. If the single-phase circuit requires a neutral, a neutral wire can be connected between the neutral terminal of the three-phase disconnect and single-phase disconnect, and between the three-phase disconnect and the transformer.

This type of circuit is used in many applications inside a commercial location where three-phase voltage is supplied to the system and you need to provide a single-phase voltage source to power a window-type air conditioner or a small refrigeration case. Once you understand that you can get single-phase voltage from the three-phase system, you will begin to see how the AC power distribution system works in industrial, commercial, and residential applications.

The remaining chapters of this book will continue to refer to the supply voltage as you will find it connected to the equipment at the line-side terminals. As a technician you will need to be aware of where this voltage comes from and how it is distributed so that you can trace the circuits if no voltage is present at the unit.

8.24 WIRING A DUPLEX RECEPTACLE AS A UTILITY OUTLET

Figure 8–26 A duplex receptacle that provides 120 V for tools and equipment.

In some applications such as rooftop air-conditioning systems, it is necessary to provide one or more duplex receptacles so that you can plug in 120-V service equipment like vacuum pumps, recovery pumps, and utility lights. It may also be necessary to provide power for electric tools such as drills or saws. Duplex receptacles provide a means to safely provide power for them. Figure 8–26 shows a duplex receptacle. When you hold a duplex receptacle in your hand, you should notice that it has gold-colored screws on one side and silver-colored screws on the other side. The L1 wire should always be connected to a gold-colored screw, and a neutral wire should be connected to the silver-colored screw. The ground wire should be connected to the green screw on the receptacle.

The openings in the face of the receptacle are different sizes. The larger opening on the left side of the receptacle face is connected to the neutral terminal, and the smaller opening on the right side of the terminal is connected to the L1 terminal. The terminal with the round opening is connected to the ground terminal. Since the openings in the face of the receptacle are different sizes, it will cause any plugs that are connected to it to be connected correctly with respect to line and neutral. If a tool such as an electric drill is made of plastic, its plug will be manufactured so that it can be plugged into the receptacle either way. If the electric drill has a metal case, the plug will have a ground terminal, and the plug will connect to the receptacle only one way.

When you are ready to connect L1 and N wires to the receptacle and disconnect, be sure the disconnect switch is open. It is also important to test the voltage from L1–N at the line side of the disconnect to ensure that you have 120 V instead of the 208 V of the high-leg delta voltage. When all the connections have been made, you can close the main disconnect switch and test the two terminals of the receptacle with your voltmeter to ensure that 120 V is present.

8.25 REVIEW

A transformer is made of two windings called the primary and secondary. The primary and secondary windings are not connected to each other in any way, so when voltage is applied to the primary winding, voltage is created in the secondary winding by induction. The induced voltage in the secondary winding will be out of phase with the primary voltage by 180°. The ratio of the number of turns of wire in the primary winding compared with the secondary winding is called the turns ratio. The turns ratio determines how much secondary voltage will be available when voltage is applied to the primary. If the turns ratio is 1:2, and 120 V is applied to the primary, 240 V will be available at the secondary windings. The current is inversely proportional to the turns ratio, so if the load connected to the secondary draws 4 A, the current in the primary winding will be 8 A. The VA rating of a transformer is similar to the power rating but is determined by multiplying the primary voltage (V) times the primary current (A).

A control transformer provides 24 V from 120 V or 240 V for the air-conditioning and heating system's control circuit. Three-phase transformers provide supply voltage to commercial or industrial customers. If the voltage is 208 V, the transformer windings are connected in a wye configuration. If the voltage is 240 V, the transformer is connected in a delta configuration. The voltage in a residential area of a community comes into the substation at the edge of town as three-phase voltage, and it is distributed throughout residential areas as single-phase voltage by using two legs of the transformer for any single-phase circuit.

Job Assignment for This Chapter

You are working on a residential air-conditioning system and you suspect the control transformer is bad. The homeowner is not sure that the transformer is bad. Your job assignment is to explain the basic parts of a transformer and explain to the homeowner what tests you will use to show whether the transformer is good or bad. You should also sketch a diagram to show the homeowner how the transformer is connected in the circuit and how it changes 240 volts to 24 volts. If you must replace the transformer, you should explain what ratings on the transformer should match exactly and which ones can be larger or smaller.

Solution to the Job Assignment

You should explain that the transformer has no moving parts; it only has a primary and secondary winding. You should draw a sketch of the primary and secondary winding of a transformer. You should explain that when AC voltage is applied to the primary

winding, voltage will be induced into the secondary winding. The ratio of the primary voltage to the secondary is determined by the ratio of the turns of wire. You can check a transformer by applying voltage to the primary and measuring the voltage to the secondary. If you do not get voltage at the secondary, you can turn power off and check the primary and secondary windings for continuity. If either winding has an open, then the transformer is faulty and must be replaced. You should also explain that if you need to exchange the transformer, the new transformer should have the identical primary and secondary voltage rating and have a VA rating that is equal to or greater than the existing one.

Assessment

QUESTIONS

1. Explain how inductance is used in the operation of a transformer.
2. Provide an example of a step-up transformer in a heating system.
3. Discuss how it is possible to have a control transformer whose primary winding can be connected to 240 or 208 V and provide 120 V at its secondary.
4. Explain how you would test for 240 and 120 V AC in a fusible disconnect on the line-side and load-side terminals.
5. Explain why a control transformer is required in some air-conditioning systems.

TRUE OR FALSE

1. If the voltage L1–L2 is 240 V and the voltage L1–N is 120 V, the transformer is connected as a delta transformer.
2. The secondary voltage is larger than the primary voltage in a step-up transformer.
3. Voltage in a transformer moves from the primary winding to the secondary winding by conductance.
4. The VA rating for a transformer is determined by multiplying the voltage at the secondary winding by the current flow in the secondary winding.
5. The 208 volts from L2 to N for a high-leg delta transformer is usable for compressor loads that require 208 V.

MULTIPLE CHOICE

1. The voltage at the secondary terminals of a _____ transformer will be larger than the applied voltage to the primary terminals.
 a. step-up
 b. step-down
 c. isolation

2. A transformer that is wired as a delta transformer has
 a. 480, 208, and 120 V.
 b. 480, 240, and 120 V.
 c. 480, 277, and 120 V.

3. A transformer that is wired as a wye transformer has
 a. 480, 208, and 120 V.
 b. 440, 277, and 125 V.
 c. 480, 240, and 120 V.

4. If a control transformer has a 2:1 turns ratio and it has 240 V applied to its primary winding, the voltage at the secondary winding will be
 a. 480 V.
 b. 240 V because the secondary voltage is strictly controlled to equal the primary voltage.
 c. 120 V.

5. The air-conditioning or refrigeration system is connected to the _____ of a fused disconnect.
 a. line-side terminals
 b. load-side terminals
 c. power company terminals

PROBLEMS

1. Calculate the secondary voltage of a transformer that has a 4:1 turns ratio and 120 V AC applied to the primary terminals.

2. Calculate the VA of a 24-V control transformer that requires 5 A.

3. Determine the turns ratio of a transformer that has 4000 turns in its primary windings and 800 turns in its secondary winding.

4. If the transformer in Problem 1 has 440 V applied to its primary winding, how many volts will be measured at its secondary?

5. Calculate the amount of secondary amps for a control transformer that is rated for 800 VA and 120 V at the secondary.

Websites

Websites change constantly. Use the key word search to locate sites that have similar content. The products that will be found at each site are indicated with the key word.

Electricity courses on line
Product: Siemens free course
Key words: Capacitive reactance, inductive reactance

Disconnect switches
Products: Disconnect switches
Key word: Disconnect

Acme Electric, Acme Transformer Division
Products: Transformers
Key word: Acme Electric

Three-phase and single-phase transformers
Products: Power transformers
Key word: Three-phase transformer

Transformer formulas
Products: Free transformer training course on line
Key word: Transformer formulas

Chapter 9

Single-Phase Open Motors

Job Assignment for This Chapter

You are on a job and find that a condenser fan is not operating. Your assignment is to explain to the customer what type of motor you have and sketch the diagram of that motor showing all the electrical parts this motor uses. You should also explain what tests you would use to determine if the motor is working correctly or if it is faulty. Finally you need to read the data plate for the motor and place an order for a new motor that will replace the condenser fan motor.

Terms You Will Learn in This Chapter

Ambient temperature rise
Ball bearing
Bushing
Capacitor-start, induction-run (CSIR) motor
Centrifugal switch
Counter EMF (CEMF)
Data plate
End plates
Frame type
Full-load amperage (FLA)
Induction motor
Locked-rotor amperage (LRA)
Motor
Multispeed motor
NEMA design

Open-type motor
Permanent split-capacitor (PSC) motor
Poles
Rotor
Run capacitor
Run winding
Service factor
Shaded-pole motor
Slip
Split-phase motor
Start capacitor
Start winding
Stator
Temperature rise

After reading this chapter, you will be able to:

1. Explain the theory of operation of a simple induction AC motor.
2. Identify the main parts of an AC motor and explain their functions.
3. Identify the parts of the centrifugal switch and explain their operations and functions.
4. Explain the data found on a typical motor data plate.

9.0 OVERVIEW OF AC SINGLE-PHASE INDUCTION MOTORS

The main loads that you will work on in most air-conditioning, heating, and refrigeration systems are motors of all types. Motors are used to move air over evaporator coils, condenser coils, and heat exchangers. They are also used as compressors to pump refrigerant. The motors that are used in compressors are sealed inside a housing so that the refrigerant does not leak out to the atmosphere, and they are called *hermetic motors* or *hermetic compressors*. The motors that are used in fan applications are called open-type motors. Figure 9–1 shows a picture of an open-type fan motor. The theory of operation, installation, and troubleshooting for open-type motors that are used to power fans, pumps, and belt-driven loads will also be explained in this chapter. The material in these sections will show that some motors require special switches and components to start and operate correctly. These devices will be explained so that when you must troubleshoot a motor you will fully understand their functions. A diagram and picture will be provided for each type of motor so that you will learn to identify each type of motor from its physical appearance and from its electrical diagram. As a technician you will be expected to identify, install, troubleshoot, and repair or replace any motors you encounter. This chapter will provide a comprehensive overview for all single-phase, open-type motors you will find in HVAC and refrigeration systems.

Figure 9–1 Open-type motor used for an evaporator fan, a condenser fan, or a furnace fan. *(Courtesy of GE Motors & Industrial Systems, Fort Wayne, Indiana)*

223

9.1 AC SPLIT-PHASE MOTOR THEORY

The AC **split-phase motor** is widely used in HVAC applications and has many of the basic parts of the other types of open motors. This motor has three basic parts: the stationary coils called the **stator**, the rotating shaft called the **rotor**, and the **end plates** that house the bearings that allow the rotor shaft to turn easily. The stationary coils are also called **poles.** Figure 9–2 shows an exploded view diagram of a capacitor-start, induction-run open-type motor so that you can see where these parts are located inside the motor. The capacitor-start, induction-run motor is a variation of the split-phase motor that has a **start capacitor** mounted on it to provide more starting torque. From this diagram you can see that the rotor is supported by the bearings in the end plates of the motor. The rotor is also mounted directly inside the stationary windings so that when voltage is applied to them, their magnetic field can be induced into the rotor. The parts of a hermetic motor are similar but they are not visible because they are housed inside the hermetic dome.

If the AC split-phase motor requires a **centrifugal switch** to help it get started, it will be mounted inside the end plate. The actuator for the centrifugal switch is mounted on the end of the rotor's shaft, and it has flyweights that swing out to open the centrifugal switch when the motor reaches 75–85% of full rpm. In the next sections of this chapter you will see how these parts operate together to provide a rotating force to the motor shaft to turn a fan or pump.

9.2 THE ROTOR

The rotating part of the motor is called the *rotor*. When voltage is applied to the coils of wire in the stator, they produce a very strong magnetic field. This magnetic field is passed to the rotor by induction in much the same way that voltage is passed from the

Figure 9–2 Exploded view of a typical capacitor-start, induction-run single-phase motor. *(Courtesy of Leeson Electric Corporation)*

Figure 9–3 Diagram of a squirrel-cage rotor for an AC motor. The rotor bars become multiple bar magnets when the rotor is magnetized.

Figure 9–4 Picture of a squirrel-cage rotor for an AC motor.

primary winding to the secondary winding of a transformer. After the rotor becomes magnetized, it begins to rotate. The speed of the rotor is determined by the number of poles and the frequency of the AC voltage applied to the motor. The rotor has a shaft that rotates to do the work of the motor, such as to turn a fan blade or to move the compressor piston to pump refrigerant. Figure 9–3 shows a diagram, and Figure 9–4 shows a picture of a rotor that is used in most AC single-phase motors. As a rule, the rotor of an AC motor does not have any wire in it (only repulsion start or some synchronous AC motors have wire in their rotors, but these motors are not used too often in HVAC applications). The rotor is made of laminated steel plates pressed on the frame. The frame of the rotor looks like the wire-frame exercise wheel that is used by hamsters. For this reason it has been named a *squirrel cage*, and when it is used in the rotor it is called the *squirrel-cage rotor*. The squirrel cage is actually the frame for the rotor, and the part that looks like the squirrel cage is actually made up of rotor bars that act exactly like multiple bar magnets. The laminated steel plates that are pressed onto the frame of the squirrel-cage rotor allow the rotor bars to be magnetized by induction when current is passed through the stator. This means that the permanent field in the stator acts like the primary of a transformer, and the rotor acts like the secondary windings of a transformer. The result is that the induction motor does not need brushes to pass current to the rotor to build its magnetic field. This is also why these types of motors are called **induction motors**. It is also important to understand that the laminated steel sections in the rotor allow it to become magnetized very easily by the AC voltage and then quickly reverse the polarity of its magnetic field as the AC voltage oscillates from positive to negative as the AC sine wave changes. The polarity of the magnetic field in the rotor of the AC motor reverses because the sine wave of the AC voltage changes from positive to negative naturally during each cycle of the sine wave. This causes the magnetic fields in the rotor to be continuous to reverse polarity and spin as long as AC voltage is applied to the motor.

9.3 LOCKED-ROTOR AMPERAGE AND FULL-LOAD AMPERAGE

When voltage is first applied to the motor, the run and start windings draw a large amount of current called **locked-rotor amperage (LRA)**. This current is called locked-rotor amps because the rotor has not started to turn at this time. The amount of current is large because the amount of resistance in the windings is very small and because the rotor has not yet started to rotate and create **counter electromotive force (CEMF)**

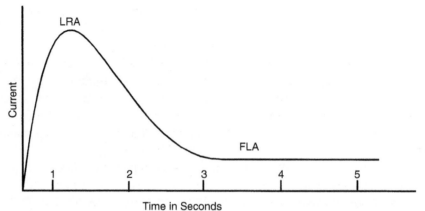

Figure 9–5 Graph of the locked-rotor amperage (LRA) that occurs when a motor is first started, and the full-load amperage (FLA) that occurs when the motor is running at full speed.

voltage. When the rotor starts to spin and comes up to full rpm, the current returns to normal and is called **full-load amperage (FLA)**. Figure 9–5 shows a graph of the large LRA current that occurs when the motor is first started and also shows the current returning to lower levels (FLA) when the motor is running at full-load speed.

The current returns to its normal value after the rotor is spinning at full speed because the rotor produces CEMF. When the rotor starts to spin, its magnetic field begins to pass the coils of wire in the run winding and generates a voltage. This generated voltage is out of phase with the applied voltage, so it is called *counter EMF* (CEMF). Counter EMF is also sometimes called *back EMF*.

The amount of CEMF will be equivalent to the speed at which the rotor is turning. If the rotor is turning at 90% of full speed, the CEMF will be approximately 90% of the applied EMF. For example, if the applied voltage is 240 V AC, and the rotor is turning at approximately 90% of full rpm, the CEMF may be 200 V AC. If the resistance of the run winding is 10 Ω, the full-load current is determined by dividing it by the difference in voltage (240–216), or 24 V AC. Thus the FLA current will be approximately 3 A.

9.4 ROTOR SLIP AND TORQUE IN AN AC INDUCTION MOTOR

The amount of LRA and FLA that an AC induction motor draws will also determine the amount of torque the motor shaft will have. *Torque* is defined as the amount of *rotating force* that the rotor shaft produces. Recall that this force is needed to pump the piston of the compressor and turn the fan blades for a condenser fan. If the motor has a large amount of LRA during starting, it will also have a lot of starting torque. If the motor draws a large FLA when the motor is running, it will have a lot of running torque.

The amount of FLA is determined by the amount of slip a motor has. **Slip** is defined as the *difference between the rated speed of a motor and its actual running speed*. For example, a four-pole motor is rated for 1800 rpm, but its actual running speed is approximately 1725 rpm. The difference of 75 rpm is called slip. At first glance you may think that slip is not good for a motor because the motor is not running as fast as it should be, but you will see that the induction motor relies on slip to create the amount of current necessary to turn the load at the end of the motor shaft. If the amount of slip is minimal and the rotor is spinning at close to the rated speed, the amount of CEMF will be large, and the difference between the CEMF and the applied EMF may be as small as 5 V AC.

In this situation the amount of current in the motor is reduced to less than 0.5 A, and consequently the amount of torque the rotor has is also reduced. It is also important to remember that the rotor gets the current required to build its magnetic fields through induction from the stator windings. The induction can occur only if the rotor is not turning at the same speed as the magnetic field is rotating in the stator.

When the rotor's torque is minimal, the shaft is not able to turn its load, and it begins to slow down, which decreases the amount of CEMF. This makes the difference between the CEMF and the applied voltage larger, which makes the motor draw more current and produce more torque. All induction motors rely on slip to operate correctly and produce sufficient torque. If it is important that the motor run at its rated speed, a special type of motor called a *synchronous motor* can be used because it runs at its rated speed with no slip. The synchronous motor is not used very often in HVAC or refrigeration applications because it requires DC voltage to be supplied to its field.

It is also possible to use a variable-frequency drive to adjust the frequency of the voltage above 60 Hz to get the motor to run at its rated speed even if it has slip. Variable-frequency drives were introduced into HVAC and refrigeration applications in the 1980s. You will learn more about variable-frequency drives in the chapter that covers electronic devices. As a technician you will encounter a variety of variable-frequency drives in newer residential air-conditioning and heating systems, where they are used to control the speed of compressors and fan motors.

9.5 END PLATES

The end plates play an important part in the operation of the motor. They house the bearings that support the rotor to allow it to spin with a minimum amount of friction, and one end plate houses the end switch that is used to help the single-phase motor get started. The bearings in a single-phase motor may be the ball-bearing type or a bushing type. The **ball bearing** is more expensive and is generally used in larger single-phase motors. The **bushing** is used in smaller and less expensive single-phase motors. Figure 9–6 shows an armature with one end plate, and Figure 9–7 shows an example of a ball bearing and several types of bushings. The ball bearing shown in Figure 9–7a uses a number of small steel balls that create a rolling surface between the inner hub and outer hub. The shaft of the motor is pressed securely into the inner hub of the bearing so that the shaft and inner hub of the bearing move together as though they were welded. As the rotor shaft spins, the inner hub rolls

Figure 9–6 An end plate shown with a rotor. The bearings are mounted in the end plates and support the rotor shaft so that it can spin freely.

Figure 9–7 a An oil-type bearing (bushing) with insulator. **b** Sealed-type bronze bushing and sealed ball bearing. **c** Pillow-block-type bearing with bronze bushing.

on the balls, and the balls make contact with the outer hub. This means that only a small portion of each ball is touching the inner and outer hubs at any time, which keeps friction to a minimum. The ball bearing for smaller motors is lubricated with grease when it is manufactured and usually is sealed for the life of the bearing, which means that motors with these types of ball bearings will not need to be lubricated. Some larger motors use bearings that must be greased periodically. A grease fitting is provided so that the ball bearing can be greased once or twice a year, depending on the number of hours the motor runs.

The bushing shown in Figure 9–7b is made of a brass or bronze alloy sleeve with a lubrication hole in it. An *oiler pad*, which is also called the *oil wick*, is mounted on the outer side of the bushing, so that oil from the pad can find its way through the hole to the inside of the bushing. The oil creates a "bearing" surface between the shaft of the rotor that spins, while the bushing remains stationary. As long as a sufficient amount of oil remains on the inside of the bushing, between it and the rotor shaft, the amount of friction will be held to a minimum. If the bushing ever runs dry, the rotor shaft will be allowed to touch the bushing surface and both surfaces will begin to wear severely. It is vitally important to ensure that the lubrication ports for the bushing-type motor are always located on the top side of the motor so that gravity can draw the lubrication through the lubrication pad into the bushing. It is also important to remember that the motor that uses a bushing must be lubricated at least once a year. Figure 9–7c shows a bushing with a pillow block to hold it in place when it is used on the shaft of a squirrel-cage fan.

9.6 THE STATOR

The stator is actually made of two separate windings called the **run winding** and the **start winding**. Each of these windings is subdivided into poles that represent the poles of a magnet. We will use a four-pole motor for our examples in this chapter. This means that the run winding will have four poles, and the start winding will have four poles. Figure 9–8 shows

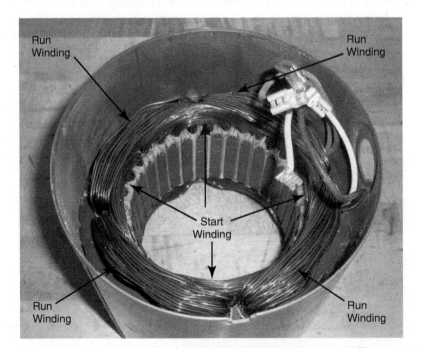

Figure 9–8 Close-up picture of start and run windings in the stator. The run winding has larger wire than the start winding, and the windings are offset from each other. This motor has four segments for the run winding and four segments for the start winding.

a picture of a stator so that you can more easily see the coils of wire that are mounted in it. From this picture you can see the four poles of the run winding and the four poles of the start winding. The windings are pressed over pole pieces that also become magnetized when the coils have current flowing in them. The pole pieces are made from laminated steel like the rotor so that the poles can quickly reverse polarity when AC voltage is applied.

The AC sine wave starts at 0 V and peaks positive, returns to 0 V and peaks negative, and returns to 0 V again. It continues to oscillate positive and negative as long as voltage is applied. This action causes the magnetic field in each winding in the stator to be alternately positive and then negative. The reversing magnetic field actually causes the magnetic field to rotate around the poles inside the stator. Since the rotor also has a magnetic field, it begins to spin and follow the rotating magnetic field in the stator. It will be easier to understand the function of the stator if you study the operation of the start winding and the run winding separately.

9.7 THE START WINDING

If the AC induction motor had only one winding, the rotor would tend not to rotate when single-phase AC voltage was first applied to the stator windings in the AC induction motor. The rotor needs a small amount of phase shift between the applied voltage and current in the magnetic field created in the stator to begin to spin. There are several ways to provide the phase shift required to get the rotor to start to spin. One way is to physically offset two windings in the stator. A second way to cause a phase shift in the stator is to put more wire in one set of windings than in the other. For this reason, the start winding is made from smaller wire than the run winding, so that it has more wire in each coil than the run winding.

Figure 9–8 shows a close-up picture of a stator that clearly shows the run and start windings. From this picture you can see that the run winding is made of larger wire and is physically offset from the start winding. Figure 9–9a shows a sketch of how the run winding and start winding are placed in the stator, offset from each other. Figure 9–9b also shows an electrical diagram of the start winding and run winding for an eight-lead AC single-phase motor, and Figure 9–9c shows a four-lead single-phase motor.

The electrical diagram in Figure 9–9a shows that the start winding is made of smaller wire and has more turns than the run winding. The start winding coil has more turns of wire than the run winding coil, so that more feet of wire are used in the start winding. Since the start winding has more feet of wire, it takes the current longer to pass through it than through the run winding. It is important to understand that when voltage is applied to an AC induction motor, voltage starts into the run winding and the start winding at the same time. Since the start winding has more wire and is offset physically from the run winding, a phase shift is created. (See Figure 9–10.)

This phase shift gives the rotor a sufficient amount of starting torque (rotational force) to get the rotor to start turning. Once the rotor begins to turn, it causes the magnetic field to continue to build as the motor comes up to full speed. The phase shift occurs between the run winding and the start winding, but it also can be described as the phase shift that occurs between the voltage waveform and the current waveform because inductors (motor windings) cause the current waveform to lag the waveform of voltage. In later sections you will see how capacitors can be added to the start windings to enhance the amount of phase shift to give the motor more starting torque.

Since the start winding is made from very small wire so that its coils can have many more turns of wire than the run winding, the start winding draws a large amount of current when voltage is applied to it. If the start winding was allowed to remain in the

(a)

Eight-Lead Motor
(b)

Four-Lead Motor
(c)

Figure 9–9 **a** Diagram showing the four-pole run winding and four-pole start winding in the stator of an AC induction motor. You should notice that the start winding is physically offset in the stator from the run winding to help provide more of a phase shift. The run winding is made of larger-gauge wire and has fewer turns than the start winding. **b** Diagram of the run winding and start winding of an eight-lead induction motor. Notice that the two sections of the run winding are identified as T1–T2 and T3–T4, and the two sections of the start winding are identified as T5–T6 and T7–T8. **c** Diagram showing a four-lead induction motor. The run winding is identified as T1–T4, and the start winding is identified as T5–T8.

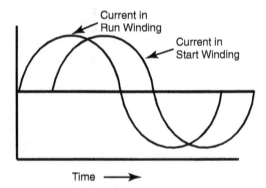

Current in
Run Winding

Current in
Start Winding

Time

Figure 9–10 Diagram of the phase shift that occurs when voltage is applied to the start and run windings of an AC motor. Since the start winding has more wire than the run winding, a phase shift is created between the magnetic fields in the two windings.

circuit continuously, this large current would damage it and eventually burn a hole in the windings. For this reason a centrifugal switch is used to disconnect the start winding after the motor's rotor has started spinning. The centrifugal switch can determine when the speed of the rotor has reached 75–85% full rpm and disconnect the start winding from the circuit until it is needed the next time the motor is started.

9.8 THE OPERATION OF THE CENTRIFUGAL SWITCH

The centrifugal switch is also called the *end switch* in a motor because it is mounted in the end plate. Even though the centrifugal switch is physically mounted in the end plates, it is important to understand that it is electrically connected to the start winding that is located in the stator. The centrifugal switch is actually part of a larger assembly that consists of an activation mechanism that has flyweights that begin to swing out when the rotor reaches 75% of full rpm. The activation mechanism is mounted on the end of the

(a) (b)

Figure 9–11 a Centrifugal switch removed from an end plate, and activation mechanism removed from the armature shaft. **b** Centrifugal switch mounted in the end plate, and the end plate mounted on the end of the motor shaft so that the activation mechanism and flyweights are touching the centrifugal switch.

rotor shaft so that it can turn when the rotor shaft is turning. The activation mechanism is mounted on the shaft in such a way that it makes contact with the centrifugal switch and holds the switch contacts in the closed position when the shaft is not rotating.

Figure 9–11 shows pictures of a centrifugal switch and the activation mechanism. Figure 9–11a shows a centrifugal switch that is removed from an end plate, and an activation mechanism removed from the armature shaft. Figure 9–11b shows a centrifugal switch as you would find it mounted in the end plate. This figure also shows the location of the activation mechanism as it touches the centrifugal switch. You can see its flyweights at rest, as you would find them when the rotor is not turning.

The centrifugal switch is a set of electrical contacts mounted on spring steel. Spring steel is used because it can be flexed so that the spring tension in the steel tends to keep the contacts in the open position. The activation mechanism that is mounted on the rotor shaft comes into contact with the spring steel of the switch when the end plate is attached to the stator. The rotor shaft goes through the bearing in the end plate, and when the end plate is drawn tight against the stator with screws, the activation mechanism touches the spring steel and causes the centrifugal switch contacts to press against each other to complete the electrical circuit through the switch.

When voltage is applied to the windings in the motor, current flows through the run winding and through the start winding, since the activation mechanism is holding the contacts of the centrifugal switch closed. As the rotor beings to spin, centrifugal force begins to pull the flyweights outward. Since the flyweights are held together with a spring on each side, they do not swing out all the way until the rotor is spinning at approximately 75–85% of full rpm. The spring tension that holds the flyweights inward determines the speed at which the flyweights will swing out completely.

When the flyweights swing out completely, they allow the spring tension on the activation mechanism to move it to a retracted position, which allows the spring tension in the contacts of the centrifugal switch to move them to the open position. When the contacts of the centrifugal switch open, current to the start winding is interrupted, and the motor continues to run with current flowing only through the run winding of the motor.

Figure 9–12a shows a diagram of the activation mechanism when the rotor is not turning. In this part of the diagram you can see that the flyweights are held in position

Figure 9–12 a Flyweights and activation device for centrifugal switch. The activation device is mounted on the end of the rotor shaft so that it can come into contact with the centrifugal switch. A spring holds the flyweights against the shaft when the rotor is not turning. **b** When the rotor is spinning at 75–85% of full rpm, the flyweights move outward from the shaft because of centrifugal force. The action of the flyweights causes the activation device to drop away from the switch contacts so that they can move to their open position.

by spring tension. When the flyweights are in this position, the activation mechanism is pressed against the switch contacts holding them in the closed position so current can flow through them. In Figure 9–12b the rotor is turning at nearly full speed. The centrifugal force that is created when the rotor shaft is spinning at full speed causes the flyweights to swing out away from the rotor shaft. When the flyweights swing out completely, the activation mechanism is pulled away from the contacts, which allows the tension of the spring steel to open the contacts.

9.9 THE RUN WINDING

The run winding is the part of the stator that remains in the motor circuit at all times. Figure 9–13 shows an electrical diagram for a single-phase motor in which you can see that the run winding is easily identified, since it uses larger wire and fewer coils in its windings. In most single-phase motors, the run winding is numbered as shown in the diagram. If the motor is a dual-voltage motor, its run winding is made in two sections, one with terminal ends T1 and T2, and the other with terminal ends T3 and T4. If the motor is wired for high voltage (240 or 208 V AC), the windings will be wired in series with each other. (See Figure 9–13a.)

Figure 9–13b shows the two sections of the run winding connected for low voltage (120 V AC). If the motor is wired for low voltage, the two sections of the run winding will be connected in parallel with each other. These two diagrams will be used in all electrical system diagrams that you will use to install and troubleshoot motors in heating, ventilating, air-conditioning, and refrigeration systems.

You can test the run winding and start winding with an ohmmeter to determine which one is which. For example, in the diagram each part of the run winding has 4 Ω of resistance, and the start winding has 12 Ω.

Figure 9–13 a Electrical diagram of an AC induction motor wired for high voltage (240 or 208 V AC). The two sections of the run winding numbered T1 and T2, and T3 and T4 are connected in series. **b** Electrical diagram for an AC induction motor wired for low voltage (120 V AC). The run winding in this diagram is shown in two sections that are connected in parallel.

9.10 ## MULTIPLE-SPEED MOTORS

The coils of wire in the run winding can be separated into sections called poles. The motor will have an even number of poles. The speed of an induction motor is determined by the number of poles and the frequency. The speed can be calculated from the following formula:

$$rpm = \frac{frequency \times 120}{poles}$$

A typical motor has two, four, six, or eight poles. The speed of each type of motor is listed in the following table.

Number of Poles	rpm
2	3600
4	1800
6	1200
8	900

Motors for HVAC and refrigeration applications need to be able to operate at a variety of speeds. This means that the motor may be manufactured with a set number of

Figure 9–14 Electrical diagram of a multitapped fan motor. The run winding is tapped so that it can provide three different choices for the number of poles that are used. The black wire is used for the high speed, the blue wire is used for the medium speed, and the red wire is used for the low speed.

windings to provide a set speed, or the motor may be reconnected in the field to give the motor a different speed.

Another way to change the speed of a motor is to tap the run winding so that more or less inductive reactance is used. This means that if you have a very long run winding, you can place taps on it just as in a multiple-tapped transformer. Each tap uses less of the total run winding; consequently, the change in inductive reactance allows the speed of the motor to change. The amount of change in rpm for each tap is approximately 15–20%. Figure 9–14 shows an example of a motor with a multitapped run winding. From the diagram you can see the taps in the run winding, and you can see the color code for the wires. If you wanted the motor to run at high speed, you would use the black wire for the run winding. If you wanted the motor to run at medium speed, you would use the blue wire, and the red wire would allow the motor to run at low speed. Generally, the multiple speeds are used for fan motors, where the high speed is used for cooling, and the low speed is used for heating. The medium speed is used when the customer wants a slightly lower speed for air conditioning, or a higher speed than the lowest speed for heating. It is important to remember that as the tap is changed for this motor it will use less of the winding to run faster, but it will also lose power.

9.11 MOTOR DATA PLATES

The motor's **data plate** lists all the pertinent data concerning the motor's operational characteristics. It is sometimes called the *name plate*. Figure 9–15 shows an example of a data plate for a typical AC motor. The data plate contains information about the ID (identification number), FR (frame and motor design), motor type, phase, horsepower

ACE MOTOR CORP.

ID		FR	TYPE	PH.	
HP	RPM	VOLTS	AMPS	HZ	SF
TIME RATING	INS. CL.	INS. TYPE	AMB. C.	NEMA DESIGN	CODE

Figure 9–15 Data plate for typical AC motor.

rating, rpm, volts, amps, frequency, service factor (SF), duty cycle (time), insulation class, ambient temperature rise, and NEMA design and code. Each of these features will be discussed in detail in the next sections.

9.11.1 Identification Number (ID)

The identification number for a motor will basically be a model and serial number. The model number will include information about the type of motor, and the serial number will be a unique number that indicates where and when the motor was manufactured. These numbers will be important when a motor is returned for warranty repairs, or if an exact replacement model is specified for exchange.

9.11.2 Frame Type

Every motor has been manufactured to specifications that are identified as a **frame type**. This information includes the distance between mounting holes in the base of the motor, the height of the shaft, and other critical data about physical dimensions. These data are then given a number such as 56. The frame number indicates that any motor that has the same frame number will have the same dimensions even though it may be made by a different company. This allows users to stock motors from more than one manufacturer or replace a motor with any other with the same frame information and be sure that it will be an exact replacement.

9.11.3 Motor Type

The motor-type category on the data plate refers to the type of ventilation the motor uses. These types include the open type, which provides flow-through ventilation from the fan mounted on the end of the rotor. In some motors that are rated for variable-speed duty (used with variable-frequency drives), the fan will be a separate motor that is built into the end of the rotor. The fan motor will be connected directly across the supply voltage, so it will maintain a constant speed to provide constant cooling regardless of the motor speed.

Another type of motor is the *enclosed type*, which is not air cooled with a fan. Instead, it is manufactured to allow heat to dissipate quickly to and from the inside of the motor outward to the frame. In most cases the frame has fins built into it on the outside to provide more area for cooling air to reach.

9.11.4 Phase

The phase of the motor will be indicated as single phase or three phase. The number may be indicated as 1 or 3, or the number may be listed by itself. Also, the symbol 1 Φ or 3 Φ may be used.

9.11.5 Horsepower Rating (hp)

The horsepower rating will be indicated as a fractional horsepower (number less than 1.0) or a larger horsepower. Fractional horsepower numbers may be listed as fractions ($\frac{1}{2}$) or decimals (0.5).

9.11.6 Speed (rpm)

The motor speed will be indicated in rpm. This will be the rated speed for the motor, and it will not account for slip. The actual speed of the motor will be less because of slip. As you know, the speed of the motor is determined by the number of poles and the

frequency of the AC voltage. Typical speed for a two-pole motor is 3600 rpm, for a four-pole motor is 1800 rpm, and for an eight-pole motor is 1200 rpm. The actual speed of the motor rated for 3600 rpm will be approximately 3450, for the 1800 rpm motor it will be approximately 1750, and for the 1200 rpm motor it will be approximately 1150. The actual amount of slip will be indicated by the *motor design letter*. The motor design letter will be explained in Section 9.11.14.

9.11.7 Volts

Voltage ratings for single-phase motors will be listed as 120 V, 208 V, or 240 V. Three-phase motors have typical voltage ratings of 208, 240, 440, 460, 480, 550, and 575. Other voltages may be specified for some special-type motors. You should always ensure that the power supply voltage rating matches the voltage rating of the motor. If the rating and the power supply voltage do not match, the motor will overheat and be damaged.

9.11.8 Amps

The amps rating is the amount of full-load current (FLA) the motor should draw when it is under load. This rating helps the designer calculate the proper wire size, fuse size, and heater size in motor starters. The supply wiring for the motor circuit should always be larger than the amps rating of the motor. The NEC (National Electrical Code) provides information to help you determine the exact fuse size and heat size for each motor application.

9.11.9 Frequency

The frequency of a motor will be listed in hertz (Hz). Typical frequency rating for motors in the United States is 60 Hz. Motors manufactured for use in some parts of Canada and all of Europe and Asia will be rated for 50 Hz. You must be sure that the frequency of the motor matches the frequency of the power supplied to the motor.

9.11.10 Service Factor (SF)

The **service factor** is a rating that indicates how much a motor can be safely overloaded. For example, a motor that has an SF of 1.15 can be safely overloaded by 15%. This means that if the motor is rated for 1 hp, it can actually carry 1.15 hp safely. To determine the overload capability of a motor, you multiply the rated horsepower by the service factor. The motor is capable of being overloaded because it is designed with features that can dissipate large amounts of heat.

9.11.11 Duty Cycle

The *duty cycle* of a motor is the amount of time the motor can be operated out of every hour. If the motor's duty cycle is listed as continuous, it means the motor can be run 24 hours a day and does not need to be turned off to cool down. If the duty cycle is rated for 20 min, it means the motor can be safely operated for 20 min before it must be shut down to be allowed to cool. A motor with this rating should be shut down for 40 min of every hour of operation to be allowed to cool.

Another way to specify the duty cycle of a motor is called the *motor rating*. The motor rating on the data plate refers to the type of duty for which the motor is rated. The types of duty include continuous duty, intermittent duty, and heavy duty, which includes

jogging and plugging duty. Continuous duty includes applications in which the motor is started and allowed to operate for hours at a time. The intermittent duty includes operations in which the motor is started and stopped frequently. This type of application allows the motor to heat up because it will draw LRA more often than will a motor rated for continuous duty.

Motors that are rated for jogging and plugging are built to withstand very large amounts of heat that will build up when the motor draws large LRA during starting and stopping. Since the motor can be reversed when it is running in the forward direction for plugging applications, it will build up excessive amounts of heat. Motors with this rating must be able to get rid of heat as much as possible to withstand the heavy-duty applications.

9.11.12 Insulation Class

The *insulation class* of a motor is a letter rating that indicates the amount of temperature rise the insulation of the motor wire can withstand. The numbers in the insulation class are listed in degrees Celsius (°C). The table in Figure 9–16 shows typical insulation classes for motors. The insulation class and other temperature-related features of a motor help determine the temperature rise the motor can withstand.

9.11.13 Ambient Temperature Rise

The *ambient temperature rise* is also called **temperature rise** or the Celsius rise, which is the amount of temperature rise the motor can withstand during normal operation. This value is listed in degrees Celsius. A typical open motor can withstand a rise of 40°C (104°F), and an enclosed motor can withstand 50°C (122°F) of rise. This means that the motor should not be exposed to environments where the temperature is 104°F above the ambient. If the ambient is considered to be 72°F, it means that the motor is limited to temperatures of 176°F. Another classification that helps determine the amount of temperature a motor will be able to withstand is the type of insulation on the motor winding. The classes of insulation that are used with motors and the amount of temperature rise these classes can handle are listed in Section 9.11.14.

9.11.14 NEMA Design

The National Electrical Manufacturers Association (NEMA) provides design ratings that may also be listed as motor design or **NEMA design**. The motor design is specified on the data plate by a letter A, B, C, or D. This designation is determined by the type of wire, insulation, and rotor that are used in the motor and is not affected by the way the motor might be connected in the field.

Class	Temperature Rise °C
A	105
B	130
F	155
H	180

Figure 9–16 Insulation class for motors. This table indicates the amount of temperature rise for which the wire's insulation is rated.

Type A motors have low rotor circuit resistance and have approximate slip of 5–10% at full load. These motors have low starting torque with a very high locked-rotor amperage (LRA). This type of motor tends to reach full speed rather rapidly.

Type B motors have low to medium starting torque and usually have slip of less than 5% at full load. These motors are generally used in fans, blowers, and centrifugal pump applications.

Type C motors have a very high starting torque per ampere rating. This means that they are capable of starting when the full load is applied for applications such as conveyors, crushers, and reciprocating compressors like air-conditioning and refrigeration compressors. These motors are rated to have a slip of less than 5%.

Type D motors have a high starting torque with a low LRA rating. This type of motor has a rotor made of brass rather than copper segments. It is rated for a slip of 10% at full load. Normally, this type of motor will require a larger frame to produce the same amount of horsepower as a type A, B, or C motor. These motors are generally used for applications with a rapid decrease of shaft acceleration, such as a punch press that has a large flywheel.

These standards are set by NEMA, and a motor must meet all the requirements of the standard to be marked as a type A, B, C, or D. This allows motors made by several manufacturers to be compared on an equal basis according to application.

9.11.15 NEMA Code Letters

NEMA code letters use letters of the alphabet to represent the amount of locked-rotor amperage (LRA) in kilovolt-amps (kVA) per horsepower a motor will draw when it is started. These code letters are listed in Figure 9–17. From this table you can see that letters at the beginning of the alphabet indicate low LRA ratings, and letters at the end of the alphabet indicate higher LRA ratings. It is important to remember that the number in the table is not the amount of LRA the motor will draw, but rather, it is the number that must be multiplied by the horsepower rating of the motor.

NEMA Code Letter	Locked-Rotor kVA/hp
A	0–3.15
B	3.15–3.55
C	3.55–4.00
D	4.00–4.50
E	4.50–5.00
F	5.00–5.60
G	5.60–6.30
H	6.30–7.10
J	7.10–8.00
K	8.00–9.00
L	9.00–10.0
M	10.0–11.2
N	11.2–12.5
P	12.5–14.0
R	14.0–16.0
S	16.0–18.0
T	18.0–20.0
U	20.0–22.4
V	22.4 and up

Figure 9–17 Locked-rotor amperage (LRA) ratings.
(Courtesy of NEMA, National Electrical Manufacturers Association)

9.12 ## WIRING SPLIT-PHASE MOTORS FOR A CHANGE OF ROTATION AND CHANGE OF VOLTAGE

The wiring diagrams for the run and start windings of the split-phase motor were shown previously in Figure 9–13. The terminals in the run and start windings can be reconnected in the field to change the rotation of the motor or to allow the motor to operate at a different voltage. Since the split-phase motor and the capacitor-start motor are very similar, these connections will be presented in Section 9.16, after you learn more about capacitor-start, induction-run motors.

9.13 ## OVERVIEW OF CAPACITOR-START, INDUCTION-RUN OPEN-TYPE MOTORS

The **capacitor-start, induction-run (CSIR) motor** is basically a split-phase induction motor that combines a capacitor with the start winding to create a larger phase shift between the start and run windings to start the motor. The CSIR open-type motor is generally used in applications where the load is connected directly to the shaft of the motor. For example, a CSIR motor can be used to turn direct-drive fans and larger pump loads. Some CSIR motors are used to drive open-type, belt-driven compressors. The torque for a CSIR motor is larger than for the split-phase motor. A typical CSIR motor is shown in Figure 9–18.

9.14 ## THEORY OF OPERATION OF A CAPACITOR-START, INDUCTION-RUN MOTOR

The operation of the capacitor-start, induction-run open-type motor is very similar to that of the AC split-phase motor explained in the previous sections. It has a rotor and a stator with a run winding and a start winding. A start capacitor is connected in series with the start winding to create a larger phase shift than that of the split-phase motor. The start capacitor is generally rated from 50 to 200 µF (microfarads). The capacitance and voltage rating for the capacitor will be listed on its case. Figure 9–19 shows a picture of two start capacitors. The start capacitor is so named because it is always connected in series with the start winding in

Figure 9–18 A capacitor-start, induction-run open-type motor used for HVAC and refrigeration applications such as direct-drive fan motors and pump motors. *(Courtesy of GE Motors & Industrial Systems, Fort Wayne, Indiana)*

Figure 9–19 The start capacitor is mounted in a round black plastic case. *(Courtesy of North American Capacitor Company)*

the CSIR motor to provide a larger phase shift, which also creates more starting torque. The start capacitor is also what gives the capacitor-start, induction-run motor its name.

The start capacitor has several physical features that make it easy to recognize when you are servicing equipment. The easiest way to recognize a start capacitor is that it is generally mounted in a black plastic case. A second characteristic is that the black plastic case for the start capacitor is generally round in shape. From the picture of the start capacitors in Figure 9–19, you can see that the start capacitor is mounted in a round black plastic case.

The start capacitor is made of two sheets of conducting foil called *plates* that are separated (sandwiched) with three sheets of insulating paper called the *dielectric*. Figure 9–20a shows an example of the foil sheets and dielectric sheets before they are rolled up. Figure 9–20b shows a diagram of the sheets rolled up so that they will fit into the round plastic case. A lead wire is soldered onto each piece of foil and brought out to the top of the capacitor to be used as the capacitor's terminals. Figure 9–20c shows the electrical symbol for the capacitor. Notice that the symbol shows the capacitor conducting plates separated by the dielectric so current cannot pass directly from one plate to the other. Figure 9–20d shows the sheets of a capacitor removed from their plastic case.

When voltage is applied to the capacitor in a circuit, current will flow as far as the plate on the left side of the capacitor. Since the capacitor plates are separated by dielectric paper, current cannot flow directly through the capacitor to the other plate, so the voltage must reverse and move backward through the circuit until it reaches the other capacitor plate. This process is referred to as *charging the capacitor and discharging the capacitor*. If the voltage that causes the current to flow is AC voltage, the AC sine wave will continually charge and discharge the capacitor, which keeps current flowing continuously through the components in the circuit.

Since it takes time for the voltage to charge up the first plate, then reverse itself and discharge the plate, the capacitor will cause a phase shift of up to 90° between the

Figure 9–20 a Two sheets of conducting foil that are separated by sheets of dielectric paper. The dielectric paper acts as an insulator between the conducting foil. The two sheets of foil are called the plates. **b** The layers of conducting foil and dielectric paper are rolled up so that they will fit into the round black plastic case. **c** The electrical symbol for the capacitor shows the two conducting plates separated by the insulator. **d** Picture of paper and foil removed from inside of a capacitor.

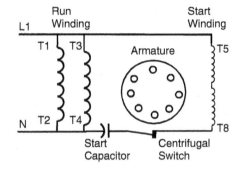

Run
Winding

Start
Winding

L1

T1 T3 Armature T5

T2 T4 T8

Start
Capacitor

Centrifugal
Switch

N

Figure 9–21 Electrical diagram of a capacitor-start, induction-run (CSIR) motor. Notice that the start capacitor (SC) is connected in series with the centrifugal switch and start winding.

voltage waveform and the current waveform in the circuit. It is important to remember that the capacitor will always cause the voltage waveform to lag the current waveform.

9.15 APPLYING VOLTAGE TO THE START WINDING AND CAPACITOR

Figure 9–21 shows an electrical diagram of a capacitor-start, induction-run motor so that you can see that the start capacitor (SC) is connected in series with the centrifugal switch and the start winding. From this diagram you can see that when voltage is first applied to the leads of the motor, current will flow from L1 through the run windings and back to L2 or N. At the same time, voltage will also be applied to the first capacitor plate. Since current cannot flow through the capacitor to the other plate, the voltage must reverse itself and move back through the run winding until it finds its way back to the start winding at the other plate of the capacitor. Since the voltage must take the long way around the motor circuit to get to the other plate, it creates a phase shift between the current flowing in the run winding and the current flowing in the start winding. This process continues as long as the centrifugal switch remains closed and current flows through the start winding. Since the capacitor causes a larger phase shift than the split-phase motor can develop with just the start winding, the capacitor-start, induction-run motor will have more starting torque than the split-phase motor.

When the motor's rpm increases to 75–85% of full rpm, the centrifugal switch opens and causes the current to stop flowing through the start winding. After the centrifugal switch opens the start winding circuit, current continues to flow through the run winding, so the motor acts as an induction motor when it is running. This is why the motor is called a *capacitor-start, induction-run motor.*

9.16 WIRING SPLIT-PHASE AND CAPACITOR-START, INDUCTION-RUN MOTORS FOR 120 V

Figure 9–21 shows the electrical diagram for a capacitor-start, induction-run motor that is connected for 120 V. This connection is also called the low-voltage connection. The diagram for the split-phase motor would be identical except it would not have a start capacitor connected to its start winding. When you must connect a split-phase motor or capacitor-start, induction-run motor for 120 V, you should connect terminals T1, T3, and T5 together with the supply voltage wire L1, and you should connect terminals T2, T4, and T8 together with the neutral supply voltage wire. Notice from the diagram that these connections place the two sections of the run winding in parallel with each other and with the start winding. You should also notice that the start capacitor is connected in series with the centrifugal switch. This means that each section of the run winding will have 120 V applied to it, and the start winding will have 120 V.

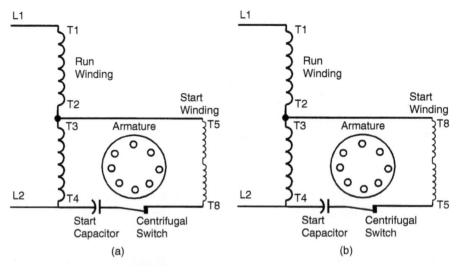

Figure 9–22 (a) Electrical diagram for a capacitor-start, induction-run motor connected for 240 V to rotate clockwise (b) T8 and T5 are reversed to reverse the rotation of the shaft

9.17 WIRING A CAPACITOR-START, INDUCTION-RUN MOTOR FOR 240 V

At times you may need to reconnect a capacitor-start, induction-run motor for a higher voltage. Figure 9–22 shows an electrical diagram of a CSIR motor that is connected for 240 V. This connection is also called the high-voltage connection. Again, the diagram for the split-phase motor would be similar except it would not have the start capacitor. From this diagram you can see that the two sections of the run winding are connected in series. This connection ensures that 120 V is applied to each winding as a voltage drop. You should understand that the insulation on the wire that is used for the run and start windings is rated for 120 V. Since the two run windings are connected in series, you are ensuring that each winding will receive only 120 V, since the 240 V will be dropped equally across each run winding. You should also notice that the capacitor is always connected to the centrifugal switch, so it will always be connected in series with the start winding.

It is important at this time to remember that the motor may be rated for 208 volts for its higher voltage rather than 240 V. The diagram for the higher voltage is the same regardless of whether the motor data plate indicates the motor is rated for 240 V or 208 V. The important point to remember is that the motor must match the exact amount of incoming voltage that is used to supply the motor. This means that you should always measure the exact amount of voltage before you connect a motor.

You should also notice that the start winding is connected so that it receives only 120 V. This means that the start winding is connected so that it is actually in parallel with the lower section of the run winding (T3–T4). One end of the centrifugal switch is connected permanently to one terminal of the start capacitor. The other terminal of the start capacitor is connected to terminal T4. This ensures that it always will be connected across the lower section of the start winding. You can see that if the motor is connected for 240 V or 120 V, one end of the centrifugal switch is always connected through the capacitor to terminal T4.

9.18 WIRING SPLIT-PHASE AND CAPACITOR-START, INDUCTION-RUN MOTORS FOR A CHANGE OF ROTATION

When you are working in the field, you will sometimes find that you must change the direction of rotation of the motor's shaft. Identify the rotation of the shaft by standing behind the motor (at the end opposite where the shaft comes out of the end plate) and

observe the direction of rotation. To change the direction of rotation of the motor shaft, you must change the direction in which current flows through the start winding. The easiest way to do this is to exchange the T5 and T8 wires.

The capacitor-start, induction-run open-type motor that is connected for high voltage can also be wired so that the direction of its shaft's rotation can be reversed. When the motor is wired for clockwise rotation, terminal T5 of the start winding is connected to terminals T2 and T3 of the run windings, and terminal T8 of the start winding is connected to terminal T4 of the run winding. Figure 9–22b shows the electrical diagram of a high-voltage CSIR motor connected for counterclockwise rotation. In this diagram you can see that terminal T8 of the start winding is connected to terminals T2 and T3 of the run winding, and terminal T5 of the start winding is connected to terminal T4 of the run winding.

9.19 CHANGING SPEEDS WITH A CAPACITOR-START, INDUCTION-RUN MOTOR

The split-phase and capacitor-start, induction-run motors are generally not designed to have their speed changed. Some more expensive CSIR motors called **multispeed motors** have additional sets of poles provided so that their speed can be changed by changing the number of poles used in each run winding, but typically this type of motor cannot be reconnected for a change of speed. This means that if you need to change the speed of the motor, you will need to replace the motor with one of a different speed. It is also important to understand that if the motor is used to drive a belt-driven fan, the speed of a fan can be altered slightly by changing the size of the pulley on the motor.

9.20 OVERVIEW OF THE PERMANENT SPLIT-CAPACITOR, OPEN-TYPE MOTOR

The **permanent split-capacitor (PSC) motor** is similar to the split-phase induction motor and capacitor-start, induction-run motor in that it has a start winding and a run winding. However, it is different in that it does not use a centrifugal switch to remove power from the start winding when the motor reaches full speed. Instead of a centrifugal switch, the PSC motor uses a **run capacitor** that is connected in series with the start winding to limit the amount of current that is allowed to flow through the start winding when the motor reaches full speed. The run capacitor is mounted in a metal container so that it can dissipate the heat more easily, which allows it to remain in the circuit even after the motor is running at full speed. The PSC open-type motor is generally used in applications where the load is connected directly to the shaft of the motor. For example, a PSC motor can be used to turn direct-drive fans such as squirrel-cage furnace fans or blade-type condenser fans, and combination evaporator/condenser fans in window air conditioners. The torque of a PSC motor is larger than that of a split-phase motor and about the same or slightly less than that of a capacitor-start, induction-run motor. A typical PSC motor is shown in Figure 9–23. You should be able to see that the run capacitor is mounted on the top of the motor. Figure 9–24 shows a special PSC motor that has a shaft extending out of both ends of the end plates. This type of PSC motor is used in window air conditioners with a blade-type fan mounted on one end to operate as the condenser fan and a squirrel-cage fan on the other end to operate as the evaporator fan.

Figure 9–23 Typical permanent split-capacitor open-type motor used for HVAC and refrigeration applications such as direct-drive squirrel-cage furnace fan motors and blade-type condenser fan motors. *(Courtesy of GE Motors & Industrial Systems, Fort Wayne, Indiana)*

Figure 9–24 A special PSC motor that has a shaft extending out each end of the motor. This type of motor is used in window air conditioners to power the condenser fan and the evaporator fan. *(Courtesy of GE Motors & Industrial Systems, Fort Wayne, Indiana)*

9.21 BASIC PARTS AND THEORY OF OPERATION OF A PERMANENT SPLIT-CAPACITOR MOTOR

The basic parts of the PSC motor include the start winding, the run winding mounted in the stator, and the end plates that support the rotor with the motor shaft on one or both ends. The operation of the PSC open-type motor is similar to that of the AC split-phase motor. It has a rotor and a stator with a run winding and a start winding. A run capacitor is connected in series with the start winding to create a larger phase shift than that of the split-phase motor. The run capacitor is generally rated from 5 to 60 µF. The capacitance and voltage rating for the capacitor will be listed on its case. Figure 9–25 shows a picture of several examples of run capacitors. The run capacitor is so named because it remains in the motor circuit even after the motor is running at full speed.

Figure 9–25 Examples of typical run capacitors. Notice that run capacitors are mounted in oval or rectangular metal containers. *(Courtesy of North American Capacitor Company)*

The run capacitor has several physical features that make it easy to recognize when you are servicing equipment. The easiest way to recognize a run capacitor is that it is generally mounted in a metal container. A second characteristic is that the metal container for the run capacitor is generally oval or rectangular in shape. The construction of the run capacitor is similar to that of the start capacitor. Two sheets of conducting foil called *plates* are separated (sandwiched) by three sheets of insulation, called the *dielectric*. The foil plates and dielectric are rolled up and then placed in the metal container and an electrical terminal is connected to each plate.

You should remember that when voltage is applied to the capacitor in a circuit, current will flow as far as the plate on the *left side* of the capacitor. In this way, energy is stored in the capacitor's plates and then released back into the circuit. Since the capacitor plates are separated by the dielectric, current cannot flow directly through the capacitor to the other plate, so the current must reverse and move backward through the circuit until it reaches the other capacitor plate. This process is referred to as *charging the capacitor and discharging the capacitor*. If the voltage that causes the current to flow is AC voltage, the AC sine wave will continuously charge and discharge the capacitor, which keeps current continuously flowing through the components in the circuit.

Since it takes time for the voltage to charge up the first plate, then reverse itself and discharge the plate, the capacitor will cause a phase shift of up to 90° between the voltage waveform and the current waveform in the circuit. It is important to remember that the capacitor will always cause the voltage waveform to lag the current waveform.

The reason a centrifugal switch is not needed to remove the start winding from the circuit after the motor is running at full speed is that the CEMF that is generated by the rotor will be present across the start winding. When this counter EMF is present, it will oppose the applied voltage, and the resulting current in the start winding will be determined by the difference between the applied voltage and CEMF. If the motor is running at near full rpm, the CEMF will be high and the amount of current in the start winding will be minimal. If the motor encounters a large load and its slip increases because the speed of its shaft slows down, the CEMF will decrease, and the difference between the applied voltage and CEMF will become larger, allowing the start winding to draw extra current, which will help provide enough additional torque to get the motor shaft back to its original speed. In this manner, the run capacitor will allow the start winding to stay in the circuit and add current when additional torque is required to get the shaft to return to its normal speed. Thus the PSC motor is able to regulate its speed. This is an important feature of the PSC motor if it is used for moving air or pumping cooling water for a chiller or pumping refrigerant as a compressor.

9.22 APPLYING VOLTAGE TO THE START WINDING AND CAPACITOR OF A PERMANENT SPLIT-CAPACITOR MOTOR

Figure 9–26 shows an electrical diagram of a typical permanent split-capacitor motor. You can see that the run capacitor (RC) is connected in series with the start winding. When voltage is first applied to the leads of the motor, current flows from L1 through the run windings and back to L2 or N. At the same time, voltage is also applied to the first capacitor plate. Since current cannot flow through the capacitor to the other plate, the voltage must reverse itself and move back through the run winding until it finds its way back to the start winding, where it reaches the other plate of the capacitor. Since the voltage must take the long way around the motor circuit to get to the other plate, it creates a phase shift between the current flowing in the run winding and the current flowing in the

Figure 9–26 Electrical diagram of a permanent split-capacitor (PSC) motor. Notice that the run capacitor (RC) is connected in series with the start winding and that this type of motor does not have a centrifugal switch.

start winding. In the PSC motor this process continues as long as voltage is applied to the motor windings. Since the run capacitor remains in the start winding circuit at all times for this motor, it is called a *permanent split-capacitor motor*.

In the PSC motor the initial current flow into the motor windings is rather large because the rotor is not turning at first. This current is called locked-rotor amperage (LRA). Since a run capacitor is mounted in series with the start winding, the current cannot flow directly through the capacitor. Instead, the voltage moves as far as the first plate when L1 voltage is applied and then stops. At this point one plate of the capacitor will have a large positive charge, which will create a potential differential across the plates of the capacitor. This potential differential causes the voltage to move from the first plate, back through the run winding and start winding until the voltage reaches the opposite plate of the capacitor. This causes a large phase shift between the current in the run winding and start winding, which is sufficient to cause the rotor to spin.

When the rotor begins to spin, it begins to generate counter EMF, which causes the current flow to diminish to a point of normal current flow called full-load amperage (FLA). The actual amount of current the motor draws will depend on the difference between the rated speed and the actual speed. This difference is called slip. The rotor will continue to rotate as long as voltage is applied to the windings.

9.23 INSTALLATION AND SERVICING PRACTICES: CHANGING ROTATION, CHANGING VOLTAGES, AND CHANGING SPEEDS OF A PERMANENT SPLIT-CAPACITOR MOTOR

When you are installing or servicing the permanent split-capacitor motor on HVAC or refrigeration equipment, you must be able to perform several important functions. These functions include reconnecting the PSC motor terminals to change its rotation, reconnecting its terminals so it will run on a different speed, and replacing the motor when it is rated for the wrong voltage. You may need to change the rotation of the motor if the fan blades the motor is driving are turning in the wrong direction or if the pump the motor is connected to is running in the wrong direction. You may need to reconnect the motor terminals to change the motor's speed if the motor is running a multispeed fan. Typically, the highest motor speed is used for air-conditioning applications and the lowest speed for heating applications. Sometimes you need to switch the low speed for an intermediate speed or a fast speed for the next slower speed if the fan is too noisy. If the PSC motor you are changing does not match the supply voltage, you will need to locate a motor that does because the PSC motor is not generally made with dual-voltage capability.

9.24 ## WIRING A PERMANENT SPLIT-CAPACITOR MOTOR FOR A CHANGE OF ROTATION

When you are changing the direction of rotation for the motor, you must change the direction in which current flows through the start winding. The easiest way to do this is to exchange the ends of the start winding wires (T5 and T8). Figure 9–27a shows the terminal connection for wiring a permanent split-capacitor motor for clockwise rotation. You can see that terminal T5 (black wire) of the start winding is connected to the terminal of the run capacitor, and T8 (white wire) of the start winding is connected to the C terminal of the motor. Figure 9–27b shows the diagram of the same motor connected for counterclockwise rotation. In this diagram you can see that terminal T8 of the start winding (white wire) is now connected to the terminal of the run capacitor, and terminal T5 of the start winding (black wire) is connected to the C terminal motor.

Some PSC motors provide a plug connection for the two leads of the start winding. When you need to reverse the start winding leads, you can simply unplug the start windings and reverse the plug, and the motor will rotate in the opposite direction. The plug makes it fairly simple to reverse the direction of rotation of the PSC motor. If the PSC motor does not have the two leads of the start winding exposed through the stator, you cannot change the rotation of the motor. Some original equipment motors are not

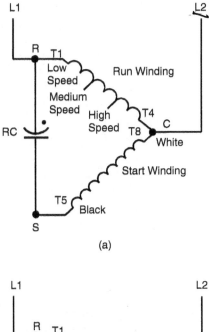

(a)

Figure 9–27 a Electrical diagram for a 240-V permanent split-capacitor open-type motor connected to run in the clockwise direction. **b** Electrical diagram of a 240-V permanent split-capacitor open-type motor connected to run in the counterclockwise direction. Notice that terminals T5 and T8 have been reversed.

(b)

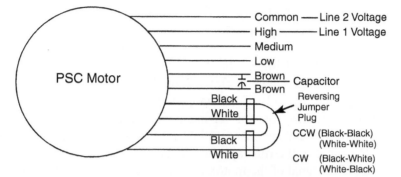

Figure 9–28 Electrical diagram that is shown on a PSC motor's data plate. This field diagram shows how the direction of rotation for the motor can be reversed by changing the start winding. This diagram shows the plug that must be reversed to reverse the start windings (black and white wires).

designed to have their rotation changed, but most replacement motors can have their rotation changed. Figure 9–28 shows an example of a diagram you would find on a PSC motor that indicates how to change the direction of rotation for the motor. In this diagram you can see that you need to remove the plug that has the start winding's black and white wires attached to it, and reverse it when you plug it in again.

9.25 CHANGING SPEEDS OF A PERMANENT SPLIT-CAPACITOR MOTOR

At times you will need to change the speed of a PSC motor, for example, that of a multi-speed furnace fan motor. The diagram on the motor's data plate will show which connections to change. Figure 9–29 shows a diagram of a multispeed motor connected for high speed. From this diagram you should notice that the run winding has multiple taps. When you use all the segments of the run winding, the motor will run at its lowest speed, and when you use just one segment of the run winding, the motor will run at its highest speed.

The wire connected to the low-speed tap is generally colored red, the wire connected to the medium-speed tap is generally colored blue, and the wire connected to the high-speed tap is generally colored black. Figure 9–30 shows a diagram of a PSC motor connected for medium speed. If the motor was connected for high speed, the high-speed terminal would be connected to L1. Each of these diagrams shows the run capacitor connected between the point where L1 is connected to the speed tap and the start winding.

Figure 9–29 Electrical diagram for a high-speed PSC motor. Notice that the middle tap is used for this motor speed.

Figure 9–30 Electrical diagram of a PSC motor connected for medium speed. Notice that the run capacitor is connected to L1 and the start winding.

Important Notice!

Some PSC motors actually have the multispeed taps in the common side of the motor rather than in the run winding. This means that when you change speeds, you must change the tap for the common terminal (L2) rather than the run terminal (L1). This also means that if you test the motor for continuity you will find that the common terminal can be used as the point of reference for some motors, and the run winding must be used as the point of reference on other motors. The main point to remember is that you must check the manufacturer's wiring diagram carefully when you change speeds. Regardless of which winding has the taps, the changing of taps will cause the motor to run at three distinct speeds.

9.26 OVERVIEW OF SHADED-POLE OPEN-TYPE MOTORS

The physical appearance of a **shaded-pole motor** is similar to that of a permanent split-capacitor (PSC) motor, but its operation is different in that it does not use a centrifugal switch to remove power from the start winding when the motor reaches full speed, and it does not use any capacitors to create its phase shift. Instead of a capacitor to create its phase shift for starting, a shaded-pole motor uses a *shading coil*. The shading coil is sometimes called the *shaded pole*, which gives the motor its name. The shaded-pole open-type motor does not have as much starting torque as the other types of open motors, so it is generally used in applications where the smaller loads are connected directly to the shaft of the motor. For example, a shaded-pole motor can be used to turn small direct-drive fans such as squirrel-cage furnace fans or blade-type condenser fans, and combination evaporator/condenser fans in window air conditioners. A typical shaded-pole motor is shown in Figure 9–31. You should notice that the shaded-pole motor looks very similar to the PSC motor except that it does not have a run capacitor. Figure 9–32 shows a shaded-pole motor with a shaft extending out of both ends of the end plates. This type of PSC motor is used in window air conditioners. A blade-type fan is mounted on one end to operate as the condenser fan, and a squirrel-cage fan is mounted on the other end to operate as the evaporator fan.

Figure 9–31 Typical shaded-pole open-type motor used for HVAC and refrigeration applications such as direct-drive squirrel-cage furnace fan motors and blade-type condenser fan motors. *(Courtesy of GE Motors & Industrial Systems, Fort Wayne, Indiana)*

Figure 9–32 A shaded-pole motor with a shaft extending out each end of the motor. The shaded-pole motor, like the PSC motor with a double shaft, is used in window air conditioners to power the condenser fan and evaporator fan. *(Courtesy of GE Motors & Industrial Systems, Fort Wayne, Indiana)*

9.27 BASIC PARTS AND ELECTRICAL DIAGRAM OF THE SHADED-POLE OPEN MOTOR

The basic parts of the shaded-pole motor are the rotor, the stator, and the end plates. The rotor is like the rotors in all the other single-phase induction motors in that it is made of laminated pieces of steel pressed onto the squirrel-cage rotor. The shaded-pole motor generally uses bushing-type bearings in its end plates. This means that it is very important to check the lubrication of the shaded-pole motor several times a year. Figure 9–33b shows an electrical diagram of a shaded-pole motor. From this diagram you can see that the stator has only a run winding. The diagram of the motor in Figure 9–33a shows the four poles of the run winding.

To understand the shading coil, you need to see exactly what it looks like. Figure 9–34a shows how the shading coil is physically mounted in the run winding, and Figure 9–34b shows how the shading coil looks when it is mounted in each of the four poles of the motor. This diagram shows a large coil made of a single turn of a solid piece of copper that is inserted inside each pole of the run windings. This coil is called the shading coil. It creates an induced magnetic field when current is flowing through the run winding. The induced magnetic field creates a small phase shift that is large enough to give the motor enough starting torque to begin to run. In some cases the shading coil is called the start winding, since it serves a purpose similar to that of the start winding of other motors. You should notice from Figure 9–34b that the shading coils are mounted only around the front part of each run winding pole.

The main problems with this type of motor are that the run winding can develop an open circuit, and the bearings can become defective, so that the rotor will not be able to rotate easily. Sometimes, the shading coil becomes loose and the motor's stator becomes overheated. You will see how to test for each of these problems in the troubleshooting section of this chapter.

9.28 THEORY OF OPERATION OF A SHADED-POLE MOTOR

The shaded-pole open-type motor has a rotor and a stator. The stator has a run winding and a shading coil that is mounted around each run winding, which acts like a start winding. When voltage is applied to the run winding, current begins flowing through the winding, which creates a very strong magnetic field. The current flowing through the run winding also creates an induced current in the shading coil. Since the shading coil is made

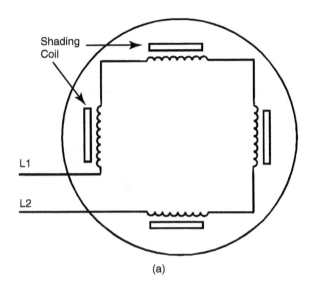

(a)

Figure 9–33 **a** A four-pole shaded-pole motor. This diagram shows that a shading pole is installed with each pole. Notice that the shading poles are not connected to the run winding or to one another in any way.
b Electrical diagram of a shaded-pole motor. The winding in the middle of the motor symbol (circle) may not be included in some symbols. The motor is generally shown as a two-wire motor.

(b)

(a)

Figure 9–34 **a** The physical location of a shading coil as you would find it mounted in the run winding. **b** The physical location of the shading coil on each of four run winding poles.

(b)

from a single-turn coil of copper, the magnetic field is out of phase with the magnetic field in the run winding. This phase shift is sufficient to cause the rotor to begin to turn.

After the rotor comes up to near full rpm, the amount of phase shift between the run winding and shading coil becomes very small. At this point, the shading coil is no longer needed, and the motor is running on just the run winding, as do all the other single-phase motors. The shading coil provides sufficient phase shift to allow the shaded-pole motor to start.

Figure 9–34b shows a cutaway diagram of the stator of a shaded-pole motor. From this diagram you can see that this motor has four poles that make up the run winding. A shading coil can be seen pressed into the windings of each pole. It is important to understand that the shading coil is not electrically connected to the run winding or to other shading coils in any way. This means that when you test the run winding with an ohmmeter, you will be able to determine the continuity of only the run winding, not the shading coil, but this will not create any problems. Since the shading coil is made of a large piece of copper, it will very seldom have any problems, since normal current flow cannot burn it out. It is also important to note that the size of the wire in the run winding of the shaded pole motor is large enough to carry the locked-rotor amperage (LRA) for a sustained period of time if the motor's rotor stalls. This is important since the shaded-pole motor is used for fans and generally uses a bushing-type bearing in the end plate. This means that if the bushing ever runs dry and the rotor stalls, the motor's run winding will tend to survive without damage.

9.29 INSTALLATION AND SERVICING PRACTICES: CHANGING ROTATION, CHANGING VOLTAGES, AND CHANGING SPEEDS OF A SHADED-POLE MOTOR

When you are installing or servicing a shaded-pole motor on HVAC or refrigeration equipment, you must be able to perform several important functions. These functions include reconnecting the shaded-pole motor terminals so it will run on a different speed, and replacing the motor when it is rated for the wrong voltage. The shaded-pole motor is generally not designed to have its direction of rotation changed. This means that if you need to change the direction of rotation of the shaft of a shaded-pole motor, you will generally need to replace the motor with one that is turning in the correct direction.

You will need to reconnect the motor terminals to change the motor's speed if the motor is used in multispeed fan applications. Typically the motor's highest speed is used for air-conditioning applications and its lowest speed for heating applications. The shaded-pole motor is generally not designed to have its windings changed if it does not match the supply voltage. Instead if you are replacing a motor that is connected to 240 V, you must be sure to have a motor with the exact voltage rating. If the motor does not match the voltage rating of the supply voltage, you will need to order a new motor that matches the voltage rating.

9.30 CHANGING SPEEDS OF A SHADED-POLE MOTOR

At times you will need to change the speed of a shaded-pole motor, for example, on a multispeed furnace fan motor. The diagram on the motor's data plate will show which connections to change. Figure 9–35 shows a diagram of a multispeed motor connected for low speed. From this diagram you should notice that the run winding has multiple taps. When you use all the segments of the run winding, the motor will run at its lowest speed, and when you use just one segment of the run winding, the motor will run at its highest speed.

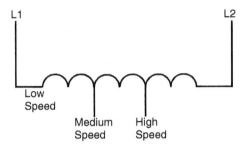

Figure 9–35 Electrical diagram for a 240 V shaded-pole open-type motor connected to run at low speed.

Figure 9–35 Electrical diagram for a 240 V shaded-pole open-type motor connected to run at low speed.

The wire connected to the low-speed tap is generally colored red, the wire connected to the medium-speed tap is generally colored blue, and the wire connected to the high-speed tap is generally colored black. When you want to change the motor's speed, you connect the L1 wire to the blue wire to make it run on medium speed, and you connect L1 to the black wire if you want the motor to run on high speed.

When the L1 wire is connected to the red wire for slow speed, the motor is using all six poles in the run winding. When you use the medium-speed tap, you are using four poles, and when you are connected to the high-speed tap, you are using only two poles. You should remember that the higher the number of poles, the slower the motor will run.

9.31 ## WIRING A SHADED-POLE MOTOR FOR A CHANGE OF ROTATION

When you are working in the field, you will sometimes have to change the direction of rotation of the motor's shaft. Since the shaded-pole motor uses the shading coils as the start winding, it is not practical to open the motor and reverse the shading coils. This means that generally you must order the shaded-pole motor as either a clockwise-rotation motor or a counterclockwise-rotation motor.

There is one other way to reverse the rotation of a shaded-pole motor's shaft. This method is not used too frequently, but it will help you to get out of trouble if you absolutely must change the direction of rotation of the motor's rotor while you are servicing equipment in the field. This method involves removing the end plates of the motor and removing the rotor and replacing it in the stator housing in the opposite direction. This means that if the rotor shaft is coming out of the left side of the stator (when you are looking directly at the motor's data plate), you will remove the rotor and place it back into the stator so the shaft is sticking out of the right side of the stator.

The reason this method may work is that the rotor bars in the squirrel-cage rotor are slightly angled rather than being square with the outside of the rotor. This means that one end of each rotor bar will pass the field coil slightly before the other end. Figure 9–36 shows the rotor bars angled for the shaded-pole rotor. Since each rotor bar becomes magnetized when current is flowing in the run winding, one end of each rotor bar will be the north magnetic pole and the other end will be the south magnetic pole. When the rotor is switched end for end in the stator, this magnetic field relationship will be reversed, and the rotor will rotate in the opposite direction.

Figure 9–36 Rotor for a shaded-pole motor. Notice that the rotor bars are angled so that the rotor can be reversed to change the direction of rotation.

Rotor Bars

Shaded-Pole Motor Rotor

If you use this method to change the direction of rotation of the rotor, you should check the motor after it has run for several minutes to see if the stator is overheating. Some shaded-pole motors were not designed to have their rotors reversed, and if they are reversed, they will cause the motor to overheat slightly.

9.32 TROUBLESHOOTING SINGLE-PHASE MOTORS

At times you will need to troubleshoot a single-phase motor. This section covers problems in which the motor tries to start but does not for some reason. Before beginning to troubleshoot a single-phase motor, you must identify its actual type, such as split phase, capacitor start, permanent split capacitor (PSC), or shaded pole. Single-phase motors can be placed into either of two categories based on their components. The first category consists of split-phase and capacitor-start, induction-run motors that have a centrifugal switch, and the second category consists of permanent split-capacitor and shaded-pole motors that do not have a starting switch such as the centrifugal switch.

The first test to perform when troubleshooting is to determine if the motor has the proper amount of voltage applied to it. If the motor has voltage but will not start, the next step is to determine if the motor "hums" or makes any noise. If the motor hums but will not start, it indicates that only one of the windings has current flowing through it. At this point you need to determine if the motor is a split-phase, capacitor-start, or PSC motor, which have a start winding and a run winding, or if it is a shaded-pole motor, which has only one winding. If the motor has a start and a run winding and hums when it tries to start, the most likely problem is that one of the windings has an open. The simplest way to troubleshoot a motor with this problem is to turn off power to the motor and test both the start winding and the run winding for continuity. Since these motors are generally fractional horsepower, you will most likely throw the faulty motor away and replace it with a new one. If you can access the terminals for the start winding and the run winding, you can use an ammeter to determine which winding has the open, because the winding with the open will not draw any current. If this is a new installation, or if the motor has just been replaced, you need to double-check that the voltage matches the motor data plate.

Another trick you can use when the motor hums but will not start is to give the shaft a twist to help the motor start. If the motor starts when you help it but does not start on its own, you can suspect that the start winding circuit has an open. If the motor has a centrifugal switch, you should suspect it. One other problem you need to consider at this time is a frozen bearing. When a motor does not have proper lubrication, the bearing on the shaft will tend to bind or seize up completely, so it can no longer rotate. When this occurs you can change the bearing or change the entire motor.

If the motor does not start, and it does not hum, you can suspect that it is not drawing any current, so the open in this motor must be in the common terminal, where the start and run winding are connected, or where the supply voltage comes into the motor. Disconnect the power from the motor and test it for continuity. If the motor feels warm to the touch, be sure to let the motor cool down prior to making a decision to replace it, because it may have overheated and tripped an internal overload. If you can determine the cause of the overload, you may be able to correct it. It is important to note, however, that most fractional horsepower single-phase motors do not have internal overloads.

The final type of problem you will have with a motor is that it will run but hum and draw excessive current after it has started. When this occurs you should suspect one of two problems. (1) If the motor has a start and a run winding, the start winding is remaining in the circuit and drawing the extra current. This can occur when the contacts of the start switch (centrifugal switch) become stuck closed. (2) The shaft does not have

proper lubrication, or the load the motor is turning is too large or lacks lubrication. You can fix some of these problems by lubricating the bearings of the motor or the load.

If the motor is a shaded-pole motor, the simplest test for troubleshooting is a continuity test, since the motor consists of four windings connected in series. A final tip is to build a test bench that has a 120- and a 240-V supply so that you can test each motor after you have removed it from a system. This will help you confirm that the problem is what you suspected when you tested the motor in the system.

9.33 REVIEW

A split-phase motor has a stator that has a start winding and a run winding. The start winding will have more resistance than the run winding and can stay in the circuit for only a few seconds while the motor is starting. The motor also has a rotor that is made of laminated steel called a squirrel-cage rotor that allows it to create a strong magnetic field. The split-phase motor has a centrifugal switch that allows current to flow through the start winding when the motor is starting and then disconnects the start winding after the motor reaches full rpm. When the motor starts, it pulls a large amount of current called locked-rotor amperage (LRA), and when the motor reaches full speed, it draws full-load amperage (FLA). The motor pulls LRA current when it starts because the rotor is not spinning. Once the rotor begins to rotate, it begins to produce back EMF, which causes the current to drop to FLA levels.

A single-phase motor will have slip when it is running at full speed. Slip is the difference between the rated speed and the actual speed of a motor. The larger the amount of slip, the more current a motor will draw, and the more torque it will produce. Torque is rotating force produced by the spinning shaft. A split-phase motor can be reversed by reversing the start winding terminals. Some split-phase motors can be connected for 120 V or 208/240 V if the run winding has two sections. The sections are connected in series for high voltage and in parallel for low voltage.

The capacitor-start, induction-run (CSIR) motor is a split-phase motor with a start capacitor added in series with the start winding. This means that all the statements about the split-phase motor are true about the CSIR motor.

The permanent split-capacitor (PSC) motor has a start and a run winding, but it uses a run capacitor connected to its start winding instead of a centrifugal switch. The start winding remains in the motor circuit when the motor is running, and the run capacitor controls the amount of current in it. The speed of the motor can be adjusted by changing the connections to the different speed taps.

The shaded-pole motor is generally used in fan applications and does not need capacitors or a centrifugal switch to start. The easiest test for a shaded-pole motor is a continuity test.

Job Assignment for This Chapter

You are on a job and find that a condenser fan is not operating. Your assignment is to explain to the customer what type of motor you have and sketch the diagram of that motor showing all the electrical parts this motor uses. You should also explain what tests you would use to determine if the motor is working correctly or if it is faulty. Finally you need to read the data plate for the motor and place an order for a new motor that will replace the condenser fan motor.

Solution to the Job Assignment

Since the condenser fan is not operating, the first thing you should do is locate the wiring diagram or schematic for the unit. The condenser motor is typically a PSC (permanent split-capacitor) motor which is shown in Figure 9–26. The motor has a run capacitor connected in series with its start winding, so you may need to test the capacitor. You will also have to test the start and run windings for resistance. The start winding should have more resistance than the run winding. You could also connect the motor to its power source and test the amount of current the start winding and run winding is drawing. You should see the start winding draw high current for a few seconds while the motor is starting and the current should drop very low (less that 1 amp) once the motor is running at full speed. You should be able to show the amount of current the run winding is pulling too, and it should be less than the FLA rating on the motor.

You should take all the data from the data plate. You should be aware of the voltage rating, rpm, shaft rotation, and hp. Also you should be aware of the capacitor size in voltage and capacitance. The motor may have a part number if it is from a brand name system.

Assessment

QUESTIONS

1. Identify the main parts of an AC motor.
2. Explain how you can identify the run winding and start winding in the stator of an AC split-phase motor.
3. Explain how a magnetic field is developed in the rotor of an AC motor.
4. Explain how the squirrel-cage rotor acts like the bar magnet's rotor in the AC motor.
5. Explain why the rotor rotates when a voltage is applied to the stator of an AC motor.

TRUE OR FALSE

1. Torque is rotational force.
2. Slip is the difference between the actual speed of a motor and its rated speed.
3. The AC split-phase motor is called an induction motor because it uses a transformer with a capacitor for starting.
4. The run winding is physically placed offset from the start winding in the stator of an AC motor to help it provide sufficient phase shift to start the rotor.
5. The start winding uses larger wire than the run winding.

MULTIPLE CHOICE

1. The start capacitor is
 a. connected in series with the run winding of a PSC motor.
 b. connected in series with the start winding of a CSIR motor.
 c. connected in series with the start winding of a shaded-pole motor.

2. The run capacitor is
 a. connected in series with the run winding of a PSC motor.
 b. connected in series with the start winding of a PSC motor.
 c. connected in series with the start winding of a CSIR motor.

3. The shaded-pole motor
 a. uses a capacitor and a centrifugal switch.
 b. does not have any capacitors or centrifugal switch.
 c. uses a capacitor but does not need a centrifugal switch.

4. The permanent split-capacitor (PSC) motor
 a. uses a run capacitor and a centrifugal switch.
 b. uses a start capacitor and a centrifugal switch.
 c. uses a run capacitor but does not have a centrifugal switch.

5. The capacitor-start, induction-run (CSIR) motor
 a. uses a start capacitor that is connected in series with its start winding.
 b. uses a start capacitor that is connected in series with its run winding.
 c. uses a run capacitor that is connected in series with its start winding.

PROBLEMS

1. Calculate the speed of a two-pole, a four-pole, a six-pole, and an eight-pole motor.
2. Draw a typical data plate (name plate) for an AC motor and identify each of the items listed in Section 9.11 of this chapter.
3. Draw the electrical diagram of a split-phase motor wired for 120 V AC.
4. Draw the electrical diagram of a capacitor-start, induction-run motor that is wired for 240 V AC.
5. Draw the electrical diagram of a PSC motor that is wired for 240 V AC.

Websites

Websites change constantly. Use the key word search to locate sites that have similar content. The products that will be found at each site are indicated with the key word.

GE Industrial Systems AC Motors
Products: AC furnace fan motors, condenser fan motors
Key word: GE Motors

Reliance Electric
Products: Electric motors
Key word: Reliance Electric

Leeson Motors
Products: AC motors
Key word: Leeson Motors

A. O. Smith
Products: Electric motors
Key word: A. O. Smith Motors

Baldor Motors
Products: Electric motors
Key word: Baldor Motors

North American Capacitor
Products: Capacitors
Key word: North American Capacitor

Chapter 10

Single-Phase Hermetic Compressors

Job Assignment for This Chapter

You are on a job at a clothing store that has one large rooftop air conditioner that cools the entire store. It is the middle of a long heat wave and customers will not come into the store to try on clothes if the air conditioning is not working. Your records show that another service person has made two service calls on the system during the three previous weeks, and each time they have replaced the main fuses for the system. You have made several tests, and you suspect that the hermetic compressor is bad. You have told the owner that you want to change out the compressor, which will take his clothing store out of business for one day while you wait on the compressor to be ordered and shipped in. You have told the owner the cost of the change out will exceed $1500.00. The owner wants you to show him how you determined that the compressor is bad and not some other part of the system before he agrees to changing out the compressor.

Terms You Will Learn in This Chapter

Bimetal element

Bimetal overload

Capacitor-start, capacitor-run (CSCR) compressor

Clamp-on ammeter

Common terminal C

Compressor shell

Compressor terminals

Current relay

Full-load amperage (FLA)

Hard-start kit

Hot wire relay

Induction

Locked-rotor amperage (LRA)

Overload

Permanent split-capacitor (PSC) compressor

Potential relay

Run capacitor

Run terminal R

Run winding

Single-phase compressor

Solid-state starting relay

Split-phase compressor

Split-phase hermetic compressor

Start capacitor

Start terminal S

Start winding

Thermal overload

Troubleshooting the compressor

After reading this chapter, you will be able to:

1. Explain the theory of operation of a single-phase hermetic compressor.

2. Identify the main parts of the split-phase hermetic compressor.

3. Explain the operation of the coil and contacts of the current relay.

4. Identify the wiring diagrams of the split-phase; capacitor-start, induction-run (CSIR); permanent split-capacitor (PSC); and capacitor-start, capacitor-run (CSCR) compressor.

5. Explain the operation of the coil and contacts of a potential relay.

6. Troubleshoot the split-phase; capacitor-start, induction-run (CSIR); permanent split-capacitor (PSC); and capacitor-start, capacitor-run (CSCR) compressor.

10.0 OVERVIEW OF THE SPLIT-PHASE HERMETIC COMPRESSOR

Single-phase hermetic compressors are used in most residential air-conditioning and refrigeration units. The main difference between the open-type motors you studied in the previous chapter and a hermetic motor is that the hermetic motor is sealed and must be cooled by the same refrigerant it is pumping. Since hermetic motors are sealed and have oil mixed with their refrigerant to lubricate their pistons and crankshafts, they cannot use a centrifugal switch to de-energize their start winding when the rotor reaches full speed. Instead, the hermetic motor must have a relay or some other type of device to de-energize the start winding, and this device must be mounted on the outside of the hermetic compressor shell. The other major difference between the open motor and the hermetic motor is that the hermetic motor must have a terminal board that is specifically designed so that the terminals are mounted through the compressor shell. Inside the shell the terminals are connected to the run winding, start winding, and common point, and on the outside of the shell, the terminals are identified as R for run, S for start, and C for common. Figure 10–1 shows examples of hermetic compressor motors you will find in residential air-conditioning and refrigeration equipment. You should remember that a motor is mounted inside each of the compressors, and the rotor of each motor is connected directly to the crankshaft of the compressor, so that when the motor runs, the compressor piston pumps refrigerant. Some compressors use different methods of pumping the refrigerant such as a scroll-type compressor or a centrifugal-type compressor. However, as far as the electric motor is concerned, they all function as one of the four types of single-phase compressor motors: split-phase; capacitor-start, induction-run; capacitor-start, capacitor-run; or permanent split-capacitor.

The single-phase motor can be wired in any one of four ways to produce the required starting torque. The *split-phase* configuration is used for low-torque applications. The *permanent split-capacitor* (PSC) is used for medium-torque applications; the *capacitor-start, induction-run* (CSIR) and *capacitor-start, capacitor-run* (CSCR) configurations are used for high-torque applications.

These wiring configurations have become so common that the name is added to the compressor when it is wired in that particular configuration. For example, when a capacitor is added to a single-phase compressor to give it more starting torque, it becomes a capacitor-start, induction-run motor and is called a capacitor-start, induction-run compressor. When a start capacitor and a run capacitor are added to a split-phase motor, it becomes a capacitor-start, capacitor-run motor.

(a) (b)

(c)

Figure 10–1 Examples of single-phase hermetic compressors. Each of these compressors has a single-phase motor mounted inside it to pump refrigerant. *(Courtesy of Copeland Corporation)*

10.1 BASIC PARTS OF A SPLIT-PHASE HERMETIC COMPRESSOR

The basic parts of a **split-phase hermetic compressor** include the rotor, whose shaft is used to turn the compressor crankshaft, the stator that houses both the **start winding** and the **run winding**, and the starting switch that de-energizes the start winding after the motor is running at 75–85% of full rpm. You should remember that the rotor becomes magnetized by **induction**, which occurs when current is induced into the rotor from the stator windings. Figure 10–2 shows a cutaway diagram of two split-phase hermetic compressors.

Internal Suction Pickup
Motor Windings
Internal Motor Overload
Motor Stacking Stator
Crankshaft
Housing
Top Main Bearing
Weld Seam
Connecting Rod
Piston
Outboard Bearing

Electrical Terminals
Discharge Muffler
Internal Pressure Relief Valve
Discharge Tube
Suction Muffler
Discharge Valve Leaf Assembly
Rubber Mounting Grommet

Figure 10–2 Cutaway pictures of two typical types of compressors. Notice that the motor in both of these compressors is mounted in the top of the compressor, and the piston is located in the bottom of the compressor. In some compressors, the motor is mounted in the bottom of the compressor. *(Courtesy of Tecumseh Products Company; and courtesy of Carlyle Compressor Company, Division of Carrier Corp.)*

The motors for each compressor are mounted in a metal container called the **compressor shell**. The motor may be mounted so that the stator is in the top of the compressor shell and the rotor shaft is pointing downward so that the piston that pumps the refrigerant is in the bottom of the shell, or it may be mounted so the piston is in the top of the shell.

Another important feature of the compressor is the way the wires are connected to the motor terminals. Since the motor is mounted inside the shell, a set of terminals is mounted through the shell in such a way as to create a seal that prevents refrigerant from leaking out of the compressor shell, yet it provides a convenient electrical connection to the motor terminals inside the shell. Figure 10–3 is a close-up cutaway picture of a compressor to show a typical terminal board on a compressor. Notice that the motor in this compressor is upside down, so that the piston is located in the top of the compressor, and the motor is located in the bottom. The terminal board in this compressor is near the bottom, and you can see how the wires from the motor are connected to the terminal board inside the compressor, and three terminal pins are located on the outside of the compressor. The terminal pins make it easy to press on terminal connectors and the current relay directly. The **current relay** takes the place of the centrifugal switch and disconnects the start winding when the motor reaches 75% of full rpm.

The terminal connections for compressors are called **compressor terminals** and they have become somewhat standardized so that the current relay from any manufacturer will fit most any compressor. You will see in the next section that the current relay must be sized for the amperage draw for each compressor motor, but the terminal numbers are standardized. Figure 10–4 shows an electrical diagram of the terminal board with the terminals identified. You should notice that the **common terminal C** is located at the top of the terminal board, and the **start terminal S** is located at the left side of the terminal board, while the **run terminal R** is located at the right side of the terminal board. The run winding is also called the main winding in some compressors, so the R terminal may also be identified as the M terminal.

Figure 10–3 Cutaway view of a compressor that shows how the motor wires are connected to the compressor terminals that are mounted through the compressor shell.
(Courtesy of Tecumseh Products Company)

* Some compressors use the letter M instead of R.

(a)

(b)

Figure 10–4 a Diagram of the terminal connections on a compressor. Notice that the C terminal is located at the top of the terminals, the S terminal is on the left, and the R terminal is on the right. The R terminal is also called the M terminal in the some compressors. **b** Close-up picture of actual terminals on a compressor. Some compressors may have different terminal locations.

10.2 THE SPLIT-PHASE MOTOR INSIDE THE COMPRESSOR

The actual split-phase motor mounted inside the compressor shell is similar to the open split-phase motor in that it has a run winding and a start winding. In some cases, the run winding is also called the main winding. Figure 10–5 shows an electrical diagram of a **split-phase compressor** motor. The run winding uses larger wire than the start winding, and the terminal end of the run winding is identified with the letter R. The start winding is made of smaller-gauge wire than the run winding, and it is identified with the letter S.

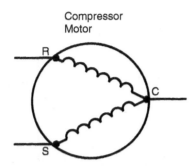

Figure 10–5 Electrical diagram of a split-phase hermetic motor. Notice that the start winding is found between the S and common terminals, and the run winding is found between the R and common terminals. You should remember that some motors have the run winding identified as the main winding.

Notice that the point where the run and start windings come together is called the **common terminal**, and it is identified with the letter C. It is also important to remember that the universal symbol for a motor is a circle with the terminals identified.

Typically the run winding will have between 4 Ω and 20 Ω of resistance, and the start winding will have between 20 Ω and 200 Ω of resistance. Recall that the start winding will have more resistance because it has more feet of smaller-gauge wire than the run winding.

10.3 THE CURRENT RELAY

The split-phase motor needs a means of de-energizing its start winding after the motor reaches 75–85% of its full rpm. In an open-type motor, a centrifugal switch is used, but it cannot be used inside the compressor shell because its open contacts may cause an explosion of the fumes from the refrigerant oil. It would also be difficult to check or replace the centrifugal switch if it were mounted inside the compressor shell, since the compressor shell is welded closed to create the hermetic seal when the compressor is manufactured.

Figure 10–6 shows the current relay connected to the motor terminals. You should notice that the current relay coil is connected in series between L1 and the M terminal of the run winding, and the current relay contacts are connected in series between L1 and the S terminal of the start winding. The M terminal on the current relay may also be identified as the R terminal. The current relay contacts are normally open until voltage is applied to the run winding of the motor. When voltage is applied to the run winding, the run winding draws a large amount of current. Since the current relay coil is connected in series with the run winding, the large current flowing through the run winding must also flow through the current relay coil. The large current that flows through the current relay coil causes it to pull its normally open contacts closed. When the contacts of the current relay are pulled closed, L1 voltage is supplied to the start winding, which begins to draw current. The current flowing through the start winding causes sufficient phase shift to cause the rotor to begin to run. When the rotor is turning at nearly full rpm, the start winding and run winding produce enough CEMF to allow the motor current to decrease to full-load amperage (FLA) levels. When the current returns to the FLA level, the current flowing through the current relay coil is not strong enough to hold the current relay contacts closed, and they return to their normally open location. When the current relay contacts open, they de-energize the start winding.

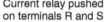

Figure 10–6 Electrical diagram showing a current relay connected to a split-phase motor. The current relay has terminals identified as 1, M, and S. The current relay coil is connected in series with the run winding (main winding), and the normally open current relay contacts are connected in series with the start winding.

Current relay pushed on terminals R and S

Figure 10–7 The diagram of a current relay mounted on the R and S terminals of a compressor. Notice the relative sizes of the compressor and current relay.

In the diagram the current relay looks as large as the compressor motor. In reality, the current relay is approximately 2 in. square, and the compressor may be as large as 18 in. tall. The important thing to remember about it is that it does the same job as the centrifugal switch in the split-phase motor. Figure 10–7 shows a typical current relay mounted directly on the compressor motor R and S terminals. You can get an idea of the relative size of the current relay and compressor motor in this diagram.

10.4 THEORY OF OPERATION OF THE SPLIT-PHASE COMPRESSOR

You should remember from Chapter 9 that the split-phase motor gets its name from the phase shift that occurs between the run and start winding currents to start the motor.

Air-conditioning and refrigeration manufacturers use the split-phase compressor in small residential air conditioners and refrigerators where a small amount of starting torque is required. Other equipment that uses the split-phase compressor includes dehumidifiers and window air conditioners.

10.5 THEORY OF OPERATION OF THE CURRENT RELAY

As you know, the start winding for a single-phase motor can stay in the circuit for only a few seconds during starting, since it will draw a large current. A switch must be added to the start winding circuit to take the start winding out of the circuit when the motor reaches 75% of full rpm. This switch is called a current relay, and it does the same job as the centrifugal switch in the open-type motor, where it de-energizes the start winding after the motor reaches 75% of full rpm. Figures 10–8a and 10–8b show examples of two types of current relays, Figure 10–8c shows the electrical diagram for a current relay, and Figure 10–8d shows a picture of a typical current relay.

You should notice that the terminals are clearly marked on each of the current relays. The current relay in Figure 10–8a has three terminals, and the current relay in Figure 10–8b has an M and an S terminal. It is important to understand that you must connect L1 power to the terminal identified as L and you must connect L2 for a 240-V single-phase compressor to the terminal identified as terminal C on the compressor motor. If the motor is a 120-V motor, the neutral power supply wire should be connected to the C terminal on the compressor.

Figure 10–8 a Example of a current relay used to disconnect the start winding in a split-phase hermetic compressor after the motor has started to run. **b** A second type of current relay. **c** A diagram of a current relay. **d** A picture of a typical current relay.

Figure 10–8c shows the electrical diagram of the current relay. Note that this is a special relay that is used only for starting single-phase compressor motors. The wire in the coil of this relay is made of very large gauge wire so that it can carry the same amount of current that the run winding will draw. You should also notice that the contacts of the current relay are normally open. They will be pulled closed when the current flowing through the coil to the run winding becomes large enough.

10.6 STARTING THE SPLIT-PHASE COMPRESSOR WITH THE CURRENT RELAY

The split-phase compressor used in window air conditioners or residential refrigerators is generally controlled by a line voltage thermostat. The line voltage thermostat is wired in series with the L1 supply voltage and terminal 1 on the current relay. Figure 10–9 shows an electrical diagram of a line voltage thermostat connected in series with a current relay and compressor motor. Whenever the area being cooled becomes too warm, the thermostat contacts will close and pass voltage through the current relay to the motor windings.

Figure 10–9 Electrical diagram of a line voltage thermostat used to control voltage to the compressor motor.

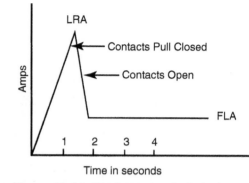

Figure 10–10 Graph showing the locked-rotor amperage (LRA) and full-load amperage (FLA) of a split-phase compressor. Notice the point where the current relay contacts are pulled closed, and where they drop out.

When voltage is first applied to the current relay and motor windings, the contacts of the current relay are open, so voltage can reach only the run winding, which immediately begins to draw current. Since the rotor is not turning, the amount of current quickly becomes very large, up to 10 times the normal full-load amperage (FLA). This current is called the **locked-rotor amperage (LRA)**, and it remains high until the rotor begins to rotate. **Full-load amperage** (FLA) is the amount of current the motor draws when it is running under normal conditions. Figure 10–10 shows the graph of the LRA and the FLA. From this graph you can see that the LRA gets very large as soon as voltage is applied to the motor windings. When the LRA is approximately 75% of its maximum, it is large enough to energize the coil of the current relay and pull the current relay contacts closed.

When the normally open contacts of the current relay close, they supply voltage to the start winding. When the voltage reaches the start winding, it begins to draw current that is out of phase with the current in the run winding. This phase shift is sufficient to provide the starting torque required to get the compressor rotor to begin turning.

When the compressor rotor is turning at 75–85% of its full rpm, the start and run windings produce enough CEMF to allow the current to drop to its FLA value. When the current flowing through the run winding drops to the FLA value, it becomes too small to keep the current relay coil energized enough to keep the current relay contacts closed. At this point, the current relay contacts return to their normally open condition and all voltage and current to the start winding is de-energized. You should notice the point in the diagram in Figure 10–10 where the contacts of the current relay are pulled in, and where they drop out.

10.7 USING THE WIRING DIAGRAM TO INSTALL THE SPLIT-PHASE COMPRESSOR

The wiring diagram for the split-phase compressor is shown in Figure 10–11, which includes all the components of the system. The main difference between the wiring diagram and the ladder diagram is that the wiring diagram shows the location and outline of all the components in the circuit. The wiring diagram is used to install the system and ensure that it is connected correctly to the power from the disconnect box, and it is used to locate components and their terminals.

Figure 10–11 Electrical wiring diagram of a split-phase compressor. Notice that the line voltage thermostat and current relay are also shown in this diagram.

For example, the current relay is shown in the wiring diagram as a square that is connected directly on the compressor. The wiring diagram also shows the locations of terminals on the current relay and the compressor as you would see them on the actual components, as well as the location of wires that are used in the circuit between the disconnect switch, the line voltage thermostat, and the compressor.

When you are ready to install a split-phase compressor in an air-conditioning or refrigeration system, you will notice that the diagram provided will look similar to the one shown in Figure 10–11. Notice that the incoming power from the disconnect and line voltage thermostat to the compressor is connected to terminal L on the left side of the current relay. The L1 wire will provide power for both the run and the start windings of the compressor from this point. Power is provided to the run winding through the coil of the current relay, and power for the start winding is provided through the contacts of the current relay. The wire for L2 power is connected to terminal C on the compressor. This provides the L2 potential voltage directly to the compressor windings at terminal C. If the compressor is a 120-V compressor, the L2 wire will be replaced with a neutral wire.

You can see why it is necessary to use the wiring diagram for installation because you may become confused when you compare the ladder diagram with the actual wires. You should remember that the ladder diagram will provide detailed information that is useful during troubleshooting, and the wiring diagram will provide information for the installation of power wires and help you locate individual components.

10.8 THE CAPACITOR-START, INDUCTION-RUN HERMETIC COMPRESSOR

At times the starting torque from the split-phase compressor will not be sufficient to start the compressor. In these cases a **start capacitor** is added to give the motor more starting torque. The start capacitor is usually added when the compressor is manufactured, but sometimes it can be added to an existing split-phase compressor motor that is having a difficult time starting. When the start capacitor is added to the split-phase motor, it is called a *capacitor-start, induction-run (CSIR) motor*.

Typical applications of the CSIR motor include larger single-phase compressors for residential and commercial air-conditioning and refrigeration equipment. Figure 10–12 shows a picture of a CSIR compressor. You should notice that the major difference

Figure 10–12 A capacitor-start, induction-run (CSIR) compressor. Notice that the start capacitor is mounted near the compressor. *(Courtesy of Copeland Corporation)*

between the CSIR compressor and the split-phase compressor is that the CSIR compressor has a start capacitor that is usually mounted on the compressor or very near it. Recall that the start capacitor is assembled in a round black plastic case. Figure 10–13 shows a diagram of the CSIR compressor. You should notice from this diagram that the CSIR compressor is basically a split-phase compressor with a start capacitor connected in series with the current relay contacts so that it is in series with the start winding.

Figure 10–13 Electrical diagram of the capacitor-start, induction-run compressor. Notice that the start capacitor is connected to the current relay contacts so that it is in series with the start winding.

R Run Winding
S Start Winding
C Common
CR Current Relay
SC Start Capacitor
OL Overload

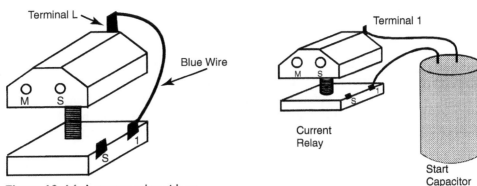

Figure 10–14 A current relay with an external jumper wire connected between terminal 1 on the base and terminal L on the top of the relay. This wire can be cut to allow connection for the start capacitor.

Figure 10–15 A start capacitor is connected to a current relay. Notice that the jumper wire has been cut so that the capacitor can be connected to the current relay.

10.9 BASIC PARTS OF A CAPACITOR-START, INDUCTION-RUN HERMETIC COMPRESSOR

The basic parts of a capacitor-start, induction-run hermetic compressor are very similar to those of the split-phase hermetic compressor that was covered in the previous section, except a start capacitor and current relay are added. The start capacitor is identical with the start capacitors used in CSIR open-type motors, and the current relay for the CSIR compressor is similar to the one used on the split-phase compressor. Since the start capacitor needs to be connected in series with the current relay contacts and the motor start winding, an open needs to be created in the circuit at the current relay. This can be accomplished by removing the jumper wire on the current relay. When the jumper is removed, the capacitor can be connected to the two terminals so it is in series with the contacts of the current relay. Figure 10–14 shows the jumper on the current relay and the capacitor connected in place of the jumper. Figure 10–15 shows the start capacitor connected to the current relay and the blue jumper removed.

10.10 THE OPERATION OF THE CURRENT RELAY

The operation of the current relay in the CSIR compressor is similar to the operation of the current relay for the split-phase compressor, since it does the same job as the centrifugal switch as it de-energizes the start winding after the motor reaches 75% of full rpm. When voltage is first applied to the motor, current flows through the coil of the current relay into the run winding. Since the contacts of the current relay are normally open, current flows only through the run winding at first. Since the current flowing through the run winding cannot get the rotor to turn, it increases to the LRA level. When the current reaches the LRA level, it is sufficient to energize the current relay coil, which pulls its contacts closed. When the current relay contacts close, they allow voltage to reach the start capacitor. After the first plate of the start capacitor charges, it discharges and allows voltage to move to the opposite capacitor plate. This action allows current to flow in the start winding, which provides the phase shift between the run and the start windings. The phase shift is sufficient to allow the rotor to spin, which generates CEMF. The increase in CEMF causes the LRA to return to the FLA level, which causes the current relay coil to reduce its magnetic field so that its contacts return to their open position.

10.11 THERMAL OVERLOAD DEVICE

The **thermal overload** is used to sense the heat buildup from excessive current used by the motor. It uses a set of normally closed contacts identified as OL. In most diagrams these contacts are located at the right of terminal C on the compressor. The overload is connected in series with the common terminal of the compressor, so that all the current in the compressor motor must go through it. Figure 10–16a shows a picture of two types of thermal overload devices. Figure 10–16b shows terminal connections for

(a)

Pin Connector

Wire Lead Assembly Screw Terminal

Screw and Lock Nut Terminal No. 3

Terminal No. 1 Terminal No. 2

Heater Contact

 Bimetal Disc

Case Contact

(b)

Open Closed

(c)

Figure 10–16 a Two types of thermal overload devices. *(Courtesy of Therm-O-Disc)* **b** Terminal connections for an overload device. **c** The contacts of an overload in the open position and in the closed position. *(Courtesy of Tecumseh Products Company)*

Figure 10–17 a Diagram of a thermal overload with its contacts closed. **b** Diagram of a thermal overload with its contacts open after the overload has been subjected to excessive current.

one type of **overload**, and Figure 10–16c shows the operation of the overload contacts in their open condition and in their closed condition.

Figure 10–17a shows an electrical diagram of the **thermal overload** with its contacts in their normally closed position. Figure 10–17b shows a diagram of the thermal overload after it has received excessive current and its contacts are open. From these diagrams you can see that the thermal overload consists of the thermal sensing element and the overload contacts. The contacts are made from copper and steel. These two metals are joined during the manufacturing process, and they are called a **bimetal element**. Copper and steel are used because they expand at different rates when they are heated. The heat causes the contacts to warp or change shape, so that the contacts do not touch when the bimetal becomes hot.

When the current flows through the overload, the thermal sensing element begins to heat up. If the current is within limits, the amount of heat is not large enough to cause the normally closed contacts to open. When the current flowing through the overload becomes excessive, the thermal element produces excess heat that causes the **bimetal overloads** to trip (warp) to their open position.

When the contacts trip to their open position, they interrupt all current flowing through them, which in turn interrupts all the current flowing through the motor. This protects the motor windings from becoming too hot due to excess current flow. When the motor has stopped running for a time and no current is flowing, the overload thermal sensing element cools down and the bimetal shrinks, which causes the contacts to return to their normally closed position. When the contacts close, current is allowed to flow through the overload, so the compressor will start to run again. Some overload devices have a manual reset that forces a technician to physically go to the compressor and press the reset button. This makes the technician check out the cause of the overload condition.

10.12 THE HOT WIRE RELAY

The **hot wire relay** is a starting switch that is used to de-energize the start winding of a hermetic compressor just like the current relay. The hot wire is not used in modern compressor applications because newer types of current relays and solid-state starting switches can do the job much better, but it is still important to understand how they work because you may see them on some older installations. Figure 10–18 shows an electrical diagram of the hot wire relay, and Figure 10–19 shows a sketch of a typical hot wire relay.

Figure 10–18 Electrical diagram of a typical hot wire relay. You should notice that its design is very similar to that of the current relay. The contacts in the start circuit open after the motor begins to run, and the overload contacts will open if the motor draws too much current.

Figure 10–19 The hot wire relay is used on older compressors. You may see one on an older compressor application. The hot wire relay will be mounted on or near the compressor.

In the electrical diagram you can see that the hot wire relay has two terminals for incoming power, which are identified as L1 and L2. The internal parts of the relay consist of a resistive wire that provides heat when current flows through it, and two sets of bimetallic contacts. One set of contacts is identified as the M contacts, and the other set is identified as the S contacts. The compressor's run winding is connected to the M contacts, and the compressor's start winding is connected to the S contacts. The common terminal C of the compressor can be connected to terminal 1, which is connected to L2. If a thermostat or other control devices are used, the C terminal can be connected to terminal 2 of the hot wire relay, and the thermostat can be connected between terminals 1 and 2 on the relay. This will ensure that the thermostat is connected in series with L2 and the common terminal of the compressor. If a starting capacitor is used, it will be connected between terminal S of the hot wire relay and terminal S of the compressor so that it is in series with the start winding, just as when it is used with the current relay.

You can see that all the current from L1 must pass through the resistive wire. After the current passes through the resistive wire, it splits and passes through either M contacts to go to the compressor's run winding, or through the S contacts to the start winding. When voltage is first applied through the resistive wire, it produces a large amount of heat, which is absorbed by the start bimetal switch and the run bimetal switch. After 2 to 3 seconds, the start bimetal switch will have absorbed enough heat to cause it to snap open. When this occurs, the start winding is de-energized from the circuit. Recall that this action is similar to the action of the current relay, which also de-energizes the start winding after 2 to 3 seconds.

After the start winding is de-energized, the total amount of current flowing through the resistive wire is reduced, and the amount of heat it produces is not large enough to cause the run bimetal switch to open. This allows the run winding to remain energized. If a problem occurs with the compressor that causes the run winding to draw excess current, the additional current will flow through the resistive wire, which will produce additional heat. The additional heat will cause the run bimetal switch to snap open and de-energize the run winding. In this way the hot wire relay is also used as an overload

device. In its day, the hot wire relay was used to great advantage as a starting switch and an overload device.

The only problem that the hot wire relay causes is that it does not allow the compressor to be started quickly a number of times. Once the compressor has started, the hot wire has transferred a lot of heat to both the start and run bimetal switches. If the compressor is turned off and restarted in just a few minutes, the leftover heat may be sufficient to cause both the start bimetal switch and the run bimetal switch to open when the starting current is sensed again, which will not allow the compressor to run. This means that the hot wire relay must be allowed to cool down for several minutes whenever it has been running. Note that you may consider replacing any hot wire relays that you find in the field with a current relay or solid-state starting device when you are troubleshooting.

10.13 THE SOLID-STATE STARTING RELAY

Since the advent of solid-state devices, newer starting switches have been designed to replace the current relay. The solid-state starting switch allows run and start current to flow through it when the motor is started, and then it de-energizes the start winding after the motor has run for 2 to 3 seconds. Figure 10–20 shows a picture of the **solid-state starting relay**. Figure 10–21 shows an electrical symbol of the solid-state starting relay. The device has two terminals that are connected in series with the power supply and terminal S on the compressor. If the device has three terminals, it should be connected to the L1 power supply terminal and terminals R and S on the compressor.

Figure 10–20 A solid-state starting relay. This starting relay is mounted near the compressor and has a starting capacitor in series with it.
(Courtesy of Watsco Components, Inc.)

Figure 10–21 Electrical diagram of a solid-state starting relay.

10.14 OVERVIEW OF THE PERMANENT SPLIT-CAPACITOR COMPRESSOR

The **permanent split-capacitor (PSC) compressor** operates like the open-type permanent split-capacitor motor. The PSC compressor uses a **run capacitor** that is connected directly in series with the start winding without using a current relay or other starting relay. Figure 10–22 shows a picture of a typical PSC compressor and a run capacitor.

The PSC configuration becomes necessary for larger compressors. The split-phase and capacitor-start, induction-run motors can be used only on smaller compressors of less than 1 hp. The reason for this is that compressors over 1 hp require larger currents that must travel through the current relay coil and contacts to get to the motor windings. The size of wire in the current relay coil is not able to handle the amount of current that larger compressors draw in their run windings. If a current relay were used on these larger compressors, the large current would cause the current relay coil to burn open or become damaged.

For this reason, manufacturers have found that the PSC motor is a more reliable method of providing voltage to the start winding of the larger compressors while still obtaining the effects of the phase shift caused by the capacitor and start winding. Figure 10–23 shows a ladder diagram of the PSC compressor. From this diagram you can see that a capacitor is connected in series with the start winding of the compressor, and no switches are used to disconnect the capacitor and voltage from the start winding. The PSC compressor, like the PSC open-type motor, does not provide a means to de-energize the start winding, so the capacitor remains in the starting circuit permanently. Thus this wiring configuration for the compressor gives it the name *permanent split-capacitor (PSC) compressor*. The word *permanent* indicates that the capacitor always remains in the start winding circuit, and the word *split* indicates that this is still a split-phase motor.

Figure 10–22 A permanent split-capacitor (PSC) compressor is shown with its run capacitor. *(Courtesy of Copeland Corporation)*

Figure 10–23 Electrical diagram of the permanent split-capacitor (PSC) compressor.

R	Run Winding
S	Start Winding
C	Common
RC	Run Capacitor
OL	Overload

Because the capacitor remains in the start winding circuit at all times, the capacitor is placed in a steel case instead of in plastic so it can dissipate more heat. The PSC compressor has become the most widely used compressor in air conditioners and refrigerators. Since there is no relay to disconnect the start winding after the compressor motor is started, there is one less component that can fail. In the PSC compressor, the capacitor is the only component external to the motor used to start the compressor.

10.15 BASIC PARTS OF THE PERMANENT SPLIT-CAPACITOR COMPRESSOR

The PSC compressor is basically a motor with a run winding and a start winding just like in the split-phase compressor and CSIR compressor. The major difference is that the PSC compressor motor is usually a larger horsepower than the split-phase compressor. The PSC compressor has the same terminal identification as the split-phase and CSIR compressors. The point where the run and start windings are connected in the compressor is called terminal C, and it is the common point. The compressor generally has an overload device connected to the common terminal and mounted on the compressor where it can sense any excess heat from the compressor shell.

The other component used with the PSC compressor is the run capacitor. Figure 10–22 shows a picture of the run capacitor with the compressor, and Figure 9–25 in the previous chapter shows pictures of several sizes of run capacitors. From these pictures you should remember that the run capacitor is housed in either an oval or a rectangular metal container. Because the run capacitor remains in the circuit the entire time the motor is running, the metal container helps the capacitor dissipate heat that builds up in it. You should notice that the run capacitor has a red dot or other means of identification on one of the terminals. It is important to be sure that the capacitor terminal that is identified is connected to the line side of the power supply and not to the compressor S terminal. The identified terminal must be connected to the line-side voltage terminal because this is the terminal connected to the capacitor plate that is nearest the metal container when the capacitor is inserted into the container. This means that if the capacitor plate was to short out the metal container, it would cause the line-side fuse to blow. If this terminal was connected to the start winding of the compressor, the short circuit would cause the large current to flow through the compressor start winding before the fuse would detect the increase in current and open the circuit. This would damage the compressor start winding.

10.16 THEORY OF OPERATION OF THE PSC COMPRESSOR

In the ladder diagram in Figure 10–23 you can see that the run capacitor in the PSC motor is connected in series with the start winding. This means the phase shift that the run capacitor provides during starting will be similar to the phase shift in the capacitor-start, induction-run compressor. The run capacitor is usually rated between 5 µf and 50 µf. Since this rating is smaller than that of the start capacitor, the PSC motor will have a little less starting torque than the capacitor-start, induction-run compressor but more than the split-phase compressor.

Once the PSC compressor starts and begins to run, the motor will again produce counter electromotive force (CEMF). The CEMF will build up to within a few volts of the applied voltage when the compressor is at full speed. The CEMF potential is present between the start and run windings. As long as the difference between the applied voltage and the CEMF is small, very little current will flow in the start winding. For example, if the applied voltage is 240 V and the CEMF is 235 V, the potential difference is 5 V, and the amount of current flowing in the start winding through the capacitor will be minimal. When the compressor loads up to pump a large quantity of refrigerant, the motor rotor will slow down slightly and the amount of CEMF will be reduced. When this occurs, the difference between the CEMF and the applied EMF will become larger and more current will flow through the capacitor. The phase shift will increase and cause more torque in the rotor shaft. You can see that this occurs at exactly the right moment to provide additional torque to move the larger load. The run capacitor is basically self-regulating, since it allows more current to pass as the difference in applied voltage and CEMF gets larger, and less current to flow when the voltage difference is small.

When the compressor is at full speed, the difference between the CEMF and applied EMF is small, and the current in the start winding is small, approximately 2 to 4 A. This small current is not enough to damage the compressor start winding and in effect is similar to having a current relay open its contacts and isolate the start winding.

Another effect of leaving the run capacitor connected to the start winding during the run cycle is that it helps regulate the compressor speed. If the compressor starts to load up (that is, more refrigerant needs to be pumped), the compressor speed tends to slow down slightly. When the rpm slows down, the CEMF becomes smaller, making a bigger voltage difference between the applied voltage and the CEMF. This causes the capacitor to allow more current to flow in the start winding, which gives the motor a little more torque, allowing the compressor to move the heavier load. As the load passes, the compressor again runs at full rpm and the current in the start winding returns to the small amount.

Some run capacitors actually have two separate capacitors mounted in the same container. One of the capacitors is smaller and is used for the PSC condenser fan, and the second is a larger capacitor for the compressor. This type of run capacitor has three terminals. One terminal is common for both capacitors, a second terminal is for the larger capacitor that is used for the compressor, and a third terminal is for the smaller capacitor for the fan. The reason two capacitors are mounted in one container is that it saves space in the unit, and it saves time and wire when the system is manufactured. Figure 10–24 shows a diagram of this type of capacitor used with the compressor and condenser fan. In this type of capacitor the terminals are identified as H, C, and F. The H terminal is connected to the compressor motor start winding, and the F terminal is connected to the start winding of the condenser fan motor. The C terminal is where power is brought into the capacitor.

Figure 10–24 A special run capacitor with two separate capacitors in one container. The terminals are identified as common, compressor, and fan. This capacitor is shown in a common circuit with a compressor and a fan.

10.17 USING THE WIRING DIAGRAM TO INSTALL A PERMANENT SPLIT-CAPACITOR COMPRESSOR

The wiring diagram for the PSC compressor is shown in Figure 10–25. From this diagram you can see all the components of the system. The main difference between the wiring diagram and the ladder diagram is that the wiring diagram shows the location and outline of all the components in the circuit. The wiring diagram is used to install the system components and to ensure that they are connected correctly to the power from the disconnect box. The wiring diagram is also used to locate components and their terminals when making voltage measurements or other tests during troubleshooting. If a component on the wiring diagram is shown near the compressor in the wiring diagram, it means that the component is physically located near the compressor.

For example, the run capacitor is shown in the wiring diagram mounted near the compressor because that is where you would physically find it. Its shape is shown as a rectangle because that is what the run capacitor physically looks like. The wiring diagram also shows the locations of terminals on the run capacitor and the compressor as you would see them on the actual components. The wiring diagram also shows the location of

Figure 10–25 Wiring diagram for a PSC compressor.

wires that are used in the circuit between the capacitor and the compressor. It is important to know the location of each individual wire in the circuit when you need to take voltage or current measurements.

When you are ready to install the PSC compressor in an air-conditioning or refrigeration system, you can use the wiring diagram that is provided with the system. In the diagram provided in Figure 10–25 notice that the incoming power to the compressor is connected to the terminal on the left side of the run capacitor. This terminal on the capacitor is identified with a *red dot*. You should recall from Section 10–15 that it is important that the L1 voltage be connected only to the capacitor terminal with the red dot, and the compressor to the other capacitor terminal. The L1 wire will provide power to both the run and start windings of the compressor from this point. The wire for L2 power is connected to terminal C on the compressor. This provides the voltage directly to the compressor.

You can see why it is necessary to use the wiring diagram for installation because you may become confused when you compare the ladder diagram with the actual wires. The ladder diagram does not necessarily show where the actual wires are connected or located. You should remember that the ladder diagram provides detailed information about the sequence of operation that is useful during troubleshooting, and the wiring diagram provides information for the installation of power wires and helps you to locate individual components.

10.18 CAPACITOR-START, CAPACITOR-RUN COMPRESSOR

Some air-conditioning and refrigeration compressors require very large amounts of starting torque. To give the single-phase compressor the most starting torque possible, both a start capacitor and a run capacitor are used. Since the compressor uses both a start capacitor and a run capacitor, it is called a **capacitor-start, capacitor-run (CSCR) compressor**. Figure 10–26 shows a wiring diagram of a CSCR compressor.

Equipment manufacturers generally use the CSCR compressor on refrigeration equipment that uses an expansion valve and on air-conditioning systems when the permanent split-capacitor (PSC) compressor has trouble starting. In some cases the start capacitor is merely added to an existing PSC compressor to give it more starting torque. If the start capacitor is added to an existing PSC system, it is called a **hard-start kit**. The start capacitor supplies the additional phase shift that provides sufficient starting torque to make the motor start more easily. The hard-start kit will be explained later in this chapter. In either case, the CSCR compressor will have both a start capacitor and a run capacitor to provide the extra starting torque.

Figure 10–26 Wiring diagram for a capacitor-start, capacitor-run compressor. Notice the start capacitor, run capacitor, and potential relay with this compressor.

10.19 COMPONENTS OF THE CSCR COMPRESSOR

Figure 10–27 shows the ladder diagram of the capacitor-start, capacitor-run compressor. From this diagram you can see that the CSCR compressor has a start capacitor like the capacitor-start, induction-run (CSIR) compressor, and it has a run capacitor like the permanent split-capacitor (PSC) compressor. The CSCR compressor also has a potential relay that disconnects the start capacitor from the start winding when the motor is running at full speed, just like the current relay. The compressor for the CSCR configuration is usually 1 to 3 hp, which is larger than those used with the other wiring configurations, because the larger compressors require the larger starting torque. If a compressor larger than 3 hp is required in the system, a three-phase compressor is generally used because it can provide even more starting torque than the CSCR compressor.

The motor used in the CSCR compressor is basically the same type of motor used in the other compressors in that it has a run winding and a start winding, so it has terminals R, S, and C like the other compressors. The run winding is connected between terminals R and C, and the start winding is connected between the S and C terminals.

You studied the start capacitor in detail in the capacitor-start, induction-run compressor, and you studied the run capacitor in the PSC compressor, so the only new component in the CSCR compressor that you need to learn about is the **potential relay**. Figure 10–28 shows a picture of a typical potential relay. From the picture you can see that the potential relay is mounted inside a black rectangular plastic case that is approximately 2.5 in. long. When you see a potential relay mounted near the CSCR compressor on an air-conditioning system, you will see only the black plastic case.

Figure 10–29a shows a sketch of the plastic case with the screw terminals identified, and Figure 10–29b shows an electrical diagram of the potential relay's normally closed contacts and its coil. The three terminal screws on the potential relay are identified as terminals 1, 2, and 5. From the diagram you can see that the potential relay coil is connected to terminals 2 and 5, and the normally closed contacts of the potential relay are connected between terminals 1 and 2. Terminal 2 acts as a common point for both the contacts and the coil. The potential relay coil is made from a large number of turns of very fine wire, so its resistance is approximately 10 kΩ to 20 kΩ. This is much higher resistance than that for the coil of a typical control relay. The potential relay is generally mounted close to the compressor or the start and run capacitors.

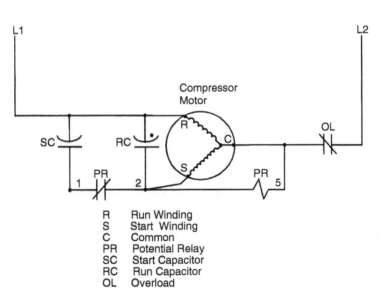

Figure 10–27 Ladder diagram of the capacitor-start, capacitor-run compressor. Notice that the wiring configuration for this compressor uses a start capacitor and a run capacitor.

R Run Winding
S Start Winding
C Common
PR Potential Relay
SC Start Capacitor
RC Run Capacitor
OL Overload

(a) (b)

Figure 10–28 a A typical potential relay with its terminals numbered. The relay looks like a black plastic cube with screw terminals on it. Typically terminals 1, 2, and 5 are used. **b** Potential relay with its cover removed.

Figure 10–29
a Potential relay with its terminals identified.
b Electrical diagram of potential relay.

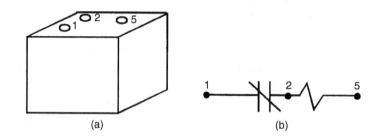

(a) (b)

10.20 THEORY OF OPERATION OF THE CSCR COMPRESSOR

From the ladder diagram in Figure 10–27, you can see that the start capacitor is in parallel with the run capacitor in the CSCR compressor. Both capacitors are connected in parallel with each other and in series with the start winding. Connecting capacitors in parallel adds their capacitive values. The formula $C_T = C_1 + C_2$ is used for adding parallel capacitors, similar to adding resistors in series.

The effect of having a start capacitor with a large capacitance value (up to 600 µf) added to the value of the run capacitor (up to 75 µf) is to cause a larger phase shift between the start and run winding voltages. Recall that the larger the phase shift, the larger the starting torque and the easier it is to start the compressor against large loads. The CSCR compressor uses a potential relay to start the motor and de-energize the start winding. In effect, the potential relay replaces the current relay that is used in the split-phase compressor and the capacitor-start, induction-run (CSIR) compressor.

In the diagram you can see that the contacts of the potential relay are connected in series with the start capacitor just as are the contacts of the current relay in the PSC compressor. From the diagram in Figure 10–27, you can see that when voltage is first applied, the potential relay contacts are normally closed, completing the circuit between the start capacitor and the start winding. As the compressor motor starts to turn, the CEMF begins to build. The CEMF will be present at terminals S and C. The coil of the potential relay (terminals 2 and 5) is connected to terminals S and C on the compressor. As the compressor reaches approximately 75% of full rpm, the CEMF will be strong enough to energize the potential relay coil and pull its contacts open. When the potential relay contacts open, the start capacitor is removed from the start winding.

The run capacitor is left in the start winding circuit, as in the PSC compressor. The combination of the start capacitor and run capacitor provides the phase shift during starting,

which gives the CSCR compressor excellent starting torque, and the run capacitor remains in the circuit, as in the PSC compressor, to give the CSCR compressor good running efficiency. The potential relay coil is connected in parallel with the S and C terminals on the compressor, which means it will last longer, since it is not connected in series with the run winding, as is the current relay, and it is not subject to large currents that tend to wear it out.

10.21 HARD-START KITS FOR COMPRESSORS

At times you will be working on a system that has a split-phase compressor or a PSC compressor that needs a little more torque to enable the compressor to start. When you encounter this problem in the field, you can add a starting capacitor to these compressors to provide them with more starting torque. If a start capacitor is added to a split-phase compressor, it technically becomes a capacitor-start, induction-run compressor. The start capacitor will provide additional phase shift, which develops the starting torque that allows the compressor to start and run. The start capacitor is connected in series with the normally open contacts of the current relay and the start winding. The parts that are used to accomplish this conversion are called a *hard-start kit*, since the result is that the capacitor will help compressors that are having a hard time starting.

One such hard-start kit contains a capacitor and solid-state starting relay in one plastic case, which has two leads protruding out one end. The plastic case is basically a traditional start-capacitor case that has been extended so the solid-state relay can fit on top of the capacitor. Figure 10–30 shows the picture and Figure 10–31 shows the electrical diagram of this type of hard-start kit. Since the solid-state starting relay is

Figure 10–30 A solid-state hard-start kit. Notice that this hard-start kit has the capacitor and solid-state relay in a single plastic case. All you need do is attach the two wires across the leads of the existing run capacitor in a PSC compressor. *(Courtesy of Watsco Components, Inc.)*

Figure 10–31 Electrical diagram of a hard-start kit connected to a compressor. Notice that the start capacitor is connected in series with the solid-state relay, and only two wires are used for connections. *(Courtesy of Watsco Components, Inc.)*

R — Run Winding
S — Start Winding
C — Common
RC — Run Capacitor
OL — Overload
Hard-Start Kit

included in this kit, it can be connected directly on the PSC compressor to make it a capacitor-start, capacitor-run compressor. The solid-state starting relay takes the place of the potential relay. In Figure 10–31 you can see the solid-state hard-start kit connected to an existing PSC motor. When the hard-start kit is connected, you can see that the compressor technically becomes a capacitor-start, capacitor-run compressor.

10.22 USING A CURRENT RELAY AND START CAPACITOR AS A HARD-START KIT

The hard-start kit is also used to add a start capacitor to a split-phase motor to provide additional torque. This is done in the field when a split-phase compressor needs additional starting torque. When the start capacitor is added to the current relay contacts and start winding, the compressor diagram will look similar to the CSIR one in Figure 10–13. It is important to remember that the current relay can be used only on smaller single-phase compressors.

10.23 ADDING A POTENTIAL RELAY AND START CAPACITOR AS A HARD-START KIT

If the PSC compressor is too large to use as a current relay, a potential relay can be used as part of the hard-start kit. The potential relay must be selected so that it matches the CEMF of the compressor. You can usually use the model number of the compressor to help you select the potential relay and start capacitor to fit the compressor. When the potential relay and start capacitor are added to the PSC compressor, it technically becomes a capacitor-start, capacitor-run compressor.

It is also possible to use a hard-start kit as a temporary repair to get a PSC or split-phase compressor that has not started for a number of months to start. Sometimes when a compressor has not started for a long period of time, the pistons become stuck in their cylinders. When this occurs, the compressor motor will hum when power is applied but the compressor will not start. You can then place a start capacitor and potential relay on the PSC or split-phase compressor to get it started. After the compressor is started, you can remove the start capacitor and potential relay and allow the motor to run as it was originally designed.

10.24	USING THE WIRING DIAGRAM TO INSTALL THE CAPACITOR-START, CAPACITOR-RUN COMPRESSOR

The wiring diagram for the CSCR compressor was shown in Figure 10–26. From this diagram you can see all the components of the system. The main difference between the wiring diagram and the ladder diagram is that the wiring diagram shows the location and outline of all the components in the circuit. The wiring diagram is used to install the system and to ensure that it is connected correctly to the power from the disconnect box, and to locate components and their terminals.

For example, the potential relay is shown in the wiring diagram as a square because that is its physical shape. The wiring diagram shows the locations of terminals 1, 2, and 5 as you would see them on the actual potential relay. The wiring diagram also shows the location of wires and jumpers that are used in the circuit between the two capacitors, potential relay, and compressor. You can see that the start capacitor and run capacitors must have terminals that can allow two wires to be connected on each terminal.

When you are ready to install a CSCR compressor in an air-conditioning or refrigeration system, you can use the diagram provided. Notice that the incoming power to the compressor is connected to a terminal on the left side of the start capacitor. This wire will provide power to both the start and run windings from this point. The wire for L2 power is connected to terminal 2 on the potential relay, which also is connected to terminal C on the compressor. This provides the voltage directly to the compressor.

Additional information about ladder diagrams and wiring diagrams was provided in Chapter 6. Recall that the ladder diagram provides detailed information that is useful during troubleshooting, and the wiring diagram provides information for the installation of power wires and helps you locate individual components.

10.25	CONTROLLING THE CSCR COMPRESSOR

Since the capacitor-start, capacitor-run compressor uses a large current to start and run, a CSCR compressor is normally controlled by a relay or contactor rather than by a thermostat (line-voltage type) connected in series with the compressor (see Figure 10–32). When a relay or contactor is used to control a compressor such as in a residential

R	Run Winding
S	Start Winding
C	Common
PR	Potential Relay
SC	Start Capacitor
RC	Run Capacitor
OL	Overload

Figure 10–32 A CSCR compressor controlled by a line-voltage thermostat.

air-conditioning system, the thermostat will be a low-voltage type and it will be connected in series with the relay coil, which draws low current and uses low voltage (24 V AC) to become energized. The contacts of the relay or contactor are connected in series with the compressor motor windings, and they open or close to control the compressor.

In this type of circuit, when power is applied, it moves as far as the open contacts of the line voltage thermostat. When the room becomes too warm, the thermostat contacts close and high voltage is applied to the run and start windings of the compressor, and a large amount of current is pulled as the compressor starts. As you know, the large starting current is called locked-rotor current or locked-rotor amperage (LRA). After the compressor is running at full speed, you will notice that the current drops slightly to its full-load amperage (FLA) level. The amount of FLA current is listed on the name plate for the air-conditioning or refrigeration system or on the compressor so that you can compare it with any measurements that you take. If the full-load current is larger than the rating, the compressor will overheat and cause the overloads to open. The main reasons for high FLA are that the compressor is overloaded and is pumping too hard, or the system is low on refrigerant, and not enough refrigerant is coming back to the compressor to keep it cool.

10.26 TROUBLESHOOTING THE SPLIT-PHASE COMPRESSOR

When a single-phase compressor fails to start in the air conditioner or refrigerator that you are working on, you must be able to troubleshoot the circuit and components quickly and accurately. The following procedures list the steps you should use to troubleshoot all four types of single-phase compressors. **Troubleshooting the compressor** involves these steps, which are broken into several parts. The first procedure should be used when power is applied to the compressor and it is trying to start and you can hear it humming. The second procedure should be used if the compressor does not try to start when power is applied. Note that the same steps are used for the split-phase, CSIR, PSC, and CSCR compressors. The only difference is that you must take into account the extra capacitors and relays that some compressors use and be sure to test them also.

10.27 TROUBLESHOOTING PROCEDURE 1

(The compressor tries to start and you can hear it humming.)

1. Put a clamp-on ammeter around the wire that is connected to terminal S to check for current flowing to the start winding, or clamp it around the wire connected to L1 to check for the current flowing to both the run and start windings. Apply power to the compressor to try to make it run by turning the disconnect to the closed position and switching the thermostat to the on position. Figure 10–33 shows the correct location to place the clamp-on ammeter to check the current the motor is drawing as you try to start the system. If the current becomes excessive and the motor will not start, turn the power off immediately and go to part 2. It is important to always try to make the compressor run even though you know it has problems because sometimes the problem will be a loose connection. If the compressor runs at certain times and not at others, you should suspect loose connections.

2. If the compressor tries to run (makes a noise) but does not start, it is indicating that it is not drawing sufficient current in either the start or run windings, or the

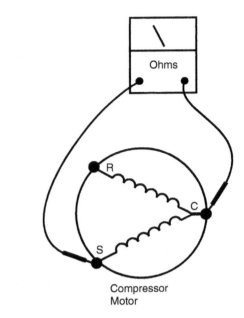

*Note: When testing any motor winding for continuity, be sure the power is turned off and the motor is isolated.

Figure 10–34 Correct location to place the ohmmeter probes to test the start winding for continuity. Be sure to turn off all power to the system and isolate the compressor terminals prior to making this test.

Figure 10–33 Correct location to place clamp-on ammeter to test the current relay and compressor start winding.

compressor piston is not able to turn due to mechanical problems. (**For this test, we will presume that the problem is not mechanical.**)

Since the motor is drawing some current but will not start, you must test to see if the system has the proper amount of voltage available at the disconnect. If the proper amount of voltage is present, and the motor is drawing some amount of current but will not start, you should suspect that sufficient current is not flowing in either the start winding or the run winding. If the current tests shown in Figure 10–33 indicates one of the windings is not drawing any current, you should suspect that winding has an open and test it for continuity with an ohmmeter as shown in Figure 10–34. Figure 10–34 shows the correct location to place the ohmmeter probes to test the start winding after voltage has been turned off. If the start winding shows some amount of resistance (5–20 Ω), it should draw current. If the compressor is a CSIR or CSCR, you should also test any of the components that are in series with the run and start windings for continuity.

When you test the start winding to determine how much current it is drawing, you may need to remove the current relay from the compressor and put jumper wires between the current relay and compressor so you have a place to clamp the ammeter around a wire. Be sure to place the current relay in an upright position so that it will operate correctly. If the current relay is operating correctly, the starting current should increase for 1 to 3 s and then return to zero after the current relay contacts open. If the current continues to be present in the start winding after 2 to 3 s, you should suspect that the current relay contacts are welded closed.

10.28 TROUBLESHOOTING PROCEDURE 2

(The compressor does not try to start when power is applied and it does not make a noise.)

Since the compressor does not try to start and no current is being drawn by the motor when power is applied, it suggests that one of several possible problems is preventing any voltage from reaching the compressor, or the compressor may have an open in a wire that is common to both the run and start windings. These problems can include no power to the system, blown fuse, bad thermostat, bad wire between power source and compressor, open at terminal C in the motor, open in the thermal overload, or open in the neutral line that provides power to the compressor.

1. Test the disconnect for proper voltage. Be sure to test for voltage at any point below the fuses at the terminals marked T1 and T2. If you do not have voltage at this point, you have lost the supply voltage.
2. If you have the proper amount of supply voltage, you should next test for voltage at terminal L of the current relay. Place one probe of the voltmeter on terminal L of the current relay and one probe on the neutral terminal of the disconnect. If voltage is present, the thermostat and all the wires connecting from the disconnect to terminal L of the current relay are good. If no voltage is present, you have an open in this side of the circuit or in the thermostat. Figure 10–35 shows the correct location to place the voltmeter probes to test for voltage at terminal L of the current relay and terminal C on the compressor.
3. If voltage is present at terminal L of the current relay and terminal C on the compressor, but the motor still does not hum or try to draw current, the problem must be in a part of the motor winding circuit that affects voltage to both windings of the motor, such as the current relay or the overload device. Turn off all power to the system and test the current relay and motor windings for continuity. You should also inspect the terminals of the compressor and current relay for dirty or loose connections. The problem you are looking for is an open in a part of the circuit that affects the run winding and the start winding. If either winding has a path for current, the compressor will try to start and you will hear it humming.

Figure 10–35 Correct location to place the voltmeter terminals to measure voltage at terminal l of the current relay and terminal C of the compressor. Remember that terminal L on the current relay may also be identified as terminal M.

Compressor
Motor

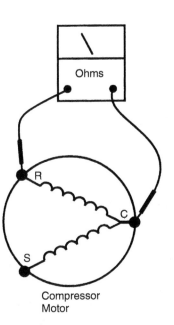

Figure 10–36 Correct location to place the ohmmeter probes to measure the continuity of the run winding of a compressor motor. Be sure to turn off all power to the system before you make a continuity test.

Ohms

Compressor Motor

*Note: When testing any motor winding for continuity, be sure the power is turned off and the motor is isolated.

Since the main symptom for this section is that the compressor does not make any noise when power is applied, the problem must be affecting both windings of the motor. This includes the current relay coil, the common terminal where the run and start windings of the motor come together, and the thermal overload. Figure 10–36 shows the correct location to place the ohmmeter probes to test the run winding for continuity. Be sure to turn off all power to the motor circuit prior to making this test. Place the probes on terminals R and C on the compressor terminals. You should also test the start winding in a similar fashion.

10.29 TESTING THE CURRENT RELAY

At times you will need to test a current relay that you suspect is malfunctioning. Use the following procedure to test your current relay. Figure 10–37 shows the proper location to place the ohmmeter probes to test the current relay. (This procedure can also be used when you are in the field troubleshooting a system and you suspect the current relay is malfunctioning.)

1. Locate the current relay you are testing. Notice that the terminals are marked S, M, and L for this example.
2. The relay coil should be located between terminals L and M. (Refer to the diagram at the bottom of Figure 10–37b.) Use an ohmmeter to test the relay coil. Set the meter at R × 1. Put one probe on the M terminal and one probe on terminal L.
 a. If the meter shows that the coil has some resistance (2–100 Ω), the coil is good. Continue to step 3.
 b. If the meter shows infinite (∞) resistance, the coil is open and the relay must be replaced.

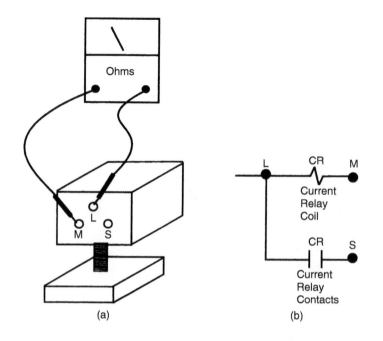

Figure 10–37 **a** Correct location to place the probes of an ohmmeter when you are testing a current relay coil. Use the diagram at the bottom to locate the contacts at terminals L and M. **b** Diagram of current relay.

3. The contacts are between terminals L and S. Use your ohmmeter to test the contacts. Place one meter probe on S and one probe on terminal L. (*Note:* This relay may be position sensitive. The arrow on the side of the relay will show in which position the relay should be mounted. The arrow should point up when the relay is in the proper position.)

a. With the arrow pointing up, the contacts should be open. Set the meter to the highest resistance scale. Measure the resistance between terminals L and S on the relay. If the reading is infinity (∞), continue to step 3b. If the contacts show some small amount of resistance, the contacts may be welded or sticking, and the relay needs to be replaced.

b. You will need to test the relay contacts also to ensure that they have low resistance when they are closed. You can perform this test by turning the relay over so the arrow points down, which will cause the contacts to close by gravity. Repeat the resistance measurement with the contacts closed. This time the contacts should show less than 1 Ω of resistance. If the resistance in the contacts is more than 1 Ω, replace the current relay. Remember, a good current relay will show low resistance in its coil, zero resistance across terminals L to S with its contacts closed, and infinite resistance across terminal L to S when its contacts are open.

10.30 TROUBLESHOOTING PROCEDURE 3

(The motor runs but draws excessive current.)

If the motor starts but draws excessive current and becomes overheated and blows a fuse or causes the overload to open, it is a good possibility that the start winding is staying energized after the motor reaches full rpm. The problem occurs when the contacts of the current relay remain closed after the current in the run winding returns to the FLA value. This can occur when the contacts become welded in the closed position or if the wrong-size current relay is being used. If the current relay has been installed on the compressor for some time, and the compressor worked correctly at one time, you can rule out the

possibility that the current relay is the wrong size and you should suspect that it is malfunctioning. If the current relay or compressor has been changed recently, you should suspect that the wrong current relay is being used.

If the current relay was working previously and the compressor is drawing excessive current after it starts, you should suspect that the contacts are welded closed. You can test for this problem by removing the current relay and testing it for continuity, or you can use a clamp-on ammeter to test it. Remember during this test that you are trying to determine if the contacts of the current relay are open. If the continuity test shows that the current relay contacts are closed when no current is flowing through the coil, the current relay is defective and should be replaced. If you use an ammeter to test the current relay, you should notice that the current through the contacts to the start winding is rather large when the compressor is started and returns to zero after 2 to 3 s when the compressor is started. If the current in the start winding does not return to zero after the compressor starts running, the current relay is faulty and needs to be replaced.

10.31 TROUBLESHOOTING THE CSCR COMPRESSOR

Troubleshooting the CSCR compressor is similar to troubleshooting the CSIR and split-phase compressors, since you can break the test into three parts: testing the circuit, testing the motor, and testing the capacitors and potential relay. The only major difference in testing the CSCR compressor is the additional testing of the potential relays and capacitor. If the problem is in the power supply voltage or the compressor windings, you can use the test described in the preceding section.

10.32 TESTING THE CSCR COMPRESSOR CIRCUIT FOR CURRENT

The CSCR compressor circuit is much easier to troubleshoot than the CSIR or the split-phase compressor because the potential relay and capacitors are mounted remotely from the compressor, which means that you can easily isolate them and get the ammeter around the necessary terminals and wires to make tests. This section will show where to place the ammeter to measure current in the CSCR compressor. Figure 10–38 shows the proper location to connect the clamp-on ammeter to test for FLA current. If the ammeter indicates that the motor is drawing some amount of current, it also means that the motor has voltage applied to its terminals and current is flowing through either the start winding or the run winding, or both. If the FLA test indicates no current flow, you should suspect the total loss of voltage somewhere between the power source at L1 or L2 or in the wire supplying this voltage to the compressor windings. Figure 10–39 shows the proper location to place the clamp-on ammeter to test for current in the start winding and the run winding. It is important to determine whether you do not have current flow through one of these windings so you will know in which part of the system to test for an open.

The previous test for current in the run winding and the start winding will indicate where the problem is located. If you have the proper amount of supply voltage and no current is flowing in the run winding, you can remove the wires from the compressor and test the run winding (R–C) for continuity, since this circuit must have an open. If the current test indicates that the start winding does not have any current flow, you can use the voltage test to test for the loss of voltage between the point where voltage is supplied to the compressor and the normally closed potential relay contacts and start winding. You can also turn off power and use a continuity test for this path. Remember that you are looking for an open in this circuit. If you have no current flow in both the start and run winding tests,

R Run Winding
S Start Winding
C Common
PR Potential Relay
SC Start Capacitor
RC Run Capacitor
OL Overload

Figure 10–38 Diagram showing the clamp-on ammeter connected around the wire to measure the amount of full-load current.

R Run Winding
S Start Winding
C Common
PR Potential Relay
SC Start Capacitor
RC Run Capacitor
OL Overload

Figure 10–39 The clamp-on ammeter should be connected around the wire that is connected between the start capacitor and terminal 1 on the potential relay to measure the current in the start winding. The meter could also be connected around the wire between terminal 2 and terminal S on the compressor.

you can suspect an open in the part of the circuit that is common to both windings, such as the overload, and you can use a continuity test to locate the problem. Remember that if the overload is open, it is a symptom of overcurrent and is not causing the overcurrent.

10.33 TESTING THE POTENTIAL RELAY OUT OF THE CIRCUIT

If you suspect that the potential relay is malfunctioning, you can remove it from the circuit and test it for continuity. The potential relay has a high-resistance coil (approximately 10 kΩ to 20 kΩ) that is energized by the counter electromotive force (CEMF)

produced by the compressor at its C and S terminals. The coil is between terminals 2 and 5 of the relay, and the normally closed contacts are between terminals 2 and 1. When you check the relay coil for continuity between terminals 2 and 5, be sure to set the meter on the R × 100k scale. The relay should have very high resistance (approximately 20 kΩ). When you test the contacts, you should use the lowest resistance range (R × 1), and the meter should indicate that the relay has closed contacts between terminals 2 and 1. If you measure high resistance or infinite (∞) resistance, the contacts are dirty or burned open. In either case, you should consider that the potential relay is not operating correctly and should be replaced.

10.34 TESTING THE POTENTIAL RELAY WHILE IT IS IN THE CIRCUIT

At times you may need to test the potential relay while it is still connected to the compressor circuit. The major reason you would consider testing the potential relay is that the compressor is having trouble getting started. When this occurs, you can check the start capacitor, run capacitor, and potential relay with a clamp-on ammeter. For the first part of this test, clamp the ammeter around the wire that is between the S terminal on the compressor and terminal 2 of the potential relay to measure the amount of current flowing to the start winding of the compressor. You will need to refer to the wiring diagram in Figure 10–38 or 10–39 to see the location of this wire. The reason you would use the wiring diagram instead of the ladder diagram is that when you look at the ladder diagram, it may not be apparent which wire to use because you will see that one end of the start capacitor and terminal 2 of the potential relay are both connected to the S terminal of the compressor through the wire that runs between the S terminal and terminal 2 of the potential relay.

If everything is operating correctly, the ammeter should indicate that 10–20 A is being used by the start winding for 1 to 5 s when it is trying to start and then the current falls off to zero. If the ammeter indicates 0 A or very low amps (less than 2 A) when power is first applied, the start capacitor, run capacitor, or potential relay is not operating correctly. If 0 A is measured, this circuit must have an open in one of the wires, or one of the components has an open. If low current is measured, it means that the wires are good, but one of the components is faulty.

Another problem that occurs with the potential relay is that an open occurs in its coil, its contacts remain closed, and the start winding remains energized after the motor has started and is running. If the start winding remains energized after the motor is running, the start winding will continue to draw large amounts of current and the compressor motor will overheat to the point at which the thermal overload will open and shut off the motor. When this happens, you will be called because the compressor will run for only a short time and then turn off.

The best way to test for this type of problem is to place the clamp-on ammeter around the wire going to the S terminal of the compressor. When power is applied to the compressor, the current in this wire should show 10–20 A for 1 to 5 s until the compressor has started and is running at full speed. When the compressor is at full speed, the current in this wire should drop to 0 A, since the potential relay coil should become energized by the CEMF and pull its normally closed contacts open. When the normally closed contacts of the potential relay open, all the current in the starting circuit should go to zero. If the current does not go to zero, it indicates that the contacts of the potential relay are still closed.

If the current test indicates that the potential relay contacts will not open, you can turn all power to the system off, remove the relay, and replace it with a known

good one and try the current test again. You should replace the potential relay because at this point all you can do is test its coil for continuity. If the coil has an open circuit, it is obviously faulty and you will need to change the relay. If the coil has the correct amount of resistance (10–20 kΩ), you can suspect that the coil is good, but the contacts of the potential relay may be welded in the closed position, so that they cannot open. In either case, you will need to replace the relay and try the current test again.

10.35 TROUBLESHOOTING THE PSC COMPRESSOR

The PSC compressor can be tested during troubleshooting by using the PSC diagram or the CSCR diagrams and procedures and omitting the start capacitor and the potential relay, since this type of motor has only a run capacitor that remains in the circuit at all times. The PSC compressor is perhaps the simplest type of compressor to troubleshoot, since it does not have a starting switch or relay. The start and run windings are identical with the windings in each of the other types of compressors, so you can use any of the preceding tests to test the windings for opens. The preceding current and voltage tests can also be used for the start and run windings.

10.36 REVIEW

The split-phase hermetic compressor is similar to the split-phase open motor in that it has a start winding and a run winding but is different from the motor in that the hermetic compressor does not have a centrifugal switch. Instead, the hermetic compressor uses a starting device that is mounted outside the compressor shell. The simplest starting switch is the current relay, which has a coil connected in series with the run winding, and a set of normally open contacts connected in series with the start winding. The current relay is a push-on type relay.

The capacitor-start, induction-run compressor has a start capacitor connected in series with the start winding to create more starting torque.

The thermal overload is an overcurrent and overtemperature device connected in series with the common terminal on the compressor. If the overload trips, you will need to wait up to 30 min for it to cool down.

The PSC compressor uses a run capacitor connected in series with the start winding. The PSC compressor does not use any starting switch with the run capacitor. The capacitor-start, capacitor-run compressor uses a potential relay that has a coil connected across terminals C and S on the compressor, and its normally closed contacts are wired in series with the start winding. When the compressor starts and comes up to full rpm, the back EMF is large enough to cause the coil of the potential relay to activate and open the contacts and stop current flow in the start winding. A hard-start kit consists of an extra start capacitor that provides extra starting torque.

The easiest way to troubleshoot a compressor is to use a clamp-on ammeter to determine which circuits in the compressor are receiving current. You can also test the windings of the compressor for continuity. The start winding should always have more resistance than the run winding.

Job Assignment for This Chapter

You are on a job at a clothing store that has one large rooftop air conditioner that cools the entire store. It is the middle of a long heat wave and customers will not come into the store to try on clothes if the air conditioning is not working. Your reords show that another service person has made two service calls on the system during the three previous weeks, and each time they have replaced the main fuses for the system. You have made several tests, and you suspect that the hermetic compressor is bad. You have told the owner that you want to change out the compressor, which will take his clothing store out of business for one day while you wait on the compressor to be ordered and shipped in. You have told the owner the cost of the change out will exceed $1500.00. The owner wants you to show him how you determined that the compressor is bad and not some other part of the system before he agrees to changing out the compressor.

Solution to the Job Assignment

Since you suspect the compressor is bad, the first step is to check the compressor data plate and the wiring diagram for the system and identify the type of compressor that you have. You should remember that single-phase compressors need additional components in their circuit to help them start. You should suspect these components are causing the problem before you change out the compressor.

Your inspection shows that the compressor is a single-phase compressor and it is wired as a PSC compressor (if it was a split-phase or capacitor-start capacitor-run, you would troubleshoot the starting components as well) so you need to turn off power and check the run capacitor and the resistance of the run and start winding. After you have checked the amount of resistance, you can turn power on again and check for the exact voltage that is applied to the compressor and also put a clamp-on ammeter on the wire to the run winding and the wire connected to the start winding. You should write down the amount of current for each winding. After you have all the data, you can see where the problem is. Since the fuses are blowing, you should pay close attention to the amounts of current the compressor is pulling, especially during starting, and then compare those values to the values listed on the data plate. Also compare the amount of current to the size of the fuses in the system.

The only time you should recommend changing the compressor is after you have ruled out problems with the amount of supply voltage, ensure all wire lugs are tightened correctly, checked for burned or bad wires and contacts, and ensured the starting components such as capacitors, contactors, etc., are all working correctly. If the start winding is open, the run winding will pull too much current and cause the fuses to blow. If either the start or run winding shows a problem with the amount of resistance, or if a megger shows problems with the windings, you should change the compressor out.

Assessment

QUESTIONS

1. Identify the windings and the terminal markings for any of the three types of single-phase hermetic compressor motors and explain why the windings are similar in each of the four types of motors.
2. Explain the operation of the capacitor-start, capacitor-run hermetic compressor and its two capacitors and potential relay.
3. Use the graph of LRA and FLA in Figure 10–10 to explain how the current relay operates.
4. Explain how you can test the run winding and start winding of a hermetic compressor.
5. Explain how the motor terminals are sealed in a hermetic compressor so refrigerant does not leak around them.

TRUE OR FALSE

1. The run winding or the start winding of a single-phase compressor will draw current if it has an open.
2. The potential relay contacts are normally closed and are opened when the counter EMF (CEMF) in the motor windings becomes large enough to energize the relay coil when the motor reaches 75% of full speed.
3. The run capacitor in the PSC compressor remains in the circuit at all times and can help regulate the speed of the compressor.
4. The contacts of the current relay are normally closed.
5. The thermal overload on the hermetic compressor can be opened by either excessive current draw or overheating of the compressor.

MULTIPLE CHOICE

1. The current relay has
 a. a coil with low resistance and normally closed contacts.
 b. a coil with high resistance and normally closed contacts.
 c. a coil with low resistance and normally open contacts.
 d. a coil with high resistance and normally open contacts.

2. The potential relay has
 a. a coil with low resistance and normally closed contacts.
 b. a coil with high resistance and normally closed contacts.
 c. a coil with low resistance and normally open contacts.
 d. a coil with high resistance and normally open contacts.

3. The split-phase hermetic compressor motor
 a. uses a current relay that has its coil connected in series with the start winding and its contacts connected in series with the run winding.
 b. uses a current relay that has its coil connected in series with the run winding and its contacts connected in series with the start winding.
 c. uses a potential relay that has its contacts connected in series with the start windings.

4. The CSIR split-phase hermetic compressor motor
 a. uses a current relay that has its coil connected in series with the run winding and its contacts connected in series with the start winding.
 b. uses a current relay that has its coil connected in series with the start winding and its contacts connected in series with the run winding.
 c. uses a potential relay that has its contacts connected in series with the start windings.

5. The CSCR split-phase hermetic compressor motor
 a. uses a potential relay that has a run capacitor connected in series with the run winding and a start capacitor connected in series with the potential relay normally open contacts and with the start winding.
 b. uses a current relay that has its coil connected in series with the start winding and its contacts connected in series with the run winding.
 c. uses a potential relay that has its contacts connected in series with the start capacitor and start winding and a run capacitor connected between the run and start windings.

PROBLEMS

1. Draw a sketch of a split-phase hermetic compressor with a current relay connected to its windings.
2. Draw a sketch of a CSIR hermetic compressor with a current relay and start capacitor connected to its windings.
3. Draw a sketch of a CSCR hermetic compressor with a potential relay, start capacitor, and run capacitor connected to its windings.
4. Draw a sketch of a PSC hermetic compressor with a run capacitor connected to its windings.
5. Draw a sketch of any of the hermetic compressors and show where the ammeter would be placed to test current in the run winding and in the start winding.

Websites

Websites change constantly. Use the key word search to locate sites that have similar content. The products that will be found at each site are indicated with the key word.

Tecumseh Products Company
Products: Compressors
Key word: Tecumseh Products Company

Copeland Compressors
Products: Compressors
Key word: Copeland Compressors

Carlyle Compressors
Products: Compressors
Key word: Carlyle Compressor

North American Capacitor
Products: Capacitors
Key word: North American Capacitor

Chapter 11

Three-Phase Open Motors and Three-Phase Hermetic Compressors

Job Assignment for This Chapter

You are on a job at a commercial location that has a large rooftop air conditioner that is supplied with three-phase voltage. The compressor and condenser fan are not running when you get to the unit. The contactor that provides voltage to the compressor and condenser fan is heavily damaged and must be replaced. Your job is to replace the contactor and test the compressor and condenser fan to see if there is any permanent damage. When the contactor is replaced, you find that the condenser fan is running backwards, and the compressor will not run. The owner wants to know what you will do next.

Terms You Will Learn in This Chapter

Changing rotation
Delta-wired motor
Hermetic compressor
Nine-lead motor
Six-lead motor
Speed

Three-phase open motor
Three-phase voltage
Torque
Variable frequency drive (VFD)
Wye-wired motor

After reading this chapter, you will be able to:

1. Explain the theory of operation of the three-phase motor.
2. Identify the main parts of the three-phase open motor.
3. Identify the main parts of the three-phase hermetic compressor.
4. Wire the three-phase open motor for high or low voltage.
5. Change the rotation of a three-phase open motor.
6. Wire the three-phase open motor for delta or wye configuration.
7. Troubleshoot a three-phase compressor.
8. Troubleshoot a three-phase open motor.

11.0 OVERVIEW OF THREE-PHASE MOTORS

When compressor or fan applications require larger motors than single-phase motors, three-phase motors are usually used. The three-phase motor can produce extremely large starting and running **torque,** which means their shaft can turn a larger load. In this section you will learn about the similarities of three-phase AC motors to single-phase AC motors. Today all HVAC technicians must be able to work on commercial jobs that utilize three-phase motors and compressors. You must be able to make small changes on three-phase motors in the field so they can operate at a different **speed,** different voltage, or different rotation. This chapter will help make the field changes, installation, and troubleshooting of three-phase motors easy to perform.

Three-phase AC voltage has three independent sources of voltage that are 120° apart. Figure 11–1 shows a diagram of **three-phase voltage,** in which three separate sine waves are 120° apart. This natural phase shift in the voltage provides the necessary shift in the magnetic field when the voltage is applied to the stationary fields (stator) of the AC motor. This means that three-phase voltage can provide the phase shift required to start a motor naturally without any capacitors or start windings.

When three-phase voltage is applied to the stator winding of a motor, the phase shift causes the magnetic field in the stator to actually rotate or move around the stator at the speed of the frequency of the AC voltage. The phase shift in the stator also causes the squirrel-cage rotor in the three-phase motor to become magnetized. When the rotor is magnetized, its field follows the rotating magnetic field in the rotor and causes the rotor to spin.

Since the three-phase voltage has a natural phase shift, it creates a very strong rotating magnetic field in the motor with the same natural phase shift. The strength of the

Figure 11–1 Example of three-phase voltage. Notice that this voltage consists of three independent sine waves.

magnetic field creates the strong torque at the shaft of the motor. This provides a means to start the motor under heavy loads such as the high pressure against which a compressor must start. The windings of a three-phase motor can be connected in a number of configurations to provide more or less starting torque and running torque, and the windings can be connected for high-voltage (480 V AC) and low-voltage (240 V AC) applications. The direction of rotation of the three-phase motor may be reversed by interchanging any two of the three supply voltage wires so that the phase relationship is reversed.

11.1 THE THREE-PHASE HERMETIC COMPRESSOR

In most air-conditioning and refrigeration systems, the compressor motor and pump are sealed in a container. When a compressor is sealed in this manner, it is called a **hermetic compressor**. The motor is sealed so that the system's refrigerant can be used to cool the motor windings without being lost to the atmosphere. Figure 11–2 shows pictures of several types of three-phase hermetic compressors and a three-phase open-type motor. Figure 11–2a shows a picture of a full hermetic compressor. The three-phase motor for this compressor is completely sealed inside the hermetic shell. The compressor in Figure 11–2b shows a semihermetic compressor. The three-phase motor in the semihermetic compressor is not sealed inside the shell; rather, it is accessible, much like an open motor. The compressor piston is the only part of the semihermetic compressor that is sealed inside the shell. In this section we will explain the wiring diagrams for only the full hermetic compressor. Since the semihermetic compressor uses an open-type motor, its diagram will be presented with the diagrams of the three-phase open motors. Figure 11–2c shows a three-phase open-type motor that is used to power large fans and pumps in HVAC systems.

As an air-conditioning and refrigeration technician you will not open the hermetic compressor for maintenance. Instead the hermetic compressor is replaced as a unit if it fails. Since the hermetic compressor cannot be opened, you must rely on cutaway diagrams and pictures to learn what is inside the motor. You can also get some idea of the basic parts in a hermetic compressor by looking at the parts of a three-phase open motor. The basic parts of an open motor are similar to those in the hermetic compressor.

11.2 BASIC PARTS OF A THREE-PHASE COMPRESSOR MOTOR

The three-phase compressor motor consists of a stator with three separate windings that are mounted out of phase from one another. Figure 11–3 shows a cutaway picture of the three-phase hermetic compressor. From this picture you can see that the windings are equally spaced around the stator, and the wire in each winding is the same gauge size. This means that all three windings in the stator are equal. The rotor is also shown in the cutaway picture of the three-phase compressor. From this picture you can see that the rotor is a squirrel-cage rotor, which is similar to the rotors in other AC motors in that it is made of laminated steel plates that are pressed on the squirrel-cage frame. The steel plates allow the rotor to become magnetized easily and to give up its magnetic field easily. This allows the magnetic bars in the rotor to become magnetized positively and negatively as the rotor spins inside the stator. Since a three-phase motor has three separate voltages, the three-phase motor will have three separate and distinct magnetic fields in the rotor at the same time, which gives the rotor optimum torque.

(c)

Figure 11–2 **a** Example of a three-phase hermetic compressor. *(Courtesy of Copeland Corporation)*
b Example of a three-phase semihermetic compressor. *(GE Motors & Industrial Systems, Fort Wayne, Indiana)* **c** Example of a three-phase open motor.

11.3 ## THEORY OF OPERATION OF THE THREE-PHASE COMPRESSOR

The theory of operation of a three-phase compressor is easy to understand. When three-phase voltage is applied to the three windings of the motor, three separate magnetic fields are developed, which immediately begin to rotate through the stator windings. Since the magnetic field rotates around the stator, the three magnetic fields in the rotor of the

Figure 11-3 Cutaway picture of a three-phase compressor. *(Courtesy of Carlyle Compressor Company, Division of Carrier Corp.)*

compressor "chase" the magnetic fields and cause the rotor to begin to spin. The speed of rotation of the magnet field is fixed and is based on the number of windings and the frequency of the applied voltage. This means that the three-phase compressor motor will run at a constant speed unless the frequency of the applied voltage is varied. In some newer high-efficiency air-conditioning and refrigeration systems, a **variable-frequency drive (VFD)** is used to control the speed of the compressor by providing less than 60 hertz at times when the amount of refrigerant that needs to be pumped is smaller.

Recall that the nature of the phase shift in three-phase voltage and the location of the three-phase windings in the stator of the compressor allow the compressor to develop the magnetic field in the rotor through induction, so no extra components or circuits such as a current relay, potential relay, or capacitors are needed. As long as voltage is applied to the stator, a magnetic field is provided in the rotor by induction and it always tries to cause the rotor to spin with the rotating magnetic field in the stator, which allows the rotor to provide torque at its shaft that can be used to pump refrigerant.

11.4 WIRING A THREE-PHASE HERMETIC COMPRESSOR

A diagram of a three-phase hermetic compressor is shown in Figure 11–4 controlled by a motor starter and a thermostat. From this diagram you can see that the motor terminals are identified as T1, T2, and T3. When you make the field wiring connections to these terminals, you can connect L1 to T1, L2 to T2, and L3 to T3. Since the motor windings are sealed inside the hermetic compressor, you cannot change the connections of the windings to change the speed of the motor or its torque. You can change its direction of rotation by switching L1 to T2 and L2 to T1. At times you may need to reverse the rotation of the

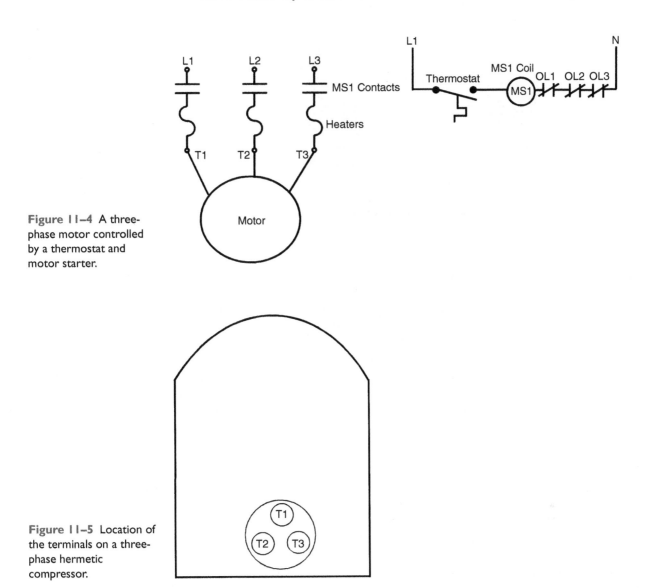

Figure 11–4 A three-phase motor controlled by a thermostat and motor starter.

Figure 11–5 Location of the terminals on a three-phase hermetic compressor.

three-phase hermetic motor to break it free if it has tied up after sitting over the winter. Normally, the compressor will pump refrigerant the same way regardless of the direction in which the compressor motor is turning. Figure 11–5 shows the location of terminals as you would find them on the side of a three-phase hermetic compressor.

11.5 CHANGING CONNECTIONS IN OPEN THREE-PHASE MOTORS TO CHANGE TORQUE, SPEED, OR VOLTAGE REQUIREMENTS

At times as a technician you will need to make minor changes to open three-phase motors that are used to turn fans, water pumps, or open-type compressors so that they will have more torque, run at different speeds, or be connected to a different voltage. Since these motors may cost several hundred dollars, it is important that you be able to make these changes in the field so a new motor does not have to be purchased and installed. After you learn about these field wiring changes, you will become confident in making these changes in the field.

Figure 11–6 Exploded view of a three-phase motor. *(Courtesy of Reliance Electric)*

11.6 THREE-PHASE OPEN-TYPE MOTORS

The internal windings of a **three-phase open motor** are similar to the three-phase hermetic compressor. The main difference in these motors is that the open-type motor has its wires brought out where the technician can make changes in its configuration. Figure 11–6 shows an exploded view of a three-phase open motor. From this picture you can see the end plates, the rotor, and the stator. The three-phase open motor does not need a centrifugal switch, capacitors, current relay, or potential relay to operate. Some open-type three-phase motors require lubrication at the bearings that are located in each end plate. The remainder of this section explains the various ways you can change the rotation, speed, and torque of a three-phase open motor.

11.7 WIRING A THREE-PHASE MOTOR IN A WYE CONFIGURATION

As you know, the three-phase motor has three equal windings, and it does not need any starting switches or capacitors connected to any of its windings. Since the motor has three equal windings, they can be connected in either a wye or a delta configuration. Figure 11–7a shows the three-phase motor windings for a **six-lead motor** connected in a typical wye configuration. This connection is also called a **wye-wired motor**. The ends of each lead are numbered so that they can be changed or reconnected for different wiring configurations. The ends of the leads for the first winding are identified as T1 and T4, the second winding leads are T2 and T5, and the third winding leads are T3 and T6. You can see that terminals T4, T5, and T6 are all connected together at one point in the center of the motor called the *wye point*. The most important point to remember about connecting a three-phase motor in a wye configuration is that it will have less starting torque than if the same motor leads are connected in a delta configuration. This means that the wye-connected motor will draw less current than the delta-connected motor.

11.7.1 Wiring a Three-Phase Motor in a Delta Configuration

The six-lead motor shown in Figure 11–7a can also be connected in a delta configuration which is also called a **Delta-wired motor**. Figure 11–7b shows the motor connected in a delta configuration. In this diagram you can see that the windings are connected in such a way that the three corners of the delta configuration are identified as T1, T2, and T3. In this configuration the incoming power supply is connected to T1, T2, and T3. You should remember that the delta-connected motor has more starting torque and draws more current than the wye-connected motor.

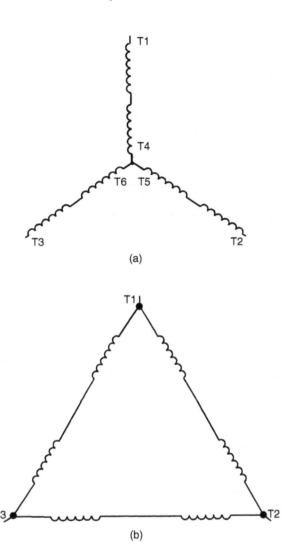

(a)

(b)

Figure 11–7 **a** A three-phase motor connected in a wye configuration. **b** A three-phase motor connected in a delta configuration.

11.7.2 Wiring a Three-Phase Six-Wire Open Motor for High Voltage or Low Voltage

The three-phase open motor is different from the three-phase hermetic motor in that all the terminal leads of the open-type motor are brought out of the stator so that the motor can be reconnected to operate with a high-voltage supply (480 V AC) or a low-voltage supply (240 V AC). Six, 9, or 12 leads may be brought out of the stator of a three-phase motor. If the motor has six leads brought out of the stator, you can use the diagram in Figure 11–8a to connect the leads for low voltage (240 V or 208 V), and you can use the diagram in Figure 11–8b to connect the leads for high voltage (480 V). It is important to understand that if the motor is a six-lead three-phase open motor, it is wired so it can be connected for high or low voltage.

Figure 11–8 **a** Connections for a low-voltage wye-connected motor. **b** Connections for a high-voltage wye-connected motor.

11.8 REWIRING A NINE-LEAD THREE-PHASE WYE-CONNECTED MOTOR FOR A CHANGE OF VOLTAGE

A number of three-phase open motors have nine leads brought out of the stator. The reason this motor has nine leads is that each of the three windings is divided into two pieces. The two sections that make up the three windings can be connected in a wye configuration or a delta configuration. Figure 11–9a shows the nine leads connected in a wye configuration. From this diagram you can see that the two ends of the first half of the first winding are identified as T1 and T4, and the terminals for the second half of the first winding are identified as T7 and common. The common point is where the three windings are connected together to form the wye point.

The terminal ends for the first half of the second winding are identified as T2 and T5, and the terminals for the second half of the second winding are identified as T8 and common. The terminal ends for the first half of the third winding are identified as T3 and T6, and the terminals for the second half of the second winding are identified as T9 and common. The diagram in Figure 11–9a shows the correct way to connect the terminals of the wye-connected motor for high voltage (480 V), and Figure 11–9b shows the correct way to connect the motor for low voltage (208 V or 240 V). It is also important to note that if the three-phase open motor brings out nine leads, it is internally connected as a wye or a delta motor. If the motor is connected as a nine-lead wye-connected motor and you need more starting torque and want to change it to the delta configuration, you will not be able to do so because the motor is internally connected so that it can be wired only as a wye-connected motor. This means that if you need a delta-connected motor, you will need to purchase a new nine-lead motor. This is a minor drawback of the three-phase motor, since you very seldom change the motor from delta to wye or wye to delta once the motor is installed.

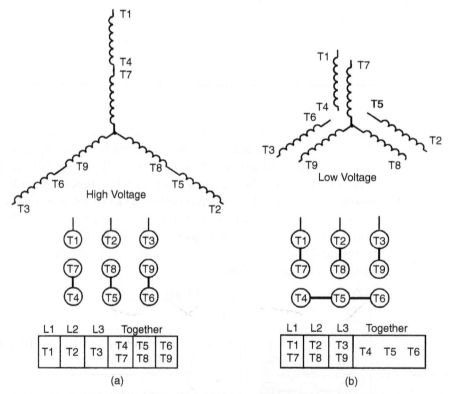

Figure 11–9 a Connections for a nine-lead wye-configured three-phase open-type motor wired for high voltage (480 V). **b** Connections for a nine-lead wye-configured three-phase open-type motor wired for low voltage (208 V or 240 V).

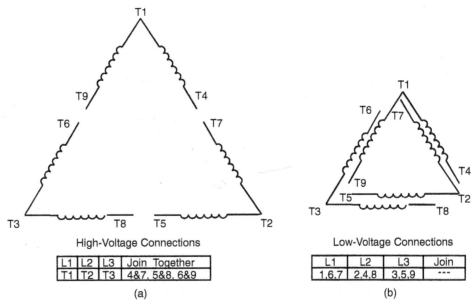

L1	L2	L3	Join Together
T1	T2	T3	4&7, 5&8, 6&9

High-Voltage Connections

(a)

L1	L2	L3	Join
1,6,7	2,4,8	3,5,9	---

Low-Voltage Connections

(b)

Figure 11–10 a Connections for a nine-lead three-phase delta-configured open-type motor wired for high voltage (480 V). **b** Connections for a nine-lead three-phase delta-configured open-type motor wired for low voltage (208 V or 240 V).

11.9 REWIRING A NINE-LEAD THREE-PHASE DELTA-CONNECTED MOTOR FOR A CHANGE OF VOLTAGE

The *nine-lead three-phase open-type motor* is also called the **nine-lead motor** and can also be rewired for high and low voltage in the field. Each winding of the nine-lead motor is divided into two sections. Figure 11–10 shows the nine leads connected in a delta configuration. From this diagram you can see that the two ends of the first half of the first winding are identified as T1 and T4, and the terminals for the second half of the first winding are identified as T1 and T9.

The terminal ends for the first half of the second winding are identified as T2 and T5, and the terminals for the second half of the second winding are identified as T2 and T7. The terminal ends for the first half of the third winding are identified as T3 and T6, and the terminals for the second half of the third winding are identified as T3 and T8. The diagram in Figure 11–10a shows the correct way to connect the terminals of the delta-connected motor for high voltage (480 V), and Figure 11–10b shows the correct connections for low voltage (208 V or 240 V). The important point to remember is that the windings are connected in series for high voltage, and in parallel for low voltage.

11.10 THE 12-LEAD THREE-PHASE MOTOR

Some larger three-phase compressors and open motors can be connected as either a wye motor or a delta motor. In order for this conversion to occur, the motor must have either 6 or 12 leads brought out of the stator. If the motor has 9 leads brought out, the motor cannot be changed from wye to delta in the field. For example, in some large refrigeration applications you must start the motor as a wye-connected motor so it does not draw too much LRA and then reconnect it as a delta motor after it is running so that it will

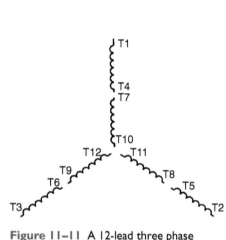

Figure 11–11 A 12-lead three phase open-type motor connected in a wye configuration.

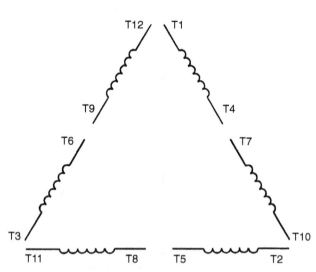

Figure 11–12 A 12-lead three-phase open-type motor connected in a delta configuration.

provide sufficient running torque. This type of motor is called a *wye-delta motor*, and the configuration is changed while the motor is running by a wye-delta motor starter. The wye-delta motor starter is a special motor starter that has two independent sets of three-phase contacts connected to it. The motor is connected in wye configuration on the first set of contacts and as a delta-configured motor on the second set of contacts. When the motor is started, the coil for the first set of contacts is energized and the motor is connected as a wye-connected motor when voltage is first applied. When the motor reaches full speed, the relay that controls the second set of contacts is energized and the first relay is de-energized so the motor is rewired as a delta-configured motor. The motor for the wye-delta starting must be a 6- or 12-lead three-phase open-type motor. Figure 11–11 shows a 12-lead motor connected as a wye motor, and Figure 11–12 shows a 12-lead motor connected as a delta motor.

11.11 WIRING A THREE-PHASE OPEN-TYPE MOTOR FOR A CHANGE OF ROTATION

At times when a three-phase open-type motor is first installed, you will need to change the direction of its rotation while it is connected to its application. This is also called **changing rotation** and it is perhaps the most common change you will make to the three-phase motor. For example, you may need to change the direction in which a pump motor runs, or you may be requested to change the direction in which a fan motor is turning. In these cases all that you need to do is exchange any two of the supply leads. For example, you can change L1 and L2 so that L2 is connected to T1 and L1 is connected to T2. Figure 11–13a shows the three-phase wye motor connected for clockwise rotation. Notice that L1 is connected to T1, and L2 is connected to T2. In Figure 11–13b you can see that the compressor is connected to operate in the counterclockwise direction. L1 and L2 are exchanged, so that L1 is connected to T2, and L2 is connected to T1. Figure 11–14a shows the delta-connected motor connected for clockwise rotation, and Figure 11–14b shows the motor connected for counterclockwise rotation.

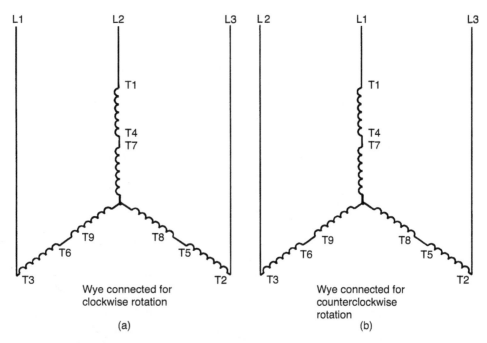

Wye connected for
clockwise rotation

(a)

Wye connected for
counterclockwise
rotation

(b)

Figure 11–13 a A three-phase wye-connected motor connected for clockwise rotation.
b A three-phase wye-connected motor connected for counterclockwise rotation. Notice that T1
is now connected to L2 and T2 is now connected to L1.

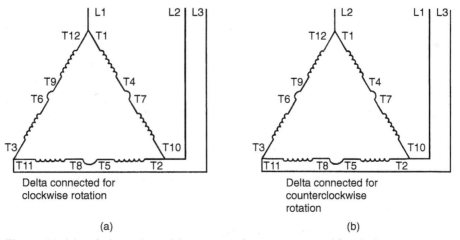

Delta connected for
clockwise rotation

(a)

Delta connected for
counterclockwise
rotation

(b)

Figure 11–14 a A three-phase delta-connected motor connected for clockwise rotation.
b A three-phase delta-connected motor connected for counterclockwise rotation. Notice
that T1 is now connected to L2 and T2 is now connected to L1.

11.12 CONTROLLING A THREE-PHASE COMPRESSOR

The three-phase compressor is generally turned on or off with a relay or contactor. The
voltage to the coil of the relay is controlled by a thermostat. When the contacts in the ther-
mostat close, voltage is applied to the coil of the relay and it draws current and becomes
magnetized and pulls the three sets of normally open contacts to their closed position. When
the contacts close, three-phase voltage is applied to the compressor and it begins to run.
When the temperature is satisfied at the thermostat, the thermostat contacts open and de-
energize the relay coil, and the three sets of contacts open and turn the compressor motor off.

11.13 CONTROLLING THE SPEED OF A THREE-PHASE MOTOR

The speed of a three-phase motor can be changed by changing the number of poles in the motor windings or by changing the frequency of the voltage supplied to the motor. Prior to the 1980s the only way that the frequency of the supply voltage could be easily changed was by changing the speed of the generator that produced the electricity. Since this was not practical, the preferred method of changing the speed of the three-phase motor was to reconnect the motor in the field so it would have a different number of poles. This basically allowed the fan motor or compressor to have two speeds.

In the 1980s more emphasis was placed on efficiency, so it became important to change the speed of three-phase fans and the speed of three-phase compressors so that they would run only at the speed necessary to provide the proper amount of cooling or heating. In the late 1970s and 1980s electronic technology began to produce products that could control larger amounts of voltage and current, so that the frequency of the voltage sent to a motor could be changed. Chapter 19 goes into detail about these devices. In this chapter all you need to know is that the variable-frequency drive is specifically designed to change the frequency of voltage that is used to power a three-phase or a single-phase motor. Since the frequency is changed, the speed of the motor can be adjusted. For example, if the motor needs to operate at a higher speed, the frequency will be increased above 60 Hz. If the speed needs to be decreased, the frequency will be lowered below 60 Hz.

Figure 11–15 shows a picture of a variable-frequency drive, and Figure 11–16 shows a block diagram of a variable-frequency drive. From the block diagram you can see that the drive is supplied with three-phase voltage. The first section of the drive is the rectifier section. In this section three sets of diodes are used to rectify the AC voltage to DC voltage. The second section of the drive is a filter section. You can see that the voltage waveform is half a sine wave as it enters the filter section and it changes to filtered pure DC voltage as it leaves the filter section. The final section of the drive is the transistor section. The transistors can be turned on or off by their triggering circuit so that the output waveform looks similar to the waveform of the original three-phase voltage that supplies this circuit. The main difference between the input voltage and the output voltage is that the frequency of the output voltage can be changed to any value between 1 Hz and 120 Hz. This means that the variable-frequency drive can create frequencies from 1% to 200% of 60 Hz, but the motor speed in practice is generally controlled at between 75% and 125% of the rated motor rpm.

Figure 11–15 A three-phase variable-frequency drive that is used to control the speed of three-phase fan motors and three-phase compressor motors.
(Courtesy of Rockwell Automation)

Figure 11–16 Block diagram of a variable-frequency drive. AC three-phase voltage enters the drive and is converted by the rectifiers to half-wave DC. The filter section converts the half-wave DC to pure DC. The transistors convert the pure DC back to variable-frequency AC. *(Courtesy of Rockwell Automation)*

11.14 TROUBLESHOOTING THE THREE-PHASE COMPRESSOR

At times in the field you will be expected to troubleshoot a three-phase motor that will not start or is malfunctioning. When you are in the field, you can test the three-phase compressor and open-type motor for voltage and current, and each of the windings for continuity. The following sections explain how to make each of these tests. The most frequent test for a three-phase motor is to test for supply voltage at the disconnect or fuse box and for voltage at each of the three motor terminals.

If the three-phase compressor fails to run when it is energized, you will need to make two basic tests. The first test involves testing each of the three lines that supply voltage to the compressor to ensure that three-phase voltage is flowing through the wires to the terminals of the motor. You can make this test by testing for voltage right at the terminals of the compressor motor. If the voltage supply for this compressor is 240 V three phase, you should measure 240 V between T1 and T2, T2 and T3, and T3 and T1. If you have voltage at two of the three terminals, it indicates that you have lost one phase of the three-phase supply voltage, and the most likely problem is a blown fuse. In some cases the compressor can continue to run for a short time when one of the phases is lost, but the compressor will not have sufficient torque to restart, and eventually the internal overload will be tripped. When the internal overload opens it will make you suspect that the winding of the compressor is open. If you test the compressor winding at this time when the internal overload is open, it may lead you to believe the compressor is bad and you will end up changing out a good compressor when the only problem is an open internal overload.

Be sure that the voltage test indicates that you have the correct amount of voltage at each of the three tests you make at the terminals on the compressor. The voltage tests are similar to the test you learned about to determine if voltage is present in a three-phase disconnect switch. Voltage tests for a three-phase open-type motor should be similar to the tests for a three-phase compressor except that some open-type motors will not have internal overloads.

If the correct amount of voltage is present at all three terminals and the motor will not run, you must suspect that one of the windings in the motor is open and has infinite (∞) resistance. (Remember that the compressor may also have an internal thermal overload that will give a similar symptom. Be sure to allow several hours for the overload to cool down before you make a decision about changing the compressor.) If you suspect that the winding is open, you will need to turn the power off to the system and disconnect all the wires from the compressor terminals and test each winding for resistance. If the three-phase motor is an open-type motor, you will use a similar test to see that the proper amount of voltage is present at its terminals. Figure 11–17 shows the proper location to place a voltmeter to make these measurements.

*Note: You must ensure that power is supplied to the motor terminals during this test. You can do this by energizing the MS1 coil so that the MS1 contacts are closed.

Figure 11–17 Proper location to place voltmeter terminals to test for three-phase voltage on a three-phase motor.

11.15 TESTING THE THREE-PHASE MOTOR FOR CONTINUITY

If the voltage test indicates that the proper amount of voltage is present at the terminals but the motor will not start and only hums or does not make any noise, you should suspect that one or more motor windings have an open and you must test the motor windings for resistance. Always remember that the voltage to the motor should be turned off for the continuity test, and the windings should be isolated. The test for resistance should be between T1 and T2, T2 and T3, and T3 and T1. If any of these windings is open, the compressor must be removed and replaced, since it cannot be repaired while it is soldered into the refrigeration system. As stated before, some compressors have an internal thermal overload built into the windings, and it may open if the internal temperature gets too hot. The motor windings may get too hot if the amount of refrigerant in the system is too low to provide cooling for the motor or if the motor is overloaded. Recall that the hermetic motor is cooled only by the extra refrigerant brought back to the motor specifically to cool the windings. If the overload is open, it will not allow current to flow through the windings and hence the motor will not run. Since the overload may be the problem, it is always a good practice to wait 1 to 2 h for the compressor to cool down and retest the windings for continuity before you change it.

It is also important to understand that in some compressors the motor windings are used as a heater when the motor is not running. This is accomplished by connecting capacitors and resistors to allow a small amount of bleed current to flow through the windings to act as a small heater. This small amount of heat keeps the windings warm enough so that the refrigerant oil does not migrate away from the compressor to a part of the system that is warmer. A problem may occur when you test the compressor if you do not make sure that this whole circuit is isolated from the motor windings when you are

Figure 11–18 Proper locations to place ohmmeter terminals to test the three-phase motor windings for continuity to check for an open circuit.

testing for an open circuit. Figure 11–18 shows the proper location to place the ohmmeter to make the continuity test.

11.16 TESTING A THREE-PHASE MOTOR FOR CURRENT

If the three-phase motor will start, you must check it for the proper amount of current to ensure that it is not overloaded. Figure 11–19 shows the proper location to place a clamp-on ammeter around one of the three-phase wires that supplies voltage to the motor. Since each of the supply wires is connected in series with the motor windings, the current measurement will indicate how much current the motor is drawing through each winding. It is important to understand that the amount of current measured in each

Figure 11–19 Proper location to place a clamp-on ammeter to measure current in the motor winding.

of the three wires should be nearly the same. If one of the leads is drawing excess current, it indicates that the motor is beginning to fail. If all three measurements are larger than the data plate rating, the motor is being overloaded and you will need to determine the cause of the overload. Sometimes three-phase compressors will draw excess current when the refrigeration system is low on refrigerant and the motor is running extremely hot. The compressor can also draw excess current if it is pumping the refrigerant as a liquid instead of as a vapor.

11.17 REVIEW

The three-phase open motor and three-phase compressor use the same basic parts and theory of operation. The motors have a stator with three identical windings, and a squirrel-cage rotor. The phase shift found naturally in three-phase voltage provides starting torque to the three identical windings.

The stator windings in a three-phase motor can be wired in a wye configuration or a delta configuration. If the motor has nine leads, it can be connected for high voltage (480 V) or low voltage (240 V).

The three-phase motor can be reversed by changing any two of the three leads. The easiest way to troubleshoot a three-phase motor is to use a clamp-on ammeter. If an open occurs in one of the three windings, the motor will hum and not start. If the motor will not start and makes no noise, you can suspect that two or more windings have an open or two or more supply voltage lines are not providing power.

Job Assignment for This Chapter

You are on a job at a commercial location that has a large rooftop air conditioner that is supplied with three-phase voltage. The compressor and condenser fan are not running when you get to the unit. The contactor that provides voltage to the compressor and condenser fan is heavily damaged and must be replaced. Your job is to replace the contactor and test the compressor and condenser fan to see if there is any permanent damage. When the contactor is replaced, you find that the condenser fan is running backwards, and the compressor will not run. The owner wants to know what you will do next.

Solution to the Job Assignment

Since the compressor and condenser fan are not running when you get to the unit, you should visually inspect all of the controls in the electrical control panel. Since the contactor that provides voltage to the compressor and condenser fan is heavily damaged, you must check the diagram to identify its voltage and current ratings. (You might be able to find a part number if you have the original manuals.) You will need the voltage and current rating to order the new contactor. After you locate the new contactor, you can remove the old one and wire the new one into place. After you have the contactor replaced, you should visually inspect the wiring and other parts of the system to see if there is any other damage. If you do not find any further damage, you should try to run the compressor and condenser fan. When you apply voltage to the condenser fan and compressor, you notice that the condenser fan is running backwards, and the compressor will not run.

Since the condenser fan is powered by three-phase voltage, you can reverse its direction of rotation by exchanging any two of the three wires supplying voltage to the fan. This will cause the three-phase motor to run in the opposite direction.

Since the compressor will not run, you will need to turn off power at the disconnect and test the compressor motor for continuity. Since it is a three-phase motor, you should find the same resistance in each of the three windings. (Be sure to isolate the terminals of the motor by removing the wires that supply voltage to it. This will keep you from reading resistance from a transformer or other motor that may be connected in parallel with the motor.)

You need to keep in mind that the voltage that makes the condenser fan run comes from the same contactor as the compressor, so since the condenser fan is running, you do not need to suspect the power supply. The problem will be the wires from the compressor to the contactor or the compressor motor windings and any internal overloads. Be sure to allow sufficient time for any overload to cool down and reset. If you test the motor and determine any of the windings are open, the compressor will have to be removed and replaced, since the windings of the hermetic compressor cannot be repaired in the field.

Note: You could also test for an open winding in the compressor by using a clamp-on ammeter or by listening to the motor when power is applied. If the motor hums loudly and tries to start and draws excessive current, you can suspect that one of the windings is open or a wire to one of the windings is open. If the motor hums, you can be certain that you have at least two of the three windings functioning correctly. Since one of the windings is not drawing current, the motor will not start but will draw locked-rotor amps (LRA), so you should not allow the motor to try to run for more that 2 to 3 seconds.

After the compressor is running again, you should be sure to check the amount of voltage to see if it is drawing less than the rated current. If the motor is drawing too much current, you will need to do further checks to determine what is causing the motor to overload. Be sure to fix the problem before you leave; otherwise, the new compressor will burn out quickly and the owner will expect the next compressor to be under warranty.

Assessment

QUESTIONS

1. Explain why a three-phase motor can start without using any start relays or capacitors.
2. Identify the main parts of a three-phase open motor and three-phase hermetic motor.
3. Explain why you would need to reconnect the terminals of a three-phase pump or fan motor to change its direction of rotation.
4. Explain why you would need to reconnect the terminals of a three-phase motor for a change of voltage.
5. The three-phase hermetic compressor you are testing for continuity indicates an open circuit. Explain why you should wait 2 hours or more before you make the decision to change the compressor.

TRUE OR FALSE

1. The direction of rotation of a three-phase motor can be reversed by exchanging any two of its three terminals.
2. One of the windings in a three-phase motor is smaller than the other two so that a phase shift can be created to help the motor get started.

3. The three-phase motor has more starting torque than an equal size single-phase motor.

4. It is possible to change the internal winding connections on a nine-lead three-phase open-type motor so that it can operate on 480 V AC or 240 V AC.

5. It is possible to change the internal windings on a hermetic three-phase compressor so it will operate at a different voltage.

MULTIPLE CHOICE

1. If you change motor terminal T1 to L2 of the supply voltage and terminal T2 to L1 of the supply voltage, the three-phase motor will
 a. not run correctly because it must be wired L1 to T1, L2 to T2, and L3 to T3.
 b. change the direction of its rotation.
 c. be able to run on 240 V AC instead of 480 V AC.

2. The three-phase motor does not require any current relays, potential relays, or a centrifugal switch because
 a. it has three equal windings instead of a start winding.
 b. it is a low-torque motor, and these relays would cause its torque to become too large.
 c. these switches are mainly used to control for overcurrent conditions.

3. The phase shift that is required to create torque for a three-phase motor is
 a. created by adding capacitors to the motor windings.
 b. found naturally in the three-phase voltage that is used to provide power for the motor.
 c. created by making each of the windings in the three-phase motor slightly different.

4. You need to know how to reconnect a three-phase motor for a change of voltage
 a. because the voltage rating of the motor that you have in stock is rated 480 V AC and the motor that it is replacing is rated for 240 V AC.
 b. because the motor may need more speed.
 c. because some three-phase motors need DC voltage instead of AC voltage.

5. If a three-phase motor runs but draws excessive current, you should suspect
 a. a faulty centrifugal switch or current relay.
 b. that the motor is running in the wrong direction, and you should exchange any two leads.
 c. that one of the motor windings is open, or one phase of the three-phase voltage is not supplying voltage.

PROBLEMS

1. Sketch a wye-connected three-phase motor.

2. Sketch a delta-connected three-phase motor.

3. Draw the connections that you would use to connect a wye-wired motor for low voltage and for high voltage.

4. Draw the connections that you would use to connect a delta-wired motor for low voltage and for high voltage.

5. Draw the electrical diagram of a nine-lead three-phase motor connected for wye configuration and delta configuration, and explain why each is used.

Websites

Websites change constantly. Use the key word search to locate sites that have similar content. The products that will be found at each site are indicated with the key word.

Tecumseh Products Company
Products: Compressors
Key word: Tecumseh Products Company

Copeland Compressors
Products: Compressors
Key words: Copeland Compressors

Carlyle Compressors
Products: Compressors
Key word: Carlyle Compressor

GE Industrial Systems AC Motors
Products: AC furnace fan motors, condenser fan motors
Key word: GE Motors

Reliance Electric
Products: Electric motors
Key word: Reliance Electric

Leeson Motors
Products: AC motors
Key word: Leeson Motors

A. O. Smith
Products: Electric motors
Key word: A. O. Smith Motors

Baldor Motors
Products: Electric motors
Key word: Baldor Motors

Chapter 12

Relays, Contactors, Solenoids, Motor Starters, and Overcurrent Controls

Job Assignment for This Chapter

You are sent to a residence to work on a split system air conditioner that has the outdoor unit supplied with 240 volts and the indoor unit (furnace) supplied with 120 volts. You find the compressor and condenser fan not running, and your preliminary tests indicate that the coil of the compressor relay is energized, but the compressor and condenser fan are not running. Your job is to decide which tests you should make next to determine what exactly is wrong with the system. You will need to explain to the home-owner what is wrong with the system and what tests you should do to demonstrate what is wrong with the system.

Terms You Will Learn in This Chapter

Armature
Bell-crank armature assembly
Clapper armature assembly
Coil
Control circuit
Double-pole, double-throw (DPDT) relay
Double-pole, single-throw (DPST) relay
Hold-in current
Horizontal-action armature
Inrush current
Load circuit

Normally closed contacts
Normally open contacts
Pull-in current
Relay
Relay coil
Relay contacts
Seal-in current
Single-pole, double-throw (SPDT) relay
Single-pole, single-throw (SPST) relay
Transformer
Vertical-action armature assembly

After reading this chapter, you will be able to:

1. Identify the contacts and coil in a relay and contactor.
2. Explain the operations of a relay and contactor.
3. Identify the components in a control circuit.
4. Select the proper size of contactor from a NEMA table.
5. Identify the basic parts of a solenoid and explain their functions.
6. Explain the operation of a motor starter.
7. Identify the main parts of a motor starter.
8. Identify the main parts of an overload and explain their operations.
9. Explain the operation of single-element and dual-element fuses.
10. Identify single-pole, two-pole, and three-pole circuit breakers and explain their operations.

12.0 OVERVIEW OF RELAYS AND CONTACTORS IN THE CONTROL CIRCUIT

The **control circuit** in air conditioning, heating, and refrigeration is the heart of the system. It consists of the control devices such as a thermostat or pressure switch to determine when to energize the system or de-energize the system. Each control circuit uses relays or contactors to control the compressor motor, evaporator fan motor, condenser fan motor, and furnace fan motor. This chapter introduces the main components that are used in the control circuits of these systems and their theory of operation. The components in the control circuit are energized and in turn they energize components such as compressors and motors in the **load circuit**. At the bottom of Figure 12–1 is an example of a typical control circuit for an air-conditioning system. The load circuit is the top part of the electrical diagram.

12.1 THE CONTROL TRANSFORMER

The control transformer was introduced in Chapter 8. In that chapter you learned that the **transformer** converts line voltage to low-voltage power for the control circuit. The load-side terminals of the control transformer are labeled R and C. In commercial air-conditioning systems and in refrigeration systems, the control transformer steps down 240 V AC to 120 V AC, and in residential air-conditioning and heating control systems it steps down 120 V AC to 24 V AC. The lower voltage is used in each case to provide power for the control circuit.

The control circuit at the bottom of the diagram in Figure 12–1 starts at the secondary side of the transformer, where it is identified by the letters R and C. The terminal that is identified as the R terminal is considered the source of power and it is called the *hot* side of the circuit. The right side of the diagram that is identified by the transformer terminal C is considered to be the *common* side of the circuit. It is important to understand that since the voltage in the transformer is AC voltage, both sides of the circuit have the same potential during every other half-cycle. The concept of hot side and common side is defined to make it easier to understand the operation of the control circuit.

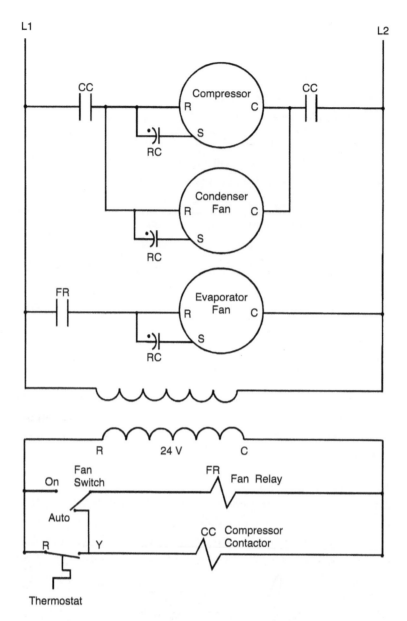

Figure 12–1 A typical control circuit for an air-conditioning system is shown at the bottom of this diagram.

12.2 THE THEORY AND OPERATION OF A RELAY

A **relay** is a magnetically controlled switch that is the main control component in air-conditioning, heating, and refrigeration systems. Figure 12–2 shows a picture of a typical relay, and Figure 12–3 shows a cutaway diagram of a typical relay, which consists of a **coil** and a *number of sets of contacts*. The coil becomes an electromagnet when it is energized, and its magnetic field causes each set of normally open contacts to close and each set of normally closed contacts to open. The contacts are basically a switch that is operated by magnetic force. The part of the relay that moves and causes the contacts to move is called the **armature**. Power is applied to the coil of the relay first, and the magnetic flux causes the armature to move and causes the contacts to change position. The coil is part of the control circuit, and the contacts are part of the load circuit.

Figure 12–4 shows the electrical diagram and symbol for the coil and contacts of a relay. The armature is not shown in the electrical symbols, since it is considered part of

Figure 12–2 A typical relay used to control an air-conditioning, heating, and refrigeration system. *(Courtesy of Honeywell)*

✪ Complete Stationary Contact Kit

Stationary Contact Block Complete with Contacts

✪ Complete Movable Contact Kit

✪ Operating Coil

Coil Housing Cover X-225897

Cross Bar Complete with Contacts

Armature & Yoke Assembly 2 & 3 Pole X-225898

4-6-8 Pole X-228033

Coil Housing Base X-225899

Figure 12–3 A cutaway diagram of a relay. Notice that the coil and contacts are mounted so that the magnetic field of the coil will make the contacts move. *(Courtesy of Rockwell Automation)*

the contacts. When you encounter a relay in the control panel of a system, you must relate the physical parts to the electrical symbols that are used to identify the components. You must also learn to envision its operation as two separate pieces, the coil and contacts, even though they are mounted near each other and operate almost simultaneously. It is important to understand that the coil must be energized first, and a split second later the magnetic field built up in the coil will cause the contacts to move.

Coil

Coil Symbols

or ⌁ or

Contacts

Contact
Symbols

Figure 12–4 Electrical diagram and symbols for the coil and contacts for a relay.

12.3 TYPES OF ARMATURE ASSEMBLIES FOR RELAYS

You will encounter many different types of relays and different name brands of relays that look like they all use different theories of operation. In reality the operation of all relays can be broken into four basic methods of operation because one of four basic armature assemblies is used in all relays. Figure 12–5 shows an example of these four basic armature assemblies. A few relays will have hybrid armatures that use some features of more than one type of armature assembly.

Figure 12–5a shows an example of a **horizontal-action armature** for a relay. In this example a set of stationary contacts is mounted in the horizontal position. A set of movable contacts is mounted on the armature assembly directly across from the stationary set. The armature is mounted inside the coil in such a way that when the coil is energized, its magnetic field causes the mass of the armature to center in the middle of the coil. Since the armature is slightly offset when it is placed in the coil, it moves to the left to center itself directly in the middle of the coil when the magnetic field is developed in the coil. The movement of the armature to the left shifts the movable contacts to the left until they come into contact with the stationary set of contacts. When the movable contacts and stationary contacts touch each other, they complete a circuit that allows large amounts of current and voltage to pass through them like a normal switch in the closed position. When the current to the coil is de-energized, a small spring causes the armature to shift back to the left to its original position.

The diagram in Figure 12–5b shows an example of a **bell-crank armature assembly**. In this example the coil is mounted above the armature. When the coil is energized, it produces a magnetic field that pulls the armature upward. The movable contacts are connected to an arm that is bent at a right angle. When the end of the arm is pulled upward by the magnet, the other end of the arm is moved to the left. This action shifts the movable contacts against the stationary contacts to complete the electrical circuit.

Figure 12–5 a Horizontal-action armature assembly. **b** Bell-crank armature assembly. **c** Clapper armature assembly. **d** Vertical-action armature assembly. *(Courtesy of Square D/Groupe Schneider. Square D/Groupe Schneider assumes to liability for accuracy of information)*

When the coil is de-energized, gravity causes the armature to drop downward away from the coil. This movement causes one end of the arm to move downward also, causing the left end of the arm to shift back to the left, which causes the contacts to move to the open position again.

The diagram in Figure 12–5c shows an example of a **clapper armature assembly**. The armature is a large arm on the right side of the coil. The armature (arm) has a pin through the bottom to act as an axis. The movable contacts are mounted to the top of the armature at the left side. When the coil is energized, it creates a magnetic field that pulls the armature to the left toward the coil. This action causes a small amount of travel at the bottom of the arm and a large amount of travel at the top of the arm, since the bottom of the arm is held in place at the axis. Since the movable contacts are mounted near the top of the arm, they will move to the left a significant distance until they come into contact with stationary contacts. When the coil is de-energized, a small spring pulls the armature to the left, which causes the contacts to return to their open position.

Figure 12–5d shows an example of the **vertical-action armature assembly**. The armature is mounted to a bracket that is shaped like the letter C. The bottom part of the C-shaped bracket is mounted directly to the armature, and a set of movable contacts is mounted directly to the top of the C-shaped bracket. When the coil is energized, it creates a magnetic field that pulls the entire bracket upward until the armature is pulled tight against the coil. This movement causes the contacts that are mounted to the top of the bracket to shift upward until they touch the stationary contacts. When the coil is de-energized, gravity causes the complete bracket to drop downward and move away from the coil, which causes the movable contacts to move away from the stationary contacts.

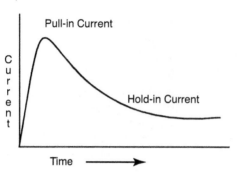

Figure 12–6 Diagram showing pull-in and hold-in currents for relay coils. Notice that the pull-in current is approximately three to five times larger than the hold-in current.

12.4 PULL-IN AND HOLD-IN CURRENT

When voltage is first applied to a coil of a relay, it draws excessive current because the coil of wire presents only resistance to the circuit when current first starts to flow. As the flow of current increases in the coil, inductive reactance begins to build, which causes the current to become lower. The current creates a strong magnetic field around the coil, which will cause the armature to move. When the armature has moved, it causes the induction in the magnetic coil to change, so that less current is required to maintain the position of the armature.

Figure 12–6 shows a diagram of the pull-in current and the hold-in current. The **pull-in current** is also called the **inrush current**, and the **hold-in current** is also called the **seal-in current**. The pull-in current is typically three to five times larger than the hold-in current. At times the supply voltage will be slightly lower in the summer. This condition is called a *brownout*, and it may cause a problem with activating relays. If the voltage is too low, it may not be able to supply sufficient current to pull the armature into place when the relay is first energized because the pull-in current is too small. If this occurs, the air-conditioning or refrigeration system will not start because the starting relay cannot close.

If the system is running when the brownout condition occurs, the relay coil will probably remain energized because the amount of hold-in current is sufficient to keep the armature in place even though the voltage is low. Since the brownout condition occurs when it is very hot outside, most air conditioners will remain running through this condition. If the air conditioner gets the room cool enough so it can cycle off, it will probably not energize again because of the low voltage and the large amount of pull-in current required.

12.5 NORMALLY OPEN AND NORMALLY CLOSED CONTACTS

A relay can have **normally open contacts** or **normally closed contacts**. It is important to understand that the word *normal* for contacts indicates the position of the contacts when no voltage is applied to the coil. The contacts can be held in their normal position by a spring or by gravity. The contacts of a relay will move from their normal position to their energized position when power is applied to its coil.

Some types of contacts can be changed or converted from normally open to normally closed in the field. Other types are manufactured in such a way that they cannot be changed. Figure 12–7 shows examples of converting normally open contacts to normally closed contacts while the relay is installed. The contacts can be converted in the field by the technician by simply removing them from the relay and turning them upside down.

Adding or Converting Contact Cartridges Having "Swingaround" Terminals

General Instructions (Specific cases below.)

1.1 Adding a contact cartridge:
As received, accessory cartridges are in the normally open mode with terminal screws adjacent to N.O. symbols. If normally closed mode is desired, convert contact as indicated in Step 1.2 below. When cartridges are inserted, the terminal screws must face the front. The clear cover may face either side. **Do not install more than 8 N.C. contacts per relay.** When installing one cartridge, locate it at an inner pole position. When installing 2 cartridges, locate both in inner or outer (balanced) positions.

1.2 Converting a contact to its alternate mode (N.O. ⇌ N.C.):
Withdraw an assemble cartridge for replacement or conversion by inserting the blade of a suitably-sized screwdriver under a terminal screw pressure plate. Slide cartridge out. See Figure 1. Back the terminal screws out of the cylindrical nuts a sufficient amount (approximately 2 turns for a fully-tightened screw) to permit rotation of each screw and nut assembly to its alternate position. See Figure 2.

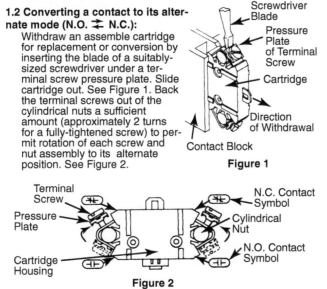

Screwdriver Blade
Pressure Plate of Terminal Screw
Cartridge
Direction of Withdrawal
Contact Block

Figure 1

Terminal Screw
Pressure Plate
Cartridge Housing
N.C. Contact Symbol
Cylindrical Nut
N.O. Contact Symbol

Figure 2

Figure 12–7 Example of changing normally open contacts to normally closed contacts in the field by turning them over. Normally closed contacts can also be converted to normally open contacts. *(Courtesy of Rockwell Automation)*

When a set of normally open contacts is inverted, the contacts become normally closed, and when normally closed contacts are inverted, they become normally open contacts. This means that a technician in the field can change the contacts in a relay to get the exact number of normally open or normally closed contacts needed for the application.

12.6 RATINGS FOR RELAY CONTACTS AND RELAY COILS

When you change a relay that is worn or broken, you must ensure that the voltage rating of the coil of the new relay matches the voltage of the control circuit exactly. This means that if the voltage for the control circuit is 24 V AC, the coil must be rated for 24 V. If the control voltage is 120 V AC, the coil must be rated for 120 V. The voltage rating for a relay coil is stamped directly on the coil. The coil may also have a color code. If the coil is rated for 24 V AC, it will be color coded black. If the coil is rated for 120 V AC, it should also be color coded red or have a red-colored stamp on the coil. If the relay coil is rated for 208 or 240 V AC, it will be color coded green or identified with a green stamp or green printing on the coil. DC coils are color coded blue. It is important to

understand that the current rating for a relay coil is seldom listed on the component. If it is important to know the current rating for the coil, you can look for it in the catalog or on the specification sheet that is shipped with the new relay. If you change a relay, you must also make sure that the rating for the contacts meets or exceeds the current rating and the voltage rating of the load to which it will be connected. For example, if the contacts of the relay are used to energize a 240-V AC fan motor that draws 3 A, the contacts must be rated for at least 240 V and 3 A. It is permissible to have the contacts in this example rated for more voltage and current, such as 600 V and 10 A.

Contact ratings are grouped by voltage and by current. The voltage ratings are generally broken into two groups, of 300 V and 600 V. This means that if you are using the contacts to control 240 or 208 V AC, you would use contacts that have a 300-V rating. If you are using the contacts to control a 480-V AC motor, you would need to use 600-V-rated contacts.

The current rating of contacts is listed in amperes or horsepower. The current rating or horsepower rating must exceed the amount of current the relay is controlling. This means that if the relay is controlling 12 A, the contacts will need to be rated for over 12 A of current. The current rating of the contacts and the voltage rating for the contacts will be printed directly on the contacts or on the side of the relay.

12.7 IDENTIFYING RELAYS BY THE ARRANGEMENT OF THEIR CONTACTS

Some types of contact arrangements for relays have become standardized so that they are easier to recognize when they are ordered for replacement or when you are trying to troubleshoot them. The diagrams in Figure 12–8 show examples of some of the standard types of relay arrangements. Figure 12–8a shows a relay with a set of normally open contacts. This type of relay can also have a single set of normally closed contacts instead of normally open contacts. Since this relay has only one contact and it can only close or open this type of relay, it is called a **single-pole, single-throw (SPST) relay**. The word *pole* in this identification refers to the number of contacts, and the word *throw* refers to the number of terminals to which the input contacts can be switched. Since the contact in this relay has one input and it can be switched to only a single output terminal, it is said to have a single throw.

Figure 12–8b shows a relay with two sets of normally open contacts. Since this relay has two sets of single contacts, it is called a **double-pole, single-throw (DPST) relay**. The *double-pole* part of the name refers to the two individual sets of normally open contacts, and the *single-throw* part of the name refers to the one output terminal for each contact. When the coil is energized, both sets of contacts will move from their normally open position to the normally closed position.

Figure 12–8c shows a relay with a set of normally open and a set of normally closed contacts that are connected on the left side. The point where this connection is made is called the *common point* and is identified with the letter C. When the relay coil is energized, the normally open part of the contacts will close, and the normally closed part of the contacts will open. Since these contacts have a common point as the input terminal, and two output terminals [one normally open (NO) and one normally closed (NC)], the relay is called a **single-pole, double-throw (SPDT) relay**. The most important feature of this relay is that the contacts have two terminals on the output side, so it is called a double-throw relay. The single-pole, double-throw relay is used where two exclusive conditions exist and you do not want them ever to occur at the same time. For example, if this relay is controlling the cooling (air conditioning) and the heating (furnace) for a residential system, you would never want both systems to be on at the same time.

Figure 12–8 a A relay with a single set of normally open contacts. This type of relay is called a single-pole, single-throw (SPST) relay. **b** A relay with two individual sets of contacts. This relay is called a double-pole, single-throw (DPST) relay. **c** A relay with two sets of contacts that are connected at one side at a point called the common (C). The output terminals are identified as normally open (NO) and normally closed (NC). This type of relay is called single-pole, double-throw (SPDT). **d** A relay with two sets of SPDT contacts. This relay is called a double-pole, double-throw (DPDT) relay. **e** A relay with multiple sets of normally open contacts. **f** A relay with a combination of normally open and normally closed contacts.

By connecting the air-conditioning system to the normally open terminal on the right side of the relay, and the furnace to the normally closed terminal, you create the condition so that the furnace and air-conditioning system cannot be on at the same time.

Figure 12–8d has two sets of single-pole, double-throw contacts, so it is called a **double-pole, double-throw (DPDT) relay**. In this case the term *double-throw* is used because two sets of normally open/normally closed contacts are provided. Each set has a common point on its left side (input side) and a terminal that is connected to the normally open (NO) set and a terminal that is connected to the normally closed (NC) set on its right side. This type of relay is used where the exclusion is needed and the loads are 208 V AC or 240 V AC where they need power from both L1 and L2. In this type of application, L1 will be connected to the common terminal (C) of one set of contacts, and L2 will be connected to the common terminal (C) of the other set. This will cause L1 and L2 to be switched the same way in both conditions.

Figure 12–8e shows multiple sets of normally open contacts. This type of relay can have any number of sets of normally open contacts. The additional sets of contacts can be added to the original contacts in some types of relays. If the original relay is manufactured with this provision, you can purchase the additional contacts and add them to the original relay by placing them on top of the original relay and tightening the mounting screws to make the additional contacts operate with the relay armature. The contacts for this type of relay can all be normally closed if the application requires it. The main feature of this type of relay is that it can have any number of contacts.

Figure 12–8f shows a relay with multiple sets of individual normally open and normally closed contacts. This type of relay is similar to the one shown in Figure 12–8e, except in this type of relay the contacts can be any combination of normally open or normally closed sets. In most cases, the contacts in this type of relay are convertible in the field. As a technician, you can add sets of contacts and change them from normally open to normally closed or vice versa as needed. In most installations, the relays are provided in the original equipment and you will need only to identify them for installation and troubleshooting purposes. Later, if the equipment is modified or if additional equipment such as electronic air cleaners, humidifiers, or other similar equipment is added, you may need to locate additional contacts on a relay to connect the add-on equipment so that it will operate correctly with the original system.

Additional Job Assignment

You have just found a relay in a residential air-conditioning system that has burned contacts. The relay is a SPST with normally open contacts. The contacts are rated for 20 amps and 300 volts, and the coil is rated for 24 volts. When you go to the truck you find that you have four relays to select from and none of them is an exact match. Your job is to select the relay that will be an acceptable replacement. The relays have the following ratings:

1. DPST relay with 24-volt coil; contact rating is 300 volts and 15 amps.
2. DPST relay with 24-volt coil; contact rating is 300 volts and 30 amps.
3. SPDT relay with 120-volt coil; contact rating is 300 volts and 30 amps.
4. SPST relay with 120-volt coil; contact rating is 300 volts and 20 amps.

12.8 EXAMPLES OF RELAYS USED IN AIR-CONDITIONING, HEATING, AND REFRIGERATION SYSTEMS

Relays are used in a variety of applications in air-conditioning, heating, and refrigeration systems. For example, one of the most common relays is the fan relay that is used in heating systems. Another common relay is the sequencing relay used in some electric furnaces. Figure 12–9 a–d shows examples of several types of relays that you will find in these systems. In some newer systems, the relays use plug-in terminals so that the relay can be removed and replaced without changing any wiring. The relay is simply plugged into a socket-type base. An example of a plug-in relay with its socket is also shown in this figure. You will begin to become familiar with relays and recognize them in the control panels of the equipment on which you work.

12.9 CURRENT RELAYS AND POTENTIAL RELAYS FOR STARTING SINGLE-PHASE COMPRESSORS

Current relays and potential relays are special relays that are designed to energize the start winding of single-phase compressor motors during starting and then immediately de-energize as soon as the motor is running. You learned about these relays in Chapter 10. The current relay has a low-impedance coil that takes advantage of pull-in and hold-in current to energize a set of normally open contacts. The coil for the current relay is

(a)

(b)

(c)

(d)

Figure 12–9 Variety of relays used in air-conditioning, heating, and refrigeration applications. *(Courtesy of Honeywell)*

connected in series with the run winding of a single-phase motor, and the contacts of the relay are connected in series with the start winding of the motor. In some cases a capacitor is also connected in series with the contacts and start winding. When voltage is applied to the motor, current can flow only through the run winding, which causes very large current to be drawn. This current is large enough to exceed the pull-in current level of the contacts, which causes them to close. When the current relay contacts close, voltage is applied to the start winding of the motor, which provides additional power to allow the motor to start and to continue running. After the motor has started and begins to run, the large initial current that is flowing through the run winding and current relay coil diminishes to a minimal level, which is below the hold-in current level. When this occurs, the current relay coil is not able to keep its contacts closed, so they open and de-energize the start winding of the motor. This action is exactly what the single-phase motor needs, since its start winding can be energized only a few seconds while the motor is starting. Figure 12–10a shows a drawing, and Figure 12–10b shows an electrical diagram of the current relay. The current relay is also available as an electronic relay. You will learn more about current relays later in this chapter.

The *potential relay* uses a high-impedance coil with a set of normally closed contacts to energize a start capacitor for a capacitor-start, capacitor-run (CSCR) single-phase compressor motor. Figure 12–11a shows a picture and Figure 12–11b shows an electrical diagram of a typical potential relay. When power is applied to the compressor motor, current flows through the run winding and through the normally closed contacts

Figure 12–10 a A sketch of a typical current relay. **b** Electrical diagram of a current relay. The run winding of a single-phase compressor motor is connected to M terminal and the start winding is connected to S terminal. **c** Picture of a typical current relay.

Figure 12–11 a A potential relay used to start single-phase compressors. **b** A potential relay connected to a CSCR compressor motor. *(Courtesy of Supco Sealed Unit Parts Co. Inc.)*

of the potential relay to the start winding. When the motor begins to run, its windings create a back voltage called back EMF. The back EMF is large enough to cause the coil of the potential relay to energize and pull its contacts open, which de-energizes the start winding. When main power is turned off to the motor, it stops running and the normally closed contacts of the potential relay return to their normally closed position, so the relay is ready to start the motor again. More information about the potential relay will be presented later in this chapter with other information about single-phase compressor motors.

12.10 THE DIFFERENCE BETWEEN A RELAY AND A CONTACTOR

A contactor is similar to a relay in that it has a coil and a number of contacts. The main difference is that the contactor is larger and its contacts can carry more current. A *relay* is generally defined as magnetically controlled contacts that carry a current of less than 15 A.

A *contactor* is defined as having contacts that are rated for 15 A or more. Some manufacturers do not follow the 15-A rating, so sometimes you will find a relay that has a current rating for its contacts in excess of 15 A, and you also may find contactors with contact ratings of less than 15 A. In general, the main difference is that a contactor is specifically designed so that its contacts can carry a larger amount of current, up to 2250 A. Contactors are rated by the National Electrical Manufacturers Association (NEMA), and their sizes range from size 00 to size 9. They may also be rated by using horsepower (hp).

In most air-conditioning and refrigeration systems the compressor motor is controlled by a contactor instead of a relay because the amount of current the compressor motor draws is usually in excess of 15 A. When you look at the compressor contactor, you will see that it looks like a very large relay. Since its contacts can carry current larger than 15 A, it will technically be called a contactor instead of a relay.

12.11 NEMA RATINGS FOR CONTACTORS

Figure 12–12 presents a table of NEMA ratings for contactors, showing that a size 00 contactor is rated to safely carry up to 9 A for a continuous load. You should also notice that the contacts are rated for up to 575 V. You should remember that the current rating for contacts depends on the size of the contacts, and the voltage rating of the contacts depends on the way the relay is manufactured so that arcs do not jump between different terminals. This means that contacts that are rated for a higher voltage have more plastic or insulating material between the sets of contacts so that arcs do not jump from the contacts to other parts of the relay or to other sets of contacts. This table also identifies the load as a maximum horsepower rating. This means that you may identify the load from its current rating or its horsepower rating and select the proper contactor size to safely control the load.

You should notice from this table that the next larger size of contactor is a size 0. This contactor is rated for up to 18 A on a continuous basis. The size 9 contactor is the largest, and its current rating is 2250 A.

NEMA Size	Continuous Ampere Rating	Maximum Horsepower Rating Full Load Current Must Not Exceed "Continuous Ampere Rating" Motor Voltage			
		200 V	230 V	50 Hz 380 V–415 V	460 V–575 V
		3 ø • 4 Power Poles • 600V AC Max			
00	9	1-1/2	1-1/2	2	2
0	18	3	3	5	5
1	27	7-1/2	7-1/2	10	10
2	45	10	15	25	25
3	90	25	30	50	50
4	135	40	50	75	100
5	270	75	100	150	200
6	540	150	200	300	400
7 ❷	810	-	300	600	600
8 ❷	1215	-	450	900	900
9	2250	-	800	1600	1600

Figure 12–12 NEMA ratings for contactors. *(Courtesy of Rockwell Automation)*

(a) (b)

Figure 12–13 a Example of NEMA size 3 contactors. **b** Example of NEMA size
5 contactors. *(Courtesy of Rockwell Automation)*

Air-conditioning, heating, and refrigeration systems usually use contactors up to a size 4. One application for contactors is to control electric heating elements for electric furnaces. These contactors are called *heating contactors*. Since heating elements may draw up to 100 A, the size 4 contactor can easily carry this current. If larger heating elements are used, a larger contactor is used. Figure 12–13a shows an example of a size 3 contactor, and Figure 12–13b shows a size 5 contactor. A size 3 contactor is approximately 5 in. tall, and a size 5 contactor is approximately 7 in. tall.

12.12 USING A RELAY TO CONTROL THE FURNACE FAN

One application of a relay in a heating system is to control the furnace fan. Since the furnace fan is called the evaporator fan if the system is also used for air conditioning, this same relay may be called the evaporator fan relay. Figure 12–14 shows an example wiring diagram of an electrical system with a fan relay. The fan relay coil is in the control system, and its contacts are in the load circuit connected in series with the load. The fan relay is identified by the letters FR. Notice that the coil is identified as FR, and the contacts are identified as FR. You should also notice that there are other sets of contacts identified as CC, which means that those contacts are controlled by the compressor contactor. The coil of the compressor contactor is identified as CC also.

When you are troubleshooting an electrical system, it is important to locate each coil in the system and see how it is identified. Then you should look around the diagram to locate all sets of contacts that have the same letters identifying them as the coil. Whenever you locate contacts and a coil in a diagram that have the same identification, such as FR, you will know they are the coils and contacts of the same relay. If you find a coil symbol that is identified as CC, and contacts that are identified as FR, you will know that the coil is part of the compressor contactor, and the contacts are part of the fan relay.

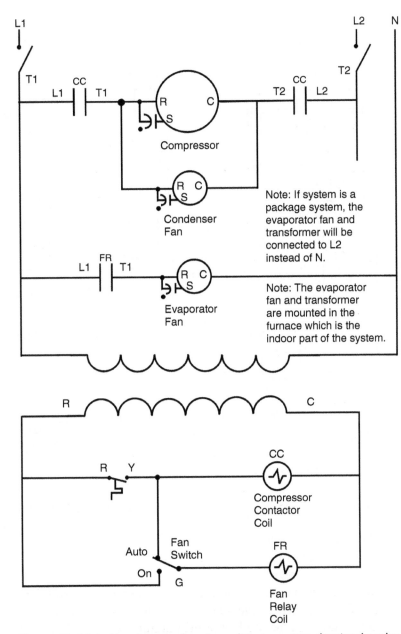

Figure 12–14 Ladder diagram of an air-conditioning system showing the relay coils in the control circuit, and the contacts of the relays in the load circuit.

It will be difficult when you first use a ladder diagram to troubleshoot an electrical system because the coils and contacts of a relay are not shown near each other. Instead the contacts are shown in the circuit they are controlling, and the coil is shown in the circuit that controls it. Figure 12–15 shows a wiring diagram of the same circuit. In this diagram you should notice that the physical outline of each relay is shown, and the contacts for each relay are shown as they would physically be found near the coil. It is important to remember that the wiring diagram is intended to show physical locations and relationships among all the controls, and the ladder diagram is intended to show the sequence of operation. See Chapter 6 for more information about ladder diagrams and wiring diagrams.

Figure 12–15 Wiring diagram of the air-conditioning system shown in Figure 12–14. Notice in the wiring diagram that the outline of each relay shows the physical relationships among the relays. This diagram also shows all contacts for a relay closed by the relay coil.

12.13 USING A CONTACTOR TO CONTROL A COMPRESSOR MOTOR

The electrical ladder diagram presented in Figure 12–14 and the wiring diagram presented in Figure 12–15 also show the compressor contactor. The coil of the compressor contactor is shown in the control circuit of the ladder diagram and is identified by the letters CC (compressor contactor). You should notice that the contacts that are identified as CC are shown connected in series with the compressor motor and condenser fan motor shown in the top line of the diagram.

The compressor contactor is larger than the fan relay and also has two sets of contacts instead of one, like the fan relay. The wiring diagram is useful for identifying the location of the compressor contactor in the control panel. According to the wiring diagram, it is located above the fan relay and to the right of the disconnect. Notice that it is rather difficult to determine the circuits and the sequence of operation in a wiring diagram because of the number of wires that are shown and the way that they cross over one another several times. Notice also that the control circuit and load circuit are not separated in the wiring diagram, which makes it more difficult to understand the operation. The wiring diagram is necessary and does serve a very important purpose in that it shows the physical relationship of all components in the circuit.

On contrast, notice how easy it is to determine the sequence of operation in the ladder diagram. For example, if you are asked to determine what conditions must exist for the compressor contactor coil to be energized, you can simply look at the ladder diagram and see that terminals R–Y on the thermostat must close to energize the coil. The main drawback of the ladder diagram is that it does not give a quick indication of the

location of components, and it does not easily indicate how many contacts a relay has. You can now begin to see why you will need to use both a wiring diagram and a ladder diagram if they are provided.

12.14 THE FUNCTION OF THE CONTROL CIRCUIT

As you have seen in this section, the function of the control circuit in air-conditioning, heating, and refrigeration systems is to control the conditions that cause the equipment to become energized or de-energized. The components in the control circuit of residential furnaces and air conditioners are generally low-voltage devices, since the control circuit operates on low voltage. The **relay coil** is located in the control section of the electrical system, and the **relay contacts** are located in the load section; they are typically connected in series with the load they are controlling. It is important to remember that the function of the control circuit is to energize the relay coil so that power can be sent through the relay contacts to the load. The action in the control circuit must always occur before the action in the load circuit.

If you must troubleshoot an electrical system for an air-conditioning, heating, and refrigeration system, you should make two checks. First, you should see if the loads are energized and operating correctly. If the loads are not energized, you should suspect that the control circuit is not energized, and move on to the second test and make checks in the control circuit. As you become more familiar with the control circuit and load circuit you will almost automatically check the control circuit to ensure that the relay coil is energized to make sure its contacts are energized. The next sections of this book will present in-depth material concerning the operation of typical loads, such as the compressor motor and the fan motors.

12.15 SOLENOIDS USED IN HVAC AND REFRIGERATION SYSTEMS

A solenoid is a device that moves something with electromagnetism. A solenoid is one type of *magnetically controlled valve*. You learned in earlier chapters how current flowing through a coil of wire causes a strong magnetic field to develop. The strong magnetic field in the solenoid valve is used to make the valve plunger move from the open to the closed position or from the closed to the open position. The valve is part of the refrigerant system plumbing or the plumbing for the water system. Solenoid valves allow the plumbing system to be interfaced with the electrical system in order to automate the plumbing system. For example, in residential heat pumps, a solenoid valve is used as a reversing valve. The reversing valve allows the system to change from a heating system to a cooling system, and it also allows the refrigerant flow to be reversed automatically during the defrost cycle to provide hot gas to the outside coil to defrost it.

12.16 BASIC PARTS OF A SOLENOID VALVE

Figure 12–16 shows typical solenoid valves, and Figure 12–17 shows a cutaway diagram of one of the solenoid valves. From these figures you can see that the solenoid valve is a plumbing-type valve with a movable plunger used to close off the valve. The valve can be designed so that the plunger is held either in the open position by spring pressure or in the closed position by spring pressure. When electric current is applied to

Figure 12–16 Typical solenoid valves. *(Courtesy of Danfoss Automatic Controls)*

4. Coil
16. Armature
18. Pilot valve plate
20. Earth screw
28. Gasket
29. Pilot orifice
40. Junction box
43. Valve cover
44. O-ring
45. Valve cover gasket
49. Valve body
51. Protective cap/Blanking plug
52. Locking button and top nut
53. Manual stem
73. Equalizing hole
74. Main passage
75. Pilot passage
76. Return spring
80. Servo piston
83. Main valve seat
84. Main valve plate

DANFOSS
A32F323.11.10

Figure 12–17 Diagram of a typical solenoid valve. *(Courtesy of Danfoss Automatic Controls)*

the coil in the valve, the strong magnetic field causes the plunger to move upward. If the valve is the type in which spring pressure is causing the plunger to stay in the closed position, the magnetic field will cause the plunger to open the valve passage. If the valve is the type in which spring pressure is causing the plunger to stay in the open position, the magnetic field will cause the plunger to close the valve passage.

The coil for the solenoid is exactly like the coils that you learned about previously for relays and contactors. Since the coil is basically a long wire that is coiled on a plastic spool, it will have two ends, and you can test the coil for continuity. Note that each type of solenoid coil has some amount of resistance that will vary from approximately

50 Ω to 1500 Ω. Remember, the exact amount of resistance does not matter because the coil is made from one long piece of wire. It is doubtful that the amount of resistance will change. The main problem a coil will have is an open somewhere along its wire.

12.17 TROUBLESHOOTING A SOLENOID VALVE

At times you will be working on a system that has one or more solenoid valves and you will suspect that the solenoid valve is not operating correctly. The first test you should make when troubleshooting a solenoid valve is to read its data plate and determine the amount of voltage the valve should have and then measure the amount of voltage that is available across the two wires that are connected to the coil. It is important to remember that the coil of the reversing valve must have the correct amount and type of voltage applied to its two wires if it is to conduct sufficient current to cause the plunger to move. This means that if the valve is rated for 120 V AC, you should measure 120 V AC across the two coil wires. If the voltage is missing or is the incorrect amount, the coil will not operate correctly. If the proper amount of voltage is not present, you will need to use troubleshooting procedures to locate the open in the circuit that is causing the loss of voltage before it reaches the coil.

If the proper amount of voltage is present and the coil will not activate, you must make several additional tests. You can test to see if the coil has developed a magnetic field by placing a metal screwdriver near the coil to see if it is attracted by the magnetic field. If the screwdriver is attracted to the coil by the magnetic field, the coil is operating correctly. If the valve is still not opening or closing, the valve seat may be frozen in place, and the valve will need to be replaced.

The other problem that the coil may have if the correct amount of voltage is present at the coil wires is that the coil may have an open in its wire. You can test the coil wire with a continuity test. Be sure to turn off all voltage to the solenoid and disconnect the coil wires so that the coil is isolated from back-feeding to the remainder of the circuit. If the continuity test indicates infinite (∞) resistance at the highest resistance range, you can suspect that the coil has an open and should be replaced. Notice that the coil can be replaced without unsoldering the valve.

12.18 REVERSING VALVE FOR HEAT PUMPS

The reversing valve is a specialized solenoid valve. Figure 12–18 shows a picture of a reversing valve, and Figures 12–19a and b show diagrams of a reversing valve in a circuit. The heat pump is basically an air-conditioning system that reverses the flow of its refrigerant in cold weather to route the hot, high-pressure refrigerant gas from the compressor through the indoor coil, where its heat is used to heat the indoor space. In this mode the outdoor coil is used as the evaporator.

In warm weather, the reversing valve switches so that the flow is returned to a normal air-conditioning system in which the compressor pumps the hot, high-pressure gas to the outdoor coil to be condensed. The indoor coil in the normal air-conditioning mode is used to evaporate the liquid refrigerant, which takes heat from the indoor space and causes it to cool.

In the diagram you can see that the indoor coil is on the right side and the outdoor coil is on the left side. The top diagram represents a typical cooling cycle. The reversing valve switches to route the refrigerant as a hot, high-pressure gas to the outdoor coil where it condenses and turns to a liquid. The liquid refrigerant in this diagram is represented by the solid line. The liquid refrigerant in the cooling cycle moves through the indoor coil, where the heat in the indoor space is used to cause the liquid to evaporate.

Figure 12-18 A reversing valve is a specialized solenoid valve that is used in heat pumps. *(Courtesy of Alco Controls Division, Emerson Electric Company)*

(a) Cooling Cycle

Figure 12-19 a A reversing valve is switched so that the heat pump is in its cooling cycle. The indoor space is cooled when the heat pump is in this mode. **b** The reversing valve is switched so that the heat pump is heating the indoor space.

(b) Heating Cycle

The bottom diagram shows the route of the refrigerant when the reversing valve is switched to the heating cycle. In this diagram you can see that the refrigerant is a hot, high-pressure gas as it comes from the compressor and is directed to the indoor coil. The heat is removed from the hot, high-pressure gas and is used to heat the indoor space. The refrigerant condenses while it is moving through the indoor coil and moves to the outdoor coil, where it evaporates. Again, the path of the liquid refrigerant is shown as a dark line.

12.19 ## THE INTERNAL OPERATION OF THE REVERSING VALVE

The reversing valve is a specialized solenoid valve known as a *pilot-operated* valve, which consists of a small solenoid-operated valve that moves a small amount of vapor and liquid to make the larger main valve switch positions. This means that the small solenoid valve switches a small amount of liquid refrigerant and vapor refrigerant directly to the main valve. The main valve is the part of the reversing valve that actually switches to change the direction of the refrigerant flow.

Figures 12–20a and b show diagrams of the internal operation of the reversing valve. In the top diagram the reversing valve is switched to the cooling cycle, and in the bottom diagram it is switched to the heating cycle. The diagrams for this figure

Figure 12–20 **a** Internal diagram of a reversing valve in the cooling cycle. **b** Internal diagram of a reversing valve in the heating cycle. *(Courtesy of Alco Controls Division, Emerson Electric Company)*

show that the main valve has one port on one side of the valve and three ports on the other side. The single port shown on the top part of the valve diagram is soldered (welded) to the compressor discharge line. The port on the left side of the diagram is soldered to the condenser coil, the middle port is soldered to the compressor suction line, and the port on the right side is connected to the evaporator coil.

The solenoid valve is shown on the right side of the diagrams, and it has three lines connected to it. The solenoid in the reversing valve is the pilot valve. This means that it takes a small amount of refrigerant and directs it to the main valve to cause the main valve to switch positions. The middle line in the solenoid valve is connected to the compressor suction line, where it takes low-pressure refrigerant that is a vapor to use as the pilot fluid. In the top diagram you can see that the reversing valve is switched for cooling, and the solenoid valve moves the pilot fluid to the left side of the main valve, causing the valve to shift to the right. You should remember that even though the pilot fluid comes from the compressor suction line, it will be at 40–70 psi, which is sufficient pressure to cause the main valve to change positions.

When the main valve is switched to the right, it makes a plumbing connection between the compressor suction line and the indoor coil. A plumbing connection is also made between the compressor discharge line and the outdoor coil. In the bottom diagram the solenoid valve has switched the pilot fluid so that it activates the right side of the main valve, which causes the main reversing valve to switch to the left side. When the main valve is switched to the left side, it makes a plumbing connection between the compressor suction line and the outdoor coil, and another plumbing connection between the compressor discharge line and the indoor coil.

The reversing valve is a pilot-operated valve because a smaller coil can be used to ensure that the small pilot part of the valve changes positions. When the pilot part of the valve switches, it can accomplish much more work from the advantage of the pressure of the pilot fluid. This ensures that the valve is as small as possible, yet it will switch positions when the electrical coil of the solenoid is energized and de-energized.

12.20 TROUBLESHOOTING RELAYS, CONTACTORS, AND SOLENOIDS

At times you will encounter a relay, contactor, or solenoid that is not operating. The first step in troubleshooting is to check for voltage across the terminals of the coil. If you do not have voltage at the terminals, the problem is in the circuit and not in the relay or solenoid. If the voltage measured at the coils matches the coil rating, then the relay or solenoid should operate correctly. If the device is a relay, you should see the contacts move as you apply voltage and remove it. If they do not move, you should suspect that the coil has an open. Turn off voltage to it, remove the wires connected to it, test it for continuity, and replace it if needed. If the coil is part of a solenoid, you can use a screwdriver or other ferrous metal tool to see if the coil has created a magnetic field. If you detect a magnetic field, the coil is operating correctly and you can continue to test the plumbing part of the solenoid.

Another problem you may encounter when you are troubleshooting a relay or contactor is that the contacts will be dirty and will not pass sufficient voltage to their load. You can test the contacts for the amount of resistance they have by removing the power applied to them and isolating them so you can get a true resistance measurement. The most important point to remember when troubleshooting relays, contactors, or solenoids is to ensure first that the coil is receiving voltage and is operating correctly before you begin to test the contacts or valve.

12.21 WHY MOTOR STARTERS ARE USED IN HVAC AND REFRIGERATION SYSTEMS

Motor starters are magnetically controlled devices that are usually used in larger commercial applications. The motor starter is a larger version of a relay or contactor and it is used to control larger motors. The motor starter also has *overcurrent protection* for motors built into it, and relays and contactors do not. This overcurrent protection is called an *overload* and is sized to trip if the amount of current drawn by the motor exceeds the designated limit.

Earlier in this chapter you learned about relays and contactors that are used to start compressors and fan motors in residential air-conditioning and heating systems. These relays and contactors are designed to close their contacts and provide current to the motors in the system. Their main function is to turn on and off to provide current to these motors. The motors in these applications are protected against overcurrent by fuses or internal overloads that are built into the winding. Since the compressor is usually the largest motor in the system, it is also protected with an internal overload or an overload that is mounted externally directly on the shell of the compressor, so it does not draw too much current.

In larger commercial systems the compressor motors and fan motors are larger and more expensive. These motors are protected by fuses or circuit breakers in the disconnect for short-circuit protection, and by the overloads in motor starters to protect against slow overcurrents. If an open motor has a problem such as loss of lubrication or overload, it will draw extra current, which will damage the motor if the overload is allowed to continue for any length of time. The overloads in the motor starter sense the excess current and trip the motor starter so that its contacts open and stop all current flow to the motor until a technician manually resets the overloads.

12.22 THE BASIC PARTS OF A MOTOR STARTER

Figure 12–21 shows a typical motor starter with all its parts identified. From this picture you can see that this is a three-pole starter that is used in three-phase circuits. The incoming voltage is connected at the top at the terminals identified as L1, L2, and L3, and the motor leads are connected to the bottom terminals, which are identified as T1, T2, and T3. The three major sets of contacts are located in the top part of the motor starter, and the overload assembly is mounted in the lower part of the motor starter. The coil is located in the middle of the motor starter, and it has an indicator that shows the word *ON* when the coil is energized, and *OFF* when the coil is de-energized.

Figure 12–22 shows a wiring diagram of a motor starter connected to a three-phase motor at the top of the diagram, and a ladder diagram of just the control circuit, which consists of the start push button, stop push button, the motor starter coil, and the motor starter overloads, in the bottom diagram. It is easy to see the parts of the motor starter in the wiring diagram because they are all in the shaded area. You can see that the coil circuit uses smaller wire than the contact circuit. The motor starter has three sets of contacts that are in series with a heater. This ensures that all the current that flows to the motor must pass through a heater. If the current flowing to the motor is normal, the heater does not provide sufficient heat to cause the overload to trip. If the motor draws excess current, that current flowing through the heater will cause it to create excess heat that will trip the normally closed overload contacts. Since the normally closed overload contacts are connected in series with the motor starter coil, current to the motor starter

Keyhole Mounting Slot – Easy to reach with large screwdriver or power driver; permits the mounting screw to be in place before installing the starter.

Lineside Power Terminals – Up-front for accessibility; self-lifting saddle clamps for ease of wiring; clearly marked in contrasting white for quick identification. Optional top wiring kit for easy connection of power factor correction capacitor ahead of the overload relay.

Arc Hood Cover Screws – Up-front for easy accessibility; allows removal of the arc hood cover for inspection of the front movable and stationary contacts, and contact springs.

Coil Cover Screws – Up-front for easy accessibility allows removal of the coil cover for coil and contact change and access to all internal components.

Auxiliary Contacts Terminals – Angled and up-front for accessibility; self-lifting pressure plates for ease of wiring; clearly marked in contrasting white for quick identification.

Coil Terminals – Up-front for accessibility; self-lifting saddle clamps for ease of wiring.

N.C. Overload Relay Contact Terminal – Self-lifting pressure plates for ease of wiring; clearly marked in contrasting white for easy identification.

Tie Point Terminal – Convenient access point for control circuit wiring; accessibility; self-lifting pressure plates for ease of wiring.

Heater Element Screws – Up-front for easy installation; allows interchangeability of heater elements for Class 10, 20, and 30 operation.

Loadside Power Terminals – Up-front for accessibility; self-lifting saddle clamps for ease of wiring; clearly marked in contrasting white for quick identification.

Straight Mounting Slots – Easy to reach with large screwdriver or power driver; permits mounting screw to be in place before installing the starter; formed corners help to retain the device base plate in the event of partial loosening of the screw.

A variety of accessories quickly and securely snap into place or easily install using only a screwdriver.

Figure 12–21 A typical motor starter with all its parts identified. *(Courtesy of Rockwell Automation)*

Figure 12–22 Electrical diagram of a motor starter and control circuit.

coil will be interrupted when the overload contacts open. A reset button on the motor starter must be manually reset to set the overload contacts back to their normally closed position.

The control circuit is shown as a ladder diagram at the bottom of this figure. In this diagram you can see that the motor starter coil is energized when the start push button is depressed. The manual push button is used for this example because it is easier to understand, but you should be aware that most motor starters in HVAC and refrigeration applications are controlled by thermostats or other types of controls.

The motor starter also has one or more additional sets of normally open contacts called *auxiliary contacts*. The auxiliary contacts are connected in parallel with the start push button. These contacts serve as a seal-in circuit after the motor starter coil is energized. The start push button is a momentary-type switch, which means that it is spring loaded in the normally open position. When the start push button is depressed, current flows from L1 through the normally closed stop push-button contacts and through the start push-button contacts to the motor starter coil. This current causes the coil to become magnetized so that it pulls in the three major sets of load contacts and the auxiliary set of contacts. When the auxiliary contacts close, they create a parallel path around the start push-button contacts so that current still flows around the start push-button contacts to the coil when the push button is released. Since the stop push button is connected in series in this circuit, the current to the coil is de-energized and all of the contacts drop out when the stop push button is depressed.

12.23 THE OPERATION OF THE OVERLOAD

The overload for a motor starter consists of two parts. The heater is the element that is connected in series with the motor, and all the motor current passes through it. The heater is actually a heating element that converts electrical current to heat. The second part of the overload device is the trip mechanism and overload contacts. The trip mechanism is sensitive to the heat, and if it detects excess heat from the heater, it trips and causes the normally closed overload contacts to open. Since the motor starter coil is connected in series with the normally closed overload contacts, all current to the coil is interrupted and the coil becomes de-energized when the overload contacts are tripped to their open position. When the coil becomes de-energized, the motor starter contacts return to their open position, and all current to the motor is interrupted. When the overload contacts open, they remain open until the overload is reset manually. This ensures that the overloaded motor stops running and cools down until someone investigates the problem and resets the overloads.

Figure 12–23 shows a typical heating element for a motor starter overload device, which is called the heater. In the cutaway diagram for this figure you can see that the heater is actually a heating element that converts electrical current into heat as it passes through the heating element. You should also notice the knob protruding from the bottom of the heater. This knob has a shaft that is held in position inside the heater, and teeth machined into the part that protrudes from the heater. The teeth are called the *ratchet mechanism*.

Figure 12–24 shows the heating element mounted in the trip mechanism. The trip mechanism consists of the ratchet from the heater and a paw. The ratchet is actually the toothed knob that protrudes from the bottom of the heater. The paw has spring pressure that tries to rotate the ratchet. Because the heater holds the shaft of the ratchet tight with solder, the paw cannot move. When the heater becomes overheated, it melts the solder that holds the ratchet in place and allows it to spin freely. When the ratchet spins, it

Figure 12–23 A typical heater assembly. Notice the ratchet that protrudes from the bottom of the heater. *(Courtesy of Rockwell Automation)*

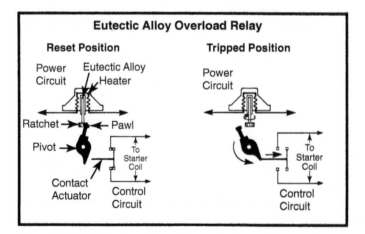

Figure 12–24 Diagram of the ratchet and paw mechanism that trips and allows the normally closed overload contacts to open. *(Courtesy of Rockwell Automation)*

allows the paw to move past it, which in turn allows the normally closed contacts to move to their open position.

After the overload condition has occurred and the overload contacts have opened, the motor starter is de-energized and the motor stops running. When the motor stops running, the heating element is allowed to cool down. After the heating element cools down for several seconds, the reset button can be depressed, which moves the paw back to its original position, and the normally closed overload contacts move back to their closed position. If the motor continues to draw excess current when it is restarted, the excess current will cause the heater to trip the overload mechanism again. If the motor current is within specification, the heaters do not produce enough heat to cause the overload mechanism to trip.

Since the overcurrent condition must last for several minutes to cause the overload mechanism to trip, the overloads allow the motor to draw high locked-rotor amperage

(LRA) during the few seconds the motor is trying to start, without tripping. If the motor has an overcurrent condition while it is running, the overloads allow the condition to last several minutes before the motor is de-energized. If the problem continues, the overloads will sense the overcurrent and trip to protect the motor. Figure 12–24 shows the overload contacts both in their normally closed or reset position and in their tripped or open condition.

12.24 EXPLODED VIEW OF A MOTOR STARTER

It is easier to see all the parts of the motor starter in an exploded view picture, as in Figure 12–25. At the far left in this picture you can see that the main contacts of the motor starter are much larger than those in a traditional relay. The coil is shown in the middle of the picture. The coil has two square holes in it that allow the magnetic yoke to be mounted through it. The magnetic yoke and coil are mounted in the movable contact carrier. When the coil is energized, it pulls the magnetic yoke upward, which causes the movable contacts to move upward until they make contact with the stationary contacts that are shown at the far left. The overload mechanism is shown at the far right in this figure. It is mounted at the lower part of the motor starter, and all current that flows through the contacts must also flow through the overload mechanism.

12.25 SIZING MOTOR STARTERS

At times you will need to select the proper size motor starter for an application. The size of motor starters is determined by the National Electrical Manufacturers Association (NEMA). The ratings refer to the amount of current the motor starter contact can safely handle. The sizes are shown in the table provided in Figure 12–26, where you can see that the smallest size starter is a size 00, which is rated for 9 A and is sufficient for a 2-hp motor connected to 480 V three phase or for a 1-hp single-phase motor connected to 240 V. You can see that the size 1 motor starter is rated for 27 A, which is sufficient for a 10-hp three-phase motor connected to 480 V or a $7\frac{1}{2}$-hp single-phase motor connected to 240 V. A size 00 motor starter is about 4 in. high, and a size 2 motor starter is approximately 8 in. high. You will typically use up to a size 3 or 4 motor starter to protect compressor motors for commercial air-conditioning and refrigeration systems. The size 4 starter will protect motors up to 100 hp. It is important to understand that the overload heaters for the motor starter can also be purchased for a specific current rating. This means that each motor starter can have a heater that is rated specifically to the amount of current the motor that is connected to it draws.

12.26 SOLID-STATE MOTOR PROTECTORS

Solid-state motor protectors, like motor starters, protect compressor motors against problems. For example, when a compressor is supplied with three-phase voltage, the solid-state motor protector provides protection against the loss of any phase of voltage, low voltage on any phase, high voltage on any phase, reversal of any phase, unbalanced voltage between any phase, and short cycling. Short cycling is a common problem that

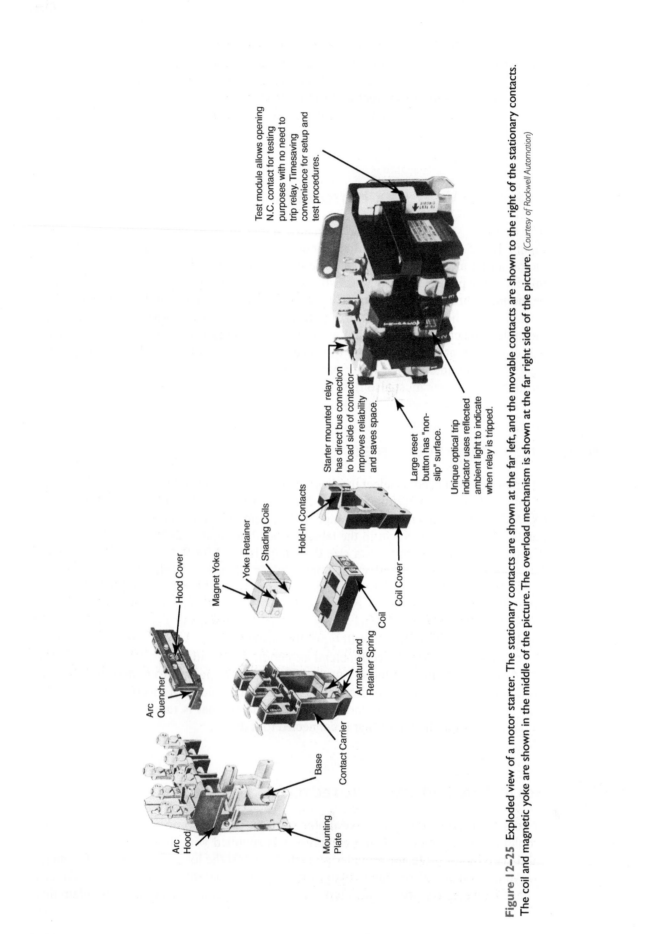

Test module allows opening N.C. contact for testing purposes with no need to trip relay. Timesaving convenience for setup and test procedures.

Starter mounted relay has direct bus connection to load side of contactor—improves reliability and saves space.

Large reset button has "non-slip" surface.

Unique optical trip indicator uses reflected ambient light to indicate when relay is tripped.

Hood Cover

Magnet Yoke

Yoke Retainer

Shading Coils

Hold-in Contacts

Coil Cover

Coil

Armature and Retainer Spring

Arc Quencher

Contact Carrier

Base

Mounting Plate

Arc Hood

Figure 12–25 Exploded view of a motor starter. The stationary contacts are shown at the far left, and the movable contacts are shown to the right of the stationary contacts. The coil and magnetic yoke are shown in the middle of the picture. The overload mechanism is shown at the far right side of the picture. (*Courtesy of Rockwell Automation*)

RATINGS FOR NEMA FULL VOLTAGE STARTERS

Nema Size	Continuous Ampere Rating	Maximum Horsepower Rating Full Load Current Must Not Exceed "Continuous Ampere Rating"			
		Motor Voltage			
				50 Hz	
		200 V	230 V	380 V–415 V	460 V–575 V
00	9	1-1/2	1-1/2	2	2
0	18	3	3	5	5
1	27	7-1/2	7-1/2	10	10
2	45	10	15	25	25
3	90	25	30	50	50
4	135	40	50	75	100
5	270	75	100	150	200
6 ❷	540	150	200	300	400
7 ❷	810	-	300	600	600
8 ❷	1215	-	450	900	900
9	2250	-	800	1600	1600

Figure 12–26 The NEMA sizes for motor starters. Notice that the smallest motor starter is a size 00, and the largest is a size 9. *(Courtesy of Rockwell Automation's Allen-Bradley Business)*

occurs when a compressor is turned off and turned on again after a few seconds before the pressure in the system has had a chance to equalize. This can occur if someone adjusts the thermostat up and then back down too quickly, or if power is interrupted due to a power outage such as occurs during a lightning storm, and then immediately comes back on. If the compressor is allowed to try to start before its pressure is equalized, it will draw large locked-rotor current and possibly not start.

Figure 12–27 shows a picture of a solid-state motor protection device, and Figure 12–28 shows a diagram of the location of the motor protection device in a compressor starting circuit. In the diagram you can see that the motor saver has a set of normally open contacts that are connected in series with the motor starter coil. The motor starter samples the voltage at L1, L2, and L3 from the three-phase voltage supplied to the compressor. If the supply voltage has a problem (low voltage, phase loss, and so on), the motor saver circuit will sense the problem, will open its contacts, and will cause current to the motor starter to be interrupted. A second circuit provides contacts for an audible alarm. When the motor saver trips, the alarm will indicate that the problem exists.

Figure 12–27 A solid-state motor protector. This device protects the motor against low voltage, loss of phase, and short cycling. *(Courtesy of SymCom, Inc.)*

Figure 12–28 Electrical diagram of a motor saver that is connected to the coil of the motor starter or contactor. *(Courtesy of SymCom, Inc.)*

Additional Job Assignment

You are sent on a job to check out a refrigeration system for a walk-in cooler that has a 5 hp compressor that is controlled by a motor starter. The motor starter overload is kicked out when you check it. You reset the overload and the compressor runs for another 20 minutes and kicks out again. The owner does not understand what a motor starter is and how it works. Your job it to explain to the owner what a motor starter is and what its basic parts are by sketching the electrical diagram of a motor starter. Next he wants you to explain the tests you will make and what you suspect is happening to the system. You should explain what you think is wrong and what tests you will use to prove what part of the system is working and what is not working correctly.

12.27 OVERCURRENT CONTROLS

Overcurrent controls include thermal overloads used in motor starters, thermal overloads used in compressors, circuit breakers, and fuses. In this section you will learn about thermal overloads for compressors, solid-state overloads for compressors, circuit breakers, and fuses. These devices are used to protect the compressor motor and other motors in the system from drawing too much current and overheating. You will find that the heat that damages compressors can come from overcurrents or from other conditions that cause the compressor to overheat when the current is within normal limits. These conditions occur when the amount of refrigerant in a system is low, and insufficient refrigerant is pumped back to the compressor to help keep it cool. Since these overheating problems can occur when the motor current is normal, additional means of detecting overheating must be used.

Figure 12–29 a Electrical diagram of thermal overload with contacts open. **b** Electrical diagram of thermal overload with contacts closed.

12.28 THERMAL OVERLOADS

Thermal overload is specifically designed to detect excessive heat that comes from overcurrent or from other physical conditions. It is mounted directly on the dome of a compressor to sense the temperature in the compressor, and it is connected in series with the compressor run and start windings to detect excessive current. Figure 12–29 shows two diagrams of thermal overload operation. Figure 12–29a shows the contacts when the overload senses too much current, and Figure 12–29b shows the overload when its contacts are closed.

All the current flowing through the compressor motor windings flows into terminal C1 at the bottom of the diagram, through the heating element, through the contacts, and out terminal C2. Since all the current for the motor must flow through the heating element, the overload must be sized for the exact amount of current the motor will draw safely. When the motor draws current in excess of this limit, the heater will produce extra heat that will cause the bimetal contacts to warp to the open position. When the overload is allowed to cool, the bimetal will snap the contacts to the closed position.

12.29 FUSES

Fuses perform a function similar to that of an overload, except a fuse uses an element that is destroyed when the overcurrent occurs. A fuse provides short-circuit protection, and the overload is not designed to protect the motor when short-circuit current occurs. The fuse provides a thermal sensing element that is capable of carrying current. When the amount of current becomes excessive, the heat that is generated is sensed by the fuse element, which melts when the temperature is high enough.

Fuses are available in a variety of sizes and shapes for different applications. Figure 12–30 shows a plug-type fuse and a cartridge-type fuse. Each fuse is sized for the amount of current it will limit. When the amount of current is exceeded, the fuse link melts and opens the fuse. Figure 12–31a shows examples of a single-element fuse in the process of blowing, and Figure 12–31b shows dual-element fuses blowing. The single-element fuse provides protection at one level. This type of fuse is generally used for noninductive loads such as heating elements or lighting applications. The dual-element fuse provides two levels of protection. The first level is called slow overcurrent (overload) protection, and it consists of a fusible link that is soldered to a contact point and attached to a spring. When a motor is started, it draws locked-rotor amperage (LRA) for several seconds. This excess current causes heat to build in the fuse, and this heat is absorbed in the slow overcurrent link. If the motor starts and the current drops to the

Figure 12–30 Plug-type and cartridge-type fuses. *(Courtesy of Bussmann)*

(a) (b)

Figure 12–31 a Single-element fuses in the process of blowing. **b** Dual-element fuses in the process of blowing. *(Courtesy of Bussmann)*

FLA level, the link will cool off, and the fuse will not open. If the LRA current continues for 30 seconds, the amount of heat generated will cause the solder that holds the slow overcurrent link to melt. When the solder melts, the spring pulls the link open and interrupts all current flowing through the fuse. The slow overcurrent link allows the fuse to sustain overcurrent conditions for short periods of time, and if the condition clears, the fuse will not open. If the condition continues to exist, the fuse will open.

The second type of element is called the short-circuit element. This element opens immediately when the amount of current exceeds the level of current the link is designed to handle. In the dual-element fuse, the short-circuit link is sized to be approximately 5 times the rating of the fuse. A short circuit is by definition any current that exceeds the full-load current rating by 5 times. (Some manufactures use the rating of 10 times.)

The single-element fuse has only a short-circuit element in it. This type of fuse is generally not used in circuits to start motors, since the motor draws locked-rotor current. If a single-element fuse is used to protect a motor circuit, it must be sized large enough to allow the motor to start, and then it is generally too large to protect the motor against an overcurrent condition of 20%, which will eventually damage the motor if it is allowed to occur for several hours.

12.30 FUSED DISCONNECT PANELS

The cartridge-type and screw-base fuses are generally mounted in a panel called a *fused disconnect*. The fused disconnect is normally mounted near the equipment it is protecting, and it serves two purposes. First, it provides a location to mount the fuses and, second, it provides a means of disconnecting the electrical supply voltage to a circuit. A three-phase disconnect is shown in Figure 12–32. The fused disconnect has a switch handle that is used to disconnect main power from the circuit and the fuses so that you can safely remove and replace or test the fuses.

Safety Notice!
It is recommended that you always use plastic fuse pullers to remove fuses from and replace them in a disconnect to protect you from electrical shock even when the fuse disconnect switch is in the off position. It is important to remember that even though the switch handle is open, line voltage is still present at the top terminal lugs in the disconnect. You should never use metal pliers or screwdrivers to remove fuses.

Figure 12–32 A three-phase fused disconnect. *(Courtesy of Challenger Electrical Equipment Corp.)*

12.31 CIRCUIT BREAKERS AND LOAD CENTERS

The *circuit breaker* is similar to the thermal overload in that it is an electromechanical device that senses both overcurrent and excess heat. Some circuit breakers also sense magnetic forces. Circuit breakers are mounted in electrical panels called *load centers*. The load center can be designed for three-phase circuits or for single-phase circuits. The single-phase panel provides 240-V AC and 120-V AC circuits. Figure 12–33 shows a typical load center without any circuit breakers mounted in it.

Circuit breakers are manufactured in three basic configurations for single-phase 120-V AC applications that require one supply wire, 240-V AC single-phase applications that require two supply wires, and three-phase applications that require three wires. The three-phase circuit breakers can be mounted only in a load center that is specifically manufactured for three-phase circuits. The two-pole and single-pole breakers can be mounted in a single-phase or three-phase load center.

The operation of the circuit breaker is similar to that of the thermal overload in that it senses excessive current that will trip its circuit after a specific amount of time. The main difference between the circuit breaker and the thermal overload is that the circuit breaker is mounted in the load center to protect both the circuit wires as well as the load. The circuit breakers are sized for the total current rating of the wire and all the loads that are connected to the wire. In some cases this means that the circuit breaker is sized for the current flowing to the compressor, condenser fan, and evaporator fan. For instance, if the compressor draws 20 A and the condenser fan and evaporator fan draw 5 A each, the circuit breaker will be rated for 30 A. A problem can occur if the compressor and condenser fan are turned off and the evaporator fan is running to circulate air and has an overload that doubles its current to 10 A. In this case the circuit breaker will not be able

Figure 12–33 A load center that is used to mount circuit breakers. The load center is sometimes called a circuit breaker panel. *(Courtesy of Challenger Electrical Equipment Corp.)*

Figure 12–34 Single-pole, two-pole, and three-pole circuit breakers for load centers. *(Courtesy of Challenger Electrical Equipment Corp.)*

to detect the additional 10 A in the evaporator fan as an overload because the total circuit current will be below the 30 A total. In these circuits it may be necessary to use a circuit breaker to protect the entire circuit, with overloads at each motor to protect them individually from overheating. This means that the main job of the circuit breaker is to protect the circuit against short-circuit conditions and to protect the entire circuit against overcurrent conditions rather than to protect individual motors against overcurrent. This is why many circuits have a combination of circuit breakers and overload devices. Figure 12–34 shows examples of single-phase and three-phase circuit breakers.

12.32

TROUBLESHOOTING MOTOR STARTERS AND OVERCURRENT CONTROLS

Troubleshooting a motor starter is similar to troubleshooting a contactor or relay. The motor starter has a single coil and three sets of normally open contacts that are rated to carry the current of the load that is connected to it. If the motor starter is not pulling its contacts closed when its coil is energized, you can make several quick tests. The first test is for voltage across the terminals of the coil. Be sure to read the voltage rating on the coil to compare it with the actual voltage applied to it. The most common problem is that the coil is not receiving any voltage. The main reason this occurs is that the normally closed overload contacts trip to the open position if the amount of current flowing through the heaters is excessive. If this occurs, the overloads can quickly be reset by depressing the reset button on the overload.

You can detect if the overload contacts are open by checking for a voltage loss through the coil circuit. You must leave one meter terminal on the common side of the circuit and move the other terminal from point to point in the coil circuit. If you detect a loss of voltage anywhere in this circuit, you have found the problem. If you have voltage directly across the coil terminals, you have determined that the circuit wiring is functioning properly and the problem must be with the coil. At this point you can try to "plug" the relay contacts to determine if they can move easily. If they can move easily, the next test is to remove power from the coil and test the coil for continuity. Be aware that the contacts may arc when you manually plug them closed.

If your test determines that the circuit is open at the overload contacts, you should have a clamp-on ammeter ready to test the load current flowing through the contacts when you reset the overload contacts. If the current measurement indicates that an overload still exists, you must take corrective action to repair this problem, or else the overloads will continue to trip. Do not oversize the overload heaters under any circumstances. If you determine that the coil is faulty, you can change the coil or change the complete motor starter.

12.33 REVIEW

A relay has one coil and one or more sets of contacts. The contacts can be normally open or normally closed. The word *normal* in this sense means the condition of the contact when power is not applied to the coil. There are four basic designs of relay armature assemblies, which move to open or close the contacts: the horizontal-action type, the bell-crank type, the clapper type, and the vertical-action type.

When a relay coil has voltage applied to it, it begins to draw current. The relay coil uses more current to pull in its coil than to hold it in. This means that if a power brownout occurs and supply voltage drops, a relay that is pulled in may not be affected, whereas a relay coil that is trying to pull in when voltage is low may not activate.

Relays have ratings for their coil and their contacts. The voltage rating for a relay coil must be matched exactly. The voltage and current rating for the contacts must not be exceeded. The typical ratings for coils are 24, 120, and 208, 240, or 480 V. Some coils also are color coded: 24-V coils are black, 120-V coils are red, and 208/240-V coils are green. Voltage ratings for contacts are 300 or 600 V.

Contacts are arranged in relays as single-pole, single-throw (SPST), single-pole, double-throw (SPDT), double-pole, single-throw (DPST), and double-pole, double-throw (DPDT).

A current relay is a specialized relay that is used for starting a compressor. It has a coil that has low resistance and a set of normally open contacts. The coil is connected in series with the run winding of a compressor, and the contacts are connected to the start winding. When the compressor is starting, the large current flowing to the run winding flows through the coil and causes the contacts to close and send voltage to the start winding.

Relays and contactors are very similar in that they both have a coil and one or more sets of contacts. The major difference between the relay and contactor is that the contactor is larger. A relay has contacts that are rated for less than 15 A, and the contactor is rated for 15 A or more. The National Electrical Manufacturers Association (NEMA) designs specifications for relays and contactors.

A solenoid has a coil like a relay, but the coil in a solenoid activates a valve rather than contacts. The valve can be normally open or normally closed. The reversing valve is a very complex solenoid that is used to change the direction of refrigerant flow.

A motor starter is a contactor with built-in heaters and overloads. The motor starter has a coil and three main contacts and one or more sets of normally open or normally closed auxiliary contacts.

The overload assembly consists of a heating element that is connected in series with each of the main contacts. The heating element converts electrical current into heat, which is transferred onto components that activate the normally closed overload contacts and cause them to open when too much current flows through the heating element. The heating element must be sized to the amount of current it is trying to protect.

Motor starters are sized by NEMA from size 00 through size 9. The table in Figure 12–26 shows the amount of current each size can safely handle.

Solid-state motor protectors act much like the overload assembly in a motor starter and can be added to any relay or contactor used in HVAC applications. Other types of overload protection are provided by fuses and circuit breakers. Single-element fuses are used to protect against overcurrent in circuits, and dual-element fuses are used to protect against overcurrent in motors.

Job Assignment for This Chapter

You are sent to a residence to work on a split system air conditioner that has the outdoor unit supplied with 240 volts and the indoor unit (furnace) supplied with 120 volts. You find the compressor and condenser fan not running, and your preliminary tests indicate that the coil of the compressor relay is energized, but the compressor and condenser fan are not running. Your job is to decide which tests you should make next to determine what exactly is wrong with the system. You will need to explain to the homeowner what is wrong with the system and what tests you should do to demonstrate what is wrong with the system.

Solution to the Job Assignment

Since your preliminary tests indicate that the coil of the compressor relay is energized, your next test should be to see if high voltage (240 volts) is present on the L1 and L2 contacts on the relay. These contacts are the input side of the contacts. If voltage is present, your next test is to measure for voltage at T1 and T2 of the contacts, which are the output terminals for the contacts. If voltage is not present there, you may have a burnt or damaged contact. If voltage is present, then you need to check the compressor motor or fan motor to see why they are not running.

If voltage is not present on the L1 and L2 contacts of the relay, then voltage is not being supplied to the relay contacts and you should suspect that the problem is the fuse box or fused disconnect or the circuit breaker that supplies high voltage to the outdoor unit. It is important to remember that the voltage for the relay coil comes from the transformer in the indoor unit (furnace) and can have 24 volts power and energize the coil, whereas the 240 volts that is supplied to the contacts comes from a different source and can be interrupted by the circuit breaker, fuse, fused disconnect, or bad wire.

Additional Job Assignment

You have just found a relay in a residential air-conditioning system that has burned contacts. The relay is a SPST with normally open contacts. The contacts are rated for 20 amps and 300 volts, and the coil is rated for 24 volts. When you go to the truck you find that you have four relays to select from and none of them is an exact match. Your job is to select the relay that will be an acceptable replacement. The relays have the following ratings:

1. DPST relay with 24-volt coil; contact rating is 300 volts and 15 amps.
2. DPST relay with 24-volt coil; contact rating is 300 volts and 30 amps.
3. SPDT relay with 120-volt coil; contact rating is 300 volts and 30 amps.
4. SPST relay with 120-volt coil; contact rating is 300 volts and 20 amps.

Solution to the Additional Job Assignment

You might complete this assignment by a process of elimination. The two relays that have 120-volt coil are automatically eliminated. The voltage for the coil must match exactly, so since the relay you are trying to replace has a 24-volt coil, the replacement must have a 24-volt coil.

The important rating is current rating for the contacts. The rating for the contacts must equal or exceed the rating of the contacts of the relay that you're trying to replace. Since the relay you are trying to replace has a 20-amp rating for its contacts, relay number 1 will not work, because its rating is only 15 amps. By process of elimination the only relay in your truck that will work is number 2, which has a 24-V coil and a contact rating of 30 amps.

Additional Job Assignment

You are sent on a job to check out a refrigeration system for a walk-in cooler that has a 5-hp compressor that is controlled by a motor starter. The motor starter overload is kicked out when you check it. You reset the overload and the compressor runs for another 20 minutes and kicks out again. The owner does not understand what a motor starter is and how it works. Your job it to explain to the owner what a motor starter is and what its basic parts are by sketching the electrical diagram of a motor starter. Next he wants you to explain the tests you will make and what you suspect is happening to the system. You should explain what you think is wrong and what tests you will use to prove what part of the system is working and what is not working correctly.

Solution to the Additional Job Assignment

Your solution should include a sketch of a motor starter with its heaters in series with the motor windings and its normally closed overload contacts in series with the motor starter coil. You should explain that the current draw of the motor causes the heaters to heat up, and when the current draw of the motor is above its ratings, the heaters will cause the overload contacts to open. Your solution should also include checking the compressor motor to see what is causing it to overheat, such as low refrigeration or other problems. You should explain that you only need excessive current in one of the three lines supplying voltage. You should also put a clamp-on ammeter on each of the windings to determine the actual current draw of each winding of the motor. If only one or two of the lines have unequal current measurements, you may suspect that the compressor motor is going bad. Another problem that could cause the unequal current is if you have low voltage in one or more of the incoming voltage lines. You can use a voltmeter to measure the amount of voltage for each winding of the supply line. You are looking for low voltage in one or more lines of the motor. You can recommend the repair to fix the problem or you can recommend more in-depth tests for the compressor if you think it is the cause of your problem.

Assessment

QUESTIONS

1. Identify the main parts of a relay and explain their operation.
2. Explain the operations of the four types of armature assemblies presented in this chapter.

3. Explain the operations of a current relay and a potential relay.

4. Discuss the differences between a contactor and a relay.

5. Identify the main parts of a solenoid and explain their operations.

6. Explain the differences and similarities between a motor starter and a contactor.

7. An overload on a motor starter has contacts and a heater. Describe the function of each.

8. Identify the parts of a thermal overload that are used to protect a compressor and explain each part's operation.

9. Discuss the differences between a single-element fuse and a dual-element fuse.

10. Explain the difference between slow overcurrent and short-circuit current.

TRUE OR FALSE

1. Plug-in relays are used in air-conditioning and refrigeration systems because they are easy to change when they become faulty.

2. The main difference between a relay and a contactor is that the contactor usually has more contacts.

3. A solenoid is similar to a relay in that it has a coil that becomes magnetic and moves the plunger of a valve.

4. It is possible to change normally open contacts to normally closed contacts in the field for some relays.

5. For a relay to operate correctly, its contacts must close first to provide current to its coil.

6. The heater is the part of the motor starter overload that opens and interrupts current flow.

7. The overload contacts on a motor starter are normally open and interrupt current when an overload condition is detected.

8. The main job of the circuit breaker is to protect the circuit against short circuits, and the entire circuit against overcurrent, rather than to protect individual motors against overcurrent.

9. A fused disconnect provides a location for mounting fuses for the system and also provides a means of disconnecting main power from the circuit.

10. An overload device may be mounted internally in a compressor or externally to detect heat on the compressor shell.

MULTIPLE CHOICE

1. The normally closed contacts of a relay
 a. pass current when the coil is not energized.
 b. pass current when the coil is energized.
 c. pass current at all times whether or not the coil is energized.

2. A current relay is
 a. a special relay designed to control only current.
 b. a relay that is designed to be used as a relay or a contactor.
 c. a relay that is designed to start single-phase compressors.

3. A solenoid is
 a. a valve that is controlled (opened and closed) by gravity and springs.
 b. a reversing relay that is used to start heat pumps.
 c. a valve that is controlled (open and closed) by springs and a magnetic field produced by a coil.

4. A reversing valve is
 a. a special type of solenoid used in heat pumps.
 b. a special relay used to reverse the direction in which the compressor motor runs.
 c. a special valve that can be used as either a fill valve or a drain valve for a heat pump.

5. An SPDT relay has
 a. one set of contacts that connects to two output terminal points.
 b. two sets of contacts that connect to one terminal point.
 c. one set of contacts and one output terminal point.

6. A motor starter
 a. is just a larger version of a relay, since it has only a coil and contacts.
 b. has a coil and contacts like a relay or contactor, and it also has overloads to provide overcurrent protection.
 c. is more like a circuit breaker than a relay in that it can detect excessive heat buildup on the outside of the compressor motor through its heaters.

7. A dual-element fuse
 a. can detect slow overcurrent and large short-circuit currents.
 b. can be used two times before it must be replaced.
 c. has two elements, so it can be used in single-phase circuits to protect both L1 and L2 supply voltage lines.

8. The overload device on a motor starter
 a. has a heater to detect excess current and overload contacts to interrupt the control circuit current.
 b. has a heater to detect excessive current and overload contacts that directly interrupt the large load current to the motor.
 c. has a heater that interrupts control circuit current and overload contacts that directly interrupt the large load current to the motor.

9. The motor starter has auxiliary contacts that
 a. are connected in parallel with a start switch to seal in the circuit when it is operating correctly and ensure the coil remains de-energized if the overload contacts trip.
 b. are connected in series with a start switch to seal in the circuit when it is operating correctly and ensure the coil remains de-energized if the overload contacts trip.
 c. can switch current to an auxiliary motor such as a condenser fan at the same time the main contacts supply current to the compressor motor.

10. Heaters for motor starters are available in different sizes so that
 a. they can provide the correct amount of current protection to match the motor that is connected to the motor starter.
 b. they can match the voltage rating of the fuses used in the circuit.
 c. they can match the temperature rating of the ambient air in which the motor will operate.

PROBLEMS

1. Draw a waveform that shows pull-in and hold-in current for a relay coil.
2. Use the NEMA table provided in Figure 12–12 to select a NEMA starter size that can control 45 continuous amps.
3. Draw an example of single-pole, double-throw (SPDT) contacts, and explain their operation.
4. Draw an example of double-pole, single-throw (DPST) contacts, and explain their operation.
5. Draw an example of double-pole, double-throw (DPDT) contacts.
6. Sketch a motor starter with its contacts connected to a three-phase motor and its coil controlled by start and stop push buttons. Be sure to show the heaters and the overload contacts in the proper circuits.
7. Sketch a thermal-type overload and explain the function of each of the parts.
8. Sketch a single-element fuse and a dual-element fuse and explain how they are different.
9. Sketch a thermal overload device connected to the outside of a compressor and explain how it protects the compressor against overheating.
10. Use the table shown in Figure 12–26 to select a motor starter for a single-phase $7\frac{1}{2}$-hp compressor that draws 27 A at 240 V.

Websites

Websites change constantly. Use the key word search to locate sites that have similar content. The products that will be found at each site are indicated with the key word.

Eaton/Cutler-Hammer
Products: Relays, motor-starters, switches
Key word: Eaton/Cutler-Hammer

Allen-Bradley Rockwell Automation
Products: Motor-starters, relays, switches, motor drives
Key word: Allen-Bradley

Square D
Products: Motor starters, relays, switches
Key word: Square D

Therm-o-Disc, Incorporated
Products: Therm-O-Discs
Key word: Therm-O-Disc

Cooper Industries, Bussmann Fuses
Products: Fuses
Key word: Bussmann Fuses

Alco Controls, a Division of Emerson
Products: Solenoid valves, reversing valves
Key word: Alco Controls

Chapter 13

Thermostats, Pressure Controls, and Timers

Job Assignment for This Chapter

You are on a service call in a residence during the heating season and the symptoms indicate the heating thermostat is not working correctly on a gas furnace. Your job is to identify several procedures and tests you would use to troubleshoot this problem. The homeowner wants to know what you think is wrong and what tests you will use to determine if the thermostat is functioning correctly.

Terms You Will Learn in This Chapter

Bimetal element
Cooling thermostat
Cut-in pressure
Cut-out pressure
Defrost timers
Dual pressure switch
Fan switch
Fixed-pressure switch
Heat anticipator
Heating thermostat
High-pressure switch
Humidistat

Limit switch
Line-voltage thermostat
Low-pressure switch
Oil pressure switch
Pilot light
Primary control
Programmable thermostat
Short-cycle protection
Thermostat
Thermostat subbase
Timer control

After reading this chapter, you will be able to:

1. Identify the major parts and circuits of a heating thermostat.
2. Identify the major parts and circuits of a cooling thermostat.
3. Explain the operation of a fan and limit switch.
4. Explain the operation and function of a high-pressure switch.
5. Explain the operation and function of a low-pressure switch.
6. Explain the operation and function of an oil pressure switch.
7. Explain the operation of a defrost timer.

13.0 OVERVIEW OF THERMOSTATS

The **heating thermostat** and the **cooling thermostat** are used to control the temperature in a conditioned space by energizing and de-energizing the furnace system and the air-conditioning system. These controls use bimetal elements and other means to sense the temperature and to activate a set of contacts for control. Some thermostats have a fan switch to control the operation of the furnace fan in manual operation.

The furnace can use electricity, gas, or oil as its main fuel. Each of these furnaces must have special controls to operate safely. For example, a gas furnace must have a gas valve with a **pilot light** to ignite the fuel safely. A fuel oil furnace must have a pump, nozzle, and main burner control to atomize the fuel oil so that it will ignite and burn. An electric furnace must have stage controls to energize each bank of electric heat in stages. All these types of furnaces need a control called a *high limit* to limit the highest temperature the system can reach if a problem such as a furnace fan failure arises. Other controls such as a *fan control* are also needed to energize the fan when the furnace is producing heat. If air conditioning is added to the furnace, additional cooling controls may be added to ensure that the fan runs when the system is in either the cooling mode or the heating mode.

13.1 SENSING TEMPERATURE

The oldest types of sensing elements used in modern times are temperature sensors. Early scientists found that different types of metals expand at different rates when they are heated. When two dissimilar metals are bonded together and heated, one of the metals will expand faster than the other and cause the two strips to curl at one end if the other end is mounted securely. Figure 13–1a shows an example of two strips of metal bonded together, and Figure 13–1b shows the bonded strips curled on one end when they are heated. Since two strips are used, this element is called a **bimetal element**.

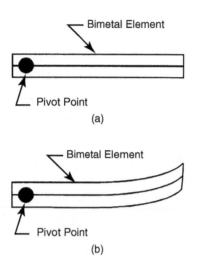

- Bimetal Element
- Pivot Point
- (a)

- Bimetal Element
- Pivot Point
- (b)

Figure 13–1 a Two strips of dissimilar metal bonded together to form a bimetal element. **b** The bimetal element curls at one end when the other end is secured and heat is added to the entire element. The bimetal element curls because one of the metal strips expands faster than the other.

13.2 USING THE BIMETAL ELEMENT TO SENSE TEMPERATURE

Figure 13–2a shows a set of electrical contacts mounted to one end of the bimetal element. One half of the contact is mounted stationary, and the second half of the contact is attached directly to the end of the bimetal element. When the element is cold, the contacts are in the normally closed position, and when the element is heated, the end of the bimetal curls and causes the contacts to open. Figure 13–2b shows the contacts when they are open. Contacts are also designed so that they are in the open position when the

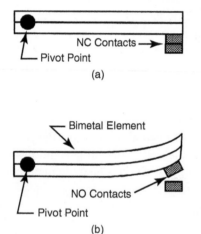

- NC Contacts →
- Pivot Point
- (a)

- Bimetal Element
- NO Contacts
- Pivot Point
- (b)

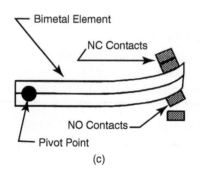

Figure 13–2 a A set of normally closed electrical contacts is mounted on a bimetal element. **b** A set of normally open electrical contacts is mounted on a bimetal strip. **c** A set of normally open and a set of normally closed contacts are mounted on a bimetal element.

- Bimetal Element
- NC Contacts
- NO Contacts
- Pivot Point
- (c)

bimetal strips are cool, and in the closed position when the strip is heated. Figure 13–2c shows two sets of contacts mounted on the bimetal element. One of the sets is in the closed position when the element is cool. This set of contacts controls the heating system. The second set of contacts closes when the bimetal strip is heated, and it controls the air-conditioning system.

13.3 AMPLIFYING THE MOVEMENT OF THE BIMETAL ELEMENT

The movement of the bimetal element is limited, so additional methods may be required to help the element become an effective temperature control switch. Figure 13–3a shows a small permanent magnet attached to the bimetal element so that when it moves close to the point where the contacts will be closed when the element is heated, the magnets will make the contacts snap to the closed position. Figure 13–3b shows the bimetal element in the shape of a spiral. When the element is lengthened and twisted into a spiral, it provides more area to sense the change in temperature and amplifies the movement caused when one metal expands faster than the other. The spiral also provides a means of moving larger sets of contacts. Figure 13–3c shows a close-up picture of a mercury bulb mounted on a bimetal spiral. Figure 13–4a shows a set of mercury contacts sealed in a glass tube and mounted to the end of the bimetal spiral. Several sets of terminals are mounted in each end of the glass tube. Mercury is a liquid at room temperature, so gravity will move it like water to the end of the tube that is down. Since mercury is a conductor, the two sets of terminals that are covered by the mercury will act like a set of closed contacts. Figure 13–4b shows the spiral expanded when heat is applied. Notice that the expansion of the bimetal element causes the mercury to move to one end. The contacts on this end of the tube are used to energize an air-conditioning system. Figure 13–4c shows the spiral when it is cooled down, and you can see the glass tube is positioned so that the mercury moves to the opposite end of the tube and covers the contacts used to energize the heating system.

Figure 13–3 **a** A set of permanent magnets is mounted to the bimetal element to make the contacts snap into the closed position when the element moves the magnets near each other. **b** The bimetal element is shaped in a spiral to amplify the movement when it is heated or cooled. **c** A mercury bulb may be used instead of contacts. The mercury bulb is mounted on the thermostat bimetal spiral, and the mercury moves to complete the electrical circuit as temperature causes the spiral to wind or unwind.

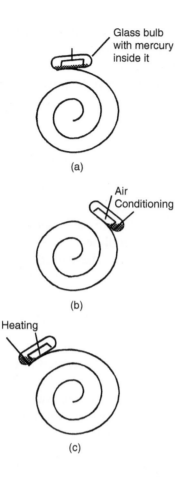

Glass bulb
with mercury
inside it

(a)

Air
Conditioning

(b)

Heating

(c)

Figure 13–4 **a** A glass tube is mounted on the bimetal spiral. The glass tube has liquid mercury in it that moves from one end to the other as the glass tube is moved. **b** The spiral is heated and expanded, which moves the glass tube so that the contacts that energize the air-conditioning system are covered by the mercury. **c** The spiral is cooled and contracted, which moves the glass tube so that the contacts that energize the heating system are covered by mercury.

13.4 THE HEATING THERMOSTAT

The **heating thermostat** has a set of contacts that close when the room temperature falls below the setpoint. The bimetal element for the heating thermostat is the part of the thermostat that senses a change in temperature and activates the contacts. The heating thermostat connects terminals R and W to the contacts. 24 V AC comes from the transformer terminal R to the thermostat terminal R. When the bimetal senses the room is cooler than the setpoint, it moves and closes the contacts and the 24 V AC moves through the thermostat and out terminal W. The gas valve, oil burner control or other furnace control is connected

(a) (b)

Figure 13–5 **a** A typical thermostat. *(Courtesy of Honeywell)* **b** T87 thermostat with cover removed.

Figure 13–6 White-Rodgers thermostat. *(Courtesy of White-Rodgers)*

to terminal W on the thermostat so that it gets voltage when the thermostat is calling for heating. It is important to understand that all thermostats operate similar to each other and most have adopted a standard terminal identification so that you can install and troubleshoot nearly any brand of thermostat. Figures 13–5, 13–6, and 13–7 show pictures of different types of thermostats you will find in residential applications. Figure 13–8 shows a diagram

Figure 13–7 Another type of White-Rodgers thermostat. *(Courtesy of White-Rodgers)*

Figure 13–8 a Electrical diagram of a subbase for a heating thermostat. **b** The internal diagram of a heating thermostat. *(Courtesy of Honeywell)*

of a thermostat. The majority of thermostats are rated to switch 24 V AC as control power. You will find some thermostats designed specifically for electric heating systems that are rated to switch 240 V AC or 120 V AC. The thermostats in the following examples are rated for 24 V AC.

Figure 13–8b shows that power is brought into the thermostat through terminal R. Terminal W is connected to the contacts that control heating, and terminal Y is connected to the contacts that control cooling. The letters R, W, and Y represent the color of wires (red, white, and yellow) that are used to connect the thermostat to the heating and cooling system. A selector switch allows the thermostat to be set to control the heating cycle or the cooling cycle or to be turned off so that neither heating nor cooling is available. Some thermostats have a position for the selector switch called *auto*, which allows both the heating and cooling to be controlled simultaneously. This function is useful where it is cool at night and a home or building requires heating, and it may be warm enough during the day to require the air-conditioning system to run.

Additional Job Assignment

You are installing a new heating thermostat on a gas furnace heating system and the instructions tell you to set the heat anticipation correctly before you complete your installation. The homeowner is looking at the installation instructions and notices that you need to set the heat anticipation to match the furnace. The homeowner asks you what exactly "heat anticipation" is. Your job is to explain to the homeowner what "heat anticipation" is and set it correctly to match the furnace gas valve.

13.5 THE THERMOSTAT SUBBASE

The thermostat mounts to a subbase that is screwed to the wall. The **thermostat subbase** has the terminals for all of the wires and also supports the fan switch and heat-off switch. The subbase is mounted to the wall and leveled so that when the thermostat is mounted to it, the thermostat will also be level. Figure 13–9a shows a typical subbase for a T87 round-type thermostat, and Figure 13–9b shows a subbase for a White-Rodgers rectangular-type thermostat. When you are installing or troubleshooting a thermostat, you will remove the thermostat from the subbase so you can see the wires that are connected to the subbase. If you are installing the thermostat, the terminal screws in the subbase are where you are connecting the wires. The terminals are identified as R, W, Y and G, which stands for the color of wires that were traditionally used in thermostat cables (R = Red, W = White, Y = Yellow, and G = Green).

13.6 THE HEAT ANTICIPATOR

When a heating thermostat is used to control a gas or oil furnace, the heat exchanger inside them stores heat energy when the fuel is being burned. When the fuel source is turned off, the heat exchanger can continue to release the stored heat for several minutes as the furnace fan blows air over it. This excess heat may cause the room to become warmer than the thermostat setpoint. One way to keep this excess heat from causing the

(a)

(b)

Figure 13–9 a A subbase T87 round-type thermostat. *(Courtesy of Honeywell)* **b** A subbase for a White-Rodgers rectangular-type thermostat. *(Courtesy of White-Rodgers)*

room to overheat is to provide an extra source of heat to the bimetal element in the thermostat to cause it to turn off 1 to 2 min prematurely.

The **heat anticipator** is shown with an arrow in Figure 13–10a and 13–10b. It is a wire-wound variable resistor that is placed on the subbase near the bimetal to provide the extra source of heat to help prevent the thermostat from allowing the system to overheat the room. The electrical resistor that is used as the heat anticipator is connected in series with the R–W circuit so that whenever the heating circuit is closed to provide voltage to the W terminal, the current flowing through this circuit to the coil of the gas valve will also flow through the resistor. The current flowing through the resistor will cause it to heat up slightly. This heat is transmitted directly to the spiral bimetal element, causing it to open the heating circuit before the room temperature reaches the setpoint. When the thermostat heating circuit opens before the room temperature reaches the setpoint, the gas valve is de-energized and the fuel is no longer provided to be burned. The residual heat that is built up in the furnace heat exchanger is then released and moved into the room by the furnace fan.

For example, if the thermostat is set for 72°F, and the room temperature drops to 70°, the thermostat's heating contacts close and energize the gas valve. When the gas is ignited, it begins to heat up the furnace heat exchanger, and the fan moves this heat into the conditioned space. The bimetal element in the thermostat senses the heat being added to the space as the room warms up to 71°. During this time the current flowing to energize the gas valve is also flowing through the heat anticipator, which causes it to add approximately 0.5°. When the room reaches 71.5°, the thermostat bimetal element actually senses 72° because the heat anticipator is adding 0.5°. This causes the heating contacts to open and de-energize the gas valve when the room is actually 71.5°. The fan continues to run for several minutes and continues to remove the residual heat that has built up in the heat exchanger. This heat adds the last 0.5° and brings the room to exactly 72°F.

Since the heat anticipator is a variable resistor, it can be adjusted to add more or less heat to the bimetal element, which will lengthen or shorten the time the gas valve is energized. This allows the resistor to anticipate how much heat is left in the heat exchanger, so that the room temperature will be controlled to the exact temperature for which the thermostat is set. You should be aware that the variable resistor (heat anticipator) can be adjusted to any amount of resistance, but if the amount of resistance is set too low, the current flowing through the resistor to the gas valve may be too large for the wattage rating of the resistor, causing a burn spot to develop on the wire.

(a) (b)

Figure 13–10 a Position of heat anticipator on White-Rodgers thermostat subbase. *(Courtesy of White-Rodgers)* **b** Position of heat anticipator on Honeywell T87 thermostat. *(Courtesy of Honeywell)*

Figure 13–11 Heat anticipator resistor wire. *(Courtesy of White-Rodgers)*

Numerical values are provided on the heat anticipator resistor adjustment to allow you to match the current flowing to the thermostat and the gas valve to a number on the heat anticipator. Figure 13–11 shows the numbers on the heat anticipator and you can see the values are 0.15 to 1.25 amps. These numbers refer to the amount of current that should flow through the resistor when it is moved to each setting. The numbers help match the amount of current flowing through the circuit to the value in the heat antici- pator to ensure that the amount of heat will cause the gas valve to shut off early.

You should use an ammeter to measure the amount of current flowing through the circuit to the gas valve, and then set the heat anticipator at the corresponding setting. The words *longer* and *shorter* are also marked on the heat anticipator adjustment so that you can extend or shorten the amount of time the gas valve is energized. You will need to use the *trial-and-error* method to find the precise setting to match the furnace to the room that is being heated. Figure 13–12 shows the instructions for setting the heat antic- ipator in the heating thermostat. The heat anticipator has current flowing through it only when the circuit between R and W is made, and the gas valve is drawing current. The adjustment on the heat anticipator can also be used to adjust the thermostat for fuel oil and electric heating systems. You may need to create a multiplier around your clamp-on ammeter to accurately measure the small amount of current in the circuit. You will know when you have the heat anticipator on the thermostat set correctly when the heating cycle is not too long and not to short and controls the room temperature within 1 degree of the setpoint.

13.7 COOLING THERMOSTAT

Figure 13–13 shows the diagram of a typical cooling thermostat. The **cooling thermo- stat** can be used exclusively for cooling or it can be added to the heating thermostat. The cooling thermostat contacts are connected between terminals R and Y. The cooling ther- mostat also needs to anticipate the changes in temperature in the conditioned space. The cooling anticipator is a fixed resistor and is wired so that it will draw current and add heat to the bimetal element when the air-conditioning contacts in the thermostat are

Setting the Heat Anticipator

When you have replaced an old thermostat with a new thermostat, you will need to set the heat anticipator on the new one to match the setting on the old one to ensure that the furnace cycles correctly. If you set the anticipator incorrectly, the furnace may short cycle and turn on and off too quickly, or the on cycle may be extended so that the room overheats. Start this process by checking the setting of the heat anticipator on the original thermostat and writing it down. Then set the anticipator on the new thermostat to the number on the old one. The settings range between 0.15 amps to 1.2 amps. You can adjust the anticipator by rotating the contact arm. The setting is indicated by the number on the base that the pointer points to. If you are not sure what setting to use, you can set the anticipator to a higher number and run the furnace through its cycle. Remember the anticipator adds heat to the thermostat sensing element as current flows through it. The anticipator is connected in series with the gas valve on the furnace, so the electrical current that flows to the gas valve solenoid will also flow through the anticipator and cause it to give off heat. If you move the anticipator to a lower number (clockwise), it will cause more heat to be developed by the same amount of electrical current flowing through it, and it will cause the heating cycle to be shortened as the anticipator will add heat and cause the sensing element to turn the thermostat earlier than it would by several seconds. If you move the anticipator to a higher number (counterclockwise), it will cause less heat to be developed by the same amount of electrical current flowing through it, and it will cause the heating cycle to be lengthened as the anticipator will not add as much heat and cause the sensing element to turn the thermostat later than it would by several seconds. Each increment of change will change the time of the cycle.

Figure 13–12 Instructions for setting the heat anticipator on a thermostat.

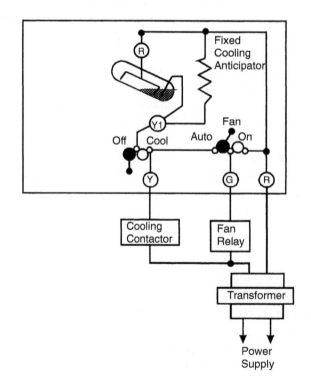

Figure 13–13 Cooling thermostat. *(Courtesy of Honeywell)*

not closed. The exact function of the cooling anticipator is to provide extra heat to the bimetal element when the air-conditioning system is not running, to cause the thermostat to close the cooling contacts a few minutes before the room temperature warms to the thermostat setpoint. This extra time allows the air conditioning to come on and begin cooling before the room gets too uncomfortable.

13.8 THE FAN SWITCH

The **fan switch** is used to provide a means of energizing the furnace fan on demand. This switch provides voltage from terminal R to terminal G when it is in the on position. The voltage at terminal G is used to energize the coil of a fan relay. Figure 13–14 shows the electrical diagram of the fan switch in the thermostat. You can see that when the switch is set to the auto position, voltage is supplied to terminal G whenever the cooling thermostat contacts close. When the switch is in the on position, voltage is supplied to terminal G at all times.

The fan switch can be used to cause the fan to run continuously for systems that have an electronic air cleaner. The fan switch can also be used to energize the fan on days when it is too warm for comfort but not warm enough to use air conditioning.

In the heating cycle, the fan motor is turned on and off by a temperature-controlled fan switch that is mounted in the furnace where it can sense the temperature inside the furnace. The switch will close its contacts when the air is warm enough to be sent into the conditioned space and not feel too cool. If the fan turned on immediately when the gas valve was energized, cool or cold air would be sent into the room and anyone in the room would notice the chilly air. The fan switch also ensures that all of the heat in the heat exchanger is moved into the air by the fan at the end of the cycle. If the fan switch on the thermostat is set to on during the heating cycle, air will continually cycle in the room, and people will not notice warm or cool areas that may stratify when air is not moving. For example, if air is not moving constantly in a room, warmer air tends to rise and cooler air tends to fall to the lowest spot in the room. This makes it feel warm for people in the upper body and face and cool around the feet, which overall is uncomfortable. When the fan runs constantly the air is mixed so that the temperature is roughly the same at all heights and levels of the room, and people will not notice when the furnace turns on and off. The only drawback to running the fan constantly is the additional expense, which is generally minimal.

13.9 HEATING AND COOLING THERMOSTATS

Figure 13–15 shows a diagram of a combination heating and cooling thermostat. This type of thermostat has a Heat-Off-Cool switch that allows you to set the thermostat for heating, cooling, or off. A fan switch is provided to allow settings for auto so that the fan will run when the heating or cooling system is called, and an on position that causes the

Figure 13–14 A diagram of the fan switch.

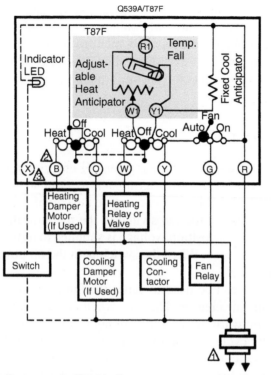

Figure 13–15 A heat–cool thermostat. *(Courtesy of Honeywell)*

⚠ Power supply. Provide disconnect means and overload protection as required.

⚠ No. 4 terminal is same as B terminal.

⚠ X terminal used on Q539 models with factory-installed indicator LED.

fan to run continuously. The electrical diagram shows that this thermostat provides two circuits for heating and cooling. Some heating and cooling thermostats provide two stages, and each stage of heating has its own independent contacts mounted in a glass bulb with mercury, and the cooling part of the thermostat has its own mercury bulb. These bulbs are interconnected with a mechanical link so that when the heating setpoint is adjusted to 70°, the cooling is mechanically set to 75° so that the thermostat does not try to energize the heating and cooling equipment at the same time.

In other types of heat–cool thermostats, a single glass bulb has a center contact that makes continuity through the mercury to the heating contact when the bulb is tipped one way, and to the cooling contact when the bulb is tipped the other way.

13.9.1 Thermostats with One Transformer

Most heating and air-conditioning systems have a single transformer that is located in the furnace (indoor unit) and supplies voltage through the thermostat to the outdoor unit (condenser). The R terminal in the thermostat supplies voltage to the air-conditioning system through the Y terminal or to the furnace through the W terminal. Figure 13–16 shows a diagram for this type of circuit. You should notice that the thermostat needs a jumper wire between the RC terminal and the RH terminal. The RC terminal is where the 24-volt supply comes from the transformer to the thermostat for the cooling functions. The RH terminal is the part on the thermostat for the 24-volt supply for the heating part of the thermostat. In most thermostats the jumper is put in at the factory and is generally in place unless you take it out.

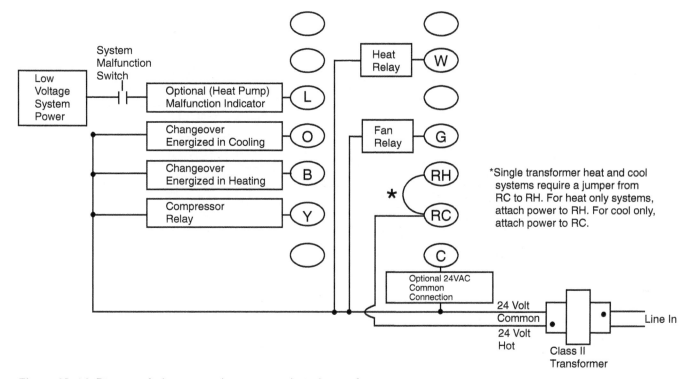

Figure 13–16 Diagram of a heat pump thermostat with single transformer. *(Courtesy of White-Rodgers)*

13.9.2 Thermostats with Two Transformers

Figure 13–17 shows a diagram with two transformers. From this diagram you can see that one transformer supplies 24 volts for the heating circuit through the terminal RH, and a second transformer is connected to the RC terminal to provide 24 volts to the cooling part of the circuit. You'll notice a black dot on one of the terminals for each transformer. This mark is used to ensure that the transformers are phased so that the hot terminal from each transformer goes to the RC and the RH terminals.

13.9.3 Staging Heating and Cooling Systems Through Thermostats

Some larger air-conditioning and heating systems have two sets of heating systems and two sets of cooling systems. The additional sets of heating and cooling are called stages. When the system needs a small amount of heat, the first stage of heating is turned on. If the temperature in the room starts to increase, the first stage of heating is allowed to run until it makes the room reach the setpoint temperature. If the room continues to get colder when the first stage of heating is energized, the thermostat will energize the second stage of heating, which will add more heat to the system.

In systems that have two stages of cooling, the second stage includes another evaporator, evaporator fan, and compressor to supply additional refrigerant, which allows the system to cool the rooms down faster. If the temperature in the room gets warmer than the thermostat setpoint, the first stage of cooling is energized. If the room begins to cool down, the first stage of cooling is all that is needed. If the heat load is too large and the room continues to get warmer, the thermostat will cause the second stage of cooling to become energized. Figure 13–18 shows a two-stage heating and two-stage cooling thermostat.

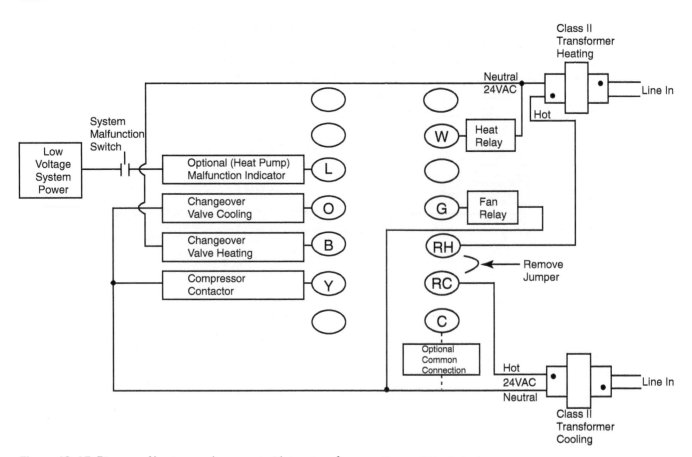

Figure 13–17 Diagram of heat pump thermostat with two transformers. *(Courtesy of White-Rodgers)*

You can see from this diagram that the first stage of cooling uses terminal Y1 and the second stage uses terminal Y2. The first stage of heating uses terminal W1 and the second stage uses W2. Any time the systems calls for heating, the first stage comes on first, and then the second stage would become energized. As the system begins to warm the

Figure 13–18 Diagram of a two-stage heating and two-stage cooling thermostat. *(Courtesy of White-Rodgers)*

room up, the second stage will turn off first and then the first stage will turn off last. For this reason, the first stage is the only stage of heating that needs a heat anticipator.

When the system calls for cooling, the first stage comes on in the beginning, and the second stage will only come on if the room is not getting colder. If the second stage is energize, it will turn off first, and the first stage will then stay on until the setpoint is met. This means that the first stage in heating and cooling are the first to come on and the last to go off.

13.10 HEAT PUMP THERMOSTATS

The bottom part of the diagram in Figure 13–19 shows a heat pump thermostat controlling a split system heat pump. Terminals R, Y1, W1, and G are similar to a regular heating or cooling thermostat. A heat pump works like a normal air conditioner in the normal cooling

Figure 13–19 Heat pump thermostat controlling a heat pump.

cycle. In the heating cycle, the refrigerant flow is reversed so that the indoor unit acts like the condenser and the outdoor unit acts like the evaporator. Since the indoor unit acts like a condenser, it will provide a small amount of heat that is moved throughout the indoor area by the indoor fan. The heat pump has a reversing valve that changes the direction of the refrigerant flow when the system calls for heating. If the refrigerant cycle cannot produce enough heat to make the indoor space warm up to its setpoint, a second stage of heating is energized. The second stage of heating can be an electrical resistance heating coil, or a gas furnace or oil furnace. The heat pump uses the refrigerant cycle for heating because it provides the lowest cost heating energy, which is lower than gas heating or pure electrical heating in some areas of the country. Chapter 16 provides in-depth information about heat pumps.

13.11 PROGRAMMABLE THERMOSTATS

The **programmable thermostat** includes a time clock with a standard-type thermostat. This type of thermostat is shown in Figure 13–20. The timer allows the thermostat to have more than one temperature setting. For example, a heating thermostat can be set for 65°F at night, and the time clock can energize the 72°F setting that will bring the temperature back up the first thing in the morning. If no one is in an office building over the weekend, the lower temperature can be programmed with the time clock to change the temperature setting to lower, more efficient temperatures during the weekend, and it can bring the temperature back to the higher temperature the first thing Monday morning.

Newer programmable thermostats have a built-in microprocessor (computer chip) to drive a digital display and allow input from a digital keypad. This type of thermostat allows multiple settings for different times of each day and for different days of the week. You should remember that these types of thermostats require both the R and C terminals to bring the "hot" and "common" of the 24-V power supply to the thermostat to run the clock and other parts of the thermostat such as LED lamps that require power. The remainder of the thermostat includes circuits for heating (terminal W), for cooling (terminal Y), and for the fan (terminal G) just like every other thermostat.

13.11.1 Touch Screen Programmable Thermostat

The newest type of programmable thermostat is called a touch screen thermostat, shown in Figure 13–21. You can see that it does not have any external buttons or switches that might wear out. Instead the touch screen provides all the menu options as icons, and you merely touch the screen to select the option that you want. For example, the screen shows the day of the week, time of day, the room temperature, the setpoint, and whether the unit is set for heating or cooling. There are also keys that allow you to increase or

Figure 13–20
Programmable thermostat.
(Courtesy of Honeywell)

Figure 13–21 Touch screen thermostat.
(Courtesy of White-Rodgers)

decrease the setpoint setting. On the lower part of the screen a menu icon is provided, and when you touch that icon the menu choices will appear.

13.11.2 Typical Program for a Touch Screen Programmable Thermostat

Figure 13–22 shows typical programming and Figure 13–23 shows typical configuration selections you would use with this thermostat. The first part of using the thermostat is the setup (steps 1–10), which lets you program in the time, date, and day of the week. Next you can program the heating cycle or the cooling cycle. You can select times of day when you are home, when you wake up, and when you go to bed. You can also select

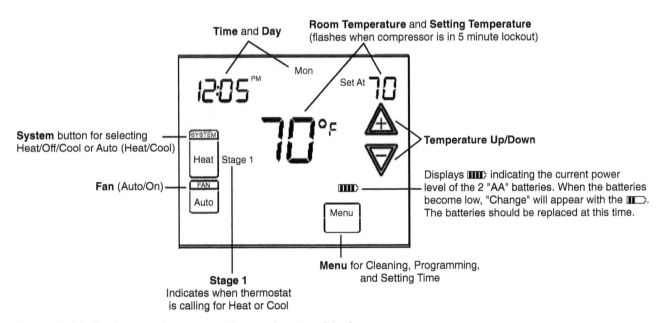

Figure 13–22 Touch screen thermostat with an explanation of the features. *(Courtesy of White-Rodgers)*

Programming and Configuration Items

1. Displays ◖ and "**Keypad Lockout**" when in keypad lockout mode.

2. Displays ◖ and "**Temperature Limit**" and "**Keypad Lockout**" when limited range is activated and locked. Displays only "**Temperature Limit**" when limited range is activated.

3. Indicates period of day being programmed.

4. RUN SCHEDULE (run program) Button.

5. SET TIME button or HOLD temperature button.

6. Displays "**Change Filter**" when the system has run for the programmed filter time period as a reminder to change or clean your filter. Displays "**Humidifier Pad**" when the system has run for the programmed filter time period as a reminder to change or clean your humidifier pad.

7. COPY button or INSTALLER CONFIG button.

8. CLEAN DISPLAY button allows 30 seconds to wipe off the display or ADVANCE DAY button for programming.

9. Used in programming to set time and in configuration menu to change selections.

10. "Hold Until" indicates the time when a temporary hold period will end.

11. "**Hours**" and "**Days**" displays during steps in installer configuration.

12. The words "**Hold At**" are displayed when the thermostat is in the **HOLD** mode. "**Temporary Hold At**" is displayed when the thermostat is in a temporary **HOLD** mode.

13. "**Call For Service**" indicates a fault in the heating/cooling system. It does not indicate a fault in the thermostat.

14. "**Stage 1**" indicates the copy program feature is being used during programming.

15. "**Copy**" indicates the copy program feature is being used during programming.

Figure 13–23 Programming and configuration for White-Rodgers thermostat.
(Courtesy of White-Rodgers)

times when you return from work if the house is vacant. The reason for these settings is to allow you to select warmer settings for air conditioning when the space is not occupied, or you can select a cooler temperature when the system is set for heating and the space is not occupied. The second part of the figure shows the screen of the thermostat and where you can find things to program.

Energy Saving Factory Preprogram

The 1F97-1271 thermostats are programmed with the energy saving settings shown in the table below for all days of the week. If this program suits your needs, simply set the thermostat clock and press the RUN button. The table below shows the factory set heating and cooling schedule for all days of the week.

	*Wake up (Morning)		Leave for Work (Day)		*Return Home (Evening)		Go to Bed (Night)	
Heating Program	6:00 AM	70°F	8:00 AM	62°F	5:00 PM	70°F	10:00 PM	62°F
Cooling Program	6:00 AM	78°F	8:00 AM	85°F	5:00 PM	78°F	10:00 PM	82°F

*You can eliminate these two program periods in the configuration menu (reference #3) if the building is occupied all day. Day will change to 6:00 AM and can be programmed as required.

Figure 13–24
Preprogram for energy saving settings on a White-Rogers thermostat.
(Courtesy of White-Rodgers)

When you set the different temperatures for each day of the week, you can press a button to make the thermostat "run" its program. When the thermostat is running its program, you will need to press the hold button to change the setting on the thermostat. The hold button allows you to determine how long the hold condition will last.

The thermostat has several indicators that will light up when the system is in heating mode or when it is in cooling mode. The thermostat also has an indicator that comes on when the system has a malfunction. For example, if the thermostat is set for 73° in the cooling mode and the room temperature continues to increase, the thermostat will presume that the air-conditioning system has a problem. The same conditions can check the furnace to ensure that the temperature in the conditioned space increases when the thermostat is calling for heat. If the room temperature does not increase in the room, then the thermostat will consider that the heating system is malfunctioning. In both of these situations, the indicator will come on at the thermostat to indicate that the system needs service.

Figure 13–24 shows the energy saving preprogram for the thermostat. This is also called the default program. If you want to use the basic settings, you can select the preprogram and it puts the setpoints shown in this figure. If the thermostat program does not function the way you want, you can manually program it to the values you prefer. Figure 13–25 shows the troubleshooting chart for the thermostat. You can use this chart to methodically work through the symptoms and isolate the problem. You will find more information about troubleshooting in later chapters.

13.12 ELECTRONIC AND DIGITAL THERMOSTATS

Some manufacturers have an electronic thermostat that has an electronic chip to control the programming features or an electronic sensing element called a thermistor. The electronic thermostat may also be called a programmable thermostat or a digital thermostat. Some electronic thermostats and temperature controllers use thermocouples, RTDs, or thermistors as the sensing device.

13.12.1 Thermocouples

The thermocouple is made of two different metals that are specifically selected and bonded together so that they will produce a small amount of voltage that is usually in the range of several millivolts, which is thousandths of a volt. The thermocouples are identified by a letter. Figure 13–26 shows the type of thermocouple, the material that it's

Troubleshooting

Symptom	Possible Cause	Corrective Action
No Heat/No Cool/ No Fan (common problems)	1. Blown fuse or tripped circuit breaker. 2. Furnace power switch to OFF. 3. Furnace blower compartment door or panel loose or not properly installed.	Replace fuse or reset breaker. Turn switch to ON. Replace door panel in proper position to engage safety interlock or door switch.
No Heat	1. Pilot light not lit. 2. System Switch not set to Heat. 3. Loose connection to thermostat or system. 4. Furnace Lock-Out Condition. Heat may also be intermittent. 5. Heating System requires service or thermostat requires replacement.	Re-light pilot. Set System Switch to Heat and raise setpoint above room temperature. Verify thermostat and system wires are securely attached. Many furnaces have safety devices that shut the system down when a lock-out condition occurs. If the heat works intermittently contact the furnace manufacturer or local service person for assistance. Diagnostic: Set System Switch to Heat and raise the setpoint above room temperature. Within a few seconds the thermostat should make a soft click sound. This sound usually indicates the thermostat is operating properly. If the thermostat does not click, try the reset operation listed above. If the thermostat does not click after being reset contact your heating and cooling service person or place of purchase for a replacement. If the thermostat clicks, contact the furnace manufacturer or a service person to verify the heating system is operating correctly.
No Cool	1. System Switch not set to Cool. 2. Loose connection to thermostat or system. 3. Cooling System requires service or thermostat requires replacement.	Set System Switch to Cool and lower setpoint below room temperature. Verify thermostat and system wires are securely attached. Same procedure as diagnostic for No Heat condition except set the thermostat to Cool and lower the setpoint below the room temperature. There may be up to a five minute delay before the thermostat clicks in Cooling if the compressor lock-out option is selected in the configuration menu (item 6).
Heat, Cool, or Fan Runs Constantly.	1. Fan Switch set to Fan On. 2. Possible short in wiring. 3. Possible short in thermostat. 4. Possible short in Heat/Cool/Fan System.	Check each wire connection to verify they are not shorted or touching together. No bare wire should stick out from under terminal screws. Try resetting the thermostat as described below. If the condition persists the manufacturer of your system or service person can instruct you on how to test the Heat/Cool system for correct operation. If the system operates correctly, replace the thermostat.
Furnace Cycles Too Fast or Too Slow (narrow or wide temperature swing)	1. The location of the thermostat and/or the size of the Heating System may be influencing the cycle rate.	Item 2 in the Configuration Menu is the adjustment that controls the cycle rate. If an acceptable cycle rate is not achieved using the FA (Fast) or SL (Slow) adjustment contact a local service person for additional suggestions.
Cooling Cycles Too Fast or Too Slow (narrow or wide temperature swing)	1. The location of the thermostat and/or the size of the Cooling System may be influencing the cycle rate.	The cycle rate for cooling is fixed and cannot be adjusted. Contact a local service person for suggestions.
Thermostat Setting and Thermometer Disagree	1. Thermostat thermometer setting requires adjustment.	The thermometer can be adjusted +/− 4 degrees as listed in item 7 of the Configuration Menu. No other adjustment is possible.
Clock Loses or Gains Time	1. Loss of power to thermostat and low batteries.	The thermostat will maintain its program in memory even with no power/no batteries but the clock time will be incorrect when power is restored. See No Heat/No Cool/No Fan (common problems) above for items to check in the system.
Heat or Cool Starts Early	1. EMR activated.	See Configuration Menu (Item 4).
Thermostat Does Not Follow Program	1. AM or PM set incorrectly in program. 2. AM or PM set incorrectly on the clock. 3. Voltage spike or static discharge.	Check current clock and program settings including the AM or PM designations for each time period. If a voltage spike or static discharge occurs use the Reset Operation listed above.
Blank Display and/ or Keypad Not Responding	1. Loss of power and dead batteries. 2. Voltage Spike or Static Discharge.	Replace batteries and check heat/cool system for proper operation. If a voltage spike or static discharge occurs use the Reset Operation listed above.

Figure 13–25 Troubleshooting procedure for White-Rodgers thermostat. *(Courtesy of White-Rodgers)*

Types of Thermocouples		
Type	Material	Range
J	Iron-constantan	−328°F to +1432°F
K	Chromel-alumel	−328°F to +2472°F
T	Copper-constantan	−328°F to +752°F
E	Chromel-constantan	−328°F to +1832°F
S	Platinum-rhodium-platinum	+32°F to +3232°F

Figure 13–26 Types of metals and temperature ranges for each type of thermocouple.

made of, and the temperature range that it works in. You can learn more about thermocouples and other temperature controls in section 20.9 in Chapter 20 in this text. The most common types of thermocouples are the J type and the K type. The J type is made of iron-constantan and can measure temperatures from −328°F to +1432°F. The K type thermocouple is made of chromel-alumel and can measure temperatures from −328°F to +2472°F. Since these sensors have such a large range of temperature measurement, they limit their measurement from approximately 0° to 100° F when they're used in thermostats or temperature controllers for residential commercial use.

Thermocouples are also used in some furnaces to detect the presence of a pilot flame. If the pilot flame is present, its heat will cause the thermocouple to produce a small amount of millivoltage, which is sent to the gas valve to determine whether it is safe to allow the gas valve to open. Since the thermocouple produces such a small amount of voltage, it is very important to ensure that its terminals are always kept clean so that resistance doesn't block any of the voltage. It is also important that the tip of the thermocouple is located properly so it can measure the source of heat.

13.12.2 Resistive Temperature Detectors (RTDs)

Another type of temperature sensor used in electronic thermostats is called a resistive temperature detector (RTD). The RTD is different from a thermocouple in that it changes its resistance when its temperature changes. Figure 13–27a shows the most common types of RTDs, what they are made of, and their temperature ranges. From this table you can see that a platinum RTD measures temperatures from −299°F to +1499°F. The platinum RTD is the most common type of RTD you will find in HVAC sensors.

13.12.3 Thermistor

Another type of temperature sensor is called a thermistor. The thermistor is a temperature-sensitive resistor much like the RTD. The major difference between the thermistor and the RTD is that the thermistor generally has a negative temperature coefficient and exhibits a greater change of resistance for each degree of temperature changed. This means that the resistance of the thermistor goes down as the temperature increases. Figure 13–27b shows the graph of the thermistor and compares it to the thermocouple and RTD. The thermistor is very accurate and makes solid-state thermostats and temperature controllers very accurate. The thermocouple, RTD, and thermistor have long lives and do not need a lot of periodic maintenance. The most likely cause of problems is that a film of oil or dirt from cooking or smoking may cover the sensor and make it less accurate.

Type	Temperature Range F	Temperature Range C	Ohms/Degree C
Platinum	−299.20° F to 1499.00° F	−184.00° C to 815.00° C	0.0039 Ω
Nickel	−99.40° F to 300.20° F	−73.00° C to 149.00° C	0.0067 Ω
Copper	−59.80° F to 300.20° F	−51.00° C to 149.00° C	0.0042 Ω
Tungsten	−99.40° F to 528.80° F	−73.00° C to 276.00° C	0.0045 Ω

(a)

(b)

Figure 13–27 a Types of RTDs and temperature ranges. **b** A graph showing the difference between an RTD, a thermistor, and a thermocouple.

13.13 LINE-VOLTAGE THERMOSTATS

Line-voltage thermostats are heavy-duty thermostats that are constructed to switch 240 V AC or 120 V AC on or off. The current rating of these thermostats is large enough to operate electric heating elements that may draw 10–20 A. They may also be used to control some refrigeration systems or commercial air-conditioning systems. Since a line-voltage thermostat can control these large currents, a relay is not needed. Figure 13–28 shows one type of line-voltage thermostat. You should notice that this thermostat has a capillary bulb that is connected by a 4- to 6-ft capillary tube. The bulb is partially filled with refrigerant and is placed where the temperature is to be measured. The heat in the space where the bulb is located will cause the refrigerant to expand and activate the switch contacts. It is important to understand that since the thermostat has 240 V AC or 120 V AC, it should be mounted in an electrical box.

Figure 13–28 a A line-voltage thermostat. **b** A line-voltage thermostat with its cover removed.
(Courtesy of Honeywell)

(a) (b)

13.14 OVERVIEW OF PRESSURE, TIMER, AND OVERCURRENT CONTROLS

As a technician, you will need to be able to identify and troubleshoot a variety of pressure controls, timer controls, and overcurrent controls. These controls are used to protect a refrigeration or air-conditioning system against conditions that will damage the system. The controls can also be used for normal operation. For example, since pressure and temperature are related, a load pressure switch can be used as a thermostat. When you are troubleshooting a system, you should be able to identify these controls and locate a diagram of the internal parts of their switches. Next you should be able to troubleshoot the system by activating the pressure, timer, or overcurrent portion of the switch.

If you find a pressure-type switch, you must be able to test the switch portion and cause the switch to activate by adjusting pressure to the switch. This means that a **low-pressure switch** must be tested by applying pressure to the switch to cause it to activate. The problem with the switch may be on the pressure side, such as a bellows that has a hole in it or a tube that is leaking, or it may be on the electrical side, where contacts are corroded. When these devices are in a system, it is important to understand that they may open their electrical contacts and shut the system down because they are working correctly. For instance, if a system has a **high-pressure switch**, it may open when pressure exceeds the normal setpoint. This can occur when the condenser fins become clogged with leaves or other matter, which reduces the air flow across the condenser and increases the condenser pressure to an unsafe level. The high-pressure switch will open and protect the system against the high pressure. If you merely test this electrical portion of the switch, you will find it is open. If you change the switch at this point without testing for the cause of the high pressure, you will merely be changing a good part. In many cases you must identify the problem the switch is reporting and repair the condition that is causing the high pressure.

You will find a wide variety of timer controls on units such as defrost timers for refrigeration units or defrost timers for heat pump systems. Timers are also used to control circulating pumps and other time-control functions. Short-duration timers are used to provide small amounts of time delay when an air-conditioning or refrigeration system is started. This type of time delay is called **short-cycle protection**. Some of the timers you encounter will be motor driven, and other types will be solid-state electronic time delay.

You will encounter a large variety of overcurrent controls, including fuses, circuit breakers, overload devices on compressors, and overload devices on motor starters.

13.15 PRESSURE CONTROLS

The most common types of pressure controls include high-pressure switches, low-pressure switches, and oil pressure switches. Figure 13–29a shows a picture of a high-pressure switch, and Figure 13–29b shows the electrical symbol for a high-pressure switch, which is a bellows that activates a contact. The contact opens upward when the high pressure is exceeded. Figure 13–30 shows the internal operation of a high-pressure switch. In this diagram you can see that the high-pressure element is a bellows made of light metal. A spring is used to adjust the amount of pressure placed on the bellows. When refrigerant pressure goes inside the bellows, it causes the bellows to expand against the spring and activate an electrical switch. The switch can be a single-pole, single-throw switch or a single-pole, double-throw switch. The switch has a screw mechanism for adjusting spring tension on the bellows. This adjustment sets the point at which the switch activates.

Figure 13–29 a A high-pressure switch. **b** Electrical symbol for a high-pressure switch. Notice the right side of the contact is shown above the terminal point to indicate this pressure switch is a high-pressure switch and it will open when pressure increases above its setpoint. *(Courtesy of Johnson Controls, Inc.)*

(a) (b)

Figure 13–30 Cutaway diagram of a high-pressure switch. *(Courtesy of Danfoss Automatic Controls)*

1/4 in. flare nut

13.16 LOW-PRESSURE SWITCHES

Figure 13–31a shows a picture of a low-pressure switch, and Figure 13–31b shows the electrical symbol for a low-pressure switch. You should notice that the low-pressure switch looks very similar to the high-pressure switch. The symbol for a low-pressure switch is different from the high-pressure switch in that the low pressure switch opens below the contact. This indicates that the switch contact opens as pressure decreases.

Figure 13–31 a A low-pressure switch.
b Electrical symbol for a low-pressure switch. Notice that the right side of this switch is drawn below the terminal point to indicate that it will open when pressure drops below the setpoint.
(Courtesy of Johnson Controls, Inc.)

(a) (b)

The low-pressure switch is constructed with a bellows that is similar to the one in a high-pressure switch, except that the low-pressure bellows is more sensitive. Again, an adjustment screw and dial are provided to show the pressure setting. Turning the adjustment screw changes the spring pressure on the bellows, which in turn adjusts the point at which the bellows will move a sufficient distance to activate the switch. The contacts in the low-pressure switch are normally closed contacts. When the pressure falls below the setpoint, the switch will activate and return to its open condition. The low-pressure switch is used to detect the loss of refrigerant or other conditions such as loss of an evaporator fan, which will cause low refrigerant pressure.

13.17 FIXED-PRESSURE SWITCHES

Figure 13–32 shows examples of **fixed-pressure switches**. These devices are nonadjustable, which means that the set point must be determined when they are purchased. They are available in a variety of pressures. This type of switch is less expensive than an

Figure 13–32 Fixed-pressure switches. *(Courtesy of Danfoss Automatic Controls)*

adjustable switch, and it is simple to install. In many cases the pressure switch can be installed in the system without making any mechanical openings. The fixed-pressure switch can be mounted on a valve stem.

13.18 DUAL PRESSURE SWITCHES

Figure 13–33a shows a picture of a **dual pressure switch**, and Figure 13–33b shows the electrical diagram of the dual pressure switch. In these figures you can see that the dual pressure switch has two connections with capillary tubes. One connection is for the high-pressure switch, and one is for the low-pressure switch. The dual pressure switch provides a high- and a low-pressure switch in one control so that it is easier to install and mount. Each switch inside the dual pressure switch is adjustable just as though it were a single high-pressure switch or individual low-pressure switch. Figure 13–34 shows a diagram of a dual pressure switch.

(a)

(b)

Figure 13–33 a Dual pressure switch. **b** The electrical symbol for a dual pressure switch. *(Courtesy of Johnson Controls, Inc.)*

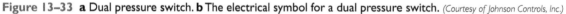

Figure 13–34 Diagram of a dual pressure switch.

(Courtesy of Danfoss Automatic Controls)

13.19 ELECTRICAL DIAGRAMS FOR PRESSURE SWITCHES

When you locate a diagram of a pressure switch as in Figures 13–29, 13–31, or 13–33 you will notice the switch symbol is provided and in some diagrams such as 13–30 the electrical terminals are numbered. Generally the switch may have an abbreviation or it might have its name added to the symbol, but if the diagram does not indicate what type of switch it is, you can examine the symbol closely to determine whether the switch will open or close when temperature increases. You should remember that an increase in pressure will cause the pressure actuator to move upwards and a decrease in pressure will cause the actuator to move downward. From there you should be able to determine if an increase in pressure will cause contacts to open or close.

Figure 13–35 shows a set of diagrams for all the pressure switches you should encounter in the field. From this diagram you can see that each type of switch is represented. You can test the switch contacts with a continuity test when no voltage is applied to them, and you can adjust the pressure below and above the switch setpoint to ensure that the pressure part of the switch is operating correctly. It is important to verify the **cut-in pressure** and **cut-out pressure** for each type of switch as well as to test the contacts to ensure that they open and close properly. You must use a set of gauges and an adjustable source of pressure to test the cut-in and cut-out pressure points of each switch.

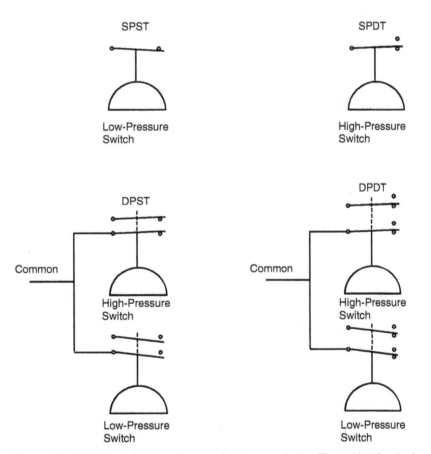

Figure 13–35 Electrical wiring diagram of pressure switches. The symbol for the low-pressure switch is shown open when pressure drops below a setpoint. The symbol for a high-pressure switch is shown open when pressure increases beyond its setpoint.

13.20 OIL PRESSURE SWITCHES

Figure 13–36 shows pictures of **oil pressure switches**, and Figure 13–37 shows the electrical diagram for the switch. This type of switch is used in refrigeration or air-conditioning systems where the compressor is some distance from the evaporator, such as in commercial buildings and malls where the compressor and condenser are mounted on the rooftop, and the evaporator coil is some distance away. When this occurs, the oil in the system tends to migrate to the warmest part of the system. In the early spring and late fall, the indoor unit (evaporator coil) may be warmer than the compressor when the system is turned off. If this occurs, the oil will migrate to the warmer coil and oil will not be available in the compressor. The oil switch is used to determine if the oil pressure has dropped or is low. When this occurs, the compressor must be allowed to run for a few seconds to get the oil back. If the oil pressure switch cannot detect the correct oil pressure, the switch will trip out and it must be manually reset.

Figure 13–36 Examples of oil pressure switches. *(Courtesy of Johnson Controls, Inc.; Danfoss Automatic Controls)*

Figure 13–37 Electrical diagram of an oil pressure switch. *(Courtesy of Johnson Controls, Inc.)*

Measuring oil pressure has two problems. The first problem is that the oil mixes with the refrigerant, so the oil pressure must be measured on top of the refrigerant pressure as a differential. The second problem is that a small amount of time delay must be provided to allow the oil to build pressure when the compressor is first started. If the oil has migrated from the compressor and the oil pressure switch started to measure the oil pressure as soon as power is applied to the compressor, it would trip and cause the compressor to shut down. For this reason, the amount of time delay is adjustable so that you can set the time delay to match the actual conditions in the refrigeration system and allow enough time to get the oil to be pumped back to the compressor. If the time is too short and the compressor trips out, you can lengthen the time slightly.

The switch contains a differential pressure mechanism and a time delay. The time-delay mechanism allows a compressor to build oil pressure and get the oil to return if it has migrated away from the compressor. The time delay is approximately 15 to 30 seconds before the switch test for the loss of oil pressure. A differential-type switch is required because the refrigerant in the compressor will provide anywhere from 30 to 60 PSI, and the oil pressure must be 30 additional pounds of pressure. This means that one side of the switch has a capillary tube connected to the low-pressure side of the refrigerant system. The other side of the switch is connected to a port that can measure oil pressure. If the refrigerant pressure in the system is 30 lb, the refrigerant side of the switch will measure 30 lb. The oil pressure measurement may indicate 65 lb, which means that 30 lb is due to the refrigerant pressure and 35 lb is due to oil pressure. The switch has a minimum and a maximum setpoint that causes its contacts to open if oil pressure drops below the lower setpoint. The second setting in the switch is to provide the amount of differential between the oil pressure and the refrigerant pressure. The time-delay circuit is operated by a smaller timer on older models, and by an *RC* time constant on newer models. The time constant is a resistor and a capacitor that are connected to a solid-state switching element such as an SCR, a triac, or a transistor. When power is applied to the circuit, voltage charges the capacitor. The resistor regulates the amount of time to charge the capacitor. If the oil pressure switch is an older-type switch, it will have an electric heating element and a bimetal strip. A variable resistor is used to help adjust the time delay on the heating strip. When current is applied to the heater element, it produces heat. When the heat builds to a sufficient level, it causes the bimetal to open and to warp, which causes its contacts to open. Both the solid-state and the heating element time delay give the compressor time to build oil pressure before the switch begins to protect it. Figure 13–37 shows the electrical diagram of the oil pressure switch. Notice that this type of switch needs a time-delay circuit to allow the compressor several seconds to build the oil pressure to a differential above the low side pressure.

If you are called to a unit where the oil pressure switch trips out occasionally, there are several other ways to keep oil migration from occurring. One way is to add a crankcase heater to the compressor, which will make it the warmest part of the system. Another type of repair for this type of problem is to add a solenoid valve in the liquid line. The solenoid valve will close immediately when the compressor turns off. This valve will trap the refrigerant and keep it from migrating to the indoor coil and keep the oil in the compressor. It is important to understand that you should not allow the compressor to run too long without oil, even if the oil pressure switch is installed. The best way to fix this problem is to lengthen the time delay slightly and, if it does cure the problem, add the crankcase heater and/or add a liquid line solenoid.

13.21 ## CUT-IN AND CUT-OUT PRESSURES

The oil pressure switch has two adjusting screws that allow you to set the cut-out point and the differential. The purpose of the switch is to protect the compressor against oil pressure failure. If the oil pressure drops below 12 PSI, the switch should open. The 12 PSI is measured in addition to low-pressure refrigerant pressure. If the oil pressure cuts out, it needs to increase to a point at which it will cut back in. Typically it is set to 15 to 20 PSI to cut in. You should adjust the cut-in point and differential on an oil pressure switch with regulated oil pressure. This means that you can set the cut-in point and differential for the oil pressure switch on the bench or when the switch is on the unit.

After you have adjusted the oil pressure switch on the bench, you can install it on the compressor and feel confident that it will protect against loss of oil pressure. When you are troubleshooting a system in the field and the oil pressure switch cuts out the compressor, you must be aware of the loss of oil pressure. Typically the loss of oil pressure will be due to a faulty oil pump in the compressor, or the migration of oil from the compressor into the warmest part of the refrigeration system. This means that many times the oil pressure switch will be open but not be broken; it is merely protecting the system against the loss of oil. This means you must correct the cause of the loss of oil rather than replace the oil pressure switch.

Additional Job Assignment

You receive a service call to a store in a shopping mall that has a 10-ton rooftop air-conditioning system with a semihermetic compressor that is not running. When you open the system and begin to check it out, you find the oil pressure control has tripped the system. The storeowner wants to know what you have found and what you are planning to do and also wants to know how the oil pressure control works and how it protects the system. Your job is to explain the operation of the oil pressure switch and identify what tests you should make to determine why the oil pressure switch continues to trip the system out.

13.22 ## TIMER CONTROLS

Timer controls are used in a variety of applications and air-conditioning and refrigeration systems. For example, most refrigerators that have a freezer compartment have some type of defrost clock or defrost timer that cycles the system into a defrost mode as needed when frost develops on the evaporator. In a heat pump system a defrost clock is used to reverse the heat pump cycle when the system is set for heating, whenever frost may develop on the evaporator. Frost can build up on the evaporator anytime the outdoor temperature is below 36°. The reason the outdoor coil can build up frost when the temperature is above freezing is that the refrigerant temperature on the coil during evaporation is well below freezing. If a defrost cycle is not actuated, frost will build up to the point that it completely covers the coil face so that air does not pass through it.

Commercial refrigeration systems also require a defrost cycle. In many cases a larger defrost timer is used to activate this cycle. Other types of timers that you will find in air-conditioning and refrigeration systems are called *short-cycle timers*. The short-cycle timer will be activated whenever the thermostat calls for cooling. Instead of starting the

compressor, the short-cycle timer will be activated to provide a short time delay of 1 to 3 min. This time delay allows time for the refrigeration pressure between the compressor and the high side to equalize to a point at which the compressor can start against a lower pressure. A problem occurs if a system is turned off and back on within 1 min. When the system is turned off and on in a short period of time, high pressure in the high side of the system will prevent the compressor from starting or will cause the compressor to draw large currents. The short-cycle timer can be a motor-driven timer, an electric heating element timer with a bimetal strip, or a solid-state *RC* time constant timer. The short-cycle timer may also be included with other compressor protection circuitry.

13.23 DEFROST TIMER

Figure 13–38 shows a typical **defrost timer**, and Figure 13–39 shows the electrical diagram of a defrost timer. In the diagram in Figure 13–39 you can see the terminals have both letters and numbers. In this explanation we will refer to the terminals by the letters that represent them. The timer has an electrical motor between terminals L1 and N, and a set of single-pole, double-throw contacts between terminal L1, NO, and NC. The normally closed set of contacts is shown between terminal L1 and NC, and the normally open contacts are shown between terminal L1 and NO. The timer motor can be connected directly across the power supply so that it runs all the time. This type of timer is called a *continuous-run timer*. The defrost timer can also be connected so that it controls the compressor through terminal NC during the defrost cycle. Figure 13–40 shows a defrost timer controlling the defrost solenoid as you would find it for a hot gas type defrost system. The compressor and evaporator fan are controlled through the system thermostat, so the compressor can continue to pump refrigerant during the defrost cycle. You can see in this diagram that during normal operation, the timer motor runs constantly and no power is sent to the defrost solenoid. When the timer motor runs to a point that a defrost cycle is required, a contact switch between L1 and NO closes a circuit to the defrost solenoid. In some systems the evaporator fan may be connected to the NC terminal of the defrost timer, and it would be turned off during the defrost cycle. The defrost cycle can be terminated by the defrost thermostat that is connected in series with the defrost solenoid, or when the timer has run for the entire length of the defrost cycle which is usually 3 to 5 minutes. At the end of the defrost cycle the contacts in the defrost control will change back to their original condition and the defrost solenoid will become de-energized again.

Figure 13–38 Typical defrost timer. *(Courtesy of Paragon Electric Company, Inc.)*

Figure 13–39 Electrical diagram of a typical defrost control. Notice the terminals may be identified with letters or numbers. *(Courtesy of Paragon Electric Company, Inc.)*

Figure 13–40 Diagram of a defrost timer connected to a compressor and evaporator fan. *(Courtesy of Paragon Electric Company, Inc.)*

You can test the timer by using an ohmmeter to measure for continuity in the timer motor. You can also test for continuity between terminals L1 and NC when the timer is in the normal position and between terminals L1 and NO when it is in the defrost position. It is possible to advance some timers by manually rotating the cam on the side of the switch with a screwdriver. The defrost timer generally has one of two problems: the timer motor stops running because its winding is open, or the defrost timer contacts stick or weld closed. You can test for these conditions by observing the timer motor to see if it is running, and you can check the contacts by manually rotating the cam and testing the contacts.

13.24 ELECTRONIC DEFROST CONTROLS

Newer types of defrost timers have solid-state electronic controls. The operation of these types of timers is similar to that of the motor-driven timers. Figure 13–41 shows an electronic defrost timer and its diagram. From the wiring diagram you can see that the electronic defrost control has the same terminal numbers as the motor driven type, so that the electronic control can replace the motor driven control. The electronic control is programmable so that you can set the day, date, and time, and then select the defrost time based on the hours the system runs or the time of day.

13.25 COMPRESSOR TIME-DELAY TIMERS

Timers called short-cycle protectors are available to create a time delay when a compressor is energized. In air-conditioning and refrigeration applications the compressor must remain de-energized for several minutes after it cycles off to allow the refrigeration pressure on its high side to equalize. When the compressor cycles off,

Figure 13–41 a An electronic defrost timer. **b** Electrical diagram of an electronic defrost timer. *(Courtesy of Paragon Electric Company, Inc.)*

(a) (b)

Figure 13–42 a A short-cycle protector for a compressor. **b** Electrical diagram of a short-cycle protector connected to a compressor contactor coil. *(Courtesy of Watsco Components, Inc.)*

(a) (b)

the refrigeration pressure may be over 150 PSI. If the compressor tries to energize too soon before the pressure is allowed to become lower on its high side, the compressor motor will not be able to start against the high pressure and will draw locked-rotor amps (LRA), which will cause its fuses to blow. The timer shown in Figure 13–42a provides a time delay of up to 10 min to allow the refrigerant pressure to equalize. Figure 13–42b shows a diagram of a time-delay timer connected in series with the coil of the compressor contactor. When the thermostat closes and sends power to the coil of the compressor contactor, the timer becomes energized and provides a time delay of up to 10 min. After the time-delay period, the timer passes power to the compressor contactor coil and the compressor is allowed to start. This type of timer is especially useful when a power outage occurs during a storm. In the summer, it is typical for lightning storms to knock out the electrical supply for a few seconds. If a compressor is running during this short power outage, it will stop and then try to start again before the refrigerant pressure has lowered enough to allow the compressor to start. When this occurs, the compressor will draw excessive current and cause its fuses to open. If the fuses are not sized properly, the compressor will be severely damaged by the excess current.

13.26 TROUBLESHOOTING PRESSURE CONTROLS AND TIMERS

At times you will be required to test pressure controls and timers in HVAC circuits. The controls themselves are rather simple because they consist of one or two sets of normally open or normally closed contacts. If you suspect that the contacts of a control are open and interrupting current flow in the circuit, you can quickly verify this with a voltage test on the circuit. If you determine that the contacts are open, you have found a symptom of the problem. If the control is a low-pressure switch, the problem is more than likely the loss of refrigerant in the system. The pressure switch contacts are open to protect the system. If you change the control as though it were bad or inoperative, the low-pressure problem will persist after you replace the switch, and the system will remain inoperative. In most cases when you find a set of open contacts in a pressure or timer control, the control is just doing its job, and you will need to troubleshoot the entire system to determine the source of the problem. For example, if the switch is a high-pressure switch, and its contacts are open, the actual problem in the system will be found in the high side of the refrigeration system, and the problem will be with air flow across the condenser. You will learn in your troubleshooting tests to distinguish between the actual problem and symptoms of the problem.

Additional Job Assignment:

You are on a service call where a new air-conditioning system has been installed at a residence that has never had air conditioning before. The complaint is that the system trips the fuses periodically when the compressor starts after it has been running and has shut off for only 2 minutes. The homeowners tell you that they frequently set the thermostat down lower if the air conditioner shuts off in the afternoon, which causes the system to come right back on after it has stopped running. Your job is to add controls to the system to prevent the homeowner from short cycling the system, and to educate the homeowner about why short cycling will cause the fuses to blow.

13.27 REVIEW

The thermostat is a device that switches its contacts open or closed in response to temperature changes. A bimetal strip activates the contacts when temperature changes. Thermostat terminals are identified as follows: R stands for red and is the terminal connected to the "hot" terminal on the transformer. W stands for white and is the terminal that is connected to the heating gas valve. Y stands for yellow and is the terminal that is connected to the cooling relay coil. G stands for green and is the terminal connected to the fan relay coil.

When connecting the thermostat to a furnace or air conditioner, you should always use colored thermostat wire and connect it to the corresponding terminal. The thermostat has a subbase at which the field wires are connected. The subbase should be leveled so that the mercury switch can operate correctly and be calibrated so that it will measure the temperature accurately.

The thermocouple is a temperature-sensitive element that produces a millivolt signal that is sent to the gas valve to indicate that a pilot flame has been established. A resistive temperature detector (RTD) is a resistive sensor that can be used to indicate that a flame is present, or it can be used as the sensor in a thermostat.

Air-conditioning and refrigeration systems use a variety of pressure controls. The low-pressure control is designed to activate its contacts when the pressure drops below a predetermined level. It may be a fixed or an adjustable-type switch. The low-pressure switch can be used as a safety control or as an operational control that can control the temperature of the system much like a thermostat.

The dual pressure switch has both a high-pressure switch and a low-pressure switch in the same control. You can adjust the cut-in and cut-out pressure separately on a dual pressure switch. An oil pressure switch provides a differential pressure sensing circuit as well as a time delay to protect the compressor against loss of oil pressure.

The timer control can be used to control defrost time or compressor time delay. The timer control can be adjustable, or it can be a solid-state-type timer.

Job Assignment for This Chapter

You are on a service call in a residence during the heating season and the symptoms indicate the heating thermostat is not working correctly on a gas furnace. Your job is to identify several procedures and tests you would use to troubleshoot this problem. The homeowner wants to know what you think is wrong and what tests you will use to determine if the thermostat is functioning correctly.

Solution to the Job Assignment

The heating thermostat closes the circuit between R and W inside the thermostat when the temperature on the thermostat is cooler than the setpoint. For example, if the thermostat setpoint is 70 degrees, and the temperature in the room is 68 degrees, the R–W circuit inside the thermostat will close and supply 24 volts to the gas valve. When you make voltage readings to see if the thermostat is working correctly, you should check from terminal R in the thermostat to the C terminal on the transformer. This is where voltage comes into the thermostat. Your next voltage measurement should be between terminal W in the thermostat and terminal C on the transformer. You are making all your measurements in reference to terminal C because that is where you will find the common terminal for the 24-volt supply. All voltage measurements should be in reference to terminal C.

You can set the heating thermostat as high as possible to ensure that it stays on. If you measure voltage from the W terminal to the C terminal, and it reads 24 volts, you can be assured that the thermostat is working correctly. If the voltage reads 0 between W and C, the thermostat is not working correctly. If the voltage between W and C reads 24 volts, then the thermostat is working correctly. If the voltage between W and C measures 24 volts and the gas valve does not energize, you can measure for voltage directly across the two terminals of the gas valve. If you have 24 volts across these terminals and the gas valve does not open, you can suspect something is wrong in the pilot or ignition circuit. If you do not have 24 volts across the gas valve terminals, but you did have voltage between W and C, you can suspect the wire between terminal W on the thermostat and where it connects to the gas valve is open and needs to be replaced. This troubleshooting sequence will help you locate where voltage is lost in the circuit.

Additional Job Assignment

You are installing a new heating thermostat on a gas furnace heating system and the instructions tell you to set the heat anticipation correctly before you complete your installation. The homeowner is looking at the installation instructions and notices that you need to set the heat anticipation to match the furnace. The homeowner asks you what exactly "heat anticipation" is. Your job is to explain to the homeowner what "heat anticipation" is and set it correctly to match the furnace gas valve.

Solution to the Additional Job Assignment

The heat anticipator is a variable resistor that produces a small amount of extra heat when current flows through it. It is mounted directly under the sensing element for the thermostat and is used to add a small amount of heat to the heating thermostat to make it turn off approximately 1 degree before the room reaches the setpoint. This is important, as the heat exchanger in the gas furnace continues to add heat into the room after the gas is turned off. The heat anticipator "anticipates" how much additional heat the furnace will add to the room after the gas is turned off. When the thermostat R–W circuit is closed, it passes current through the heat anticipator to the gas valve, and this current causes the heat anticipator resistor to add a small amount of heat to the thermostat sensing element. This additional heat should make the furnace turn off just before the room reaches setpoint and furnace fan will continue to add heat so the room will heat up to exactly the setpoint with out overshooting.

The numbers on the heat anticipator indicate the amount of amperes the gas valve will draw, and it will produce the approximate amount of heat needed. You can find the amperage of the gas valve indicated on some systems. If the amperage rating is not shown, then you can measure the amperage with a clamp-on ammeter, and it should be less than one amp. You might need to wrap the wire around one of the claws of the clamp-on ammeter to create a multiplier. After you have determined the amperage, you can set the heat anticipator to the same number.

When you match the heat anticipator to the amperage that the gas valve draws, you should test the system by letting it run and reach temperature several times. If the furnace shuts off too soon, you can adjust the variable resistor to provide a longer cycle by moving it in the direction of the arrow that indicates "longer cycle." If the thermostat is overshooting, you can adjust the heat anticipator slightly to create a shorter cycle.

Additional Job Assignment

You are on a service call where a new air-conditioning system has been installed at a residence that has never had air conditioning before. The complaint is that the system trips the fuses periodically when the compressor starts after it has been running and has shut off for only 2 minutes. The homeowners tell you that they frequently set the thermostat down lower if the air conditioner shuts off in the afternoon, which causes the system to come right back on after it has stopped running. Your job is to add controls to the system to prevent the homeowners from short cycling the system, and to educate the homeowner about why short cycling will cause the fuses to blow.

Solution to the Additional Job Assignment

Your first check for the system is to see if the current draw is near the rating of the circuit breakers or fuses. If the current measurement is close to their rating, you may need to increase their size to the maximum rating of the wire. If the fuse and circuit breaker are already at the maximum rating of the wire, you may need to change the wire for the next larger size. Do not increase the fuse or wire size above the manufacturer's specifications. If the current measurement is substantially lower than the fuse and circuit breaker rating, you should suspect the system is short cycling, which means it is trying to turn on too soon after it turns off, and the pressure is not allowed to equalize before the compressor tries to start again. If the compressor tries to start against the higher refrigerant pressure, it will cause it to draw too much current. Another problem you should investigate is that since the homeowners had just started using the new air conditioner, they may not understand that you cannot turn the unit back on immediately after it cycles off. You may start the service call discussing this matter with the homeowners and explain that the system needs time to let the pressure equalize. Another solution to this problem is to put a time-delay relay into the system to prevent the compressor contactor coil from energizing for 3 to 5 minutes. If this fails to solve the problem, you may need to check the starting capacitor to ensure that it is working correctly. If the capacitor is not functioning properly, the compressor will draw excessive current and cause fuses to blow.

Additional Job Assignment

You receive a service call to a store in a shopping mall that has a 10-ton rooftop air-conditioning system with a semihermetic compressor that is not running. When you open the system and begin to checkout it out, you find the oil pressure control has tripped the system. The storeowner wants to know what you have found and what you are planning to do and also wants to know how the oil pressure control works and how it protects the system. Your job is to explain the operation of the oil pressure switch and identify what tests you should make to determine why the oil pressure switch continues to trip the system out.

Solution to the Additional Job Assignment

If you are called to a unit where the oil pressure switch trips out occasionally, there are several things you should check for. First try to determine the amount of time delay the oil pressure switch has. If it is a very short time, 10–12 seconds, you can lengthen the time slightly if the switch is an adjustable type. If the oil pressure switch is fixed, you must look to other ways to fix the oil migration problem. One way is to add a crankcase heater to the compressor, which will make it the warmest part of the system. Another repair for this type of problem is to add a solenoid valve in the liquid line. The solenoid valve will close immediately when the compressor turns off. This valve will trap the refrigerant and keep it from migrating to the indoor coil and keep the oil in the compressor. It is important to understand that you should not allow the compressor to run too long without oil, even if the oil pressure switch is installed. The best way to fix this problem is to lengthen the time delay slightly, and if it does cure the problem, add the crankcase heater and/or add a liquid line solenoid. Lengthening the time indicates you have not lost oil, rather it has combined with the refrigerant, and adding heat will fix the problem.

You might also check the system for leaks on the liquid lines that could allow oil to escape and leave an oil stain. If this occurs, you should fix the liquid line leak first and then add oil to the system when you add refrigerant to the system.

Assessment

QUESTIONS

1. Explain how two strips of metal that form a bimetal element sense a change in temperature.
2. Explain how a spiral bimetal element reacts to changes in temperature.
3. Explain the function and operation of a heat anticipator.
4. Explain the operation of a thermocouple and tell where you would find a thermocouple.
5. Explain the operation of a fan switch on a thermostat.
6. Explain why an oil pressure switch needs a time-delay period.
7. Explain the difference between a fixed pressure switch and an adjustable pressure switch.
8. Identify the basic parts of a high-pressure switch and explain the conditions a high-pressure switch will protect against in an air-conditioning system.
9. Identify the basic parts of a low-pressure switch and explain the conditions a low-pressure switch will protect against in a refrigeration system.
10. Explain the operation of a defrost timer.

TRUE OR FALSE

1. The contacts of the heating thermostat go closed when temperature falls below the setpoint.
2. The gas valve is basically a solenoid valve with a safety circuit to ensure that the pilot light is burning.
3. The oil burner control uses 24 V to provide the spark to the ignition points.
4. The fan and limit switch provides an operational control to energize the fan, and a safety circuit to de-energize the gas valve if the temperature gets too high.
5. The heat anticipator produces heat while the heating thermostat contacts are in the open position.
6. The oil pressure switch requires a time delay to allow the oil pressure to build up when the compressor is first turned on.
7. A fixed-pressure switch can be adjusted a small amount around a fixed-pressure setpoint.
8. The cut-in and cut-out pressure settings on an oil pressure switch will be the same pressure.
9. A defrost timer is normally used to stop the compressor periodically and energize a defrost heating element.
10. A commercial timer can be used to cycle off large commercial air-conditioning systems on weekends and other times when no one is in a building.

MULTIPLE CHOICE

1. The fan switch in the thermostat has
 a. two positions (auto and on).
 b. three positions (auto, off, on).
 c. two positions (off and on).

2. The gas valve has a button that must be depressed when the pilot light is being lighted to
 a. bypass the high-temperature safety circuit.
 b. bypass the low-temperature safety circuit.
 c. overcome the pull-in current of the solenoid in the valve until the thermocouple can produce sufficient current to hold in the solenoid.

3. The cooling anticipator in the cooling thermostat produces
 a. heat when the contacts of the cooling thermostat are open.
 b. heat when the contacts of the cooling thermostat are closed.
 c. a cooling effect when the contacts of the cooling thermostat are open.

4. The fan switch
 a. has an auto and an on position.
 b. sends voltage to the compressor contactor.
 c. controls the condenser fan.

5. The thermostat subbase
 a. has the fan switch mounted in it.
 b. is where you would connect all the wires to the thermostat.
 c. both a and b.

6. An oil pressure switch compares
 a. oil pressure with high pressure.
 b. oil pressure with low pressure.
 c. oil pressure with a setpoint of 15 PSI.

7. An oil pressure switch needs
 a. a small amount of time delay to allow the compressor to build up oil pressure.
 b. a small amount of time delay to allow the compressor to build up its high pressure.
 c. no time delay because the oil pressure must be detected immediately or the switch will shut the compressor off so that no damage occurs if the oil pressure is low on startup.

8. A low-pressure switch connects the compressor or the coil of the compressor contactor to terminal
 a. the normally closed contacts.
 b. the normally open contacts.
 c. for the time delay.

9. When you are troubleshooting an air-conditioning system and find that it is shut down because its low-pressure switch is open, you should
 a. test the switch for continuity a second time and replace it if it is open.
 b. test the switch for continuity a second time and replace it if it is closed.
 c. look for the cause of the low pressure.

10. If a defrost timer does not cycle the system through a defrost cycle periodically, you should suspect that
 a. the timer motor is not running continuously or is not running at all.
 b. the defrost heater is bad.
 c. the compressor cannot shut off.

PROBLEMS

1. Sketch the mercury-type switch that is contained in a glass element, and explain how it works when it is attached to a spiral bimetal element.
2. Use the diagram in Figure 13–8b to identify the terminals that are used in the heating circuit for a heating thermostat.
3. Use the diagram in Figure 13–13 to identify the terminals that are used in the cooling circuit for a cooling thermostat.
4. Sketch the fan switch circuit in a heat–cool thermostat and explain its operation in the auto and the on settings.
5. Use Figure 13–21 and sketch the face of the thermostat. Explain the operation of each part on the display.
6. Use the diagram in Figure 13–14 to sketch the circuit of a fan switch.
7. Sketch a high-pressure switch including its electrical diagram and explain its function.
8. Sketch a low-pressure switch including its electrical diagram and explain its function.
9. Sketch a dual pressure switch including its electrical diagram and explain its operation.
10. Sketch an oil pressure switch including its electrical diagram and explain its operation.
11. Sketch a defrost timer including its electrical diagram and explain its operation.

Websites

Websites change constantly. Use the key word search to locate sites that have similar content. The products that will be found at each site are indicated with the key word.

White-Rodgers
Products: Thermostats, controls
Key word: White-Rodgers Control

Lennox
Products: Thermostats, furnaces, air conditioners
Key word: Lennox

Ranco Controls
Products: Air-conditioning and refrigeration controls
Key word: Ranco Controls

Johnson Controls
Products: Air-conditioning components and controls
Key word: Johnson Controls

Danfoss Automatic Controls
Products: Air-conditioning and refrigeration controls
Key words: Danforth Automatic Controls, Penn Controls, Johnson Controls, Honeywell Controls

Honeywell Controls
Products: Thermostats and solid-state HVAC controls
Key word: Honeywell Controls

Robertshaw Controls
Products: Solid-state HVAC controls
Key word: Robertshaw Controls

Chapter 14

Controls for Gas, Electric, and Oil Heating Systems

Job Assignment for This Chapter

You are on a service call to a residence where the homeowner complains that the furnace blows cold air at the end of its cycle. Your job is to identify the problem and explain to the homeowner what tests you will make and what you are trying to prove. If you plan to replace a part or adjust a control, you should be able to explain to the homeowner what you are trying to do to fix the problem.

Terms You Will Learn in This Chapter

Boiler
CAD cell
Cam stat
Direct-spark ignition
Electric furnace
Electric heating coils
Electric sequencer relay
Fan and limit switch
Fan control
Fan switch
Flame rod

Flame rollout switch
Gas valve
Heat exchanger
High efficiency furnace
Hot surface igniter
Oil burner
Oil burner control
Protectorelay
Thermal fuse
Thermocouple

After reading this chapter, you will be able to:

1. Explain the operation of a direct-spark ignition system.
2. Explain the operation of a high-efficiency furnace.
3. Explain the basic operation of a gas valve.
4. Explain the operation of an oil burner control.
5. Explain the operation of a fan and limit switch.
6. Explain the operation of a two-stage electric furnace.

14.0 FUNDAMENTALS OF HEATING SYSTEMS

In previous chapters we studied air-conditioning system controls and motors. In this chapter you will study the different types of fuels or energy sources and the controls that are used to make heating systems operate safely and efficiently. Heating systems use gas, oil, or electricity to heat air or water, which is used to heat residences and commercial buildings. Gas and oil type furnaces have a **heat exchanger** where the fuel is combusted (burned). The heat exchanger allows the fuel to be burned and exhausted, and provides a means for air around it to be warmed. In a **boiler**, the fuel is burned and heats water that flows through the boiler. In **electrical furnaces**, electric coils are heated until red-hot, and air is blown directly over them.

14.1 TYPES OF HEATING SYSTEMS: ELECTRIC, GAS, AND OIL

The gas type furnace, shown in Figure 14–1, can use natural gas or liquid petroleum (LP) type gas, depending on where the furnace is located. Major parts of the United States have natural gas lines running through them and utilties provide this gas directly to the home in a continuous supply. In other parts of the country the natural gas lines have not been run, so LP gas is used and is stored in a tank right on the property.

Over the years, different areas of the country have used natural gas, LP gas, fuel oil, or electricity to heat homes. As the price of these different fuels varied, different heating systems became popular. You may find large areas of homes and commercial establishments that use a similar energy source for the heating system.

Many years ago some parts of the country burned coal for heating systems, but these systems have largely been converted to natural gas or LP gas. In some areas of the country, electricity is the least expensive energy source, so electric furnaces and electric baseboard heating is the most common type of heating system. Electricity is also used in the southeast and southwest parts of the country where winters are milder and a small amount of heat is needed.

Today the emphasis for heating systems is to become as efficient as possible. For this reason furnaces have become much more complex and include electronic controls for gas valves and variable speed fan motors.

When natural gas and LP gas became scarce and expensive in the 1970s and 1980s, heat pumps became popular. The heat pump uses a traditional air-conditioning system

Figure 14–1 Typical gas furnace (Infinity 96 Furnace) *(Courtesy of Carrier)*

and reverses the refrigerant flow so that the hot gas from the compressor is sent to the indoor unit in the wintertime. Heat pumps will be covered in detail in Chapter 16. In this chapter you'll learn the basic part of each type of heating system and be able to troubleshoot typical problems found in them. Cutaway diagrams and electrical diagrams will be provided to help you understand how to troubleshoot these heating systems.

14.2 GAS FURNACE CONTROLS

Gas furnaces have changed radically in the past 20 years to meet changing efficiency standards. Early gas furnaces used a standing pilot light that burned a small flame continuously. Figure 14–2 shows a standing type pilot. The pilot light flame is available to ignite gas for the main burner whenever the thermostat energizes the main gas valve. In this figure you can see that a thermocouple is mounted so that it is in the tip of the pilot flame. The pilot light must be lit manually to ignite the pilot flame. The pilot is supplied with a small amount of gas continuously so that the pilot flame burns continuously. The thermocouple generates a small amount of voltage (millivolts) when it is heated, and this millivolts signal is sent to the gas valve to indicate a pilot flame is burning. If the pilot flame goes out, the thermocouple stops generating voltage, and the gas valve will not open, even if it gets an electrical signal from the heating thermostat. Since the thermocouple generates a very small voltage signal, it is important that it is located exactly in the tip of the pilot flame, so it receives the most heat possible. It is also important that the terminal end of the thermocouple, which is the end that is not in the flame, is mounted into the gas valve tightly and that the terminal is very clean. If the terminal end is loose or dirty, it will not make good enough contact for the millivolts signal to energize the gas valve. If you find a furnace that has the pilot flame burning, but the gas valve will not open when it is energized, you should suspect the thermocouple and you should take it apart and clean the terminal end and ensure that the tip is located directly in the pilot flame.

Newer, **high-efficiency furnaces** do not use standing pilot lights; rather, they ignite the pilot light just prior to igniting the main burner. In some furnaces an electric spark is used to ignite the pilot light; in other furnaces, a glow coil (hot surface igniter) is used in place of the pilot light to ignite the furnace.

Figure 14–2 A typical pilot assembly for a gas-fired furnace.
(Courtesy of Honeywell)

Figure 14–3 Electrical diagram of a typical standing pilot furnace.

14.3 ## MAIN PARTS OF THE STANDING PILOT TYPE GAS FURNACE

The main parts of the standing pilot type gas furnace include the pilot flame and **thermocouple**, **gas valve**, transformer, heating thermostat, and **fan and limit switch**. The **fan switch** turns the furnace fan on and off, and the limit switch, also called the high limit switch, is a normally closed, temperature-controlled switch that opens if the temperature in the furnace gets out of control and becomes too hot. Figure 14–3 shows an electrical diagram of a standing pilot type furnace. In the diagram you can see that the tip of the thermocouple is located in the pilot flame and the terminal end is mounted in the gas valve. The thermostat and gas valve are part of the control circuit, and the pilot light and thermocouple are also included. The fan switch, high limit switch, and furnace fan are all part of the load circuit at the top of the diagram.

14.4 ## NORMAL OPERATION OF THE STANDING PILOT TYPE GAS FURNACE

Before we began to discuss the normal operation of the standing pilot type gas furnace, you should notice from the diagram how the loads and controls are wired. The transformer primary winding is connected to the 120-V power supply and has the high limit switch connected in series with it. The secondary winding of the transformer provides power to the R terminal of the thermostat, and power goes through the W terminal of the thermostat to the gas valve coil. You can also see the pilot flame send heat to the thermocouple, which generates a small amount of voltage that is sent to the gas valve to

indicate that a pilot flame is present. When the thermostat calls for heat, voltage is sent to the gas valve coil and it opens and sends gas to the main burner. Heat from the main burner begins to warm the heat exchanger, and when the temperature reaches the set-point of the fan switch, the furnace fan is turned on.

The first step in this sequence is that the pilot light must be lit from an external source of fire, such as a match or lighter. When you are lighting the pilot for the first time, you must manually turn the gas valve control knob to the pilot position and then press down the button on the gas valve that enables gas to flow to the pilot assembly. This button actually provides pull-in function for the gas valve pilot solenoid. When the thermocouple begins producing millivoltage, the amount of millivoltage will be sufficient to keep the pilot solenoid energized. If you let up on the pilot button on the gas valve before the thermocouple begins producing sufficient millivoltage, the pilot solenoid will de-energize and cut off gas flow to the pilot assembly.

Once the flame in the pilot is established and the thermocouple begins generating millivoltage, the gas furnace is ready to accept a signal from the heating thermostat indicating that the furnace should be turned on. When the room where the thermostat is located becomes too cool, it will cause the R–W circuit to be completed and 24 V AC is sent to the gas valve main coil. When the gas valve's main solenoid coil receives 24 V, it opens the gas valve completely, so that gas is supplied to the main burners. When gas reaches the main burners, it passes over the pilot flame where it is ignited. The burners are mounted in such a way that gas is distributed equally to each of them and they ignite one after another.

After all the burners have ignited, heat is added to the heat exchanger. This heat continues to build until the temperature on the fan switch reaches setpoint, and then it turns the furnace fan on. The fan and limit switch is a combination switch that works off of the same temperature actuator, which is mounted so that it senses the temperature near the heat exchanger. The actuator is made of bimetal and moves as temperature increases. When the furnace fan turns on, it begins to blow air through the ductwork into the conditioned space, and the room begins to warm up. When the temperature in the room reaches the setpoint of the thermostat, the thermostat opens the R–W circuit, current to the main gas valve coil is turned off, and the gas valve closes. This stops the gas going to the burner assembly, and the furnace begins to cool down. When the furnace cools down to the lower setpoint on the fan switch, the furnace fan will turn off and the furnace waits for the cycle to begin again. Once the pilot light is lit, it will remain burning until the gas valve is manually turned off.

If for some reason the furnace fan fails to turn on, the heat in the heat exchanger will continue to build to a dangerous level where a fire could occur. In this case, the high limit switch part of the fan and limit switch would open and turn off voltage to the primary of the transformer. This would shut down the 24 volts on the secondary, which will turn off power to the main gas valve solenoid, and main gas flow to the furnace would shut down. When the high limit switch is activated and opens because of overtemperature, it will remain open until it is manually reset. This prevents the furnace from continually cycling on and off due to the overtemperature condition.

14.5 PILOT ASSEMBLY AND THERMOCOUPLE

The pilot and thermocouple were shown in Figure 14–2 and they provide the function to establish a pilot flame to ignite the main burner. The thermocouple is used as a safety device to indicate that the pilot light is burning correctly. The pilot light is a small orifice through which gas flows. This gas must be ignited with a match or other source of

fire when the flow of gas is turned on to the pilot. The gas for the pilot comes through the main gas valve. The pilot circuit in the gas valve is controlled by a solenoid that is energized by the small voltage produced by the thermocouple. Since the thermocouple produces a small amount of voltage (15–30 mV DC) that is not sufficient to pull the armature of the solenoid open initially, the solenoid must be opened manually by depressing the valve by hand. After the solenoid is opened manually, the small voltage is able to keep the coil energized so that it keeps the pilot valve open.

14.6 THE GAS VALVE

The gas valve that uses a standing pilot is shown in Figure 14–4, and a cutaway diagram of the gas valve is shown in Figure 14–5. In the cutaway diagram you can see that it has a main valve that is operated by a solenoid coil, and a pilot valve that is operated by a separate solenoid. The main valve is spring loaded in the closed position, and when its solenoid coil receives voltage from the thermostat, it becomes energized and pulls the valve open. The pilot valve is controlled by voltage from the thermocouple.

The body of the gas valve has one inlet on the left side and two outlets on the right side. The small outlet supplies gas to the pilot assembly, and the large outlet supplies gas to the main burners. You should notice that the flow of gas to the main valve must pass through the pilot valve. This means that if the pilot valve is not open, gas cannot flow to the main valve, so the pilot valve acts as a safety device for the main valve. The electrical voltage that energizes the pilot solenoid comes from the thermocouple and is very small. To activate the pilot valve, you must depress the pilot valve manually and hold it down, which provides the *pull-in* function for the solenoid valve. After the valve is pulled in, the small voltage from the thermocouple is sufficient to "hold in" the solenoid. If the pilot light fails, the temperature at the thermocouple drops to a point at which the voltage cannot hold the pilot valve open, and a spring forces the valve closed. This action creates a failsafe control for the gas valve and pilot assembly, so that the gas valve does not open and allow gas to flow into the combustion chamber of the furnace if the pilot light goes out.

Figure 14–4 A typical gas valve. *(Courtesy of Honeywell)*

Figure 14–5 Cutaway view of a typical gas valve. *(Courtesy of Honeywell)*

14.7 TROUBLESHOOTING THE STANDING PILOT TYPE GAS FURNACE

When you are called to troubleshoot a gas furnace, the first thing you should do is set the thermostat at a high temperature well above room temperature and try to operate the furnace. When the thermostat closes its heating circuit (R–W), the furnace should begin its operation by opening the main gas valve. A furnace that does not operate correctly can have problems in two basic areas: the fuel system and the furnace fan.

If the furnace does not ignite the main burners when you turn the thermostat up, you should begin your troubleshooting procedure with an inspection of the pilot light to ensure that it is burning. If the pilot light is not burning, you would manually turn the gas valve to the off position and start the pilot lighting procedure from the beginning. You may need to check the instructions on the furnace if they are available and follow them. Generally you start lighting the pilot by manually turning the gas valve to the pilot position and depressing the pilot button. This should allow the supply of pilot gas, and you should provide a fire from a match or source of flame. If the gas does not ignite immediately, you should check for the proper supply of gas. Sometimes an insect or dirt will block the small pilot line or orifice. You may need to turn the gas off and remove the tubing and orifice and blow air through it. If you can determine it is clear, you can check to make sure the main gas supply is available. Sometimes you will get a call for "no heat" and find that the main gas supply has been turned off because the bill is not being paid. You may need to check the main gas valve to ensure that gas is available. Sometimes you can check other gas appliances in the house such as a gas range in the kitchen to ensure that it has gas.

If the pilot light will ignite, but does not stay on, you will need to check the thermocouple and ensure that its terminal end is clean, that it is seated tightly in the gas valve, and that the tip is correctly located in the flame. If the pilot flame does not stay lit, the thermocouple may need to be replaced.

If the pilot flame is burning and the main gas valve does not open, you can begin the troubleshooting process by taking the following voltage measurements. Refer to the diagram in Figure 14–3. One additional note to remember before you begin to troubleshoot

and take voltage measurements: Make all of your 24-volt measurements in reference to terminal C and all 120-volt AC measurements in reference to N. This means that you should leave your meter leads on terminal C for 24-volt measurements and leave your meter leads on N (neutral) for 120-volt measurements.

1. Check terminals R to C at the furnace terminal board for 24 V AC. _____

 a. If you have 24 volts at terminals R–C, move to step 2; if not, move to step 1b.
 b. If you do not have 24 volts at R–C, check for voltage at the primary terminals of the transformer and move to step c.
 c. If you have voltage at the primary terminals, you should suspect the transformer is bad and turn off power, remove it, and test its primary and secondary windings for continuity. If either of the windings has an open, you will need to install a new transformer.
 d. If the primary terminals do not have 120 volts, you should suspect the high limit switch is open and find the reset button on it. If the furnace starts after you depress the limit reset button, you should be sure to determine what made the overtemperature condition occur.

2. Check from terminal C to terminal W on the terminal board of the furnace control for 24 V AC to test whether the thermostat is calling for heat. _____
 a. If you measure 24 volts at terminals W–C, the thermostat is calling for heat and you should move to step 3.
 b. If you do not measure 24 volts from W–C, you should suspect that the thermostat is not set high enough, that it is not working correctly, or that a wire between the thermostat and the furnace is open.

3. Check directly across the terminals on the gas valve for 24 V AC to test whether the gas valve is getting full voltage. _____
 a. If you measure 24 volts directly across the terminals of the gas valve, it should open, the burners should ignite, the furnace should start to warm up, and you can move to step 4. If you do not measure 24 volts, you should move to step 3b.
 b. If you do not measure 24 volts across the terminals of the gas valve, you should suspect the wiring between the terminal board where your last test was made in step 2 and where the wires connect to the gas valve. Replace any bad wires and continue with trying to make the furnace run.

4. If the gas valve opens and the burners ignite, the temperature should begin to rise. If the furnace fan comes on, the furnace is now working correctly. If the furnace fan does not turn on, you should move to step 5. If the furnace does come on correctly, you can turn the thermostat down and ensure that the furnace turns off. You should cycle the furnace one more time before you leave the service call.

5. If the furnace fan does not cycle on when the temperature increases, you should check the voltage through the fan switch. You can check for 120 volts at the terminals of the furnace fan motor. If you have 120 volts and the motor does not run, you should suspect the motor has a problem. If you do not have 120 volts at the motor terminals, you should suspect that the fan switch is not closed. Move to step 5a. _____
 a. If you do not have voltage at the furnace fan motor terminals, you should check for voltage where voltage is supplied to the fan switch. If you have voltage at the input of the switch, but not at the output, you should turn off power, remove the fan switch, and check it for correct operation. If the fan switch is faulty, replace it and cycle the furnace several times to ensure that it is operating correctly.

14.8 THE MAIN PARTS OF THE ELECTRONIC IGNITION TYPE GAS FURNACE

Newer models of furnaces are designed to be more energy efficient. The first step in this process was to create a pilot light that only burns gas for a few seconds when the main burners are turned on, rather than burn gas constantly. This effort brought about two types of pilot ignition systems. One is the **direct-spark ignition** system, which has a spark igniter similar to a spark plug, and a flame rod. The spark igniter creates a high-voltage spark directly in the pilot flame, which will cause it to ignite. The **flame rod** does not create a voltage like the thermocouple; rather it uses a unique principal that the resistance of a burning flame is lower than the resistance of air. The flame rod is part of an electronic circuit that has its resistance lowered when the pilot flame has been established. When this occurs, the pilot flame completes the circuit and causes a lower resistance path that allows enough current to flow to activate the electronic circuit. You will learn more about the flame rod later in this chapter, and you will learn more about electronics in Chapter 19. Figure 14–6 shows a high-efficiency furnace that has a direct-spark ignition system that ignites the pilot light on demand.

The second type of ignition system is called a hot plate igniter. It has a ceramic plate that receives voltage, which causes it to glow bright red at a set temperature and ignite the pilot gas. This type of system is sometimes called a glow plug.

Another feature that is used to make furnaces more efficient is to take more heat out of the combustion process, which lowers the temperature of the exhaust (combustion) gases. In traditional furnaces, a natural vent is used whereby the high temperature of the combustion gases creates a large temperature differential that causes a "draft" so that exhaust gases flow up a flue pipe or chimney naturally. The combustion chamber must be kept at a very high temperature for the exhaust gases to draft properly. This higher temperature causes the furnace to be very inefficient, since a lot of heat from the combustion process goes up the chimney instead of into the heated space.

One way to combat this waste is to lower the combustion temperature and ensure that more of the heat from the combustion process gets into the heated space. The problem this creates is that the exhaust gases have so much heat removed from them

1. Heat exchanger

2. Variable speed fan

3A. Gas valve

3B. Manifold

4. Quiet draft inducer

5. Direct spark ignition

6. Modulating ICM2 + blower motor

7. Easy-access washable filter

8. Integrated furnace control

Figure 14–6 Typical high-efficiency furnace with direct-spark ignition. *(Courtesy of Rheem Air Conditioning Division)*

that they no longer draft naturally. Newer furnaces now have a combustion fan that is specifically designed to vent the cooler exhaust gases. The temperature of the exhaust gases in newer high-efficiency furnaces is so low that these gases can now be vented through plastic (PVC) pipe, rather than through a metal vent pipe or masonry chimney. Some very high efficient furnaces also have two-stage gas valves. The first stage is used when only a small amount of heat is needed, and the second stage is opened when maximum heat is required, such as when the outdoor temperature is extremely cold. Another time the second stage will come on is when the temperature differential between the desired temperature and the actual temperature in the heated space is very large, such as when the temperature has been turned down for a period of time, and then set back up when people return to the space.

The newer high-efficiency furnaces also try to eliminate waste in the electrical system. In a traditional furnace, the furnace fan comes on and runs at one speed, which is inefficient. Newer high-efficiency furnaces use a variable speed drive that controls the furnace fan at the slowest speed required to keep the heated space comfortable. The improvement of electronics allows these furnaces to have an electronic circuit board that controls the motor speed and checks for ignition and other safety circuits in the furnace.

The main parts of high-efficiency furnace includes an on-demand pilot system, a combustion blower that creates an induced draft, a variable speed furnace fan motor controller, and an electronic circuit board that controls and checks all the systems.

14.9 NORMAL OPERATION OF THE ELECTRONIC IGNITION TYPE GAS FURNACE

The basic operation of an electronic ignition type gas furnace begins when the thermostat calls for heating. When this occurs a signal is sent from the thermostat to the W terminal on the furnace control board. When the signal is received, the process to light the pilot begins. First a combustion blower turns on, and a sensor verifies that a draft has been established by measuring the amount of air that is moving out the vent pipe. When the control board has determined that a proper draft has been created, a signal is sent to the pilot ignition system. If the system is a direct-spark type system, a source of high voltage is sent to the spark igniter, and it begins sparking. The gas for the pilot flame is turned on so that it flows over the spark, and ignition will occur.

If the system uses a hot surface igniter, it will receive voltage and began to glow brightly. When the pilot gas passes over it, it will ignite and established the pilot flame. Both systems have a flame rod or other sensor that will determine that a pilot flame has been established and send a signal back to the control board. At that point the control board will send 24 V to the main gas valve and cause it to open and began to ignite the burners. The combustion blower will constantly force the combustion gases out the vent.

When the heat exchanger begins to warm up, the control board will receive a signal from the fan switch and start the fan motor. If the fan motor is variable speed, it will start on low speed and increase RPM to the speed necessary to ensure that the conditioned space is warming up.

When the conditioned space reaches the setpoint on the thermostat, the thermostat turns off voltage to the W terminal, and the furnace begins the shutdown process. The first thing that occurs is voltage to the main gas valve is turned off, and the main burners are extinguished. The combustion blower will turn off as soon as the main burners are extinguished. The furnace fan will continue to run until all of the heat in the heat exchanger has been removed. When the furnace fan shuts off, the system is ready to start the cycle from the beginning.

14.10 DIRECT-SPARK IGNITION GAS VALVES

Newer gas valves use a pilot light that does not stay on continuously. Instead, the pilot light is ignited only when the furnace is starting its cycle. The pilot light is ignited by a series of sparks generated by the spark igniter, which is similar to a spark plug in that high voltage jumps a gap between the tip of the spark igniter and the metal around it, creating a high-voltage spark. The high voltage is created in the direct-spark ignition module as a series of pulses and sent to the spark igniter. Figure 14–7 shows the parts of this system. Figure 14–7a shows the igniter module, which is called the smart ignition module. This module contains electronic circuits that create the high-voltage source that is sent to the igniter plug to create a spark, and an electronic circuit that uses the flame or rod or other sensor to prove that the pilot flame has been established. Figure 14–7b shows a typical flame rod that is mounted in the pilot flame so it can detect that the pilot flame has been established. The flame rod is different from a thermocouple in that it does not generate voltage, rather it completes the electrical circuit through the gas flame. The flame rod uses a principle that the electrical resistance in a flame (heated air) is less that the resistance of air at room temperature. If the pilot flame is established, a low resistance circuit path is created between the flame and the rod, and electrical current will begin to flow in the electronic circuit in the control module. If the flame is not ignited,

(a)　　　　　　　　　　　　　　(b)

Figure 14–7 **a** A direct-spark ignition module for electronic ignition. **b** The flame rod for electronic ignition. **c** The spark igniter for electronic ignition. **d** Electronic spark igniter with flame sensor. *(Courtesy of Honeywell)*

(c)　　　　　　　　　　　　　　(d)

and gas flows out the pilot orifice, the resistance between the rod in the gas is too high and current cannot flow in the electronic circuit. The electronic control is designed to see the higher resistance as a "no flame" condition and the low resistance as a "flame established" condition. The flame rod tends to last much longer than a typical thermocouple.

Figure 14–7c shows a typical spark igniter, and as you can see in this one, the high-voltage electrical signal will jump the gap between the tip of the igniter and the grounding case that is just above the tip. Figure 14–7d shows the location of the flame rod in relation to the spark igniter.

After the pilot flame is established, the resistance between the flame rod and flame is reduced and current flow begins in the electronic circuit that tells the main valve it is safe to open. In this type of gas valve assembly, the 24-V AC signal from thermostat terminal W starts the pilot ignition cycle, and heat from the pilot flame heats a sensor that allows the main valve to open and provide fuel to the main burner.

14.11 IGNITER AND FLAME ROD

The main parts of the direct-spark ignition gas valve are the spark igniter and flame rod. In Figure 14–7c you can see that the igniter looks like a large spark plug with a gap between its main electrode and ground. The shell of the igniter is made of ceramic material so that it can withstand the 30,000–40,000-V electrical voltage that is supplied to it to produce the spark at the gap. The control module uses an oscillator signal to provide three to four sparks per second. The control module also provides a voltage signal to the coil of the pilot solenoid so that pilot gas can flow to the pilot assembly, where the spark from the igniter can cause it to start burning. Figure 14–8 shows a diagram of the

Figure 14–8 Electrical diagram of a direct-spark ignition system. *(Courtesy of Honeywell)*

1. Power supply, provide disconnect means and overload protection as required.
2. Alternate limit controller location.
3. Maximum wire length 3 ft (.9 m).
4. Controls in 24V circuit must not be in ground leg to transformer.
5. For module with TH-W terminal and vent damper plug, connect thermostat to the TH-W. Leave 24V open. Do not remove vent damper plug.

direct-spark ignition system, and you can see how the spark igniter is grounded to complete the circuit. A high-voltage pulse is sent to the center electrode, which makes it jump between the electrode and the tip that is grounded. This causes a spark to occur between the electrode and the tip, and the spark is hot enough to ignite the pilot gas.

After the pilot flame is established, the electrical resistance between the flame and the flame rod becomes low enough that current flow is established that completes the electronic circuit. When the resistance is reduced, the current flow in the circuit in the control module indicates the pilot flame is established and it is safe to allow the main valve to open so that gas can flow to the main burner. When the main gas comes into contact with the pilot flame, it ignites and establishes the main fire. When the main burners are ignited, enough heat is produced to cause the fan switch to close and energize the furnace fan motor. The furnace fan moves air across the heat exchanger and into the conditioned space. When the desired temperature is reached, the thermostat is satisfied, and it opens its contacts between R and W, which de-energizes the direct-spark ignition control module. This de-energizes the main gas valve and pilot valve so that they close and shut off the flow of gas. When the thermostat calls for heating again, the pilot ignition process starts again.

14.12 HOT SURFACE IGNITER

Another system that is used to ignite the pilot flame is a **hot surface igniter**. Hot surface igniters are made of high-purity recrystallized silicon carbide. They act like a large resistor that heats up when voltage is applied and current flows through it. Figure 14–9 shows an example of a hot surface igniter glowing red hot with current flowing through it, and Figure 14–10 shows an example of a hot surface igniter when it does not have voltage applied and it is cooled. In this figure you can see the hot surface igniter has a tip made of silicon carbide, a ceramic holder, and wires that are attached to allow current flow through the silicon material. Typical hot surface igniters are available for 12-V, 24-V, 120-V, and 240-V supplies. They are designed for long-term reliability and service,

Figure 14–9 Typical hot surface igniter. *(Courtesy of Saint-Gobain Norton Igniters)*

Figure 14–10 Hot surface igniter. *(Courtesy of Hot Surface Igniters)*

and they will reach ignition temperatures of 2100° to 2400°F within 20 to 40 seconds. The hot surface igniter provides positive ignition over a wide range of gas and air mixtures and conditions. They are used in all types of heating systems from furnaces and boilers to rooftop heaters, unit heaters, infrared burners, water heaters, and every other type of gas appliance.

When you are handling a hot surface igniter during installation or testing, you should always handle it by touching only the ceramic holder when possible. You must also be careful not to drop it or bump the tip as it may crack, which will cause an open in its electrical circuit and it will no longer work. The hot surface igniter is used to ignite the pilot flame. It is located near the main burner, where the flow of gas from the pilot orifice is directed. When the control circuit in the heating unit determines that the heat should be turned on, the proper amount of voltage is sent to the hot surface igniter and it begins to warm up until it is glowing red hot. At the same time, the pilot gas is allowed to flow over the hot surface of the igniter and the hot surface is hot enough to cause the gas to ignite and the pilot flame is established. The pilot flame must produce sufficient heat to show the control system that the pilot flame is established. When the sensor indicates that the pilot flame is established, the main gas valve is energized and full gas flow is allowed to flow to the main burner. When the main burner flame is established, the voltage to the hot surface igniter is turned off, and the hot surface igniter begins to cool down. If the main burner flame is not detected, the main gas valve will be de-energized and the main gas flow will be shut off. The system may go into a lockout mode or it may be allowed to repeat the ignition process for a specific number of retries before it is locked out.

If you are asked to troubleshoot a hot surface igniter, you should check to determine if it is receiving the proper amount of voltage. If the proper amount of voltage is present at the two terminals of the hot surface igniter, it should begin to warm up immediately. If it does not warm up to a red-hot glow, you can turn off power and replace it. If you replace the hot surface igniter, be sure to use one with identical voltage rating and size. You should also test the igniter several times for proper operation and test the lockout feature by turning the manual shutoff valve to the off position.

14.13 FAN AND LIMIT CONTROLS FOR FURNACES

Each type of furnace needs a control to protect it against overheating. If the furnace fan fails to operate correctly, the fire in the heat exchanger will make the furnace overheat and could cause a fire. A high-limit switch is used to sense the overtemperature, which opens its contacts. The contacts for the switch are usually wired in series with the voltage that is supplied to the gas valve or oil burner pump or other source of heat. When the high-limit contacts open, they de-energize the source of heat and cause the furnace to cool down safely. The limit is usually a snap-action switch that must be manually reset. The manual reset function causes someone to come to the furnace to depress the reset button. This person must also check the furnace to learn the cause of the overtemperature. Figure 14–11a shows an example of a high-limit switch with a reset button, and Figure 14–11b shows the switch with its cover removed. Figure 14–11c shows the fan and limit as you would see it mounted in a furnace. Figure 14–12 shows the high-limit switch in an electrical diagram. Notice the fan switch is connected to the furnace fan, and the high-limit switch is connected in series with the primary side of the control transformer.

Fan
Preset

Limit
Preset

Reset
Button

(a)

(b)

Figure 14-11
a Insertion type
adjustable fan and limit
control that is used to
control the fan and
protect against
overtemperature. *(Courtesy
of Honeywell)* **b** Location of
reset, limit preset, and fan
preset. **c** The part of the
temperature-sensitive
spiral that is located
inside the furnace heat
exchanger.

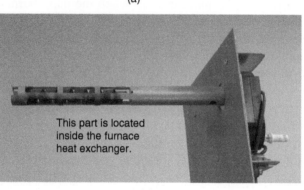

This part is located
inside the furnace
heat exchanger.

(c)

Figure 14-12 A typical
electrical diagram for a
limit switch.

L1 Fan
 Switch Fan N

 R C
 S

 Transformer
 Primary Winding

High-Limit Fan and Limit Switch
Switch

14.14 THE FAN SWITCH PART OF THE FAN AND LIMIT CONTROL

The furnace also needs a temperature control that controls the fan. When the gas valve or oil burner is first energized, the heat exchanger is cool. If the fan was energized at this time, it would blow cool air into the conditioned space. The **fan control** senses the temperature in the heat exchanger and energizes the fan when the temperature reaches approximately 110°F. When the source of heat is de-energized, the fan continues to run to allow the residual heat that is built up in the furnace to be moved into the room. The fan switch allows the heat exchanger to cool down to approximately 100°F before its contacts open

Combination Fan & Limit Control
• FAL SPDT

This Cam-Stat® Combination Fan and Limit Control is designed to replace OEM FAL models which contain SPDT fan switch or SPDT limit switch or both.

The fan portion incorporates an adjustable fan "OFF" to enable the installer to match the settings of the old control.

The limit portion will limit the temperature of the furnace to a safe operating temperature. A limit temperature adjustment is provided to match the setting of the old device being replaced.

Fan Terminals	Volts AC	Full-Load Amps. (Inductive)	Locked-Rotor Amps
2 & 4 N.O. –[SPDT Switch]– 1 & 4 N.C.	120	13.8 (3/4 HP)	82.8
	240	8.0 (1HP)	48.0
	120	5.8 (1/4 HP)	34.8
	240	2.9 (1/4 HP)	17.4
Limit Terminals			
5 & 6 N.C. –[SPDT Switch]– 6 & 7 N.O.	120	4.4 (1/6 HP)	26.4
	240	2.2 (1/6 HP)	13.2
	30	40 VA	N/A
	120	125 VA	N/A

NOTE: May be substituted for FAL 7C control by using terminals 2 & 4 (FAN) and 5 & 6 (Limit).

Model Number	Element Length	Adjustable Settings (°F)		
		Fan Off Temperature	Limit Temperature	Fan & Limit Fixed Directionals
FAL 7C-20BT-120-AB	7 in.	90 to 120	150 to 250	25° to 30°

Figure 14–13 Combination fan and limit control. *(Courtesy of Cam-Stat)*

and de-energize the fan. Figure 14–13 shows another type of fan and limit switch. This switch has the trade name **Cam stat**. The turn-on and turn-off temperatures for the fan switch are adjustable on some fan switches, and they are fixed on others. Some fan switches use a time-delay element in addition to the temperature-sensing element to provide sufficient time delay for the fan to cool the heat exchanger at the end of a cycle. The time delay can be used to provide time delay at the beginning of the cycle as well.

14.14.1 Time-Temperature Start Fan Control Switch

Another type of switch that is used to control the furnace fan is called the time-temperature start fan control switch. An example of this type of switch is shown in Figure 14–14a. This type of switch incorporates the best features of both the temperature-actuated switch and the time-delay control to start the furnace fan. When the thermostat sends a signal to the furnace and calls for heat, the time-delay part of the switch becomes energized. If the temperature-sensing part of the switch does not reach its setpoint temperature, the time-delay feature will turn on the furnace fan. If the temperature rises to the setpoint first, the furnace fan will be energized by the temperature-controlled part of the switch.

Figure 14–14b shows a typical wiring diagram for this type of switch. From the diagram you can see that the switch uses a resistive type heater that is wired in parallel with

Figure 14–14 a Picture of a Cam-Stat used as a replacement for a fan and limit switch. **b** Electrical diagram of a Cam-Stat used to control fan and high limit on a gas furnace. *(Courtesy of Cam-Stat)*

(a) (b)

the gas valve. When the W terminal of the room thermostat sends 24 volts to the gas valve, the heater of the switch also gets voltage. The contacts of the switch are located at terminals 2 and 4, and they are wired in series with the 120-volt power supply and the furnace fan. When the heater of the switch gets warm enough, it closes the contacts and energizes the furnace fan. When the switch setting is changed, the amount of resistance is increased or decreased, which changes the amount of heat that is generated and in turn the amount of time delay. The contacts are also activated by the temperature-activated element that is mounted in the furnace where it can sense the change of temperature caused by the burners.

When the furnace gas valve turns off, the fan continues to blow air over the heat exchanger, which starts to cool down the furnace. When the furnace reaches the low set-point temperature of the fan switch, the switch contacts open and turn off the fan. This type of switch can have its time delay or its turn-on setpoint adjusted to ensure the furnace fan always comes on at the most comfortable and efficient point.

14.15 THE HIGH-LIMIT CONTROL PART OF THE FAN AND LIMIT SWITCH

The fan and limit switch can be operated off the same temperature-sensing element and can be set at different temperatures. Figure 14–15a shows the typical adjustable fan and limit switch that works from the same temperature-sensing element, and Figure 14–15b shows the switch with the cover removed. Figure 14–15c shows an electrical diagram of a fan and limit switch. This control has two separate switches that are activated by two separate cams that are mounted to the temperature-sensing bimetal element. The bimetal element is placed in the top of the heat exchanger, where it can sense the warmest temperature. When the heat exchanger begins to warm up, the spiral bimetal element begins to move, and the cam for the fan switch moves across the fan switch and activates it. This causes the normally open fan switch contacts to close and energize the fan. The fan moves cool air from the conditioned space over the heat exchanger and keeps the heat exchanger at a constant temperature. When the source of heat is de-energized, the heat exchanger cools down, and the bimetal element returns to its original position, allowing the fan switch to return to its open position.

If the fan fails to operate correctly, the temperature in the heat exchanger will continue to increase, and the bimetal element will continue to move. When the temperature of the heat exchanger reaches approximately 130°F, the cam for the limit switch on the bimetal element causes the limit switch part of the fan and limit to activate. When this occurs, the normally closed contacts of the limit switch open and de-energize the source of heat for the furnace. When the furnace cools down, the bimetal element returns to its normal position, the limit contacts return to their closed position, and the heat source is energized again.

The furnace will continue to cycle on the high temperature until the temperature exceeds the fixed high-temperature limit or until someone notices that the furnace is cycling on the fan and limit switch.

14.15.1 Fixed-Temperature High-Limit Switches

In some furnaces a fixed-temperature high-limit switch is used. Figure 14–16 shows one type of fixed-temperature high-limit switch. The fixed temperature has one advantage in that it cannot be set too high. Some of the fixed-temperature high-limit switches have a reset button that requires them to be manually reset. The switch is

(a) (b)

(c)

Figure 14–15 a A high-limit switch that automatically resets itself when the overtemperature condition returns to normal. **b** A high-limit switch with its cover removed. **c** Diagram of a high-limit switch connected in a circuit. *(Courtesy of Honeywell)*

Figure 14–16 Fixed-temperature switches. *(Courtesy of Therm-o-Disc Incorporated)*

mounted in the top of the furnace where it can measure the warmest temperature and open if the temperature gets out of control. The fixed limit switch is wired in the system exactly as the variable switch. The only difference is that the temperature setting cannot be adjusted.

14.15.2 Flame Rollout Type Limit Switch

Figure 14–17 shows a typical **flame rollout switch**. This is a temperature switch that looks similar to the fixed-temperature high-limit switch. The main difference with this switch is where it is mounted. The flame rollout switch is mounted near where the burner takes in air in a standing pilot type furnace. Since this type of furnace uses a natural draft to move combustion gases up the flue, it is possible that on a windy day, air is forced back down the chimney or flue pipe. If this occurs, the flame can be forced out the front of the furnace and a fire can be started in the front of the furnace or in the room where the furnace is located. The flame rollout switch senses the excessive temperature that occurs with this condition and turns the gas valve off.

Since the flame rollout switch is connected in series with the gas valve, it will turn off all power to the gas valve. The switch has a feature that requires it to be manually reset. If you are on a service call and find the flame rollout switch has been tripped, you should be very careful to ensure the condition that caused the switch to trip has been repaired. A vent cap that has come loose on the top of the flue pipe may cause this condition, or it may have been rusted through or blown off. In any case, you must thoroughly inspect the flue pipe system to ensure that it is working correctly before you reset the switch and leave the service call.

14.16 THE MAIN PARTS OF THE HIGH-EFFICIENCY GAS FURNACE

Older furnaces used natural draft vents, which included a long vent pipe that usually went from the furnace and extended through the roof. These furnaces were designed to use excess heat from the combustion process to keep the vent pipe warm enough to make the exhaust gases rise through them quickly. This meant that a lot of heat was wasted to make the exhaust gas move. Some newer furnaces incorporate a combustion blower to assist in removing exhaust gases. This allows the furnace to remove more heat from the exhaust gas and use it to heat the conditioned space. Figures 14–18, 14–19, and 14–20 show examples of high-efficiency furnaces. In these figures, you can see all the main parts of the different models of high-efficiency furnaces.

The high-efficiency furnaces may also use an electronic type ignition for the pilot light as well as electronic control of the furnace fan. Some of the models will use a two-stage gas valve and some have two separate heat exchangers, such as the model in Figure 14–19. The first heat exchanger removes approximately 80% of the heat and the second one removes the remaining 12–16% of the heat, which makes the furnace very efficient. Since these types of furnaces take as much heat out of the combustion gas as possible, a combustion blower must be used to ensure that the exhaust gas is removed safely. This combustion blower may be called an *induced-draft blower*. The furnace uses several controls and an electronic control board to ensure that the combustion fan is working prior to the

Duralok® Heat Exchanger – Made of patented ArmorTuf™ steel. Life-tested to 2-1/2 times agency requirements (ANSI Standard Z21.47) for long life and durability.

Two-Stage Gas Valve – Quietly adjusts the amount of gas to accommodate two levels of heating for better temperature control.

Special Sound-Absorbing Insulation – Further reduces sound for even quieter operation.

Patented Sound Enclosure – Gas burners are carefully contained, minimizing sound.

Variable Speed Motor – Efficient motor quietly controls flow of air for added humidity control, indoor air quality and comfort.

Sure®Light Control Board – Controls furnace operation to ensure high reliability and efficiency.

Durable Steel Cabinet – Made to last with an attractive, high quality, textured paint finish.

Figure 14–18 A high-efficiency furnace. *(Courtesy of Lennox International Inc.)*

Figure 14–19 A high-efficiency furnace.

(Courtesy of Carrier.)

Remote Sensor

Direct Spark Ignition & Remote Sensor

Primary and Secondary Heat Exchanger

In-Shot Burners

Gas Valve and Manifold

Pressure Switches

Draft Inducer

Integrated Furnace Control

Figure 14–20 A high-efficiency furnace that shows the heat exchanger, direct-spark ignition and remote sensor, induced-draft system, integrated furnace control, gas valve and manifold, pressure switches, and primary and secondary heat exchangers.
(Courtesy of Rudd Air Conditioning Division)

pilot light starting and the main gas valve opening. Figure 14–20 shows a high-efficiency furnace that has a modulating gas valve. The gas valve is controlled by an electronic circuit that modulates it between 40% and 100% to ensure that the correct amount of fuel is supplied to the system at all times to ensure the highest efficiency. This furnace also has a direct-spark ignition and an induced-draft system that uses a combustion blower to quietly vent the combustion gas. The integrated furnace control includes a seven-segment LED display that provides troubleshooting codes to help you diagnose problems.

14.17 TROUBLESHOOTING THE HIGH-EFFICIENCY GAS FURNACE

When you are called to troubleshoot a gas furnace, the first thing you should try to do is to set the thermostat at a high temperature well above room temperature and try to operate the furnace. When the thermostat closes its heating circuit (R–W), the furnace should begin its ignition process and you should hear the combustion blower begin to operate. A furnace that does not operate correctly can have problems in two basic areas: the fuel system and the furnace fan.

If the pilot did not ignite and the main gas valve did not open, you can begin the troubleshooting process by taking the following voltage measurements. Refer to the diagram in Figure 14–21. You could also refer to the fault status light conditions shown in Figure 14–22 if you were actually in front of the furnace and could observe the lights.

Figure 14–21 Electrical diagram of a residential gas furnace with a variable-speed fan. *(Courtesy of Nordyne)*

Status Red Light	Fault Condition
On	Power on
1 Flash	Limit circuit open or external load on "W"
2 Flashes	Pressure switch is open with inducer on
3 Flashes	Pressure switch is closed with inducer off
4 Flashes	Ignition failure (check ground)
5 Flashes	120 VAC & neutral reversed or no ground
Continuous	False flame or gas valve relay shorted
Yellow Flame Light	**Fault Condition**
Continuous flash	Low flame sensor signal
On	Flame present

Figure 14–22 Fault conditions and fault lights.

1. Test terminals R to C at the terminal board of the integrated furnace control for 24 V AC. _____
2. Test terminals W to C at the terminal board of the integrated furnace control for 24 V AC to test if the contacts of the thermostat are closed. _____

3. Test terminal C to terminal 3 of the terminal board of the integrated furnace control for 24 V AC to see if the supply air limit switch, flame rollout switch, and vent safety switches are all closed. _____

4. Test terminal C to terminal 5 at the terminal board of the integrated furnace control for 24 V AC to see if the air pressure switch is closed. _____

5. Test terminals 3 to 6 on the six-pin connector for 120 V AC going to the hot surface igniter (HSI), which is the glow plug. _____

6. Test terminals C to 7 on the terminal board of the integrated furnace control for the sensor signal to the gas valve. _____

7. After the pilot has been established, test terminals 4 to 2 on the terminal board of the integrated furnace control for 24 V AC to the main gas valve. _____

8. Measure the voltage directly across the two terminals of the gas valve where the brown and yellow wires are connected. _____

If voltage is not present at the values indicated at each point, you can begin to determine the problem by identifying the last test that has the proper amount of voltage and the first test point where voltage is lost. For example, if you do not have 24 V at the first test, it indicates that the transformer is not providing the 24 V AC used as the supply of control voltage. This means that you should test for 120 V AC at the black and white wires connecting the transformer primary terminals at L1 and neutral on the integrated furnace control board. If 120 V AC is not present at these terminals, you will need to test for L1 voltage through the door switch. You can do this by connecting one voltmeter terminal to the neutral terminal on the integrated furnace control board and the other terminal to T2 on the six-pin connector. If you do not measure 120 V AC, either the door is not closed correctly or the door switch is faulty. Another possibility is that the main 120 V AC power to the furnace at terminals 5 and 2 on the six-pin connector does not have power, and you will need to check the supply voltage to the furnace to see why it is lost. (For example, the circuit breaker is open or a fuse is blown.)

If you have voltage at test point 1 but not at test point 2, you should suspect that the high-limit switch is faulty. You can turn off power and isolate the switch by removing its wire from terminal 8 and testing it for continuity. If the high limit is open, be sure to check whether a high temperature has occurred or the switch has failed. If a high temperature has occurred, be sure to fix the cause. If the switch is faulty, change the switch. It is important to understand that if the high limit is open because of an overtemperature condition, you will not solve the problem by changing the switch. You must fix the problem that is causing the overtemperature condition.

If you have voltage at test point terminal R but not at test point terminal W, you should suspect the thermostat. Be sure it is set to heating, and that the setpoint is significantly above the room temperature. If you still do not have voltage at test point W, you should remove the thermostat and test the R–W circuit for continuity. You can also visually inspect the thermostat to see if the heat anticipator resistor is burned open. If you cannot get continuity through the thermostat, you should change it and try to start the furnace again.

If you do not have voltage at test point terminal 5, you should suspect that the air flow switch, the flame rollout switch, or the vent safety switch are open. It is important to understand that the air flow switch is measuring the difference of air pressure at two different points in the combustion air circuit. If the combustion fan is not moving sufficient air, the air switch contacts will be open. You must correct any problems that inhibit the proper amount of airflow. It is also important to understand that the small plastic tubing that connects the switch to the two test points in the combustion air circuit may become clogged, so that the switch cannot sample the differential pressure correctly. You should

also visually inspect the intake and the exhaust air pipes to see that they are clear. It is possible for them to be partially clogged with snow, or a small animal or debris may obstruct the pipes.

You should be able to look into the combustion chamber and visually confirm that the pilot flame has been established. If you do not have a pilot flame, you can test for 120 V AC at the hot surface igniter (glow plug) at test point 6, and you should also visually confirm that the glow plug is glowing red. If it has voltage and is not glowing, you should remove voltage from the igniter and test it for continuity. If it is open, you will need to change it. If you have 24 V AC at both coils of the gas valve and it will not open, you can suspect that the gas valve is faulty. Remove and replace the gas valve if needed and try to operate the furnace again.

14.18 ## TROUBLESHOOTING THE COMBUSTION BLOWER AND FURNACE FAN

The combustion blower and the furnace fan are both powered by 120 V AC. Each motor is controlled by the contacts of a relay. The furnace fan is a PSC motor, and it is controlled by the fan control circuit board. The speed circuit determines whether the fan runs at high or low speed. The circuit board receives the signal that comes from the R–G circuit in the thermostat and by voltage from terminals G and C on the terminal board of the integrated furnace control.

The combustion blower is also called the "inducer" or induced-blower motor in the wiring diagram. It receives voltage from the six-pin connector at terminals 1 and 4. When you find that the furnace fan will not turn on, you can use steps 1 to 3 to troubleshoot the fault. If you find that the combustion blower will not turn on, you can use steps 1 to 4 of the second set of measurements to troubleshoot it.

Test points for the furnace fan:

1. Terminal L1 to N (black and white wire) for 120 V AC _____
2. Terminal 5–4 on the motor speed terminal board for 120 V AC _____
3. 24 volts will be provided to terminal G to energize the fan circuit and then 120 V AC will be sent to the motor _____

If you do not have 120 V AC at L1 and N (black and white wire) at the furnace, you should test the incoming voltage at the fuse box or circuit breaker. You should also test to see that the door switch is operating correctly. If you have voltage at test point R to C, you should test to see that testing for 24 V AC at terminals G–C energizes the fan circuit.

Test points for combustion blower (inducer motor):

1. Terminal L1 to N for 120 V AC _____
2. Terminal W–C for 24 V AC to ensure the thermostat is calling for heating _____
3. Terminal 1 to terminal 4 on the six-pin connector for 120 volts _____
4. Terminal R–C directly across the terminals on the induced blower for 120 V AC _____

If you do not have voltage at test 1, use the same inspections as for step 1 for the furnace fan. If you have voltage at test point 1 move to step 2.

If the thermostat is not calling for heat, set the thermostat higher to bring in voltage at terminals W–C.

If you get the 24 volts at the W–C terminals, but the combustion inducer blower is not running, you should suspect that the control board is not sending power to the blower motor or the blower motor is bad. To determine which problem you have, check step 3 for voltage from terminal 4–6 on the six-pin terminal connector. If voltage is present at terminals 4–6, you should move on to terminals R–C directly on the motor. If you do not have voltage at terminals 4–6, then you should suspect the board has a problem.

If you have 120-volt power at terminals R–C on the blower and it is not running, then you should suspect the motor is bad. Remove power and isolate the motor leads so that it can be tested for continuity. Replace the motor if it is found to be faulty. After you have found the problem and corrected it, you should operate the furnace through several cycles to ensure that you have fixed the problem.

You can see additional furnace diagrams in Figures 14–23 and 14–24. These diagrams are similar and the components can be tested with these same procedures.

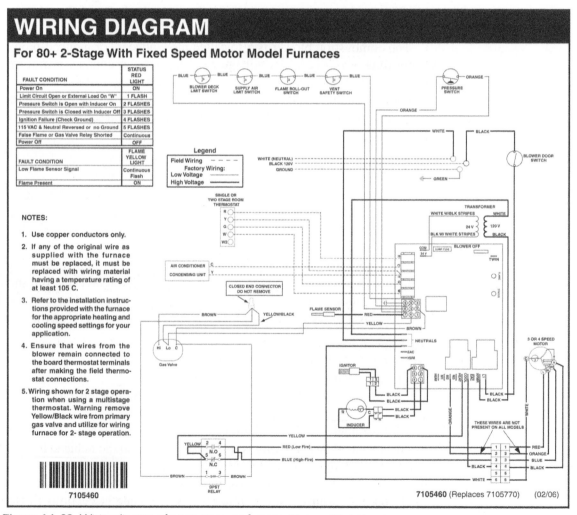

Figure 14–23 Wiring diagram of a two-stage gas furnace. *(Courtesy of Nordyne)*

Figure 14–24 Wiring diagram of a 90+ model gas furnace. *(Courtesy of Nordyne)*

14.19 THE OIL HEATING SYSTEM

The main difference between the oil heating system and the gas heating system is that in the oil heating system oil must be pumped from a storage tank to the nozzle in the end of the **oil burner** under high pressure to ensure that the oil is atomized as it leaves the nozzle. When the oil atomizes, it becomes easy to ignite and burn. Figure 14–25 shows a typical oil furnace. From this diagram you can see the main parts of the oil furnace are the furnace fan, the primary ignition control, the oil pump motor, the transformer, the fan and limit switch, and the high limit.

14.20 NORMAL OPERATION OF THE OIL FURNACE

Figure 14–26 shows a wiring diagram of a typical oil furnace, and Figure 14–27 shows a ladder diagram of the oil furnace. When the oil furnace is operating correctly, it will receive a 24-V AC signal from the thermostat on terminals W–C. This signal will cause the primary control for the oil burner to become energized and will cause the pump to pressurize the oil to the nozzle and cause a continuous high-voltage spark for ignition at

Figure 14–25 Typical oil furnace. *(Courtesy of Carrier)*

Figure 14–26 Wiring diagram of an oil furnace.

Figure 14–27 Ladder diagram of an oil furnace.

the tip of the burner. When the spark ignites the atomized oil, it creates a high-temperature flame in the heat exchanger. The increase in heat causes the fan switch to close and energize the furnace fan. When the thermostat is satisfied, it will open its contacts and de-energize the furnace primary control. If the furnace develops an overtemperature problem, it will cause the high limit to open and de-energize the primary control. Figure 14–28

Figure 14–28 A wiring diagram and schematic (ladder) diagram of an oil furnace. *(Courtesy of Nordyne)*

shows an additional wiring diagram and schematic (ladder) diagram of a different brand of oil furnace. You can see from this diagram that the furnace manufacturer prints both diagrams on the same label and attaches it to the furnace door so you can find it when you are troubleshooting.

14.21 OIL BURNER CONTROLS

Fuel oil furnaces have been used for many years, and the **oil burner controls** have not changed significantly during this time. This means that if you understand the main parts of this system, you will be able to troubleshoot newer and older systems. The oil burner is shown in Figure 14–29a, and the oil burner control is shown in Figure 14–29b.

The main parts of the system are the primary transformer, the igniter electrodes, the cadmium sulfide cell (**CAD cell**), the primary controls, the oil pump motor, and the limit switches that shut down the system in the event of overtemperature. The basic function of these parts is to compress the fuel oil to a high pressure with a pump so that when the oil passes through a small nozzle, it will atomize completely so that it can burn easily. When fuel oil is in a liquid form, it does not burn well. When it is atomized into a vapor, it ignites easily and burns completely. The ignition system provides a continuous spark across a gap that is created by placing two electrodes close to each other at the tip of the pump nozzle, so that the atomized fuel comes into contact with the spark that jumps across the electrodes. The following sections explain some of the basic oil burner controls. Figure 14–30 shows a picture of the stack-mounted **Protectorelay®** control from Honeywell. This control is used to protect the oil burner system. Figure 14–31 shows a burner control from Figure 14–29b mounted on the oil burner along with a typical diagram for the control.

(a) (b)

Figure 14–29 a An oil burner. **b** An oil burner control. *(Courtesy of R. W. Beckett Corporation, Honeywell)*

Figure 14–30 RA117 Protectorelay control. One-piece, stack-mounted oil burner primary controls that cycle the burner on and shuts down the burner on flame loss or system malfunction.
(Courtesy of Honeywell)

Figure 14–31 Honeywell S7184 solid-state oil burner igniter. *(Courtesy of Honeywell)*

14.22 THE IGNITION COMPONENTS

The ignition system uses a step-up transformer. The primary side of the transformer is connected to a 120-V AC source, and the secondary side of the transformer produces a voltage in the range of 10,000 V. The voltage is applied to the electrodes, where it jumps the gap between the electrodes to create a high-voltage spark. The gap in the electrodes is placed directly in the spray pattern of the nozzle so that the spark can ignite the atomized fuel. The control module provides supply voltage whenever the pump is running. Note that the spark from the electrodes will occur at the frequency of the 60-Hz supply voltage.

14.23 THE PRIMARY CONTROL AND CAD CELL

The primary control contains the safety circuit for the oil burner and relays to energize the various parts of the system. An electrical diagram of this control is shown in Figure 14–32. The main problem that occurs with the fuel oil furnace is that the burner

Power supply. Provide disconnect means and overload protection as required.

To use R8184C with line voltage controller. Jumper T-T terminals and connect line voltage thermostat in series with limit controller.

On models with remote alarm indication, remove plug and route violet leadwire through hole in base of R8184.

Remote alarm dry contacts (available on some models) make to turn on alarm when safety switch locks out.

Figure 14–32 Electrical diagram of the R184G flame control. *(Courtesy of Honeywell)*

(a) (b)

Figure 14–33 a A diagram of a cadmium sulfide cell connected to the oil burner relay. **b** Typical cadmium sulfide cell for oil burner. *(Courtesy of Honeywell)*

flame is interrupted, and fuel continues to be pumped into the burner chamber. If flame is established later, the excess fuel that is in the chamber will ignite and burn uncontrolled. This excess fuel may also vaporize, which can cause a dangerous explosion. The cadmium sulfide cell (CAD cell) shown in Figure 14–33 is a sensor whose resistance changes as light strikes its surface. Since the flame of burning fuel oil produces light, the CAD cell can determine whether the flame is burning safely. If the flame diminishes or goes out completely, the CAD cell can detect this condition, and its resistance changes. This change in resistance causes the main burner control to de-energize relays that control voltage to the pump motor. You can see in the electrical diagram that the burner control has a bilateral switch to provide a signal to the base of the triac. The triac is a device that switches AC voltage. If you want more information about the operation of electronic devices, you can refer to Chapter 19. The triac provides a current path through its main terminals when a voltage signal is applied to its base.

14.24 ## THE OPERATIONAL AND SAFETY CIRCUIT

When the thermostat calls for heating, a 24-V AC signal is sent to terminal T on the burner primary control. This 24-V signal originates from the secondary of the control transformer that is inside the primary control. When the thermostat calls for heating, its contacts close, which completes the circuit to the coil of the 1K relay. The 1K relay contacts close to supply 120 V AC to the pump and to the ignition system. When the pump runs, fuel is pressurized so that it comes through the nozzle as a fine mist. When the oil mist comes into contact with the sparks at the electrodes, it ignites and a flame develops. The flame is directed into a ceramic firebox and its heat begins to warm the heat exchanger. The furnace fan blows air across the outside of the heat exchanger into a warm air duct and through registers into the room that is being heated.

The burner control requires a safety circuit to ensure that the flame is established when the pump begins to run and remains burning as long as the pump is running. Figure 14–34 shows the safety circuit highlighted on the diagram. From this diagram you can see that the circuit uses a transformer with a center-tapped secondary winding. The primary side of the transformer is energized by 120 V AC, which also powers the burner motor and furnace fan. The secondary voltage of the transformer is 24 V AC. This voltage is switched on and off through the R–W terminals of the thermostat. When the thermostat calls for heating, the voltage is supplied to the coil of the 1K-relay coil. The contacts of the 1K relay control voltage to the burner motor in the high-voltage side of the circuit and also interlock their own relay coil on the secondary side of the circuit. Voltage is also provided to the heating element of the safety switch SS1 through terminals MT1 and MT2 of the triac. When a flame is established, the light from the flame causes the resistance in the CAD cell to decrease, which changes the voltage that is used to provide a signal to the bilateral switch. Since the bilateral switch provides the pulse for the triac, it turns the triac off when a flame is detected.

△1 Power supply. Provide disconnect means and overload protection as required.

△2 To use R8184C with line voltage controller. Jumper T-T terminals and connect line voltage thermostat in series with limit controller.

△3 On models with remote alarm indication, remove plug and route violet leadwire through hole in base of R8184.

△4 Remote alarm dry contacts (available on some models) make to turn on alarm when safety switch locks out.

Figure 14–34 The current flow through the oil furnace safety circuit when a flame is detected. *(Courtesy of Honeywell)*

Figure 14–35 Current path through the safety control when a flame is not detected and the heating element of the SSI switch is energized. (Courtesy of Honeywell)

When the triac is off, it interrupts the current flow to the heater of the safety switch SS1. Figure 14–35 shows the current path when a flame is detected and everything is normal.

If a flame is not detected, the resistance in the CAD cell remains high and the bilateral switch receives sufficient voltage through R1 and R2 to provide a pulse to the gate of the triac. When the gate of the triac is pulsed, it completes the circuit through the main terminals of the triac, which allows current to flow to the heating element of the SS1. The heat from the element of SS1 causes the normally closed SS1 contacts to open after 15 to 20 s. When SS1 "trips" to the open position, it activates an external button to indicate that the system has tripped. When the contacts of SS1 open, the secondary voltage to the thermostat and the coil of 1K is interrupted. The coil of 1K is de-energized, which opens the 120-V AC circuit to the pump motor. The SS1 safety switch must be manually reset to get the pump to operate again. It is important to check the furnace to determine why loss of flame occurred.

14.25 TROUBLESHOOTING THE COMBUSTION CONTROL OF AN OIL FURNACE

This section explains how to troubleshoot some of the basic oil burner controls shown previously in Figures 14–29 through 14–31. If the oil furnace does not operate correctly, you will need to make several voltage measurements to determine the location of the fault. Make the following voltage measurements in the diagram shown in Figure 14–36 and record the voltage. Figure 14–37 shows another list that is usable for troubleshooting oil furnaces.

1. L1 to N for 120 V AC _____
2. T1 (black wire side) of limit switch inside fan/limit control to N for 120 V AC _____

Figure 14–36 Diagram of the RA117 Protectorelay control.

(Courtesy of Honeywell)

⚠️ POWER SUPPLY, PROVIDE DISCONNECT MEANS AND OVERLOAD PROTECTION AS REQUIRED.

⚠️ IF USING A TWO-WIRE THERMOSTAT, TAPE LOOSE ENDS OF RED WIRE (IF NECESSARY).

⚠️ USE GREEN TERMINAL TO CONNECT CONTROL CASE TO GROUND.

 3. T1 (yellow wire side) of upper limit control to N for 120 V AC _____
 4. T1 (black wire side) of control transformer to N for 120 V AC _____
 5. T1 (black wire side) of burner assembly to N for 120 V AC _____
 6. Terminals R to C on secondary of transformer for 24 V AC _____
 7. Terminals W–C furnace terminal board for 24 V AC _____
 8. Terminals 1–3 on the heating relay for 24 V AC _____
 9. Black wire to white wire on oil burner pump for 120 V AC _____

 If you do not have voltage at L1 to N at test point 1, you do not have incoming voltage to the oil furnace and you need to check the power source for any fuse or circuit breaker that may be tripped or disconnected. If you have voltage at test point 1 but not at test point 2, you should suspect the door switch. Turn power off and test for continuity in the door switch and the two wires connected to it. If you have voltage at test point 2 but not at test point 3, the limit switch parts of the fan/limit control or the two wires connected to it are open. Turn power off and isolate the switch and wires and test them for continuity. If the limit is open, it may be functioning correctly to protect the furnace against an overtemperature condition. If the furnace is overheated, you must find the source of the problem and correct it. For example, the furnace fan may not be operational and it is causing the furnace to overheat. If the furnace is cool and the limit is open, it may be defective and you should change it. If you have voltage at test point 3 but not at test point 4, you should suspect the upper limit control. Again, if the furnace is overheated, this limit switch will be open as well as the limit switch in the fan/limit control. If the furnace is cool, you should suspect the upper limit or the wires connected to it of being open. Turn off power and test them for continuity.

 If you have voltage at test point 4 but not a test point 5, you should suspect the black and white wire that connects the burner assembly to the main furnace controls. Since

Problem	Possible Cause	Remedy
Furnace will not start.	Thermostat not calling for heat.	Check thermostat and adjust. Also, check thermostat for accuracy; if it is a mercury switch type, it might be off level.
	No power to furnace.	Check furnace switch, main electrical panel furnace fuse, or circuit breaker. Also look for any other hand-operated switch, such as an old poorly located furnace switch, which was not removed during furnace replacement.
	Thermostat faulty.	Check reset button on protector relay. Remove thermostat wires from protector relay terminals T T. Place a jumper across T T. If furnace starts, replace thermostat, thermostat subbase (if equipped), or both.
	Protector relay faulty.	Check reset button on protector relay. Remove thermostat wires from protector relay terminals T T. Check for 24 V across T T. If no voltage is present, check for 115 V to protector relay. If 115 V is present, replace protector relay.
	Photo cell wiring shorted or room light leaking into photo cell compartment.	Check photo cell (CAD cell) wiring for short circuits. Also, check for room light leaking into CAD cell compartment. Repair light leak if necessary.
	Open safety switch.	Check for open limit or auxiliary limit, open door switch (if equipped). Also, check internal wiring connections, loose connectors, etc.
Furnace will not start without first pushing protector relay reset button. (Happens on frequent basis)	No fuel oil.	Check fuel oil supply. Check that all hand-operated fuel oil valves are in the open position. Fill oil storage tank if necessary.
	Clogged nozzle.	Replace nozzle with high-quality replacement. Use rating plate.
	Clogged oil filter.	Replace oil tank filter or in-line filter if used.
	Low oil pump pressure.	Connect pressure gauge to oil pump. Adjust pump pressure, or replace oil pump if necessary. Ensure that erratic pressure readings are not caused by defective fuel oil line.
	Air getting into fuel oil lines, or fuel oil line dirty, clogged, or in some manner defective.	Check fuel oil lines. Replace any compression fittings found with high-quality flared fittings. Check for any signs of oil leaks. Any oil leak is a potential source of air or contaminants.
	Defective burner motor.	Check burner motor. If burner motor is cutting out on overload, determine why. Replace if necessary.

Figure 14–37 A troubleshooting procedure for an oil furnace.

Problem	Possible Cause	Corrective Action
Furnace starts, but cuts out requiring manually resetting the oil protector reset button.	Photo cell (CAD cell) defective.	If CAD cell is dirty, clean it. (Determine why CAD cell is getting dirty.) If CAD cell is poorly aimed, realign it. NOTE: The photo cell should have a resistance of 100K Ω in absence of light, a maximum of 1500 Ω in the presence of light. Ensure that room light is not leaking into the CAD cell compartment.
	No fuel oil.	Check fuel oil supply. Check that all hand-operated fuel oil valves are in the open position. Fill oil storage tank if necessary.
	Clogged nozzle.	Replace nozzle with high-quality replacement. Use rating plate.
	Clogged oil filter.	Replace oil tank filter or in-line filter if used.
	Low oil pump pressure.	Connect pressure gauge to oil pump. Adjust pump pressure, or replace oil pump if necessary. Ensure that erratic pressure readings are not caused by defective fuel oil line.
	Air getting into fuel oil lines, or fuel oil line dirty, clogged, or in some manner defective.	Check fuel oil lines. Replace any compression fittings found with high-quality flared fittings. Check for any signs of oil leaks. Any oil leak is a potential source of air or contaminants.
	Defective burner motor.	Check burner motor. If burner motor is cutting out on overload, determine why. Replace if necessary.
	Water or contaminants in oil.	Drain fuel oil storage tank, replace fuel oil. (Consult with fuel oil supplier.)
	Frozen oil line.	Gently warm oil line. Insulate oil line. (Outdoor piping size may require increased diameter.)
Oil burner sputtering at nozzle	Electrodes out of adjustment or defective.	Check electrode settings. Check electrodes for dirt buildup or cracks in porcelain.
	Poor transformer high voltage connections or defective transformer.	Check contacts between transformer and electrodes. If OK, replace transformer.
	Fuel oil filter clogged.	Replace fuel oil storage tank filter and/or fuel oil in-line filter.
	Defective oil pump.	Check burner motor/fuel oil pump coupling. Check oil pump pressure. Replace fuel oil pump if necessary.
	Fuel oil line partially clogged or contains air.	Bleed air from oil line. If problem persists, replace oil line.
Excessive fuel oil consumption.	System temperature rise too high.	System temperature rise should not exceed 85°F. Check for clogged air filters. Check blower fan for excess dirt buildup or debris. Speed up blower fan if necessary.
	Blower fan control out of adjustment (fan stops too soon).	Check fan control settings. The fan control is adjusted with a duct thermometer in the supply air plenum take-off or first few inches of the supply air trunk duct. The "fan off" setting should be 90°–100°F. Once set, the "fan on" setting is normally adjusted 10°–30°F higher than the "fan off" setting.
	Fuel oil leak.	Check fuel oil line for leaks. Repair or replace if necessary.
	Stack temperature too high.	Check stack temperature. Stack temperatures will normally range from 350° to 575°F. Check draft regulator. Draft should be set to 0.02 in. w.c.

Figure 14–37 *(continued)*

these wires are connected to the other wires in the furnace controls with wire nuts, it is possible that you have a loose connection. You can also turn off power and test the wires for continuity. If you do not have voltage at terminal 6, you should suspect the transformer of being faulty. If the voltage at test point 4 indicates that you have voltage at the primary side of the transformer, you should also have 24 V AC on the secondary side, or the transformer may have an open circuit. Turn off power and test the transformer for continuity.

If you have 24 V AC at test point 6 but not at test point 7, you should suspect that the room thermostat is set incorrectly, or it has an open in the heating circuit or the two wires connected to it. If you have voltage at test point 7 but not at test point 8, you should suspect the low-voltage wires between the 24-V terminal board on the furnace and terminals of the heating relay. Turn off power and test these two wires for continuity. If you have voltage at test point 8 but not at test point 9, you should suspect the coil or contacts of the heating relay. You can turn off power and test the coil and contacts of this relay for continuity.

If you have voltage at test point 8 but not at test point 9, you should suspect the wires connected to the heating relay or the contacts of the heating relay. If you have voltage at test point 9 and the pump motor and ignition probes do not provide a spark, you will need to troubleshoot the burner assembly itself. You may need to replace the pump motor or high-voltage transformer that provides the spark. If it is running and the fuel does not ignite, you may have problems with getting fuel to the nozzle and in getting a good spray pattern to ignite properly.

14.26 TROUBLESHOOTING THE OIL FURNACE FAN

If the burner control operates but the furnace fan does not turn on, you will need to troubleshoot the furnace fan. Use the following test points to help you locate the problem, and record the voltage:

1. L1 to N at the voltage terminal board for 120 V AC _____
2. Terminals G–C on the low-voltage terminal board for 24 V AC _____
3. Terminal T2 (brown wire) at the fan relay to N for 120 V AC _____
4. Terminal T2 (red wire) at the fan switch to N for 120 V AC _____
5. Terminals 2–5 on the furnace fan terminal board for 120 V AC _____
6. Terminal T3 (red wire) at the fan relay to N for 120 V AC _____
7. Terminals 1–5 on the furnace fan terminal board for 120 V AC _____

If you do not have voltage at test point 1, use the same troubleshooting procedure as in the first set of tests, since this indicates you have lost incoming voltage. If you have voltage at test point 1, you can continue to test point 2. If you have voltage at test point 2, it indicates that the thermostat is sending a signal to the furnace to operate the furnace fan at high speed. If you do not have voltage at test point 2, the heating relay will allow voltage to pass through its normally closed contacts to the fan switch part of the fan/limit switch. This is considered the heating mode for the fan, and it will run at low speed when the fan switch closes as the furnace heats up. You can continue to follow the voltage that makes the fan run at low speed.

If voltage is available at test point 3, it means that voltage is available to the fan switch. If voltage is not available at test point 4, you must determine if the temperature in the heat exchanger is warm enough to cause the fan switch to close.

If voltage is available at test point 4 but not at test point 5, you should suspect the fan switch or the wires connected to it. If you have voltage at test point 5 and the motor does not run, you should suspect that the motor has an open. Turn off power and test the motor for continuity. You should recognize the motor as a PSC motor. If the thermostat turns on the furnace fan by sending a 24-V AC signal to terminals G–C on the furnace control board, the fan relay should energize and turn the fan on for high speed. This will be used if the furnace fan is used as the evaporator fan when an air-conditioning system is added to the furnace. If you do not have voltage at test point 2, and the furnace fan switch is set to the *on* position, you should suspect the transformer secondary voltage (24 V AC) or the thermostat fan switch or the wires between the thermostat and fan relay (R wire for hot and G wire for fan). If you want the fan to run in high speed, you can omit test points 3 to 6 because these are part of the fan's heating circuit that cause the motor to run at low speed. When the fan switch is on in the thermostat, the furnace fan motor should run at high speed.

If you have voltage at test point 2 but not at test point 7, you should suspect the normally open contacts or the coil of the fan relay. Turn off power and test the coil and the contacts for continuity. You should also test the wires connected to these terminals for continuity. If you have voltage at test point 7, you should suspect the fan motor itself, and use the tests for the PSC-type motor that are found in Chapter 9. If you do not have voltage at test point 7, you should suspect the 120-V AC wire (red wire) between the fan relay and the motor terminal board terminal T1. When you have found the problem, change the parts or wires as needed and then cycle the furnace several times to make sure it is running correctly.

14.27 BOILERS

Boilers are heating systems that circulate water or other fluid. The source of the heat can be natural gas, propane, fuel oil or electric. Figure 14–38 shows a typical gas fired boiler. The operation of the gas fired burner and its controls are very similar to the gas

Figure 14–38 Gas fired boiler.

warm air furnace except it has additional controls for pumping the hot water throughout the system. If the boiler was fuel oil fired, it would use a control similar to the one shown in Figure 14–39. This figure shows an example of the wiring diagram for Honeywell S7184 solid state burner control that is specifically designed for use with boilers. You can see the control is very similar to the one for the warm air furnace, and that it has a place in the circuit for an aquastat controller. If you need to troubleshoot a boiler, you can use the same steps as shown in the previous troubleshooting sequence.

14.28 ELECTRIC HEATING SYSTEM

Perhaps the simplest type of heating system is the electric heating system. **Electric heating coils** can act as the only source of heat for a system, or they can be an add-on for heat pumps or solar heating systems. Figure 14–40 shows a picture of typical electric furnaces. Figure 14–41a shows a typical set of electric heating coils used in an electric furnace, and Figure 14–41b shows a cutaway diagram of an electric furnace. The furnace has a blower that moves air over two separate sets of electric heating coils. You will be able to identify all the parts of the system and the electrical controls from the wiring diagram and ladder diagram. Figure 14–42a shows the wiring diagram. Figure 14–42b shows a picture of a typical electric furnace sequence relay, and Figure 14–43 shows the ladder diagram for the furnace. The wiring diagram will help you locate the parts of the furnace and identify wire colors and terminal points. The ladder diagram shows the sequence of operation of the system. Both

Figure 14–40 Typical electric furnace.
(Courtesy of United Technologies Carrier Corporation)

Figure 14–39 Diagram of the Honeywell S7184 solid-state oil burner control. This control can also be used with an oil burner boiler. *(Courtesy of Honeywell)*

Figure 14–41 **a** Example of electric heating coils for an electric furnace. **b** Cutaway diagram of an electric furnace.

Figure 14–42 **a** Wiring diagram for a typical electric furnace. **b** Picture of electric furnace sequence relay.

diagrams provide a legend showing the symbol and the abbreviation used to identify each of the parts.

Figures 14–44 and 14–45 show two additional wiring diagrams for electric furnaces.

14.29 THE MAIN PARTS OF THE SYSTEM

In the electrical diagram you can see that the main parts of the electric heating system are the electric heating coils, the high-temperature limit control, and the fan control. Two heating contactors called **electric sequencer relays** are used with sequencers to provide a means of staging the heating coils by turning one heating element on at a time. A thermostat is used to control the point at which the heat is turned on or off

Figure 14–43 Ladder diagram showing the sequence of operation for an electric furnace.

Legend

FR Fan Relay
SEQ1 Heating Sequencer 1
SEQ2 Heating Sequencer 2

during the first stage of heat, and the sequencer energizes the second stage on time delay. Some heating coils have a **thermal fuse** that protects the electric furnace for over temperature or overcurrent.

14.30 NORMAL OPERATION OF THE ELECTRIC HEATING SYSTEM

In normal operation the thermostat is mounted on an inside wall in the space that is to be heated, and it is set to a temperature such as 70°. When the temperature inside the space drops below 70°, the thermostat closes the circuit between the R and W contacts, which provides 24 V to the coil of the first heating contactor HC1. When the coil of HC1 is energized, its contacts close and provide 240 V AC single-phase voltage to the first heating coil.

The current flowing through the heating coil produces a large amount of heat, and this heat causes the fan switch to close its contacts and send voltage to the furnace fan. The furnace fan in this diagram is rated for 240 V AC, but in some electric furnaces the fan can be powered with 120 V AC. When the furnace fan is energized,

Figure 14–44 Wiring diagram of electric furnace. (*Courtesy of Nordyne*)

Figure 14–45 Wiring diagram of an air handler with factory installed electric heat. (*Courtesy of Nordyne*)

it turns a squirrel-cage blower that moves air through the ductwork into the controlled space. When the temperature in the controlled space increases to the setpoint and exceeds it, the thermostat opens its contacts and the coil of the heating contactor is de-energized.

If the controlled space continues to get colder and its temperature drops 2° below the setpoint, the thermostat will close a set of second-stage contacts, which will energize heating contactor HC2. The contacts of HC2 will close and provide 240 V AC to a second set of heating coils, which will double the amount of heat provided for the room. As the temperature in the controlled space increases back to the setpoint, the second-stage heating will drop out when the temperature difference at the thermostat becomes less than 2°, and when the temperature reaches the setpoint, both of the heating contactors will be de-energized. In some cases, both heating contactors are controlled by the first stage of the thermostat, which makes the two heating coils act as one larger coil. If the temperature in the furnace becomes too hot, the high-limit control will open its contacts and cause the voltage to the heating contactors to be interrupted. In most cases, the only way the temperature can get too hot is if the furnace fan fails to come on and remove the heat from the furnace. When this occurs, the high-limit control provides protection against overheating.

14.31 TROUBLESHOOTING THE ELECTRIC HEATING FURNACE

If the electric heating system fails to provide heat, you will need to troubleshoot it to locate the source of the fault. We will refer back to the diagrams shown in Figures 14–42 and 14–43. You will see that in this simple system either the wiring diagram or the ladder diagram can be used to trace wires and locate faults. As systems become more complex you will find that the wiring diagram will be more difficult to follow for troubleshooting purposes and that it will be used primarily to locate parts and to identify wire colors and terminals. The ladder diagram will always provide the sequence of operation.

When the system does not work properly, you will need to conduct several voltage tests to determine whether the fault exists in the wiring, in the controls, or in the major load of the circuit. For example, if a wire between the thermostat and the coil of heating contactor HC1 is broken, voltage cannot get to the coil and energize the HC1 coil. Another type of problem can occur if the thermostat is faulty, so voltage cannot pass through its contacts to the coil of HC1. Yet another problem can be caused if the coil of HC1 is open, or if an open develops in the main heating coils themselves. In each of these conditions you can see that the result is the same in that the furnace will not produce any heat, and you will need to trace the circuit and test the wires and components to locate the fault.

The first step you should perform when you are asked to troubleshoot a system is to try and make the system run. You must remember that the person setting the thermostat or controlling the furnace may not understand how to set the thermostat and may not understand that the circuit breaker for the furnace must be turned on for the system to operate properly. By trying to make the system run, you may find that no fault actually exists and that the problem is in the way the furnace is being operated. If the furnace does not run when you try to make it operate, you will need to use a voltmeter to take several measurements. The checklists in the following sections will provide a detailed picture of how far voltage is moving through the system and where the fault exists. You will find by making these voltage measurements that the fault will be in the wires and controls or that the motor or heating coils are faulty.

14.32 TROUBLESHOOTING THE HEATING COILS

Use the following tests to determine whether voltage is reaching the main heating coils. You can also use a clamp-on ammeter to take a current test to determine the amount of current flowing through the main coils. Refer to the diagrams in Figures 14–42 and 14–43.

1. Test at L1 and L2 for 240 V AC _____
2. Test across the primary terminals of transformer _____
3. Test for 24 V AC at R to C (secondary of transformer) _____
4. Test from W to C for 24 V AC _____
5. Test across coil terminals of HC1 for 24 V AC _____
6. Test from T1 on HC1 to L2 for 240 V AC _____
7. Test from T2 on HC1 to L1 for 240 V AC _____
8. Test directly across heating coil for 240 V AC _____

You should write the amount of voltage down for each test point so that you can look at the list and basically determine the location of the fault by determining the last point that has voltage and the first point where voltage is lost. This will be the area where the fault exists.

If the voltage at test 1 is not 240 V, it indicates that the supply voltage to the system has a fault. Locate the source of voltage for the system and check to see that the circuit breakers are not tripped. If you have voltage at test point 1 but not at test point 2, it means that you lost voltage in the wires between the primary terminal of the transformer and the point where voltage comes into the system at L1 and L2. You should notice in the wiring diagram that after voltage is provided to the incoming terminal lugs, it is sent immediately to the L1 and L2 terminals of HC1.

If you have voltage at the first two test points but not at test point 3, you have determined that the transformer is faulty. If no voltage is present at the secondary terminals of the transformer, you can turn power off to the system and remove the transformer and test it for continuity. You can also refer to earlier chapters in this text that explain the operation of the transformer in more detail. If voltage is present at test point 3 but not at test point 4, you should suspect that the thermostat is not set correctly, or that it is faulty. If voltage is present at test point 4 but not at test point 5, you should suspect the wiring between the W and C terminals on the terminal board and the point where the wires connect to the coil of HC1.

Voltage tests at test points 5, 6, and 7 should help you to determine if the coil of HC1 is energized and if voltage is passing through the contacts. If voltage is present at the coil and the contacts do not close, you should suspect that the coil is faulty. You can turn power off to the system and remove HC1 and test the coil for continuity. If the coil causes the contacts to close, but the contacts do not provide voltage at test point 6 or at test point 7, you should suspect faulty contacts in the heating contactor. You can visually inspect these contacts, or turn off power to the system and isolate the contacts so that you can test them for continuity. If you have 240 V AC at test point 8 and the heating coils do not produce any heat, you should suspect that the heating elements are open, and you can turn off all power to the system and test them for continuity. As you locate any faulty parts, you will need to change them out and continue testing the system until it operates correctly.

14.33 TESTING THE FAN MOTOR

If the heating element produces heat but the fan motor does not turn to move the air, you must test the motor circuit for faults to determine if the problem is in the wiring, in the controls, or in the fan motor itself. If the fan does not operate, use the following test points to determine the location of the fault.

1. Test L1 to L2 for 240 V _____
2. Test T1 on fan switch to L2 _____

Note: The temperature must be high enough to cause the contacts of the fan switch to close.

3. Test T2 on fan switch to L2 _____
4. Test across the fan motor _____

You can see that if you have voltage at all four steps, the circuit wires are good and the fan control is working properly, and you should suspect that the problem is in the fan motor. If you have lost voltage at any point along the test, you can suspect the point between the last place at which voltage is measured and where it is lost. Be sure to turn off voltage to the furnace and isolate any components that you test for continuity. It is also important to understand that some electric furnaces use a relay to control the fan in addition to the fan limit. The relay allows smaller voltage and current to be controlled by the fan switch, and the 240 V for the fan is controlled by the relay contacts. You should also remember to test the fan for bearing failure, which may cause the fan motor to draw excessive current.

14.34 ## REVIEW

Heating systems include electric furnaces, gas furnaces, and oil furnaces. Electric furnaces have one or more banks of electric heating coils that are turned on in stages with heating relays. If stages are used, a thermostat with stage control must be used, or sequence control to take advantage of the stages. If sequence control is used, a time delay is built into sequencer 1 to provide a short time delay before stage 2 is brought on line. The electric furnace has a high limit to protect it against overheating.

When you are working in the field you will find gas furnaces with either a standing pilot or an electronic pilot. If the furnace has a standing pilot, it will remain burning continuously during the heating season. If the furnace has an electronic pilot, it becomes energized when the thermostat calls for heating. Once the pilot is safely established, the main gas valve is allowed to open to ignite the main burner. The gas furnace has a high-limit switch to protect against overtemperature. If the furnace is a high-efficiency type that uses a combustion blower, the blower must come on prior to ignition and establish a pressure drop across the combustion chamber that indicates that sufficient air is being exhausted. A pressure switch is used to detect the proper amount of pressure drop.

The oil furnace is different from the gas furnace in that the former must provide a pump to move its fuel to the burner. The burner control provides a relay that controls the pump, and provides current to the igniter. The igniter consists of two tips that are mounted a short distance apart and receive current from an ignition transformer. It is important to position the tips accurately and to ensure no carbon deposits build up between them. The oil furnace has an upper limit switch to protect against overheating.

An oil burner control uses a CAD cell to indicate the presence of a flame in the oil furnace. The CAD cell shuts the oil burner control off if it detects that the flame has failed.

The fan and limit control has a normally open set of contacts that activate to the closed position to cause the furnace fan to run, and it has a set of normally closed contacts that open if the furnace gets too hot.

Job Assignment for This Chapter

You are on a service call to a residence where the homeowner complains that the furnace blows cold air at the end of its cycle. Your job is to identify the problem and explain to the homeowner what tests you will make and what you are trying to prove. If you plan to replace a part or adjust a control, you should be able to explain to the homeowner what you are trying to do to fix the problem.

Solution to the Job Assignment

The fan and limit switch controls the furnace fan motor. There are two settings on the fan switch: the low temperature is where the switch turns off, and the higher temperature is the point where the temperature turns on the furnace fan. If the lower temperature is set too low, the furnace fan will stay on too long and blow cool air at the end of the cycle. Your solution to this problem is to increase the lower temperature on the switch by 2° or 3°, which will cause the furnace to turn off sooner and keep the unit from blowing cold air at the end of the cycle. If this does not fix the problem completely, you may need to increase the temperature setting on the switch a little higher.

Assessment

QUESTIONS

1. Identify the major parts of an electric furnace and explain their operations.
2. Explain what must occur for the heating elements to be energized in an electric furnace.
3. Explain what causes the furnace fan to turn on in an electric furnace.
4. How much voltage do the electric heating elements require in an electric furnace?
5. Identify the major parts of a gas furnace and explain their operations.
6. Identify the components of a direct-spark ignition system and explain their operations.
7. Explain the operation of the direct-spark ignition system.
8. Explain the operation of the hot plate igniter ignition system.
9. Explain the operation of the Protectorelay device on the oil burner.
10. Explain the differences between a standing pilot type furnace and a high-efficiency furnace.

TRUE OR FALSE

1. The hot surface igniter creates sparks that ignite the pilot flame.
2. The gas valve is basically a solenoid valve with a safety circuit to ensure that the pilot light is burning.
3. The oil burner control uses 24 V to provide the spark to the ignition points.

4. The fan and limit switch provides an operational control to energize the fan, and a safety circuit to de-energize the gas valve if the temperature gets too high.

5. The electric furnace uses an electronic direct-spark ignition.

6. The combustion blower in a high-efficiency furnace turns on after the gas valve opens and the main burner flame is established.

7. The spark for the ignition system on the oil furnace turns off after the flame is established.

8. After the flame is established in the gas furnace, the temperature increases until the fan switch is enabled and energizes the furnace fan.

9. The pilot light must be established before the thermostat can energize the gas valve in the gas furnace.

10. The differential air pressure switch in the high-efficiency gas furnace closes when the exhaust fan is moving a sufficient amount of air.

MULTIPLE CHOICE

1. If the main heating coils of an electric furnace are not heating up, you should suspect that
 a. the coil of the fan relay is open or the fan relay contacts are faulty.
 b. the fan switch contacts are not closed or the fan switch is faulty.
 c. the system has an open fuse, the contacts of the first sequencer are not closed, or the main heating element is open.

2. If the main gas valve of a standing pilot gas furnace is not energized and allowing gas to flow to the main burner, you should suspect that
 a. the heat anticipator of the thermostat is set incorrectly.
 b. the pilot flame is not established, the thermocouple is faulty, or the coil in the gas valve is open.
 c. the fan switch is open, the fan is faulty, or the filter may be dirty.

3. The sequence of operation of a high-efficiency furnace is
 a. the thermostat calls for heat, the combustion fan turns on, the pilot is established, the main burner comes on.
 b. the thermostat calls for heat, the pilot is established, the combustion fan turns on, the main burner comes on.
 c. the thermostat calls for heat, the pilot is established, the main burner comes on, the combustion fan turns on.

4. The sequence of operation for an oil burner is
 a. the thermostat calls for heat, the ignition spark is established, the furnace fan comes on, the main flame is established.
 b. the thermostat calls for heat, the ignition spark is established, the main flame is established, the furnace fan comes on.
 c. the thermostat calls for heat, the furnace fan comes on, the ignition spark is established, the main flame is established.

5. The sequence of operation of a standing pilot gas furnace is
 a. the thermostat calls for heat, the pilot is ignited, the main gas valve is energized, the furnace fan comes on.
 b. the thermostat calls for heat, the furnace fan comes on, the pilot light is ignited, the main gas valve comes on.

 c. the pilot light is ignited and remains on continuously, the thermostat calls for heat, the main gas valve is energized, the furnace fan comes on.

6. The direct-spark ignition system
 a. provides a spark to start the pilot flame.
 b. provides a spark to start the main flame.
 c. provides a spark to the thermocouple.

7. The flame rod in the direct-spark ignition system
 a. detects the main flame and pilot flame.
 b. detects only the pilot flame.
 c. detects only the main flame.

PROBLEMS

1. Sketch the electrical diagram of the control circuit that includes the thermostat for a standing pilot gas furnace.

2. Sketch the electrical diagram of the control and load circuit for an oil furnace.

3. Sketch the electrical diagram of the control and load circuit for an electric furnace.

4. Sketch the electrical diagram of the control and load circuit for a high-efficiency gas furnace.

5. Sketch the electrical diagram of the fan and limit control that controls the furnace fan.

Websites

Websites change constantly. Use the key word search to locate sites that have similar content. The products that will be found at each site are indicated with the key word.

York International Corporation
Products: Air conditioners, heat pumps, furnaces
Key word: York International

R.W. Beckett Corporation
Products: Oil burners
Key word: R.W. Beckett

Carrier Corporation
Products: Furnaces, air conditioners
Key word: Carrier Corporation

Bard Furnaces and Air Conditioners
Products: Furnaces, air conditioners, and heat pumps
Key word: Bard Manufacturing

Armstrong Air Conditioners
Products: Air conditioners, furnaces, and heat pumps
Key word: Armstrong Air Conditioners

Heat Controller, Inc.
Products: Century air conditioners, Comfort-Aire air conditioners
Key word: Heat Controller

Johnson Controls
Products: Solid-state relays and controls
Key word: Johnson Controls

Ranco Controls
Products: Solid-state controls
Key words: Ranco Controls

Honeywell Controls
Products: Thermostats and solid-state HVAC controls
Key word: Honeywell Controls

Chapter 15

Controls for Air-Conditioning Systems

Job Assignment for This Chapter

You are sent on a service call to a residence that has no cooling. When you get there you find the indoor (furnace) fan running, but the compressor and condenser fan are not running. The homeowner wants to know what could make the indoor fan motor run and why the outdoor fan and compressor are not running. Your job is to explain what you think could be wrong and what test you would make.

Terms You Will Learn in This Chapter

Air-conditioning system

Compressor

Condenser fan motor

Evaporator fan motor

Packaged air-conditioning system

Split-system air conditioner

Window air conditioner

After reading this chapter, you will be able to:

1. Identify the basic electrical components of an air-conditioning system and explain their operations.

2. Explain why a window-type air conditioner has one fan motor for both the evaporator fan and the condenser fan.

3. Sketch a split-type air-conditioning system and explain which parts are in the outdoor unit and which parts are in the indoor unit.

4. Explain the difference between a split-type air-conditioning system and a packaged-type air-conditioning system.

15.0 OVERVIEW OF AIR-CONDITIONING SYSTEMS

As a technician you will be required to work on a variety of **air-conditioning systems**. These systems will include rather small window-type air conditioners, split-system residential air conditioners that are part of the heating system, and larger commercial package systems. This chapter will help you identify all the components in each system and develop a troubleshooting strategy to help you determine the location of faulty parts quickly. The main point you should remember in this chapter is that all the parts in the different types of air-conditioning systems are very similar to one another. In many cases, components such as **compressors** and fans are just different sizes in the different units, and they may be powered with single-phase voltage in the smaller units and with three-phase voltage in the larger units. You should also remember that you have learned how all the components in the system operate individually, and now you will see how they operate together as a system. You will also get experience reading both the ladder diagram and the wiring diagram for each system. Remember, if you need help in moving from a ladder diagram to the wiring diagram, you can refer to Chapter 6. Note that sometimes only one of the two types of diagrams will be provided, and you may need to sketch the other type of diagram to get sufficient information to understand the system.

15.1 ROOM AIR CONDITIONERS (WINDOW AIR CONDITIONERS)

The simplest type of system to troubleshoot is a **window air conditioner**, which is also called a room air conditioner. Figure 15–1 shows a cutaway picture of a typical window air conditioner. Figure 15–2 shows a wiring diagram of a window air conditioner, and Figure 15–3 shows a ladder diagram of a window air conditioner. Recall that the wiring diagram shows the relative location of each of the components in the system. We will use this diagram to identify all the components in the system.

Figure 15–1 A typical window air conditioner. *(Courtesy of Heat Controller Inc.)*

Figure 15–2 Wiring diagram of a window air conditioner.

Figure 15–3 Schematic (ladder) diagram of a window air conditioner.

15.2 THE MAIN COMPONENTS OF THE SYSTEM

You can see in the electrical diagrams that the window air conditioner is powered by 120 V AC and has a single-phase compressor that is wired as a PSC motor. It also has a single fan with a double shaft that has the evaporator squirrel-cage blower mounted on one end and a blade fan mounted on the other end for the condenser fan. This fan is a three-speed PSC motor. You can review the troubleshooting operation of these motors in previous chapters.

The controls for this system include the fan switch, which controls the speed of the fan and whether the compressor runs with the system or the fan runs alone. The compressor is turned on and off with the thermostat connected in series with it. The other components in this system include the external overload and the run capacitor for the compressor.

15.3 NORMAL OPERATION

In normal operation, the compressor will run and pump refrigerant through the system whenever the contacts in the thermostat close and send it power. Since the evaporator fan and condenser fan are the same motor, it is called the *fan motor*, and it will run anytime the compressor is energized. The fan motor can also run when the compressor is turned off. This means that the system is fairly simple to understand, since the controls only have to supply power to the compressor and fan motor to get them to run. If voltage cannot get to either motor, or if a motor is faulty, the system will not run correctly. When this occurs, you can use either the wiring diagram or the ladder diagram to troubleshoot the circuit.

You can see that it is more difficult to trace the route of voltage to each component in the wiring diagram than in the ladder diagram, so we will use the ladder diagram to trace voltage to each system. In the ladder diagram you can see that all the L1 voltage comes from the 120 V AC plug and directly goes into the main control switch. When this switch is set to the fan only position, the fan receives power and the compressor does not. This allows the fan to circulate air in the room for a moderate cooling effect. When the switch is adjusted to the low cooling setting, the fan receives voltage to the low speed, and voltage is sent to the thermostat. If the room is warm enough, the thermostat contacts close and send voltage to the C terminal on the compressor. The S and R terminals are connected directly to the other neutral potential through the run capacitor, so that when voltage comes from the thermostat, the compressor starts and runs. When the room has cooled sufficiently, the thermostat opens and stops the voltage supply to the C terminal of the compressor. It is important to note that some manufacturers connect the C terminal of the compressor to the neutral potential in the circuit, and the R and S terminals to the L1 side of the circuit.

15.4 TROUBLESHOOTING THE SYSTEM

At times you will encounter a window air conditioner that does not run correctly when power is applied. You will need to determine the exact nature of the fault in order to correct the problem and get the system running again. You will be able to use some of the same troubleshooting tests you used previously to locate faults. The first test that you should always make when you start to work on a system is to try to make the system run. This sounds rather simple, but you must remember that the customer may not be familiar with the terminology on the operating switches on the front of the unit and may not know that the system must be plugged into a wall outlet, the switch must be set to cooling, and the thermostat must call for cooling for the compressor to run.

The next test you should do is to determine if any motors run or try to start or if power is interrupted completely. If either the compressor or fan runs without the other, you at least know that the system has some voltage and the likely problem is in the wiring or in the components inside the window air conditioner. If both the compressor or fan fail to run when power is applied, the next test should be for supply voltage. You must use a voltmeter and measure the exact voltage available at the wall outlet. This is also a good time to feel the power cord and wall outlet to determine if they are warm or hot, which could cause a fire. You must remember that some older homes are not wired for the larger current required by the window air conditioner. If either the power cord or the wall outlet is hot, you must ensure that it is at room temperature when you have completed your repairs. If it continues to be warm, you will need to advise the homeowner of

the problem. Remember that if the unit is trying to start and draws large amounts of current, the wall outlet may be warm until you repair the problem.

15.5 TESTING THE FAN MOTOR

One problem that can occur with the system is that the fan motor or the circuit that supplies voltage to it is faulty. The simplest way to test for a faulty motor or the loss of voltage is to place one terminal of the voltmeter at the locations identified here and record the voltage at each point. If you have voltage at the motor, you must suspect that the motor is faulty, and if you do not have voltage, you must suspect an open in the circuit that supplies voltage to the motor. Refer to Figure 15-2.

1. Terminal 1 of the fan switch to N _____
2. Terminal Low of the fan switch to N _____
3. Terminal Low of the fan to N _____
4. Terminal Low to N directly on the fan _____

If you have 120 V AC at each of these tests, the circuit is complete and the controls that supply voltage to the motor are good, and the motor should run. Be sure that the fan blades can rotate freely. If the blades do not turn freely, you may need to lubricate the fan bearing. If the motor does not run, you can change the motor at this time or use additional troubleshooting tests found in Chapter 9 and determine which part of the motor is faulty, but you will still need to replace the motor.

If voltage is present at the supply but not at the motor at reading 4, voltage has been interrupted at the switch or in the wiring, and the measurements you have for tests 1, 2, and 3 will locate the fault. If you have full voltage at test 1 but not at test 2, the switch is faulty. If you have full voltage at test 1 and test 2, but not at test 3, the wire between the switch and the motor or between the motor and C is faulty. Make repairs as needed and try to start the system.

If the fan runs on one speed but not on others, you can use a similar test to determine if the problem is in the switch, the motor, or the wire between the switch and motor. Repeat the test with the fan switch set for high or medium speed and place the voltmeter probes on the high- and low-speed terminals instead of on the low-speed terminals.

15.6 TESTING THE COMPRESSOR

You can use these troubleshooting tests or the ones found in Chapter 10 to locate faults. When you test the compressor, you should break the test into two parts: Does the compressor try to start and make a noise, or is it completely quiet? If the compressor tries to start, it indicates that voltage is making its way to the compressor, and the fault is in the starting winding or the capacitor, and you can use the troubleshooting procedure provided in Chapter 10 for permanent split-capacitor (PSC) compressors.

If the compressor is not trying to start and it does not make any noise, it means that voltage to the compressor has been interrupted by the controls or an open in the motor itself. Use the following voltage tests to identify the location of the open circuit. Refer to Figure 15-3.

1. L1 in switch to N _____
2. Output terminal in switch to N _____

3. Incoming terminal of thermostat to N _____
4. Outgoing terminal of thermostat to N _____
5. Incoming terminal of overload to N _____
6. Outgoing terminal of overload to N _____
7. Incoming terminal A of capacitor to C of compressor _____
8. Outgoing terminal B of capacitor to C of compressor _____
9. Terminal R of compressor to C of compressor _____
10. Terminal S of compressor to C of compressor _____

If the circuit is providing voltage to the compressor correctly, you should have voltage at each of the 11 points listed. If you lose the voltage at any of the points, you can back up one test and the open circuit is between the last place you have voltage and the point where voltage is lost. This is a very simple testing procedure that will help you determine where the open in the circuit has occurred. If the open occurs in a component such as the thermostat, you may need to return to the chapter that explains the operation and troubleshooting of the individual component. If you have voltage at test points 9, 10, and 11, the circuit is operating correctly and the problem is in the compressor motor itself. Be sure to consult Chapter 10 about the PSC compressor motor for more troubleshooting details.

15.7 SPLIT-SYSTEM RESIDENTIAL AIR-CONDITIONING SYSTEMS

The split-system residential air-conditioning system adds a condenser, which includes the compressor and condenser fan to the existing furnace. Figure 15–4 shows a picture of the components of a typical **split-system air conditioner**. Figure 15–4a shows the

| (a) | (b) | (c) |

Figure 15–4 a The condenser unit is part of a split-system residential air conditioner and is located outside the house. **b** The furnace is the part of the split-system residential air conditioner that is located inside the house. **c** A cutaway picture of the condenser unit.

(Courtesy of Armstrong Air Conditioning Inc.)

condenser that is outside the house and contains the compressor, condenser fan, and controls. Figure 15–4b shows the furnace part of the system that is inside the house and has the evaporator fan and the transformer to provide control voltage to the coil of both the evaporator fan relay and compressor contactor. Figure 15–4c shows a cutaway picture of the condenser. The thermostat that controls the system is mounted on an interior wall in the house.

15.8 THE MAIN COMPONENTS OF THE SYSTEM

Figure 15–5 shows a wiring diagram of the split-system air conditioner, and Figure 15–6 shows a ladder diagram of the system. You can see in the electrical diagrams that the condensing unit is powered by 240 V AC, and the control circuit for the system is powered by 24 V AC. The top part of the ladder diagram shows the components that are mounted in the condensing unit. The system uses a single-phase compressor that is wired as a PSC motor. You can also see that a current relay can be used with a start capacitor if the compressor requires additional starting torque. (Note: The diagram shows where the start capacitor and relay go if they are used, otherwise the compressor is connected as a PSC motor.) The **condenser fan motor** is also powered with 240 V and is also wired as a PSC motor. You can review the troubleshooting operation of PSC open motors and compressor motors in previous chapters. Since the compressor for this unit sits outdoors, it has a crankcase heater to keep the compressor pump warm so that oil does not migrate into the evaporator coil inside the house.

The main control for this system is the compressor contactor whose contacts close to provide voltage to the compressor. The 24-V coil of the compressor contactor is connected to the low-voltage control circuit, where it is energized by the thermostat connected in series with it. Other components in this system include the external overload

Figure 15–5 A wiring diagram for the outdoor section of a split-system air-conditioning unit with the evaporator fan in the furnace, and the compressor and condenser fan in the condensing unit outside the house.

R Run Winding
S Start Winding
C Common
PR Potential Relay
SC Start Capacitor
RC Run Capacitor
OL Overload

Figure 15–6 Ladder diagram for the outdoor section of split-system air-conditioning unit.

and run capacitor for the compressor. Figure 15–7 shows a wiring diagram and schematic (ladder) diagram that you would find posted on the door of the outdoor unit. You can see the diagram is similar to the ones in Figures 15–5 and 15–6.

15.9 NORMAL OPERATION

In normal operation the temperature in the room increases above the setting on the thermostat and the contacts in the thermostat close and energize the coil of the compressor contactor and cause its contacts to close. When the compressor contacts close, the compressor motor runs and pumps refrigerant through the system. The condenser fan is also connected to the contactor and it runs whenever the compressor runs. The evaporator fan is mounted in the furnace and is controlled by a fan relay. As you know, the thermostat can be set to run the fan continuously or in the auto position to run only when the compressor is energized. This system, like the window air conditioner, is fairly simple to understand, since the controls only have to supply power to each of the motors to get them to run. If voltage cannot get to a motor or if a motor is faulty, the system will not

Split System Air Conditioner (Outdoor Section)

Single Phase

NOTES:

1. Disconnect all power before servicing.
2. For supply connections use copper conductors only.
3. Not suitable on systems that exceed 150 volts to ground.
4. For replacement wires use conductors suitable for 105° C.
5. For ampacities and overcurrent protection, see unit rating plate.
6. Connect to 24 vac/40va/class 2 circuit. See furnace/air handler installation instructions for control circuit and optional relay/transformer kits.
7. Anti-Short Cycle Timer (ASCT) may or may not be installed in the unit. If desired, ASCT is factory installed on select models only or may be field installed as shown using manufacturer's approved kit. If not present, connect Yellow and Black wires per Note 6.

1. Couper le courant avant de faire le'entretien.
2. Employez uniquement des conducteurs en cuivre.
3. Ne convient pas aux installations de plus de 150 volt a la terre.

7103880

CC - Contactor Coil
CCH - Crankcase Heater
HPS-High Pressure Switch

Legend

Field Wiring	
Factory Wiring:	
Low Voltage	
High Voltage	

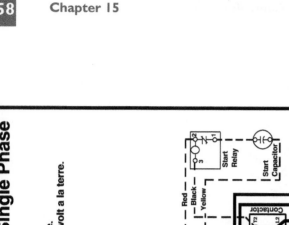

Figure 15–7 Wiring diagram and schematic (ladder) diagram for the outdoor section of a split-system air conditioner. (*Courtesy of Nordyne*)

operate correctly. When this occurs, you can use either the wiring diagram or the ladder diagram to troubleshoot the circuit.

15.10 USING THE WIRING DIAGRAM TO TROUBLESHOOT

The wiring diagram is provided in Figure 15–5 and you can use it to locate the parts of the system. You can see that because the diagram of this system has several additional components and more wires than the simple window air conditioner, the wiring diagram is more difficult to use to trace the circuit. The wiring diagram is better to use to locate the components, identify terminals, and locate the points where actual wire connections are made. For example, you can see that the supply voltage for the system is actually connected at the L1 and L2 terminals of the compressor contactor. You can also see that a wire is connected directly to the C terminal on the compressor, and the other end of that wire is connected to the T1 terminal on the compressor contactor. The wiring diagram also shows that the run capacitor for the compressor motor and the fan motor is located in the top right part of the diagram, and its terminals are identified as common (COMM), fan, and hermetic (HERM).

15.11 USING THE LADDER DIAGRAM TO TROUBLESHOOT

Because it is more difficult to trace the route of voltage to each component in the wiring diagram than in the ladder diagram, we will use the ladder diagram to trace voltage to each part of the system. In the ladder diagram you can see that all the supply voltage comes from L1 and L2, which are the terminals for the 240 V AC supply. As you know, this voltage must come from a disconnect switch or a circuit breaker.

When the disconnect or circuit breaker is turned on, 240 V AC is supplied directly to terminals L1 and L2 on the compressor contactor. Since the sump heater is also connected to these same two terminals, it will always draw current whenever power is applied to the system. The ladder diagram clearly shows that both the compressor and condenser fan motors are controlled by the compressor contactor that is located between terminals L1 and T1 on the compressor contactor.

15.12 TROUBLESHOOTING THE SYSTEM

At times you will encounter a split-system air conditioner that does not run correctly when power is applied. You will need to determine the exact nature of the fault in order to correct the problem and get the system running. Again, the first test that you should always make when you start to work on a system is to try to make the system run. Remember, the customer may not be familiar with the terminology on the thermostat and may not know that the disconnect or circuit breaker for the system must be turned on, that the thermostat must be set to cooling, and the temperature in the room must be above the setting on the thermostat for the compressor to run.

The next test you should do is to determine if any motors run or try to start, or if power is interrupted completely. If the compressor and either fan try to start or run, you at least know that the system has some voltage, and the problem likely is in the

wiring or components. If the compressor and both fans fail to run when power is applied, the next test should be to test for supply voltage. You must use a voltmeter and measure the exact voltage available at the L1 and L2 terminals at the compressor contactor.

15.13 TESTING THE CONDENSER FAN MOTOR

One problem that can occur with the system is that the condenser fan motor or the circuit that supplies voltage to it is faulty. The simplest way to test for a faulty motor or the loss of voltage is to place one terminal of the voltmeter at the following locations and record the voltage at each point. If the motor has voltage supplied to it, you should suspect the motor of being faulty. If you do not have voltage to the motor, you must suspect an open in the wiring that supplies voltage to the motor. Remember that you must always use the terminals shown for each reading because they provide a reference for each measurement to the opposite voltage potential. Refer to Figure 15-5 and 15-6 .

1. Terminal T1 of the compressor contactor to L2 _____
2. Terminal C of the motor (black wire) to L2 _____
3. Red incoming terminal on the capacitor to L1 _____
4. Yellow wire (terminal R) on the fan to L1 _____
5. Red wire (terminal S) on the fan to L1 _____

If you have 240 V AC at each of these tests, the circuit is complete, the controls that supply voltage to the motor are good, and the motor should run. If the fan does not run, it may have an open or it may have a mechanical problem. You can remove the wires to the fan and test the fan windings for continuity to see if one of the wires or windings has an open circuit. Since the problem may also be mechanical you should try to turn the fan blades to see if they can rotate freely. If the blades do not turn freely, you may need to lubricate the fan bearing. If the motor does not run, you can change the motor at this time or use additional troubleshooting tests found in Chapter 9 to determine which part of the motor is faulty, but you will still need to replace the motor.

If voltage is present at the supply but is not present at one of the other points, it has been interrupted by an open in the wiring. The measurement that indicates 0 V also indicates the point that has the problem. You can also use a clamp-on ammeter to test for current in the wire for the start winding, run winding, and common. If the wire connected to the start or run winding does not draw current, it is an indication of an open circuit in the wire that supplies voltage to that winding, or the winding itself has an open.

15.14 TESTING THE COMPRESSOR

The test for the compressor in the split system is very similar to the tests for the compressor in the window unit. The main difference is that the compressor in the split system is usually powered with 240 V AC rather than 120 V AC. The compressor motor may also be connected as a PSC motor or CSCR motor depending on

its horsepower. You can use the same two tests as before with the other compressor when you determine whether the compressor tries to start and makes a noise or if it is completely quiet. If the compressor tries to start, it indicates that voltage is making its way to the compressor, and the fault is in the starting winding or the capacitor, and you can use the troubleshooting procedure provided in Chapter 9 for PSC compressors.

If the compressor does not try to start and does not make any noise, it means that voltage to the compressor has been interrupted by the controls or an open in the motor itself. Use the following voltage tests to identify the location of the open circuit:

1. L1 on the compressor contactor to L2 _____
2. T1 on the compressor contactor to L2 _____
3. Terminal C on the compressor to L2 _____
4. Terminal R on the compressor to L2 _____
5. Capacitor (incoming) RED terminal to L1 _____
6. Terminal S on the compressor to L1 _____

If the circuit is providing voltage to the compressor correctly, you should have voltage at each of the six points listed. If you lose the voltage at any of the points, you can back up one test, and the open circuit is between the last place you have voltage and the point where voltage is lost. This is a very simple testing procedure that will help you to determine where the open in the circuit has occurred. If you have voltage at test points 3, 4, 5, and 6, you have determined that the circuit wires are doing their job and they are supplying voltage directly to the motor, and you can suspect that the motor has an open.

If the open occurs between terminals L1 and T1 on the compressor contact, it indicates that the contacts are not closed or they are faulty and are not passing voltage. Be sure to test the two terminals of the coil of the compressor contactor for 24 volts. If 24 volts is present, you must test the coil and the contacts for continuity. Be sure to turn off all power to the system before you make any continuity tests.

If 24 V AC is not present at the coil terminals, you must test the low-voltage control circuit for voltage. Use the following tests (remember you should measure 24 V AC at each of these test points):

1. R and C terminals on the transformer in the furnace _____
2. Terminal Y to C on the terminal board in the furnace _____
3. Terminal Y to C in the condensing unit _____
4. Incoming terminal on the contactor coil to C _____
5. Incoming terminal on the high-pressure switch to Y _____
6. Outgoing terminal of the high-pressure switch to Y _____
7. Across both terminals of the compressor contactor coil _____

Some technicians find it is easier to start at step 7 and work their way back to the voltage source at the transformer. You are trying to locate the point where voltage is lost. If voltage is not present at the R and C terminals of the furnace transformer, be sure that power to the furnace is turned on. If voltage is present on the 120-V primary side of the transformer and not at the secondary, you can disconnect the transformer and test it for continuity.

If you have lost voltage at the thermostat, make sure that it is set so that the room is warmer than the setting. Remember that the R–Y circuit of the thermostat must have continuity for it to pass voltage. If you have voltage inside the house on the furnace terminal board but lose it when it comes outside to the condensing unit terminals, you should suspect broken wires between the furnace and condenser. You should find that this system is a fairly easy way to locate the source of the open circuit.

15.15 COMMERCIAL PACKAGED AIR-CONDITIONING SYSTEMS

The commercial **packaged air-conditioning system** is similar to the split-system air conditioner and the window air conditioner in that it has a compressor, condenser fan motors, and **evaporator fan motor**. It also has a room thermostat that controls when the compressor and evaporator fan should run. The packaged air-conditioning system usually is mounted on the roof of a commercial building, and the conditioned air and return air are both ducted through the roof into the attic, where ducts distribute the air to each part of the building through ceiling registers. Since this type of unit is mounted on a commercial building, it will most likely be powered with three-phase voltage, and it may have an air economizer circuit.

Figures 15-8a and 15-8b show a picture of a typical package-type system. Figure 15–9a shows a wiring diagram, and Figure 15–9b shows a ladder diagram of the electrical system. In the electrical diagram you can see that this type of unit has components similar to those of the previous air-conditioning systems. The major difference in this system is that the unit is powered with 480/575 three-phase V AC. The compressor motor and crankcase heater are connected to this three-phase voltage. The primary winding of a step-down transformer is connected across L1 and L2 of the three-phase voltage, and it drops the voltage to single-phase 240 V AC. The 240 V is used to power the condenser fan and evaporator fan. Since the condenser is powered with different voltage than the compressor, a condenser fan relay is used

(a) (b)

Figure 15–8 a A typical packaged air-conditioning system. This type of system may also be called a rooftop system. **b** A cutaway picture of a packaged air-conditioning system. *(Courtesy of Armstrong Air Conditioning Inc.)*

Connection Diagram
Three Phase
(460V, 575V-3-60)

(a)

Figure I5–9a **a** Wiring diagram for a packaged air-conditioning system. *(Courtesy of Armstrong Air Conditioning Inc.)*

to energize and de-energize the condenser fan. Figure 15–10 shows a wiring diagram and schematic (ladder) diagram that you would find posted in the door of the unit. You can use this diagram or the ones shown in Figure 15–9 when you are troubleshooting.

A primary winding of a second transformer is connected across the terminals of the step-down transformer. This transformer converts 240 V to 24 V AC for the control voltage that is used to power the coil of the contactor relay, the condenser fan relay, and the evaporator fan blower control. The thermostat is used to control the voltage that energizes and de-energizes the coils of these relays. You can use the procedures provided earlier in this chapter to troubleshoot this type of system.

Figure 15–9b continued b Ladder diagram for packaged air-conditioning system. *(Courtesy of Armstrong Air Conditioning Inc.)*

15.16 REVIEW

Air-conditioning systems consist of small window units, split-system residential systems, and packaged systems that are used in residential and commercial applications. The main components in any air-conditioning system include the compressor, the condenser fan, and the evaporator fan. In the window unit, a single motor with a shaft on each end is used to drive the evaporator fan and condenser fan. This means that if the motor becomes faulty, neither fans' blades will turn. The fan for a window unit usually has two or three speeds for user comfort.

In the split system the compressor and condenser fans are located in the unit outside the house and the evaporator fan in the unit inside the house. The control transformer for the system is located in the inside unit, and it provides 24 V to control the fan relay and the compressor relay through the thermostat. The supply voltage for the outdoor unit is controlled through a disconnect that may be fused. If the fuses blow, the indoor unit may operate correctly while the outdoor unit is inoperative. If the indoor unit loses power, neither the indoor nor the outdoor unit will operate.

Figure 15–10 Wiring diagram and schematic (ladder) diagram for the outdoor section of a packaged air conditioner. (*Courtesy of Nordyne*)

In the packaged system the evaporator fan, the condenser fan, and the compressor are in one unit. The power supply to these three devices comes from the same source, so if a fuse is blown, all the components will be inoperable. The packaged unit has a transformer to provide 24-V control that is used with a low-voltage thermostat. A commercial packaged system will be located on the roof, whereas a residential application will be situated outside at ground level.

Job Assignment for This Chapter

You are sent on a service call to a residence that has no cooling. When you get there you find the indoor (furnace) fan running, but the compressor and condenser fan are not running. The homeowner wants to know what could make the indoor fan motor run and why the outdoor fan and compressor are not running. Your job is to explain what you think could be wrong and what test you would make.

Solution to the Job Assignment

The indoor fan is mounted in the furnace and runs off of 120 volts. The outdoor unit has 240 volts supplied to it from a fused disconnect. Since the indoor fan is running and the outdoor fan and compressor are not running, you should start at the disconnect and 240-volt supply for the outdoor unit. It is not likely that both motors are bad at the same time, rather the voltage supplying them is probably interrupted at some point. Your first test should be at the top of the disconnect. If you do not have voltage at the top of the disconnect, you should check the circuit breaker for the outdoor unit in the circuit breaker box and see if it has tripped. The next test should be at the bottom of the fuses. If you do not have voltage at the bottom of the fuses but you did at the input terminals of the disconnect, you should turn off power and test the fuses for continuity to determine which fuse is bad.

If you have voltage at the bottom of the fuses, your next check should be at the terminals on the compressor contactor in the electric box of the outdoor unit. If you have voltage there you should check to see if the coil for the compressor contactor is pulled in. If it is not, you should troubleshoot the cooling thermostat, wiring, and the coil of the compressor contactor.

If the compressor contactor is pulled in, you should check the contacts to see if they are dirty or pitted, and also check the capacitor for both motors.

Assessment

QUESTIONS

1. Explain the function of the electrical components of a window air conditioner.
2. Identify the electrical components of a split-system air conditioner and explain their functions.
3. Explain why a window air conditioner has only one fan motor.
4. Identify the electrical components of a packaged-type air-conditioning system and explain their functions.

5. Describe which electrical components are in the indoor unit and which electrical components are in the outdoor unit of a split-system air-conditioning system.

TRUE OR FALSE

1. The Y terminal on the thermostat is used to control the air-conditioning outdoor system.
2. In the split-system air conditioner the condenser fan can run without the compressor running.
3. The evaporator fan can run without the compressor running.
4. The evaporator fan is located in the outdoor unit in a split-system air-conditioner.
5. The G terminal on the thermostat is used to control the evaporator fan relay.

MULTIPLE CHOICE

1. The evaporator fan for a split-system air conditioner is a
 a. squirrel-cage fan.
 b. blade fan.
 c. shaded-pole fan.

2. If the compressor tries to start but does not make any noise, you should suspect that
 a. the run winding is good and drawing current but the start winding is open.
 b. the main fuse is blown or the compressor is not receiving voltage.
 c. the start winding is good and drawing current but the run winding is open.

3. If the condenser fan in a split system is running and the compressor will not run and is humming, you should suspect that
 a. the main fuse is blown or the main power to the outdoor section is not receiving power.
 b. the indoor temperature is at the thermostat setpoint, so the compressor does not need to run.
 c. either the start or run winding is open or the components connected to the start windings of the condenser fan are malfunctioning.

4. If the evaporator fan in a window unit is operating correctly and little or no air is moving across the condenser fins, you should suspect that
 a. the condenser fan is faulty or not running at all.
 b. the condenser fan blade may be loose or the condenser fins may be covered with dirt, since the condenser and evaporator use the same fan motor.
 c. the fuse to the condenser fan is probably open.

5. If the condenser fan in a packaged-type air-conditioning system is not running and the compressor and evaporator fan are running, you should suspect that
 a. the condenser fan has an open winding or a wire to the fan is open.
 b. one of the main fuses to the packaged unit is open.

c. there is a short in the compressor motor, which is causing low voltage in the condenser motor.

PROBLEMS

1. Sketch the wiring diagram or ladder diagram that includes all the components in a window air conditioner and explain what controls the voltage to each of the motors.

2. Sketch the wiring diagram or ladder diagram that includes all the components in a split-system air conditioner and explain what controls the voltage to each of the motors.

3. Sketch the wiring diagram or ladder diagram that includes all the components in a packaged unit air conditioner and explain what controls the voltage to each of the motors.

4. Identify the steps you would use to troubleshoot the compressor of a split-system air conditioner that does not try to run when power is applied.

5. Identify the steps you would use to troubleshoot the condenser fan of a packaged air conditioner that does not try to run when power is applied.

Websites

Websites change constantly. Use the key word search to locate sites that have similar content. The products that will be found at each site are indicated with the key word.

Carrier Corporation
Products: Furnaces, air conditioners
Key word: Carrier Corporation

Bard Furnaces and Air Conditioners
Products: Furnaces, air conditioners, and heat pumps
Key word: Bard Manufacturing

Armstrong Air Conditioners
Products: Air conditioners, furnaces, and heat pumps
Key word: Armstrong Air Conditioners

York International Corporation
Products: Air conditioners, heat pumps, and furnaces
Key word: York International

Heat Controller, Inc
Products: Century air conditioners, Comfort-Aire air conditioners
Key word: Heat controller

Johnson Controls
Products: Solid-state relays and controls
Key word: Johnson Controls

Ranco Controls
Products: Solid-state controls
Key word: Ranco Controls

Honeywell Controls
Products: Thermostats and solid-state HVAC controls
Key word: Honeywell Controls

Robertshaw Controls
Products: Solid-state HVAC controls
Key word: Robertshaw Controls

Chapter 16

Electrical Control of Heat Pump Systems

Job Assignment for This Chapter

You are called to a residence that has a heat pump as its heating source. The second-stage heating is electrical heat. The customer complaint is that the utility bills seem much higher than last year even though the temperature has not been as cold as last year. The system is still heating the home. Your job is to check out the system and explain to the customer why the system may be costing more to run this year.

Terms You Will Learn in This Chapter

Backup heat
Cooling cycle
Heat pump

Heating cycle
Reversing valve
Second-stage heating

After reading this chapter, you will be able to:

1. Identify the basic parts of a heat pump system when it is in the cooling mode.
2. Explain how the reversing valve changes the system from cooling to heating.
3. Identify the basic parts of a heat pump system when it is in the heating mode.
4. Explain the defrost cycle of a heat pump.
5. Troubleshoot a heat pump system and locate problems in the heating and cooling cycles.

16.0 OVERVIEW OF HEAT PUMP SYSTEMS

A **heat pump** operates like a normal air-conditioning system in the summer to provide cool air to the conditioned space. In the winter a reversing valve is energized to reverse the refrigerant flow so that the indoor coil becomes the condenser. In the heating cycle, the indoor fan moves air across the indoor coil, which moves heat from the coil to the indoor space like a regular furnace. The heat pump gets its name from the way it takes heat from outside air, which may be as cool as 10°F, and pumps it into the indoor coil in the winter, where it is converted to temperatures of 80°F by increasing refrigerant pressures. In the summer it pumps heat from inside the conditioned space to the outdoor coil.

It is sometimes difficult to understand how the heat pump gets heat out of very cold air in the winter until you remember that the heat pump system is a basic refrigeration cycle, and it is possible to get the refrigerant to evaporate in the outside coil even at very low temperatures. When this occurs, the heat that causes the refrigerant to evaporate is absorbed by the refrigerant, and the compressor pumps the refrigerant in vapor form and compresses it into a hot gas. The hot gas is pumped to the indoor coil, where air from the conditioned space is moved over it to condense the refrigerant. The condensing process allows the refrigerant to give up the heat it picked up in the outdoor coil to the indoor air as air is moved over the coil by the furnace fan. The amount of heat that is available at the indoor coil is sufficient to heat the indoor space, but the temperature of the air will not be over 100° as it moves over the indoor coil, so sometimes people inside the conditioned space will complain that it feels as if cool air is coming out of the air registers. Basically the heat pump operates like a regular air conditioner in the summer, and in the winter it reverses its cycle and provides warm air like a furnace. Figure 16–1a shows a picture of the indoor unit for a split-system heat pump. Figure 16–1b shows a picture of the outdoor unit for a split-system heat pump, and Figure 16–2 shows a cutaway picture of the outdoor unit for a split-system heat pump.

The indoor coil acts as the evaporator when the heat pump operates like an air conditioner, and it acts like the condenser for the system in the winter when the refrigerant flow is reversed. A *reversing valve* changes the flow of the refrigerant when the heat pump is switched from cooling mode to heating mode. The heat pump requires a two-circuit thermostat; one of the circuits is for the heating cycle, and the other circuit is for the cooling cycle. The remainder of this chapter will explain several other modes, including the defrost mode, second-stage heating mode, and emergency heating mode, which the heat pump must utilize to operate efficiently. You will need to determine if the heat

Figure 16–1 **a** The indoor unit for a split-system heat pump. **b** The outdoor unit for a split-system heat pump. *(Courtesy of Bard Manufacturing Co.)*

(a) (b)

Figure 16–2 A cutaway picture of the outdoor unit for a split-system heat pump. *(Courtesy of Armstrong Air Conditioning Inc.)*

pump is trying to operate in its heating cycle or in its cooling cycle when it is in these modes so that you will know which part of the system to troubleshoot when it is faulty.

16.1 THE MAIN PARTS OF A SPLIT-SYSTEM HEAT PUMP

The main parts of the heat pump are discussed in relation to the indoor part of the system and the outdoor part of the system. The main parts of the indoor system include the furnace fan and the indoor coil. The indoor section also will have some type of alternative heating source such as an electric heating coil, a gas valve and heat

exchanger, or an oil burner and heat exchanger. The alternative heating source can be used as a second stage of heating if the first stage does not provide enough heat, and it is also used to temper the indoor air when the heat pump is in its defrost cycle. The main parts of the outdoor system include the outdoor coil, the outdoor fan, the compressor, the reversing valve, the defrost timer, and the overpressure controls.

16.2 REVERSING VALVE FOR HEAT PUMPS

The **reversing valve** is a specialized solenoid valve. Figure 16–3 shows a cutaway picture of a reversing valve, and Figures 16–4a and 16–4b show diagrams of a reversing valve in a circuit. The heat pump is basically an air-conditioning system that reverses the flow of its refrigerant in cold weather to route the hot, high-pressure refrigerant gas from the compressor through the indoor coil, where its heat is used to heat the indoor space. In this mode the outdoor coil is used as the evaporator.

In warm weather, the reversing valve switches so that the flow is returned to a normal air-conditioning system in which the compressor pumps the hot, high-pressure gas to the outdoor coil to be condensed. The indoor coil in the normal air-conditioning mode is used to evaporate the liquid refrigerant, which takes heat from the indoor space and causes it to cool.

In the diagram you can see that the indoor coil is on the right side and the outdoor coil is on the left side. The top diagram represents a typical **cooling cycle**. The reversing valve switches to route the refrigerant as a hot, high-pressure gas to the outdoor coil where it condenses and turns to a liquid. The liquid refrigerant in this diagram is represented by the solid line. The liquid refrigerant in the cooling cycle moves through the indoor coil, where the heat in the indoor space is used to cause the liquid to evaporate.

The bottom diagram shows the route of the refrigerant when the reversing valve is switched to the heating cycle. In this diagram you can see that the refrigerant is a hot, high-pressure gas as it comes from the compressor and is directed to the indoor coil. The heat is removed from the hot, high-pressure gas and is used to heat the indoor space. The refrigerant condenses while it is moving through the indoor coil and moves to the outdoor coil, where it evaporates. Again, the path of the liquid refrigerant is shown as a dark line.

Figure 16–3 A reversing valve is a specialized solenoid valve that is used in heat pumps. *(Courtesy of Alco Controls Division, Emerson Electric Company)*

(a) Cooling Cycle

(b) Heating Cycle

Figure 16–4 a A reversing valve is switched so that the heat pump is in its cooling cycle. The indoor space is cooled when the heat pump is in this mode. **b** The reversing valve is switched so that the heat pump is heating the indoor space.

16.3 THE INTERNAL OPERATION OF THE REVERSING VALVE

The reversing valve is a specialized solenoid valve known as a *pilot-operated* valve, which consists of a small solenoid-operated valve that moves a small amount of vapor and liquid to make the larger main valve switch positions. This means that the small solenoid valve switches a small amount of liquid refrigerant and vapor refrigerant directly to the main valve. The main valve is the part of the reversing valve that actually switches to change the direction of the refrigerant flow.

Figures 16–5a and 16–5b show diagrams of the internal operation of the reversing valve. In the top diagram the reversing valve has switched to the cooling cycle, and in the bottom diagram it is switched to the heating cycle. The diagrams for this figure show that the main valve has one port on one side of the valve and three ports on the other side. The single port shown on the top part of the valve diagram is soldered (welded) to the compressor discharge line. The port on the left side of the diagram is soldered to the condenser coil, the middle port is soldered to the compressor suction line, and the port on the right side is connected to the evaporator coil.

The solenoid valve is shown on the right side of the diagrams, and it has three lines connected to it. The solenoid in the reversing valve is the pilot valve. This means that it takes a small amount of refrigerant and directs it to the main valve to cause the main valve to switch positions. The middle line in the solenoid valve is connected to the compressor suction line, where it takes low-pressure refrigerant that is a vapor to use as the pilot fluid. In the top diagram you can see that the reversing valve is switched for cooling, and the solenoid valve moves the pilot fluid to the left side of the main valve, causing the valve to shift to the right. You should remember that even though the pilot fluid comes from the compressor suction line, it will be at 40–70 psi, which is sufficient pressure to cause the main valve to change positions.

Figure 16–5 a Internal diagram of a reversing valve in the cooling cycle. b Internal diagram of a reversing valve in the heating cycle. *(Courtesy of Alco Controls Division, Emerson Electric Company)*

When the main valve is switched to the right, it makes a plumbing connection between the compressor suction line and the indoor coil. A plumbing connection is also made between the compressor discharge line and the outdoor coil. In the bottom diagram the solenoid valve has switched the pilot fluid so that it activates the right side of the main valve, which causes the main reversing valve to switch to the left side. When the main valve is switched to the left side, it makes a plumbing connection between the compressor suction line and the outdoor coil, and another plumbing connection between the compressor discharge line and the indoor coil.

The reason the reversing valve is a pilot-operated valve is that a smaller coil can be used to ensure that the small pilot part of the valve changes positions. When the pilot part of the valve switches, it can accomplish much more work from the advantage of the pressure of the pilot fluid. This ensures that the valve is as small as possible, yet it will switch positions when the electrical coil of the solenoid is energized and de-energized.

16.4 NORMAL OPERATION IN COOLING

When the heat pump is in the cooling mode, the indoor coil acts as the evaporator coil and the outdoor coil acts as the condenser coil. The furnace fan acts as the evaporator fan, and the outdoor fan acts as the condenser fan. The compressor pumps the refrigerant as a hot vapor to the outside coil, where it is condensed and flows as a liquid to the indoor coil. The furnace fan circulates inside air across the indoor coil and causes the refrigerant to evaporate and cool the air. Figure 16–6 shows a diagram of the heat pump system. You can see in each diagram that the control circuit is shown in relationship to the heat pump thermostat. This will make it easier to understand the operation based on the setting of the thermostat.

You can see that the wiring diagram looks very complex for the heat pump system and that the ladder diagram simplifies the sequence of operation somewhat. The problem with the ladder diagram from this manufacturer is that it tries to combine the wiring diagram of the heat pump control module with the ladder diagram. Whenever you try to show part of the diagram as a wiring diagram and part of the same diagram as a ladder diagram, it leads to extreme confusion. Figure 16–7 shows the manufacturer's schematic (ladder) diagram and a wiring diagram, as you would see it posted on the door to the outdoor unit. It is important to understand that the schematic (ladder) diagram in this figure does not try to show the relationship between the location of terminals for the heat pump control module. Rather, it shows where each switch in the heat pump control module is located in the ladder diagram. You should also notice that the components in the furnace

Figure 16–6 A ladder diagram of a split-system heat pump.

Figure 16–8 Simplified ladder diagram of a split-system heat pump. This diagram does not try to show the relationship between the terminals on the heat pump control module.

(indoor unit) and the control circuit are shown separately from the high-voltage circuit at the top of the ladder diagram. Figure 16–8 shows all the circuits of the heat pump. Figure 16–9 shows all the circuits that are energized in the cooling mode.

In normal cooling operation the thermostat circuit R–Y is closed, which sends voltage to terminal Y of the heat pump control module and energizes it. When the heat pump control module is energized, it closes its contacts LRI–B, which provides voltage to energize the coil of the compressor contactor (CR1) in the first line of the ladder diagram. When the contacts of CR1 close, it energizes the compressor and the condenser fan in the outdoor unit. The thermostat also provides voltage to terminal G, which causes the fan relay to be energized. The contacts of the fan relay close and provide voltage to the furnace fan. Voltage from terminal Y of the thermostat also provides voltage to the coil of the defrost relay (CR3), but at this point it does not affect anything until the heating cycle. The remainder of the heat pump circuits are utilized for heating and defrost circuits, so they do not provide any function at this time.

16.5 NORMAL OPERATION IN THE HEATING CYCLE

Figure 16–10 shows the schematic (ladder) diagram of a heat pump when it is in the **heating cycle**. All the components that are energized during this cycle are shown in bold. When the heat pump is in normal operation in the heating cycle, the compressor

Figure 16–9 Diagram of heat pump with cooling mode circuit in bold.

runs and the reversing valve is energized so that the hot gas from the compressor moves to the indoor coil. The furnace fan is energized to blow the indoor air over the indoor coil to heat the air. The outdoor fan is also energized with the compressor to move sufficient outdoor air over the outdoor coil so that heat can be removed from it and moved to the indoor coil. The heating part of the thermostat uses two or three stages to provide sufficient heat. When the conditioned space is 1° to 2° below the setpoint of the thermostat, the stage-one circuit of the thermostat is closed and 24 V AC is sent to the W1 terminal, which is jumpered to the B terminal. This causes the reversing valve and the heat pump module to become energized. The heat pump module causes the CR2 (compressor contactor) to energize. When the contacts of the CR2 contactor close, they provide 240 V AC to the compressor and outdoor fan. During the heating cycle the outdoor coil and outdoor fan act like the evaporator part of the refrigeration cycle. An outdoor thermostat is connected across terminals Y and Y1. In normal operation it provides a circuit until the outdoor temperature drops below 10°, which is the lowest temperature at which the heat pump can efficiently remove heat from the outdoor air. If the temperature drops below this point, it will cost more to run the compressor than it will to get the heat from the electric heating coil.

The heat pump control module also has an outdoor thermostat that is connected to two terminals identified as *sensor*. This sensor is mounted close to the outdoor coil and it is used during the defrost cycle to bring the system out of the defrost cycle when the

Figure 16–10 Heating circuit for heat pump is shown in bold.

temperature reaches above 57°. If the temperature inside the conditioned space drops more than 2° below the setpoint, a second stage of heating is energized. In this system in the second stage of heating the thermostat energizes the W3 terminal, which closes and sends 24 V AC to the CR4 coil. The CR4 contacts close and provide 240 V AC to the electric heating coil. The electric heating coil can provide additional heat, which also causes the temperature of the indoor air to increase several degrees and make the air coming through the warm air vents feel warmer. This also causes the conditioned space to warm up quickly. When the indoor temperature increases to within 2° of the setpoint, the second-stage contacts of the thermostat open and de-energize the CR4 contacts.

16.6 NORMAL OPERATION DURING THE DEFROST CYCLE

Figure 16–11 shows the heat pump when it is in its defrost cycle. The lines of the circuit that are energized during the defrost cycle are shown in bold. If the heat pump runs long enough during the heating cycle, the outdoor coil begins to form a frost over it, which causes the outdoor coil to become less efficient as the frost blocks the air passages through the coil. At a point where the frost completely covers the coil, the temperature drops sufficiently to bring in the defrost circuit. During the defrost cycle the heat pump is switched back to air-conditioning mode, in which the outdoor coil becomes the

Figure 16–11 Defrost cycle of a heat pump is in bold.

condenser, and the compressor pumps hot gas to it. During the defrost cycle the outdoor fan is de-energized so that the outdoor coils can receive maximum heat from the hot gas flowing through the outdoor coil. This heat causes frost to melt from the outdoor coils. When a sufficient amount of frost has melted from the coil, it begins to warm up quickly from the hot gas flowing through it because the outdoor fan is not running. The defrost thermostat senses the increase in temperature and switches the system out of the defrost cycle and back to the heating cycle. Some heat pump systems use a defrost timer or a differential air pressure switch to bring the system into and out of the defrost cycle. If a defrost timer is used, it automatically sends the heat pump into defrost at timed intervals. You should understand that the heat pump must be set up differently for each area of the country. For example, in the southern part of the country, the outdoor temperature routinely gets above 45° every day or two. This temperature is warm enough to help the outdoor coil defrost without using the defrost cycle. In the northern part of the country, the temperature can stay below 32° for several weeks at a time. When this occurs, the outdoor coil will continue to add frost, and a defrost cycle must be used to keep the outdoor coil free of frost so that the heat pump can run efficiently. If a pressure differential switch is used to determine the need for a defrost cycle, it samples the pressure differential that is created as air tries to move across the coil as it becomes blocked by frost. As more frost builds up, the pressure differential of the air moving across the coil becomes larger.

When the heat pump switches to its defrost cycle, the reversing valve is de-energized, so that the system runs in its air-conditioning mode and liquid refrigerant is pumped to the inside coil, where it removes heat from the inside air. The heat is pumped to the outdoor coil, where it melts the frost, but it also causes the system to blow cold air from the indoor vents during the time the unit is in defrost. Since this is not desirable, the second stage of heating is energized during the defrost cycle to provide additional heat to the indoor air. In the circuit, you can see that a jumper is provided from the DH terminal to the CR2 contacts to ensure that the second-stage heating relay coil (CR4) is energized during defrost.

16.7 TWO-STAGE HEAT PUMPS

A two-stage heat pump has a second source of heating beyond the refrigeration cycle. Figure 16–12 shows a wiring diagram and schematic (ladder) diagram that shows the operation of a two-stage heat pump. The **second-stage heating** is also called **backup heat**. For some systems the second stage is electrical heat. This heat source is used anytime the system needs more heat than the heat pump can provide through its refrigeration cycle. The second stage of heat also comes on when the system is in its defrost cycle so the conditioned space does not get cool air in the winter.

The second stage of heat can also be natural gas, LP, or fuel oil depending on availability and costs. The idea of the heat pump is to provide the most cost effective source of heating without losing comfort. The second stage of heating is brought in at a set of conditions that consist of determining the temperature in the home (conditioned space) where the heat pump can no longer provide sufficient heat to keep the room comfortable. In this case the thermostat will automatically have a 2° or 3° differential, where the second stage will automatically come on. This means that if the setpoint on the thermostat is 70°, the second-stage heat will automatically come on when the indoor room temperature reaches 68° or below. The second stage of heat is also brought in anytime the thermostat setpoint is increased more than 2° or 3° higher than the room temperature. You should remind the homeowners that they should be aware to try and keep the heat pump running on the first stage of heating as much as possible, as the second stage will typically be more expensive to operate.

You can quickly determine if the thermostat is on second-stage heat, as it will illuminate a small light or show an icon on the programmable thermostat. You may need to set the system to backup heat or constant second-stage heat if the refrigeration cycle has a problem such as a compressor that has gone bad. The system may also go to constant second-stage heat if the outdoor temperature drops too low. Every heat pump has a temperature called the *balance point* where the temperature gets too cold for the refrigeration cycle to create enough heat to make it economical to run the heat pump refrigeration cycle any longer. The exact temperature of the balance point depends on the efficiency of the heat pump and the heat loss of the house. This means that a well-insulated home in the southern states may seldom have days where the temperature is below its balance point and can run off its refrigeration cycle in its most efficient mode most of the time. A home in northern states may have 2 or 3 months where the temperature is too cold to run the heat pump efficiently, so the backup heat source is used as the only source of heat for weeks at a time.

16.8 THE HEAT PUMP THERMOSTAT

The heat pump thermostat is slightly more complex than a normal heating and cooling thermostat. Figure 16–13 shows a picture of a manual heat pump thermostat, and Figure 16–14 shows a picture of a programmable thermostat. Figure 16–15 shows a wiring diagram of a

Figure 16–12 Wiring diagram and ladder diagram of the outdoor section of a two-stage split-system heat pump. (*Courtesy of Nordyne*)

Figure 16–13 Manual heat pump thermostat.
(Courtesy of Honeywell)

Figure 16–14 A typical programmable heat pump thermostat. This model has a touch screen.
(Courtesy of White-Rodgers)

typical heat pump thermostat. In this diagram you can see that the heat pump has an R terminal, where all the 24-V AC power source originates. The R lead of the transformer is connected directly to the R lead of the thermostat. The system switch determines if the thermostat is set to heating mode or to cooling mode. When the thermostat is set to cooling mode, the reversing valve is energized directly from the R–B terminal. When the thermostat system switch is set to the heating mode, the reversing valve is de-energized. Some heat pumps have the reversing valve energized during the heating cycle rather than in the cooling cycle. If the heat pump has the reversing valve energized during the heating cycle, you would use terminal B. If the reversing valve were powered during the cooling cycle, you would use terminal O. Figure 16–16 shows a table that identifies all the terminals used in a heat pump.

Some thermostats have two stages of heating: W1 and W3. Some units use terminal O instead of W1. The W1 or O circuit powers the heat pump during the first stage of heating. It is connected to the Y1 circuit through internal circuitry in the thermostat so that whenever the circuit between R and W1 is made, 24 V AC is also sent to the Y1 terminal to ensure that the compressor contactor is energized.

The second stage of heating circuit (W3) is used to energize the auxiliary heat. In most cases the auxiliary heat is an electric heating coil that is energized by a heating contactor. Some heat pump thermostats provide three stages, so you will find a W1, W2, and W3. The second and third stages of heating will each be a bank of electric heat. In some parts of the country where natural gas or fuel oil is available, the second stage of heating can be a gas furnace or an oil furnace.

The E terminal in the thermostat provides a signal to the emergency heat circuit. The emergency heat circuit will be utilized if a problem occurs in the outdoor unit and it needs to be shut off. The emergency heat circuit activates the coil of a control relay

Figure 16–15 Internal wiring of a heat pump thermostat. *(Courtesy of White-Rodgers)*

Terminal Designation	Description
L	Heat pump malfunction indicator for systems with malfunction connection
O	Changeover valve for heat pump energized constantly in cooling
B	Changeover valve for heat pump energized constantly in heating
Y	Compressor relay
Y2	Second-stage compressor
W/E	Heat relay/emergency heat relay (stage 1)
W2	Second-stage heat (third-stage heat in HP2)
G	Fan relay
RH	Power for heating
RC	Power for cooling
C	Common wire from secondary side of cooling system transformer or heat only system transformer

Figure 16–16 Terminal designation descriptions for heat pump systems. *(Courtesy of White-Rodgers)*

whose contacts energize the auxiliary heating system directly and de-energizes the compressor and outdoor fan by opening the circuit to the compressor contactor. When the thermostat is switched to emergency heat, an indicator illuminates on the thermostat to remind the person operating the thermostat that the outdoor unit requires service.

The cooling circuit in the thermostat is controlled through the Y1 and Y terminals. When the thermostat is switched to automatic or cooling, power is switched from the R terminal to the Y1 terminal. Power from the Y1 terminal is used to energize the heat pump control module, which closes a set of contacts that provides voltage to energize the compressor contactor coil. The contacts of the compressor contactor provide high voltage to the compressor and outdoor fan.

The Y terminal is provided to allow an outside thermostat to de-energize the heat pump cycle if the outdoor temperature drops below a specified temperature called the *balance point*. The balance point is determined by calculating the cost of electricity versus the cost of the auxiliary heating energy. The outdoor thermostat is set to the temperature at which the efficiency of the heat pump becomes too low to make the heat pump cost effective. Whenever the outdoor temperature is below this temperature, the outdoor thermostat de-energizes the heat pump cycle, and the auxiliary heat is used to provide all the heat for the system. When the outdoor temperature returns above this temperature, the heat pump becomes the primary heating source for the system and the auxiliary heat comes on only during defrost or when the indoor temperature drops more than 2° below the thermostat setpoint.

The thermostat also has a set of contacts called DH, which receives power from the outdoor unit when it goes into the defrost cycle. If the heat pump is operating when the outdoor temperature drops below 40°, the surface of the outdoor coil will be below 32°. This allows any moisture in the air to freeze on the surface of the coil and begin a process of building up and blocking the air flow across the coil. The defrost cycle can be energized by time and temperature or by sensing the pressure of the refrigerant in the outdoor coil, which can be used to calculate the temperature on the surface of the coil. Some brands of heat pumps also use various methods of measuring the pressure drop of air flowing across the coil to help determine when the coil needs to be defrosted.

The G terminal on the thermostat provides voltage to the coil of the indoor fan relay. The G terminal receives power from the R terminal when the fan switch is on. When the fan switch is set to auto, the G terminal receives power from the Y terminal when the thermostat is calling for cooling or heating. The fan relay energizes the fan to run at high speeds because the heat pump needs to move a lot of air, since the temperature of the air is not as warm as the air coming from a gas or oil furnace.

The thermostat also has a C terminal to provide the return potential for the 24-V AC source. The C terminal on the thermostat is connected to the C terminal on the transformer. This provides a connection to the common side of the 24-V circuit for any indicators on the thermostat such as the emergency heat indicator and the second-stage heating indicator if one is used. When you suspect that the thermostat is not operating correctly, you can disconnect power and test each of the circuits for continuity with respect to the R terminal. Be sure the thermostat is set correctly and that the temperature is above or below the setpoint as needed to activate the heating or cooling circuits. This means that during the troubleshooting period you may need to set the thermostat substantially higher or lower than normal.

16.9 PROGRAMMING THE PROGRAMMABLE HEAT PUMP THERMOSTAT

The programming screen for the programmable heat pump thermostat is shown in Figure 16–17. On the screen you can see that the time and day of the week are shown in the upper left. The room temperature and the setpoint temperature is shown in the middle and the top right corner. A + and − button allows you to increase or decrease the setpoint temperature. The fan switch and the condition of the fan setting (auto or on) is

Figure 16–17 Typical home screen display for White-Rodgers heat pump thermostat. *(Courtesy of White-Rodgers)*

shown in the bottom left. The system status, heat, cool, or auto is shown in the system button, and you can change the system to heating, cooling, or off with this screen button. The instructions for programming and making changes to the heat pump thermostat is shown in Figure 16–18. You should use the instructions when setting up a thermostat or making changes. You should explain to customers that this type of thermostat requires a battery and they should change the batteries if the indicator comes on. This type of thermostat has a touch screen rather than any electrical switches or buttons that might wear out. You can simply touch the screen where the buttons (icons) are displayed.

Programming and Configuration Items

1. Displays 🔒 and "**Keypad Lockout**" when in keypad lockout mode.
 Displays 🔒 and "**Temperature Limit**" and "**Keypad Lockout**" when limited range is activated and locked.
 Displays only "**Temperature Limit**" when limited range is activated.

2. Indicates period of day being programmed.

3. RUN SCHEDULE (run program) Button.

4. SET TIME button or HOLD temperature button.

5. Displays "**Change Filter**" when the system has run for the programmed filter time period as a reminder to change or clean your filter. Displays "**Humidifier Pad**" when the system has run for the programmed filter time period as a reminder to change or clean your humidifier pad.

6. COPY button or INSTALLER CONFIG button.

7. CLEAN DISPLAY button allows 30 seconds to wipe off the display or ADVANCE DAY button for programming.

8. Used in programming to set time and in configuration menu to change selections.

9. "**Hold Until**" indicates the time when a temporary hold period will end.

10. "**Hours**" and "**Days**" displays during steps in installer configuration.

11. The words "**Hold At**" are displayed when the thermostat is in the **HOLD** mode. "**Temporary Hold At**" is displayed when the thermostat is in a temporary **HOLD** mode.

12. "**Call For Service**" indicates a fault in the heating/cooling system. It does not indicate a fault in the thermostat.

13. "**Stage 1**" indicates when heating or cooling stage is energized. "**+2**" also indicates when a second stage is energized.

14. "**Copy**" indicates the copy program feature is being used during programming.

Figure 16–18 Programming and configuration items for White-Rodgers heat pump thermostat. *(Courtesy of White-Rodgers)*

16.10 TROUBLESHOOTING THE HEAT PUMP

When you troubleshoot a heat pump system, you must be sure which mode you are trying to make it run in. For example, if it is summer and you are trying to make the system run in the cooling mode, you will need to make only the compressor, outdoor fan, and indoor fan run. If you are trying to make the heat pump run in winter, you will also need to check to ensure that the second stage of heating and the defrost cycle are operating correctly. This means that the customer complaint may be with the second stage of heating and not the heat pump, or the heat pump may be working correctly to provide heat, but it may not be able to go into defrost mode. In all these cases you will need to refer to the diagram and make the correct tests.

16.11 TROUBLESHOOTING THE HEAT PUMP IN THE COOLING MODE

When you are requested to troubleshoot the heat pump in the cooling mode, you should set the thermostat for a temperature that is lower than the room temperature and try to make the system run. If the system does not run, you should make the following voltage tests. Use Figure 16–9 for these tests.

1. L1–L2 for 240/208 V _____
2. R to C on the thermostat or transformer (24 V AC) _____
3. Y to C on heat pump terminal board (24 V AC) _____
4. T1 of the high-pressure switch to C (24 V AC) _____
5. T2 of the high-pressure switch to C (24 V AC) _____
6. Coil of CR2 to C (24 V AC) _____
7. L1 terminal of CR2 contacts to L2 (240 V AC) _____
8. T1 terminal of CR2 contacts to L2 (240 V AC) _____
9. R–C on the compressor (240 V AC) _____
10. S–C on the compressor (240 V AC) _____
11. R–C on the condenser fan (240 V AC) _____
12. S–C on the condenser fan (240 V AC) _____
13. Across the terminals of the reversing valve (24 V AC) _____

If you do not have 240 V AC at L1–L2, you have lost the source voltage for the system, and you should suspect that the circuit breaker is tripped, the fuses are blown, or the disconnect is not on. Locate the source of the problem and correct it. If you have 240 V at L1–L2, but you do not have 24 V AC at terminals R and C inside the unit, you should suspect that the transformer or the wire supplying voltage to it is not operating correctly. You can make a voltage test directly across the terminals of the primary side of the transformer for 240 V AC if it is connected to L1–L2 or 120 V AC if it is connected to L1–N. If you do not have voltage directly across the primary terminals, but you have supply voltage to the unit, you should suspect that one of the wires connecting the transformer to the power supply is open.

If the primary side of the transformer has the proper amount of voltage, but the secondary does not, you should suspect that one of the windings in the transformer is open. Turn power off and test the primary and secondary sides of the transformer for continuity.

If you have voltage at the R–C terminals but do not have voltage at Y–C on the heat pump terminal, you should suspect that the thermostat is not supplying voltage to the heat pump. Check the thermostat to ensure that it is set for cooling and that it is set well below

the room temperature. If the thermostat is set correctly and you still do not have voltage at Y–C, you should suspect that the thermostat or the wire connecting the thermostat to the heat pump is faulty. You can place a jumper across the Y–R terminals in the thermostat to see if the problem is in the thermostat, or you can remove power from the system and test the thermostat cooling circuit for continuity at the R–Y wires at the furnace. Be sure to turn off power and remove the wires from the terminal board so that they are isolated. If you get voltage at terminals Y–C in the heat pump when the jumper is placed on terminals R–Y in the thermostat, you can suspect that the thermostat is faulty and you should replace it. If you still do not have voltage when the jumper is in place, you should suspect the wires between the thermostat and the furnace, or the wires between the furnace and the heat pump. Remember that the wires run from the thermostat and the furnace, and then from the furnace to the outdoor unit. Visually inspect the terminals in the thermostat, furnace, and outdoor unit to ensure that the wires are connected securely. You may have to test each wire involved for continuity until you locate the faulty wire and replace it.

If you have 24 V AC at test point 2 but do not have voltage at test point 4 (terminal T1 of the high-pressure switch and C), you should suspect the wire between the R terminal in the heat pump terminal board, and the high-pressure switch. Turn off power and test this wire for continuity. If the wire is faulty, replace it, and try the system again.

If you have voltage at test point 4 but not at test point 5, you should suspect that the high-pressure switch is open. You must determine if the high-pressure switch is open because the pressure is too high or because the switch is broken. This means that you must put gauges on the system to determine the pressure and compare it with the pressure chart for the refrigerant that is used and the ambient temperature at the coil. If the pressure is actually too high, you should suspect that the outdoor unit is not moving air across its coil correctly, and you should check the outdoor fan to see if it is operating correctly. If the fan is running, the outdoor coil may be plugged with debris such as grass or leaves so that sufficient air cannot pass through it. If the pressure is high, you must correct the problem that is causing it. If the pressure is normal, you should suspect that the switch is faulty, and you can test it and replace it if necessary.

If you have voltage at test point 5 but not at test point 6, you may have a faulty heat pump control module, or the wire between the LRI and B contacts and the coil may be open. The heat pump control module causes contacts LRI–B to close when a 24-V signal is received at terminal Y1 on the module. You determine if the contacts in the module are closed by testing for 24 V AC from the common terminal (C) to the terminal where voltage comes in and then test a second time for voltage on the terminal where the wire is connected to the coil. If you have voltage at the input terminal and not at the output terminal, you can suspect that the contacts are not closed, and the problem is in the module or the signal coming from the Y1 terminal. If you have 24 V AC at both the incoming terminal and the outgoing terminal, the contacts are closed and passing power correctly.

If you do not have voltage at the coil, you can suspect that the wire between the LRI and B contacts and the coil is open. Turn off power and test this wire for continuity and replace the wire if necessary. If you have voltage (24 V AC) across the coil of the compressor contactor (CR2), its contacts should be closed. If its contacts do not close, you should suspect that the contactor is faulty. You can turn off all power and test the coil for continuity. If the contacts close, you should move to test points 7 and 8 to determine if 240 V AC is passing through the contacts. You can also notice if the outdoor fan is running. If the outdoor fan is running, you have determined that the contacts are passing voltage, and you can move on to the next test. If you do not have 240 V AC at test point 7 (L1 to L2), you should suspect a problem with the fuse disconnect or the wires between the disconnect and the terminal box in the outdoor unit. You really should have noticed this loss of voltage with test 1 of this procedure. If you have 240 V AC at test

point 7 but not at test point 8, it indicates that the compressor contacts are not closed or they are malfunctioning. Turn the power off and test the contacts for continuity by physically activating (pressing the contacts closed) the contactor. If you do not have low resistance, remove and replace the contactor.

If you have voltage at test point 8 but not at test point 9, you should suspect the wire between the compressor contactor and the R terminal on the compressor. It is important to note that the compressor is wired as a permanent split-capacitor (PSC) motor, and the terminal connection for this wire may be on the run capacitor where voltage is supplied for both the compressor and the outdoor fan. You can check this wire with a continuity test or by using a jumper wire.

If you have voltage at test point 9, you should hear the compressor trying to start, since voltage at the R terminal should cause current flow through the compressor run winding to terminal C. You can also use a clamp-on ammeter to measure the exact amount of current in this circuit. If the compressor does not make a noise or if no current flow is measured, you should suspect that the compressor has an open in its run winding, its overload is open, or the wire from terminal C on the compressor to L2 may be open. Turn the power off and test each of these parts of the circuit for continuity. If the compressor is "humming" or trying to start, you have determined that there is a complete path from either the R terminal to L2 or the S terminal to L2. This means that the overload is operating correctly and that you have voltage.

If you have voltage at test point 9 but not at test point 10, you should suspect that the capacitor is faulty, since the voltage for test point 10 must come through the run capacitor. You should notice that the capacitor in this diagram is a two-circuit capacitor, which means it has one common terminal. It also has a terminal for the capacitor for the compressor, and a second capacitor on the outdoor fan. You can test the capacitor or you can try a different capacitor to make the system run. If you have voltage at test point 10, you should suspect that the start winding is open. At this point you can confirm if the problem is in the start winding by measuring the current in the start winding circuit. If this winding does have current flow, you should suspect that the start winding is open. Turn off power and test the start winding for continuity.

If the compressor is running and the outdoor fan is not running, you should test the circuit of the outdoor fan. Remember that the outdoor fan is connected as a PSC motor and receives voltage from the same source as the compressor. This means you do not need to test the voltage source if the compressor is running and the fan is not running. If the fan is not running, you should suspect the fan itself, its run capacitor, or the normally closed LRI–1 contacts. These contacts are opened when the system is placed in the defrost mode. If they are open, you will have voltage from R–L2 but not at R–C on the motor. You should suspect that the LRI–1 contacts are open and you can check the heat pump control module for proper operation. If you have voltage at the R–C terminals (test point 11) and at S–C terminals (test point 12), you should suspect that the motor is faulty. Turn off all power and test the motor windings for continuity.

The final test for voltage across the terminals of the reversing valve will need to be made only if the system is producing warm air at the indoor coil and cool air at the outdoor coil. This condition indicates that the reversing valve is not energized. Test for 24 V AC, and if it is present and the valve is in the wrong position, you will need to remove power and test the heat pump reversing valve solenoid coil for continuity. If the voltage is not present, you will need to make several other voltage tests to locate the loss of voltage. Remember, the voltage should come directly from the R terminal to the B terminal when the system switch on the thermostat is set to cooling. When you have completed all tests, and replaced the appropriate parts as necessary, you should cycle the system several times through its cooling cycle to ensure that it is operating correctly.

16.12 TROUBLESHOOTING THE HEAT PUMP IN THE HEATING MODE

When the heat pump is placed in the heating mode, the signal to energize its cycle comes from the W1 terminal of the thermostat. You should notice in the diagram that the W1 circuit is jumpered to the B terminal, and the B terminal controls voltage to the reversing valve. Voltage is also sent to the heat pump control module to energize the CR2 coil, which ensures that the compressor and outdoor fan are powered. Voltage is also sent to the FR1 fan relay coil to ensure that the indoor fan is also powered. Figure 16–10 shows the diagram of the heat pump with the heating circuit highlighted. If you must troubleshoot a heat pump system that will not operate in the heating mode, you can use the previous set of tests to ensure that the compressor, outdoor fan, and indoor fan are running, and you will use the following tests to determine the location of the faults if the reversing valve is not de-energized and if the second stage of heating is not energized.

16.13 TROUBLESHOOTING THE HEAT PUMP IN THE DEFROST MODE

When the heat pump is in its heating cycle and it goes into defrost mode, it is basically changing back to the cooling mode, and the second stage of heating is turned on. When this occurs, the reversing valve is powered again to cause the refrigeration cycle to switch. The types of problems you will get with the heat pump defrost cycle are that it will not go into defrost, it will not come out of the cycle correctly, or it will not provide backup heat during the defrost cycle. If the system does not go into defrost often enough, you can suspect that the defrost timer is set incorrectly. Recall that the defrost timer uses a clock-type timer to set the amount of time between defrost cycles. This is activated by the outdoor thermostat, which indicates that it is cold enough to cause frost to form. If the system does not go into defrost, you should test for voltage through the circuit that has the defrost relay coil. If the defrost cycle does not come out of the defrost cycle correctly, you can suspect that the defrost timer is faulty if the cycle is terminated by time, or the thermostat for the defrost cycle is faulty if the cycle is terminated by temperature.

If the second stage of heating does not energize when the system is in defrost, you should test for voltage across the second-stage heating coil for 24 V AC. If the coil has 24 V AC, and the second stage of heating is not energized, you can use the test found in the previous section. If the coil of the heating relay does not have 24 V AC, you should suspect the contacts of the defrost relay. You can test for a voltage loss in this circuit and test for continuity when you find the suspect parts or wire. Figure 16–11 shows an electrical diagram of the heat pump in the defrost mode.

16.14 PACKAGE-TYPE HEAT PUMPS

Figure 16–19 shows an example of a package-type heat pump, which is similar to the split-system heat pump except that in the package system the indoor fan is in the same cabinet as the outdoor equipment. In the package system the wiring between the indoor and outdoor sections is all completed through a single terminal board, since both sections are in the same cabinet. The other main difference is that all the fan motors in the package system are powered with 240 V AC, whereas the indoor fan in a split system is usually powered with 120 V AC. You can use the troubleshooting tests for cooling and heating cycles found in the previous sections to find similar problems in a package

Figure 16–19 A package-type heat pump. *(Courtesy of Bard Manufacturing Co.)*

system. Figures 16–20 and 16–21 show the electrical diagram of a package system heat pump.

REVIEW

A heat pump can be a split system or a package system. The main feature of the heat pump is that it has a reversing valve that allows the refrigerant flow to be reversed so that the refrigerant flows from the compressor directly to the indoor coil when the system is set for heating. This makes the indoor coil act like a condenser and give up heat. In this configuration the outdoor coil acts like an evaporator and absorbs heat. Since the outdoor temperature is usually cool during the heating season, the outdoor coil will frost over, so it periodically needs a defrost cycle to keep the system running efficiently.

In normal operation during the summer, the reversing valve directs refrigerant flow so that the system operates exactly like a typical air conditioner, in which the indoor coil is the evaporator and the outdoor coil is the condenser.

Another feature of the heat pump is the electric heating elements that are used to add heat when the outdoor temperature gets too cold or when the system goes into defrost mode. The electric heat can be staged to add only the amount of heat necessary. The system is most efficient when it is running in heat pump mode. Its efficiency begins to fall when it adds electric heat. The electric heat is also called the emergency heat because it is sufficient to heat the entire space if the compressor or any other part of the heat pump becomes defective, or if the outdoor temperature drops below the point where it produces enough heat for the system to heat the space effectively. The secondary heat in some systems is a gas furnace or an oil burner. The type of fuel used will depend on its price.

The heat pump thermostat has several extra terminals. The O terminal is used to control the reversing valve directly. Terminal L is used to control emergency heat. Terminal Y controls the compressor, and terminal G controls the indoor fan. The other major feature of the heat pump thermostat is that it has a small lamp that indicates when the backup heat is running.

When the heat pump goes into the defrost mode, it must use a timer or other sensor such as a temperature sensor to bring the heat pump out of the defrost cycle. Since the system is least efficient during the time it is running in defrost, it is important to limit the defrost time.

Figure 16–20 Electrical diagram of a package system heat pump. (*Courtesy of Bard Manufacturing Co.*)

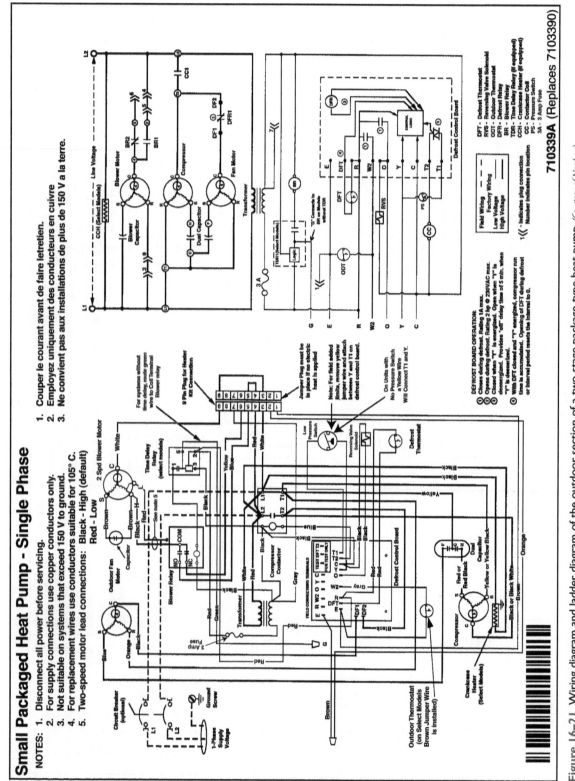

Figure 16–21 Wiring diagram and ladder diagram of the outdoor section of a two-stage package-type heat pump. *(Courtesy of Nordyne)*

Job Assignment for This Chapter

You are called to a residence that has a heat pump as its heating source. The second-stage heating is electrical heat. The customer complaint is that the utility bills seem much higher than last year even though the temperature has not been as cold as last year. The system is still heating the home. Your job is to check out the system and explain to the customer why the system may be costing more to run this year.

Solution to the Job Assignment

Since the customer complaint is that the utility bills seem higher than last year and the system is still heating the house, you should check to see if the system is running on the second-stage (or backup) heat more frequently instead of using the heat pump refrigeration cycle.

The possible causes of are a faulty thermostat, a refrigeration cycle that is low on refrigerant or other similar refrigerant cycle problem, a problem with the heating sequence relay which is keeping the electric heat on too much, the system going into defrost too often, or the outdoor balance point thermostat set too high.

You can start your tests by turning the thermostat 1° to 2° above the room temperature and seeing if the second stage of heat comes on. The second-stage heat should not come on until the heat pump refrigeration system provides as much heat as possible. Remember the refrigeration cycle of the heat pump should be the more cost effective system to run than the second-stage heat.

You can also check the heat pump cycle by itself by turning off the circuit breaker to the electric heating coils or ensuring that the heating relay does not become energized. If the heat pump cycle does not produce sufficient heat, you should check the system for the proper amount of charge and determine if the outdoor (balance point) thermostat is set to the correct temperature. If it is set too warm, the heat pump will turn off before the temperature gets too cold, and this will cause the system to run on backup heat more often.

You can check the defrost cycle by observing what brings the heat pump into the defrost cycle and ensuring that the system actually needs a defrost cycle. You can manually force the system into defrost and observe the system operation. If all these things check out, you can show the customer how to keep the system in first-stage heating as much as possible to make sure the system is running efficiently.

Assessment

QUESTIONS

1. List the parts of a heat pump when it is in the cooling cycle.
2. List the parts of a heat pump when it is in the heating cycle.
3. Explain how the defrost cycle starts and ends in the heat pump. Be sure to include the condition that must occur to get the heat pump to go into the defrost cycle.
4. Identify the ways a heat pump can provide second-stage heating.
5. Explain the difference between a package-type heat pump and a split-system heat pump.

TRUE OR FALSE

1. When the heat pump is in the cooling cycle, the indoor coil will be cool.
2. When the heat pump is in the defrost cycle, the outdoor coil will be cool.
3. When the heat pump is in the defrost cycle, the auxiliary heat will be on.
4. If the reversing valve has an open in its coil and it becomes inoperative, the heat pump will remain in the heating cycle.
5. When the heat pump is in the defrost cycle, the outdoor fan will be de-energized, and the indoor fan will be running.

MULTIPLE CHOICE

1. When the heat pump is in the cooling cycle, _____ will be warm.
 a. the indoor coil
 b. the outdoor coil
 c. both coils

2. When the heat pump is in the defrost cycle, _____ will be warm.
 a. the indoor coil
 b. the outdoor coil
 c. both coils

3. When the heat pump is in the emergency heat cycle, the
 a. compressor will be turned off and the auxiliary heat will be turned on.
 b. outdoor section will be energized, and the indoor section will be turned off.
 c. reversing valve must be in the cooling mode.

4. The auxiliary heat will be energized
 a. only when the heat pump is in the defrost mode.
 b. only when the heat pump is in the heating mode and second-stage heating is needed.
 c. when the heat pump is in the defrost mode or when it needs additional second-stage heating in the heating mode.

5. The balance point is
 a. the outdoor temperature at which it is no longer economical to operate the heat pump in the heating cycle, and the auxiliary heat is used when the temperature is below this point.
 b. the indoor temperature at which it is no longer economical to operate the heat pump in the heating cycle, and the auxiliary heat is used when the temperature is below this point.
 c. the comparison of the indoor temperature and the outdoor temperature at which the heat pump is the most efficient.

PROBLEMS

1. Use the diagrams found in this chapter to sketch an electrical diagram of the components that are energized when the heat pump is in the cooling mode.
2. Use the diagrams found in this chapter to sketch an electrical diagram of the components that are energized when the heat pump is in the heating mode.

3. Use the diagrams found in this chapter to sketch an electrical diagram of the components that are energized when the heat pump is in the defrost mode.

4. Use the diagrams found in this chapter to sketch an electrical diagram of the heat pump thermostat, and explain which part is energized during the heating cycle, cooling cycle, defrost cycle, and emergency heat cycle.

5. Use the diagrams found in this chapter to sketch an electrical diagram of the heat pump reversing valve, and indicate when it is energized or de-energized during the heating cycle, cooling cycle, defrost cycle, and emergency heat cycle.

6. Explain several problems that could occur that may make the heat pump run less efficiently.

Websites

Websites change constantly. Use the key word search to locate sites that have similar content. The products that will be found at each site are indicated with the key word.

Carrier Corporation
Products: Furnaces, air conditioners
Key word: Carrier Corporation

Bard Furnaces and Air Conditioners
Products: Furnaces, air conditioners, and heat pumps
Key word: Bard Manufacturing

Armstrong Air Conditioners
Products: Air conditioners, furnaces, and heat pumps
Key word: Armstrong Air Conditioners

York International Corporation
Products: Air conditioners, heat pumps, and furnaces
Key word: York International

Heat Controller, Inc.
Products: Century air conditioners, Comfort-Aire air conditioners
Key word: Heat Controller

Johnson Controls
Products: Solid-state relays and controls
Key word: Johnson Controls

Ranco Controls
Products: Solid-state controls
Key word: Ranco Controls

Honeywell Controls
Products: Thermostats and solid-state HVAC controls
Key word: Honeywell Controls

Robertshaw Controls
Products: Solid-state HVAC controls
Key word: Robertshaw Controls

Chapter 17

Electrical Control
of Refrigeration Systems

Job Assignment for This Chapter

You are called to a convenience store that has a reach-in freezer case. The customer complaint is that the temperature in the case is getting too warm and the product is not staying frozen. The customer wants you to take a look at the system and give an estimate of what is wrong and what it will take to repair it before you begin your repair. Your initial check indicates that the evaporator has an excessive amount of frost on it. Your job is to look at the system and identify possible problems and report back to the store owner your estimate of what needs to be done.

Problems with refrigeration systems tend to be complex and be a combination of refrigeration system problems as well as electrical problems. Since this text covers more electrical problems, you should consider this problem as a pure electrical problem when you create your solution.

Terms You Will Learn in This Chapter

Condenser
Defrost control
Door heater
Electronic defrost control
Electronic refrigeration control
Evaporator
Freezer display case

Ice maker
Oil pressure switch
Pressure control
Reach-in cooler
Temperature control
Timed defrost

After reading this chapter, you will be able to:

1. Identify the main parts of a reach-in cooler and explain their functions.
2. Identify the main parts of a reach-in freezer and explain their functions.
3. Identify the main parts of a package-system ice maker and explain their functions.
4. Identify the main parts of a split-system ice maker and explain their functions.
5. Troubleshoot the electrical circuit of a reach-in cooler, a reach-in freezer, a package-type ice maker, and a split-system ice maker.

17.0 OVERVIEW OF REFRIGERATION SYSTEMS

You will find different types of refrigeration systems in use as you make service calls. This chapter will discuss reach-in and walk-in freezers and coolers as well as commercial ice makers. You will find that the basic parts of each of these systems are similar, and that troubleshooting them will be very similar. All commercial refrigeration systems have a compressor and a condenser fan motor. Some systems have a plate-type evaporator, so that no fan is used, and other systems use an evaporator fan to help move the cool air. The thermostat for this system can be a temperature-type control or a pressure-type control, since the low pressure can be used to determine the temperature in a system. Most refrigeration systems also utilize a defrost system to remove unwanted frost from the evaporator when it builds up. The **evaporator** for a refrigerator case can be a plate type that is mounted in the floor or shelves of a display case, or the evaporator can be a tube-and-fin type mounted in the display case with a fan.

This chapter will present a reach-in cooler, a reach-in freezer, and two examples of commercial ice makers. One of the ice makers has a self-contained **condenser**, and the other has a split condenser that is placed outside the building. The chapter will also explain the functions of a variety of switches such as low-pressure switches, high-pressure switches, and **oil pressure switches** that are used to protect the system against all types of high- and low-pressure problems or the loss of oil pressure. You will learn the theory and sequence of operation and the relationships among the parts of each system. You will also learn the proper sequence of voltage and current tests to help you isolate problems when you troubleshoot the system.

17.1 REACH-IN COOLERS AND REFRIGERATION CASES

The **reach-in cooler** is commonly used in grocery stores, convenience stores, and floral shops. Figure 17–1 shows a typical reach-in cooler. Some coolers have a fin-type evaporator mounted on a wall or in the ceiling of the cooler, and a fan is built into the evaporator housing to move air over the fins. The condenser fan and compressor can be mounted in the bottom of the case or on top of the case, or they can be located remotely, which means that they are placed outside the building.

Figure 17–2 shows an electrical diagram of a reach-in cooler. From this diagram you can see that the main electrical parts of the reach-in cooler are two compressors,

Figure 17–1 A reach-in cooler used in grocery stores or convenience stores. *(Courtesy of Tyler Refrigeration Corp.)*

Figure 17–2 Electrical diagram of a reach-in refrigeration cooler.

(Courtesy of Manitowoc Equipment Works)

two condenser fans, two evaporator fans, three door heaters, and a cabinet light. The compressors in this system are capacitor-start, induction-run compressors, and each has a current relay, starting relay, and a thermal overload. The condenser fans and the evaporator fans are shaded-pole motors. The compressors and condenser fans are energized through the contacts of a relay with a thermostat switch that controls the temperature in the cooler. A high-pressure cut-out switch is used to protect the compressor and condenser fan. The evaporator fans run whenever all the doors are closed. The system can be powered with 120 V AC (L1–N) or with 240 V AC (L1–L2).

17.2 NORMAL OPERATION FOR THE REACH-IN COOLER

During normal operation the evaporator fans run whenever the disconnect switch is closed for the unit. The heaters in the door seals are also powered anytime the disconnect is closed. These heaters keep moisture from building on the door seals. The normally closed contacts of the high-pressure switch provide voltage to the coil of the compressor relay. Voltage for the compressors and condenser fans is supplied through the contacts of the compressor relay and through the contacts of the thermostat when the temperature in the case is above the setpoint of the thermostat.

When voltage is first supplied to the compressors, it causes current to flow through the run winding to the C terminal. Since the coil of the current relay is connected in series with the run winding, this large current causes the coil of the current relay to become energized and pull in its contacts. The current relay contacts are connected in series with the start capacitor to supply voltage to the start winding. When the current relay contacts close, it allows voltage to flow to the start winding, which causes a current to flow through the compressor start winding. When the compressor begins to run, its counter EMF causes the current in the start winding to return to lower levels, which allows the current relay coil to return its normally open contacts to the open position. The compressor continues to run with current flowing only through its run winding. The shaded-pole condenser fan begins to run as soon as power is applied to its winding.

When the temperature in the case cools down and reaches the setpoint, the contacts in the thermostat open and de-energize the compressor and the condenser fan. The evaporator fan runs continuously to keep air circulating in the case. Whenever the temperature increases above the setpoint, the thermostat contacts close, the compressor and condenser fan are energized again, and the temperature in the cooler begins to decrease.

17.3 TROUBLESHOOTING THE REFRIGERATION SYSTEM

When you must troubleshoot the refrigeration case, you should make the following tests to determine where voltage loss has occurred. Be sure to start at the source of voltage for the case.

1. L1–L2 for 240 V AC or L1–N for 120 V AC _____
2. T4 on the compressor relay to L2 for 240 V AC _____
3. T6 on the compressor relay to L2 for 240 V AC _____
4. T1 thermostat to L2 for 240 V AC _____
5. T2 thermostat to L2 for 240 V AC _____
6. C on the compressor to L2 for 240 V AC _____
7. 2 on the potential relay to L1 for 240 V AC _____

8. R on the compressor to L1 for 240 V AC _____
9. Across the condenser fan motor terminals for 240 V AC _____
10. Across the evaporator fan motor terminals for 240 V AC _____
11. Across the door heaters for 240 V AC _____

Figure 17–3 shows the part of the electrical diagram where current flows for test points 1–6. From this diagram you can see that all voltage measurements must be made in reference to the L2 terminal. If you do not have 240 V AC at L1–L2 for a system connected for high voltage or 120 V AC at L1–N for a system that is connected for low voltage for the first test point, you do not have supply voltage and you should see if the disconnect or circuit breaker is set to the on position. If you have voltage at test point 1 but not at test point 2 (terminal T4 on the compressor relay), you should suspect the wire between L1 and terminal T4 on the relay. Turn power off and test the wire for continuity.

If you have voltage at test point 2 but not at test point 3 on the compressor relay, you should suspect that the high-pressure switch is open, the coil of the relay is bad, or the contacts of the relay are bad. You can test across the coil of the relay for 240 V AC. If you have voltage, you should suspect that the coil or contacts of the relay are bad. You can turn off power and test the coil and contacts for continuity. If you do not have voltage across the coil, you should suspect that the high-pressure switch is open. When the pressure switch is open, you must verify the system refrigeration pressure to see if it is too high. Many times, the high-pressure switch is open because of high pressure, which can be caused by a dirty condenser coil, faulty condenser fans, or blocked air supply to the condenser. If refrigeration pressure is normal and the switch is open, you will need to replace the switch.

Figure 17–3 Electrical diagram with current path for test points 1–6 *(Courtesy of Manitowoc Equipment Works)*

If you have voltage at test point 3 but not at test point 4, you should suspect the wire between terminal 6 on the relay and terminal 1 on the thermostat. Turn power off and test the wire for continuity. If you have voltage at terminal point 4 but not at terminal point 5, you should suspect that the thermostat is open. You will need to verify whether the temperature in the cooler is above the setpoint of the thermostat. You should remember that if the temperature in the cooler is below the setpoint of the thermostat, the contacts of the thermostat are supposed to be open, and it is operating correctly. You can temporarily set the temperature of the thermostat to the lowest setting while you troubleshoot the system and then set it back to its normal setting when you are finished with your tests. If the contacts of the thermostat are still open after you have it set at its lowest temperature, you should turn off power and test the thermostat for continuity. Change the thermostat if it is faulty. If you have voltage at test point 5 but not at test point 6, you should suspect that the thermal overload is open. You should determine if the compressor has been trying to start. If it has an overtemperature condition, you must wait for the compressor to cool down. If the compressor is cool and the overload does not return to its normally closed condition, you should replace the thermal overload.

If you have voltage at test point 6 but not at test point 7, it indicates that you have lost voltage between the contacts of the start relay and the S terminal of the compressor. You should suspect either the capacitor or the wire between the capacitor and the S terminal on the compressor. Turn off power and test the capacitor for proper operation, and test the wire for continuity.

If you have voltage at test point 7 but not at test point 8, you should suspect the wire between the start relay and the R terminal on the capacitor or the coil of the start relay. Turn off power and test each of these for continuity.

If you have voltage at test point 9 across the terminal of the condenser fan motor and the condenser fan is not running, you should suspect that the condenser fan is faulty. Turn off power and test the condenser fan for continuity. Since this is a shaded-pole motor, it should run if the coil has continuity. Be sure and test to see if the fan shaft can turn freely, or if it needs lubrication.

If you have voltage at test point 10 across the evaporator fan motor terminals, the fan motors should be running. If they are not running, turn off power and test them for continuity. The evaporator fan motors are also shaded-pole motors, which means that they should turn whenever power is applied to the windings. If you do not have voltage at test point 10, you should suspect that one of the door switches is open. Notice that the door switches are connected in series, so that if one of them is opened, power will be lost to the evaporator fan.

The last test is for voltage at the door heaters. You should suspect that the door heaters are not functioning properly if you notice moisture around the door seals. The job of the **door heater** is to provide sufficient heating on the door seals to prevent moisture from forming on them. It is important to understand that the three door heaters are wired in parallel with one another, which means that one of them can have a problem and the other two can be operating correctly. The one that has a problem will be the one that has moisture on it. You can also test each door heater for the amount of current it is drawing.

17.4 FREEZER CASES

Freezer cases are used in grocery stores and convenience stores to display frozen foods. The main difference between a freezer case and a refrigeration case is that the evaporator on the freezer case will reach temperatures below freezing, so a method must be provided for a defrost cycle. The defrost cycle is controlled by a defrost timer control.

Figure 17–4 **a** A display-type freezer case. **b** A deep-type freezer case. *(Courtesy of Tyler Refrigeration Corp.)*

The remainder of the components of a freezer case are very similar to those of the refrigeration case. The compressor for the freezer case may be mounted directly below the case, or it may be part of a remote condensing unit. Figures 17–4a and 17–4b show two typical freezer cases, and Figure 17–5 shows an electrical diagram of a freezer case. In this diagram you can see that the freezer case has a compressor, condenser fan, evaporator fan, door heaters, and a defrost heater. A defrost timer is also used to make the system go into defrost.

17.4.1 Normal Operation of a Freezer Case

In normal operation, the freezer case energizes its compressor with a thermostat. When the temperature in the freezer case rises above the setpoint of the thermostat, the thermostat closes its contacts, which energizes the compressor and condenser fan. The evaporator fan in the freezer case runs continuously just as in the refrigeration case. The operation of the freezer case is exactly like the operation of the refrigeration case, except for the defrost cycle. When the temperature in the freezer case drops below the setpoint, the thermostat contacts open and the compressor and the condenser fan are turned off.

17.4.2 Operation of the Freezer Case During Defrost Cycle

When the **defrost control** mechanism senses frost on the evaporator coil, or when time is expired on the defrost cycle time clock, the contacts in the defrost cycle timer close and cause the system to go into a defrost cycle. During the defrost cycle, the compressor and the condenser are de-energized, and the defrost heater is energized. The defrost heater provides heat directly on the evaporator coil, which causes the frost to melt. The defrost cycle lasts for the duration of the timer, or until the system determines that the frost has been eliminated. In some systems hot gas from the condenser is used to melt

Figure 17–5 Electrical diagram for a freezer case. *(Courtesy of Manitowoc Equipment Works)*

the frost in much the same way as a heat pump cycle is reversed. This type of defrost is called a *hot gas bypass defrost*. The diagram for this freezer shows an electric heater defrost.

17.5 TROUBLESHOOTING THE FREEZER CASE

When you must troubleshoot a freezer case, you can use the following tests to help you determine the location of the problem. These tests are very similar to the ones for the refrigeration case, except the freezer case has a defrost cycle. It is also important to understand that the compressor for this system is larger than the ones in smaller refrigeration cases and it is wired as a capacitor-start, capacitor-run (CSCR) motor. Notice that the CSCR compressor uses a potential relay to energize the start winding. Figure 17–6 shows the ladder diagram for the freezer case. You can use either the wiring diagram or the ladder diagram for this troubleshooting exercise.

1. L1–L2 for 240 V AC _____
2. T4 on the compressor relay to L2 for 240 V AC _____
3. T6 on the compressor relay to L2 for 240 V AC _____

Figure 17–6 Electrical ladder diagram for the freezer case.

4. T1 thermostat to L2 for 240 V AC _____
5. T2 thermostat to L2 for 240 V AC _____
6. Terminal 2 to L2 on compressor TB (240 V AC) _____
7. C on the compressor to L2 for 240 V AC _____
8. R on the compressor to L1 for 240 V AC _____
9. Terminal 1 on the potential relay to L1 for (240 V AC) _____
10. Terminal 2 on the potential relay to L1 for 240 V AC _____
11. Across the condenser fan motor terminals for 240 V AC _____
12. Across the evaporator fan motor terminals for 240 V AC _____
13. Across the door heaters for 120 V AC _____
14. Across the defrost heater for 240 V AC _____

If you don't have voltage at test point 1, you have lost the supply voltage. Check that the main disconnect is closed and that voltage is coming through the fuses. If you have voltage at test point 1 but not at test point 2, you should check the high-pressure cutout.

Be sure that the refrigeration pressure is normal. Remember if the refrigeration pressure is high, the high-pressure compressor cutout will be open. Turn off power and test the high-pressure cutout for continuity. If you have voltage at test point 2 but not at test point 3, you should suspect that the compressor relay has not been energized. You can test the coil or the contacts for continuity. If you have voltage across the compressor relay coil, its contacts should be closed. If you have voltage at test point 3 but not at test point 4, you should suspect the wire between the compressor relay and the **temperature control**. The temperature control is the device that turns the system on and off at the temperature setpoint. Turn off power and test this wire for continuity.

If you have voltage at test point 4 but not at test point 5, you should suspect the temperature control. Set the temperature control to the lowest possible temperature and try to make the system run. If the compressor still does not try to start, you can turn off power and test the temperature control for continuity. If you have voltage at test point 5 but not at test point 6, you should suspect the wire between the temperature control and the compressor terminal 2. Turn off power and test this wire for continuity. If you have voltage at test point 6 but not at test point 7, you should suspect that the thermal overload on the compressor is open. Feel the area where the compressor overload is mounted to see if the compressor is warm to the touch. If the compressor has tried to start several times and has not started, the thermal overload will be opened due to excessive current. Allow the thermal overload to cool down and try to start the compressor again. If the contacts of the thermal overload will not close, replace the thermal overload. If you have voltage at terminal 7 but not at terminal 8, you should suspect voltage coming through the start capacitor called line 2. Turn off power and test the wires connected to the start capacitor for continuity, and test the capacitor for normal operation.

Figure 17–7 shows the voltage path for the next tests. In this diagram you can see that voltage must make its way from L1 through the contacts of the defrost switch, the contacts of the compressor relay, and then through the overload device until it reaches the C terminal on the compressor. If you do not have voltage at L1, you have lost incoming voltage and you should check to see that the disconnect is turned on and that the fuses are good. If you have voltage at L1 but not at the defrost switch, you have lost voltage in the contacts of the defrost switch or in the wires that connect to the defrost switch and to the compressor relay. You will need to verify whether the system is in a defrost cycle. If the system is in a defrost cycle, the compressor is supposed to be de-energized. You can advance the defrost timer to bring the system out of defrost so that you can continue to make tests. If you do not have voltage through the defrost contacts, you can turn off power and test the timer contacts for continuity and the defrost timer for proper operation. You can also test the wires between L1 and the contacts of the defrost timer and the wire between the contacts of the defrost timer and the compressor contactor for continuity.

If you have voltage at defrost contacts but not at terminal 6 of compressor contactor, you should suspect that the contacts of the compressor contactor are not closed. You should test to see if the coil of the contactor has 120 V AC across its terminals. If voltage is present, the contactor may be faulty and you can test its coil for continuity. If voltage is not present, you can use Figure 17–7 to trace the path of the voltage that energizes the compressor contactor coil. You can see from this diagram that voltage comes from L1, through the contacts of the high-pressure cut-out switch, to the coil of the contactor, and back to N. Leave one meter probe on N terminal and test for voltage at the input of the high-pressure switch. If you do not have voltage at this point, the wire between L1 and the switch is open. If you have voltage at the input of the high-pressure switch but not at the output, you should suspect that the high-pressure switch is open. Be sure to test for high refrigerant pressure and correct its causes if necessary. If you have voltage at the output of the high-pressure

Figure 17–7 Electrical diagram of voltage path from L1 to coil of the compressor contactor highlighted.

switch but not at the coil, you have an open in the wire between the switch and the coil. When you find the cause of the open circuit, make repairs as necessary and try to start the system.

If you have voltage at terminal 6 but not at T1, you should suspect the wire between the compressor contactor and the thermostat. If you have voltage at T1 but not at T2, you should suspect that the thermostat is open. Be sure it is set to its lowest position. If it still does not pass voltage, you can turn off power and test it for continuity. If you have voltage at T2 but not the OL, you should suspect the wire between the thermostat and the OL. If you have voltage at the OL but do not have voltage at terminal C on the compressor, you should suspect that the thermal overload on the compressor is open. Recall that the overload will protect the compressor if it tries to start and draws excessive current because one of its windings is open or if one of its capacitors is faulty. Sometimes the compressor continues to try and start with faulty conditions and stresses the overload device until it is damaged. If you change the overload, be sure and continue to test the

Figure 17–8 Electrical diagram of voltage path for condenser fan, evaporator fan, and door heaters highlighted.

compressor circuit until you are certain that the compressor will start and draw normal current. Figure 17–8 shows the circuit where L1 voltage is provided to the condenser fan, evaporator fan, and door heaters.

If you have voltage at terminal C of the compressor, it means that the path for the L1 side of the circuit is good. You can leave one meter probe on terminal L1 as a reference for tests 8–10. Figure 17–9 shows the path of voltage from L2 to the compressor motor. From this diagram you can see that voltage on the L2 side goes directly to terminal R of the compressor. If you have voltage at L1–L2 but not at L1 to R, you can suspect the wire from L2 to R. Turn off power and test it for continuity.

If you have voltage at terminal R–C on the compressor, it should hum when it tries to start, since this is a complete circuit through the run winding. If the compressor does not hum and try to start, you can test this circuit for current. If no current is being drawn, you can suspect that the compressor has an open in its run winding. Turn off power and remove all wires from the compressor terminals and test the compressor run and start windings for continuity. If the compressor hums and tries to start, you can test for a

Figure 17–9 Electrical diagram of the path of voltage on the L2 side of the compressor highlighted.

voltage loss in the start winding circuit, which includes the start capacitor and potential relay. If you have voltage at L2 but not at terminal 5 on the compressor, you can suspect the start capacitor or the normally closed contacts of the potential relay. Recall that the contacts of the potential relay will open after the compressor is running at approximately 75% and sufficient counter EMF is developed to energize the potential relay coil. Turn off power and test these components for continuity, and test the capacitor for proper operation.

The condenser fan motors, evaporator fan motors, and heaters can also be tested. If you suspect that a condenser fan or evaporator fan is not running, you can physically test to see if their fan blades are turning. Recall that the evaporator fan should run whenever the doors are closed, and the condenser fan should run whenever the compressor is running. Since these are shaded-pole motors, you can test for the presence of voltage. If you have voltage and the motor does not run, you can test it for an open. If you do not have voltage, you can use voltage tests or continuity tests to locate the open in the wiring.

17.6 TROUBLESHOOTING THE DEFROST CYCLE

Since the freezer case has evaporator temperatures below 32°F, frost can build up on the fins of the evaporator. The system can be sent into a defrost cycle by timer or if the temperature in the coil is too low. The frequency and the duration of the defrost cycle can be set in the timer. When the system goes into a defrost cycle, 240 V AC is sent to the defrost heating coil. The defrost coil is mounted near the evaporator so that heat from it can melt frost from the evaporator coil. When the coil is covered with frost, it activates the defrost thermostat to move its contact from the normally closed position to the normally open position. This interrupts voltage to the evaporator fan while the system is in a defrost cycle. When sufficient frost has been removed, the thermostat returns to its normally closed position, which energizes the evaporator fan again. It also ends the defrost cycle.

The coil of the timer motor receives voltage whenever the system disconnect switch is closed. Figure 17–9 shows the electrical diagram of the defrost cycle that includes the defrost timer and the defrost heater coils. You should notice that the defrost heater is a simple electric heating coil that is located on the evaporator coil so that the heat it produces will melt the frost. The main problem that occurs with the defrost cycle is that it will not start or it will not end. If the defrost cycle does not start, you should test for 240 V AC across the timer motor at terminals.

17.7 ICE MAKERS

Ice makers can be either flake type or cube type, and they are found in hotels, motels, commercial kitchens, and cafeterias. They provide a continuous source of ice cubes for the food industry. Figure 17–10 shows typical cube-type ice machines, and Figure 17–11 shows an electrical diagram of a typical cube-type ice machine. From this diagram you can see that the main components of the cube-type ice machine include a compressor, a

Figure 17–10
a A cutaway picture of a cube-type ice machine.
b A cube-type ice machine. *(Courtesy of Scotsman®)*

(a) (b)

Air Cooled, 60 Hz, Single Phase Schematic Diagram

Figure 17–11 Electrical diagram of ice maker.
(Courtesy of Scotsman®)

condenser fan, evaporator plates that make the ice cubes, a water pump, an inlet water valve, and a hot-gas solenoid valve. The typical controls for a cube-type ice machine include an ice-size sensor, a low-pressure cutout, a high-pressure cutout, and a bin control to turn the machine on and off. Control systems vary greatly from brand to brand.

17.8 NORMAL OPERATION OF AN ICE MAKER

During normal operation the ice maker is connected to 240 V at line 1 and line 2. When the system is making ice, the master switch, the high-temperature cut-out switch, the high-pressure cut-out switch, and the bin thermostat are closed, which provides power to the coil of the compressor contactor. The compressor is wired as a capacitor-start, capacitor-run motor and it gets voltage for its start winding through a *potential relay*. The ice-making part of this system has a water pump motor that is energized and a water

solenoid valve that is energized to the open position to allow water to flood the evaporator plate. The ice bin sensor control turns the compressor contactor relay coil on or off, and the contracts of this relay open to de-energize L1 voltage to the compressor motor when the level of ice in the bin is full. The ice sensor also regulates the amount of time the water pump and solenoid stay energized to provide the correct amount of water to ensure the ice cubes are full size. As water flows into the evaporator it begins to freeze on the evaporator plate and form ice cubes.

When the ice has frozen solid, the sensor starts the *harvest cycle*. During the harvest cycle, the hot-gas solenoid and finish relay are energized to redirect the hot gas from the compressor to the evaporator plate to allow the ice to melt slightly so that it falls free of the evaporator plate and into the storage bin. When the ice falls free of the evaporator, the harvest cycle is complete, and the freezing cycle starts again. The water pump and solenoid are energized to allow water to flood the evaporator again. If the bin that contains the ice cubes becomes full, the bin level sensor opens and de-energizes the primary voltage to the transformer, which prevents water from flowing to the evaporator plate. The refrigeration cycle continues for several seconds after this point until the low-pressure switch opens. The pressure in the evaporator drops until it reaches its preset pressure, since no water is available to freeze.

17.8.1 Troubleshooting the Ice Maker

When you must troubleshoot the ice maker you can use the following list to make voltage measurements at specific points in the circuit. After you have made the measurements, you can analyze them to determine where the problem lies. If the compressor is not running, you should use tests 1–11 to find the problem. If the compressor is running but the condenser fan is not running, you should use tests 12 and 13 to determine the problem. If the compressor and condenser are running but the machine is not making ice, you should use tests 14 and 15 to determine the problem.

1. L1–L2 for 240 V AC _____
2. Master switch contacts to L2 for 240 V AC _____
3. High-temperature cut-out contacts to L2 for 240 V AC _____
4. High-pressure contacts to L2 for 240 V AC _____
5. Bin thermometer contacts to L2 for 240 V AC _____
6. Contactor coil to L2 for 240 V AC _____
7. Compressor switch contacts to L2 for 240 V AC _____
8. Start capacitor to L1 for 240 V AC _____
9. Potential relay contacts to L1 for 240 V AC _____
10. Terminal S on the compressor to L1 for 240 V AC _____
11. L1 to L2 on the compressor for 240 V AC _____
12. High-pressure fan control to L2 for 240 V AC _____
13. Across the condenser fan terminals for 240 V AC _____
14. Across the water pump for 240 V AC _____
15. Across the water solenoid for 240 V AC _____

If you do not have 240 V AC at test point 1, you should check the service box to determine if you have a blown fuse or an open circuit breaker. If you have voltage at test point 1 but not at test point 2, you should suspect that the master switch contacts are open or faulty. If you have voltage at test point 2 but not at test point 3, you should suspect that the high-temperature cut-out switch is open. The problem may be that the temperature is too warm. If you have voltage at test point 3 but not at test point 4, you should suspect

that the high-pressure contacts are open. Be sure to determine if the pressure in the system is high from a problem such as an inoperative condenser fan, or a condenser coil covered with dirt so that air cannot get through the coils to help remove heat. If the refrigerant pressure is at the correct level, and the high-pressure switch is still open, you should suspect that the switch itself is faulty.

If you have voltage at test point 4 but not at test point 5, you should suspect that the bin thermostat switch is open. This switch may be open because the bin is full. If you have a problem with any of these switches, you must be sure to check the condition that may cause the switch to open. You must remember that you may have to correct the condition that is causing the problem rather than changing the switch. For example, if the pressure is not at the correct level, you may have to correct the pressure problem before the system will operate correctly. If the pressure is at the correct level and the switch is still open, you should then suspect the switch itself.

If you have voltage at test point 5 but not at test point 6, you should suspect that the coil that closes the compressor contacts is not energized. If voltage is present at the coil but the contacts are not closed, you may need to test the relay coil to ensure that it has continuity. If you have voltage at test point 6, and the contactor coil is energized, you should move to test point 7 and test to see if voltage is passing through the contacts of the contactor. If the contacts are closed and the compressor is trying to start but it cannot run, it will cause excessive current, which will open the overload after a short period of time. Other problems that may keep the compressor from running and allow it to draw excessive current include high refrigerant pressure, which causes the compressor to work too hard, or low refrigerant pressure, which causes the compressor to run hot, since not enough refrigerant is coming back to the compressor in liquid droplets to keep it cool. If all the conditions in the compressor are normal and the overload is still open, you should suspect the overload itself.

If you have voltage at test point 8 but not at test point 9, you should suspect that the wire between terminal C on the compressor and terminal 3 on the overload is open. If you have voltage at test point 9 but not at test point 10, you should suspect that the wire that provides L2 voltage to the circuit through the start capacitor is open. If you have voltage at test point 10 but not at test point 11, you should suspect that the contacts of the potential relay are faulty. If you have voltage at test point 11, the circuit is functioning properly and you should suspect the start winding or run winding of the compressor motor. Turn off power and test each winding for continuity.

Another problem that causes the ice maker to fail is that the compressor is running, but the condenser fan is not running. When this occurs you should use test 12. This test is for voltage across the contacts of the high-pressure fan control. If you have voltage at this point, you should use test 13, which tests for voltage directly across the terminals of the fan motor. If you have voltage and the motor does not turn, you should suspect that the motor is faulty and test it for continuity. The condenser fan motor is a shaded-pole motor. If you do not have voltage at the motor terminals, you should suspect the wires that connect the motor directly to the voltage.

If the compressor and condenser fan are both running but the system is not making ice, you should suspect the water and ice control circuit. You can start with test 14. If you do not have voltage at test point 14, you can suspect that the bin switch is open. Be sure to determine whether the level of ice in the bin is keeping the switch open, or if the switch is faulty or not adjusted correctly. If the pump is running but you do not have water flowing, you should suspect the water solenoid and check for voltage directly at the terminals of the water solenoid. If you have voltage at these terminals but the solenoid is not energized, you should turn off power and test the solenoid coil for continuity. You should also remember that sometimes the water solenoid valve corrodes and cannot operate correctly even though the electrical control part of the valve is functioning.

If the machine is making ice but does not harvest it correctly, you should suspect the ice sensor control, which is also known as the bridge thickness sensor, or the hot-gas solenoid. If it has voltage but does not operate, you should suspect the valve coil and test it for continuity.

You will find that troubleshooting a system is a process of determining what is operating correctly and what is not. You must trace voltage to determine how far it is moving through a circuit, and know where to find information about motors and controls in prior chapters for more detailed troubleshooting. The biggest mistake that most starting technicians make when troubleshooting is finding a control such as a high-pressure control that is open and thinking that the control is faulty when, in fact, the control is simply open because it is doing its job correctly by detecting a high-pressure condition. You must determine the cause of the high-pressure condition and correct it rather than change the high-pressure switch.

17.9 ELECTRONIC CONTROLS FOR REFRIGERATION SYSTEMS

Figure 17–12 shows an electronic refrigeration control that contains an **electronic defrost control**. The electronic board inside this control allows selectable **timed defrost** intervals from 2 to 48 hours. The amount of time the control keeps the system in defrost mode lasts between 1 and 63 minutes. It also has a mode that allows the system to be set on timed interval or compressor run time. The control provides three types of defrost termination, including temperature only, which requires an external thermostat; **pressure control**, which provides an external pressure switch; and time only, which stops the defrost cycle strictly by a predetermined amount of time. Manual defrost cycle is provided through a pushbutton on the control.

Figure 17–12 Paragon ERC 2 electronic refrigeration control. This is a microprocessor-based electronic controller designed to control both the temperature and the defrost function of a refrigeration system. *(Courtesy of Paragon)*

The **electronic refrigeration control** is a microprocessor-based electronic controller designed to control the system temperature in the cooled space as well as all defrost functions. The control has four relay outputs to control the compressor, the defrost controls, the evaporator fan, and any alarm system components. The controller has a digital display that provides readouts for temperature, time, and diagnostics. The display can be mounted locally on the control, or it can be mounted remotely from the unit in a more accessible location. A keypad is provided with the display for simple programming. The electronic controls provide the same control as older electric motor driven controllers as well as additional functions such as diagnostics.

17.10 REVIEW

Refrigeration systems consist of an evaporator, condenser, and compressor. The condenser and compressor can be located directly under the display case, or they can be located outdoors. The evaporator for a refrigeration case can be a plate type that is mounted in the floor or shelves of a display case, or the evaporator can be a tube-and-fin type that is mounted in the display case with a fan.

If the refrigeration system uses temperatures below freezing or near freezing, it will need a defrost cycle to ensure that frost does not build up on the evaporator. Refrigeration cases or freezer cases also require door heaters if the units have doors. The door heaters provide a small amount of heat to the door frame to keep moisture from gathering at the point where moist room air meets the cool door frame. If the system is a freezer, the door heaters prevent the door gasket from freezing to the door frame.

A thermostat is used to turn the compressor on and off to control the temperature. In some commercial units, a low-pressure control is used to control temperature based on the inverse temperature and pressure relationship.

The components and controls of an ice maker are similar to those of a refrigeration system. The major difference is that the ice maker has mechanical parts to harvest the ice. A hot-gas bypass solenoid is used to route a small amount of refrigerant in gas form through the evaporator plate to loosen the ice when it is time to harvest it. The ice maker also has a water solenoid to control the amount of water that is allowed to flow on the evaporator plate to make the cubes.

Troubleshooting the refrigeration system or freezer case is similar to troubleshooting other air-conditioning systems. If the compressor is not running, you can follow the path of voltage and check for a loss of voltage by keeping one voltmeter lead on the L2 or N terminal as a reference and moving the other terminal around the circuit following the voltage path to the compressor. If you find a component that you suspect is faulty, you can find more information about troubleshooting it in an earlier chapter.

Job Assignment for This Chapter

You are called to a convenience store that has a reach-in freezer case. The customer complaint is that the temperature in the case is getting too warm and the product is not staying frozen. The customer wants you to take a look at the system and give an estimate of what is wrong and what it will take to repair it before you begin your repair. Your initial check indicates that the evaporator has an excessive amount of frost on it.

Your job is to look at the system and identify possible problems and report back to the store owner your estimate of what needs to be done.

Problems with refrigeration systems tend to be complex and be a combination of refrigeration system problems as well as electrical problems. Since this text covers more electrical problems, you should consider this problem as a pure electrical problem when you create your solution.

Solution to the Job Assignment

Since we are limiting our problem to a pure electrical problem, and you know that the evaporator has an excessive amount of frost, we will focus our test on the defrost control. Your first test should be to see if the defrost control has power. Next you should check to see the amount of time and temperature on the control and see if they are set correctly.

Next you can put the system into a manual defrost cycle. If the system does not begin a defrost cycle, you need to check to see if they are using a solenoid to create a hot-gas defrost or if they are turning off the system using another type of defrost. If the defrost cycle does not start immediately, you should suspect that the solenoid or defrost system is not starting. You can troubleshoot the defrost control to see if it is putting out the correct amount of voltage to the defrost controls. If it is, then you should suspect the controls such as the solenoid. If it is not providing an output voltage, you should check out the defrost controller and see if it is actually in a defrost cycle. If the control is faulty, you should install a new one and set it up for defrost conditions.

Assessment

QUESTIONS

1. Identify the main parts of a reach-in cooler and explain their functions.
2. Compare the parts of a cooler with the parts of a freezer and explain the basic differences.
3. Identify the controls for a freezer or cooler and explain their operations.
4. Explain why a freezer requires a defrost cycle and a reach-in cooler does not.
5. Identify the main parts of an ice maker and explain which parts control the refrigeration cycle and which parts control the water flow.

TRUE OR FALSE

1. A cooler or freezer can use a low-pressure control as a thermostat to control the temperature in the display case.
2. The door heaters on a reach-in cooler are needed because the door handles are too cold when people touch them.
3. In most reach-in coolers the evaporator fan and condenser fans are three-phase motors.

4. Freezers require a defrost heater because frost will build on evaporator fins and block air flow.
5. The ice maker shown in the electrical diagram in Figure 17–11 uses a hot-gas solenoid to warm the evaporator plate during the harvest cycle.

MULTIPLE CHOICE

1. If you are troubleshooting the reach-in cooler in Figure 17–2 because its compressor is not running, you should
 a. test the door heaters because an open in a heater will keep voltage from reaching the compressor.
 b. test the supply voltage, compressor relay, and thermostat to see that voltage is passing through these components and reaching the compressor.
 c. test the door switches to ensure that voltage is reaching the compressor.

2. The compressor in the electrical diagram for the reach-in cooler in Figure 17–2 is connected as a
 a. capacitor-start, induction-run (CSIR) compressor.
 b. permanent split-capacitor (PSC) compressor.
 c. split-phase compressor with a current relay.

3. The condenser fan motor in the electrical diagram for the reach-in cooler in Figure 17–2 is connected as a
 a. shaded-pole motor.
 b. capacitor-start, induction-run (CSIR) motor.
 c. permanent split-capacitor (PSC) motor.

4. The compressor and condenser fan for the reach-in cooler in Figure 17–1 is physically located
 a. outside, since this is a split-system type unit.
 b. below the cooler case toward the back of the unit.
 c. on top of the cooler case.
 d. all the above can be used.

5. If the defrost cycle for the freezer in Figure 17–9 will not activate, you should suspect that
 a. the defrost timer motor has stopped.
 b. one of the door switches is faulty.
 c. the drain line heater has failed, and the drain is frozen shut.

PROBLEMS

1. Use the diagram in Figure 17–2 and sketch the electrical diagram of the circuit for the evaporator fan.
2. Use the diagram in Figure 17–2 and sketch the electrical diagram of the circuit for the compressor. Be sure to include the control circuit and all controls for the compressor relay.
3. Use the diagram in Figure 17–2 and sketch the electrical diagram of the circuit for the condenser fan.
4. Use the diagram in Figure 17–11 of the ice maker and sketch the electrical diagram of the circuit for the defrost hot-gas solenoid and all the controls for defrosting.
5. Use the diagram in Figure 17–11 of the ice maker and sketch the electrical diagram of the circuit for the compressor.

Websites

Websites change constantly. Use the key word search to locate sites that have similar content. The products that will be found at each site are indicated with the key word.

Carrier Corporation
Products: Refrigeration
Key word: Carrier Corporation

Tyler Refrigeration
Products: Refrigeration equipment
Key word: Tyler Refrigeration

Scotsman
Products: Ice makers
Key word: Scotsman

Johnson Controls
Products: Solid-state relays and controls
Key word: Johnson Controls

Ranco Controls
Products: Solid-state controls
Key words: Ranco Controls

Honeywell Controls
Products: Thermostats and solid-state HVAC controls
Key word: Honeywell Controls

Robertshaw Controls
Products: Solid-state HVAC controls
Key word: Robertshaw Controls

Chapter 18

Troubleshooting Heating, Air-Conditioning, and Refrigeration Systems

Job Assignment for This Chapter

You receive a work order to troubleshoot a packaged air-conditioning system that has gas heat. It is summer and the unit is in the air-conditioning mode. The complaint is that the unit is not cooling, and your job is to create a process to troubleshoot the problem. Part of your process should include identifying what parts of the system you can ignore and where you should focus your troubleshooting. You need to explain to the homeowner what tests you will make and what troubleshooting processes you will use. You will also need to explain why you can ignore parts of the system.

Terms You Will Learn in This Chapter

Anti short cycle timer (ASCT)
Circuit breaker
Common point
Compressor contactor
Continuity test
Megger
Ohmmeter

Run winding
Start capacitor
Start winding
Troubleshooting
Voltage drop method of troubleshooting
Voltmeter

After reading this chapter, you will be able to:

1. Explain the fundamentals of troubleshooting.

2. Explain the difference between a symptom and a problem.

3. Develop a procedure for locating a loss of voltage in a circuit.

18.0 INTRODUCTION TO TROUBLESHOOTING

As an HVAC and refrigeration technician, you will be requested to troubleshoot equipment that has stopped running or is not operating correctly. Many people fear this task because it looks so complex. This chapter will help you gain the skills that make **troubleshooting** much easier and more predictive. The most important point that you must understand is that technicians who have been in the field for many years have developed tests to verify the condition of every component that you may find in an HVAC or refrigeration electrical control or load circuit. You must believe in the value of these tests and strictly follow their procedures. You can review these tests and their procedures in the previous chapters of this book. After reading this chapter, you will have the skills to find problems in any system that you are requested to troubleshoot. You will also learn troubleshooting processes that can be used every time you troubleshoot something, and you will be able to verify the condition of the components you suspect are working correctly or the components you suspect are faulty.

Troubleshooting can be broken into multiple scenarios. You may be asked to troubleshoot an HVAC or refrigeration system that is already in operation and has a wiring diagram and a ladder diagram that explains the sequence of operation of the HVAC or refrigeration system. In this case you would use all of the diagrams to test the operation of the HVAC or refrigeration system and determine which parts are inoperative.

In some cases you may be asked to troubleshoot a system that has a wiring diagram but does not have a ladder diagram. In this case you may need to convert all or part of the wiring diagram to a ladder diagram in order to determine the sequence of operation for the system. If the HVAC or refrigeration system has only a ladder diagram but does not have a wiring diagram, you must convert the ladder diagram into a wiring diagram so that you can determine the location of the components. In some cases you may be able to open the cabinet door and identify where the components are located. This chapter will explain how to troubleshoot an HVAC or refrigeration system when you have both a wiring diagram and a ladder diagram.

18.1 WORKING SAFELY

The most important point to remember when you are troubleshooting is to work safely and follow all safety procedures provided in the previous chapters. At times it is important that you turn off all electricity to a system when you are removing and replacing parts or testing for ohms. You should understand that some tests require that these systems be fully energized to complete the test. In these cases be sure that you are following all safety rules and are fully aware of the potential for injury when you are working around live circuits.

You should also work as though each circuit and power system is live and active anytime you are working around an HVAC or refrigeration system.

18.2 WHAT IS TROUBLESHOOTING?

Many HVAC and refrigeration technicians believe that troubleshooting is a process to determine what is broken on HVAC or refrigeration systems. This chapter will show you that it is actually easier to identify what is working correctly on an HVAC or refrigeration system and, through a process of elimination, not work on the part that is working correctly, rather than determine what is not working correctly and only focusing on that part of the system. This is a little-known secret that the best troubleshooting technicians have learned. This section will explain the basic things you should do when you are trying to troubleshoot HVAC or refrigeration systems.

Another problem that a number of technicians have when troubleshooting HVAC or refrigeration systems is that they do not use a consistent procedure each time. Instead they try different things each time they troubleshoot a different HVAC or refrigeration system, and sometimes they basically try to remember what was broken on the HVAC or refrigeration system the last time and work from there as though the problem is simply repeating itself. In this chapter, you will learn to develop a procedure that you can use every time you troubleshoot an HVAC or refrigeration system.

18.3 THE DIFFERENCE BETWEEN SYMPTOMS AND PROBLEMS

When you are first learning to troubleshoot an HVAC or refrigeration system, you may find what appears to be a problem but instead is only a symptom of the problem. For example, you may find a blown fuse on an HVAC or refrigeration system and think you have found the problem. In fact if you replace the fuse, the HVAC or refrigeration system will begin to operate again. What is hard to understand is that a motor in the HVAC or refrigeration system is overheating and overworking for some reason, which is causing the fuse to blow. Each time you replace the fuse, the motor will start and run until it draws enough overcurrent and blows the fuse again. If you continue to replace fuses, the motor will eventually overheat to the point at which it is totally destroyed. You will learn in this chapter how to determine what are problems and what are symptoms. You may need to revisit some of the early chapters to determine how components such as relays, transformers, and motors operate.

18.4 STARTING THE TROUBLESHOOTING PROCEDURE

The first step in troubleshooting an HVAC or refrigeration system should be to determine what the HVAC or refrigeration system was doing prior to stopping. Sometimes this can be done by simply talking to the HVAC or refrigeration system owner. You can ask a series of questions to determine if the HVAC or refrigeration system was operating correctly prior to this malfunction or if the problem has been occurring for some time. You may also ask the customer to try and operate different parts of the HVAC or refrigeration system in the mode that is not working correctly. Sometimes you will find that

the HVAC or refrigeration system can operate partially, or that one part of the HVAC or refrigeration system, such as the indoor fan, can operate, whereas another part, such as the outdoor unit, cannot. This will allow you to focus on just the part of the system that is not functioning correctly.

18.5 OBSERVING THE HVAC OR REFRIGERATION SYSTEM AND USING ALL OF YOUR SENSES

When you first begin to troubleshoot an HVAC or refrigeration system, you may feel that it is a large job that is very complex. As you work with more experienced technicians, you'll find that they use a number of methods to determine what is wrong with the HVAC or refrigeration system. For example, they may spend a few minutes just looking at the HVAC or refrigeration system and trying to get an idea of what components are running or turned on. They may also try to identify what power systems the HVAC or refrigeration system has, such as the type of voltage each section gets. Another important part of this process is to identify what types and how many motors the HVAC or refrigeration system has. Do the motors have variable-frequency drives (VFDs) or other types of acceleration and speed controls or are they connected directly to a motor starter? Finally you may take a minute to determine whether the HVAC or refrigeration system uses regular hardwired controls or a direct digital controller (DDC) or programmable logic controller (PLC).

A second step in the process of troubleshooting includes using your other senses beyond vision. For example, you may want to listen for problems such as motors that are under extreme loads or belts and pulleys that are squeaking, which indicates that they are running improperly and causing friction. Another sound that you should listen for is motors that hum but do not start. These clues will help you later when you are trying to determine what is wrong with the HVAC or refrigeration system. Another sense that is usually overlooked is smell. In some cases you will be able to smell a motor that is overheating before you're able to determine from some other test that it has a problem.

18.6 SWAPPING PARTS

When you have tested the HVAC or refrigeration system and have identified a faulty component, you may want to swap a part to ensure that the part you suspect is faulty is actually bad. Swapping parts may involve going to the truck, locating a good part, and swapping it with the suspected part. However, this procedure has several drawbacks. One problem is that some technicians defer to the component swap method before thoroughly testing the part; they basically guess which part is bad and may waste a lot of time changing parts.

The advantage of swapping a part is that if the part is expensive or has a two- or three-day shipping time, you can ensure that you have actually determined the part is faulty before you order a new one and wait for it to be shipped.

The basic rule of thumb to follow when swapping parts is to ensure that you have completely tested the part while power is applied and you have completed an alternative test with power off. If you have completed these tests and are sure the part is faulty, you can safely change or swap the part to verify your suspicions.

18.7 WHEN TO USE A VOLTMETER TO TROUBLESHOOT

The **voltmeter** is the meter of choice for most technicians. The voltmeter allows the technician to make tests in the circuit with power applied. It also allows technicians to make a large number of preliminary tests without disconnecting any wiring or components. If you are called to troubleshoot an HVAC or refrigeration system that has stopped completely, it more than likely has an open circuit, where voltage is interrupted. The first test that you should make with the voltmeter is to measure the source voltage. You should always set the meter to the highest voltage range until you have determined the actual voltage. After measuring the source voltage, you will need to use the ladder diagram for the HVAC or refrigeration system and measure voltage throughout the circuit wherever you suspect the voltage has stopped. When you find the loss of voltage, you can turn power off to the HVAC or refrigeration system, lock it out, disconnect the wire or component that you suspect is faulty, and test it with an ohmmeter. One of the best ways to determine where you have a problem when a control circuit does not energize the coil of the **compressor contactor** is to use a voltmeter to locate the point where voltage has been lost. Later in this chapter we will use the control circuit shown in Figure 18–1 to make several voltage tests to determine where voltage has been lost. You should remember that the schematic (ladder) diagram indicates the sequence of operation for the HVAC or refrigeration system. The wiring diagram indicates the location of wire terminals on the components—where you would find them if you were looking in the electrical cabinet.

You also should be aware of some problems that you might encounter when measuring voltage. The first problem is voltage feeding backwards through a circuit. This usually occurs in circuits that have one blown fuse. If you have one blown fuse in a circuit, it is possible for voltage from the remaining supply line that has a good fuse to feed through a transformer winding or motor winding and flow backwards up to the bottom of the blown fuse. When you test the output side of each fuse, your meter will measure voltage and it will appear that all of the fuses are good. If you are aware this can happen, you can use a fuse puller and pull all three of the fuses from their holders and test each fuse for continuity with an ohmmeter. You should also remember that if a circuit has one blown fuse, you cannot have any current flow, so you could check for current flow too. You will also get feedback voltage through a compressor or condenser motor when either the run or the start winding is open and the other winding is good. All you need to do is be aware that feedback voltage does occur, so you can double check the source of voltage if you suspect that the voltage you are reading may be coming from a feedback source.

Figure 18–1 Schematic (ladder) diagram of control circuit for air conditioner.

Another problem is that the electronic controls on high-efficiency furnaces may cause odd values of voltage to show up in a circuit. If a circuit has a variable-frequency drive, the drive can produce voltages that are above or below 60 Hz. Since most digital and analog voltmeters are designed for 60 Hz, the voltage reading on the meter could be slightly more or less than it actually is, making the technician feel that the circuit is faulty when in fact it is operating correctly. You can use a true RMS reading voltmeter to get an accurate reading in these types of circuits if it is important, but normally you can just remember that the voltage may be higher or lower in an electronic circuit and everything is still working correctly.

18.8 WHEN TO USE AN OHMMETER TO TROUBLESHOOT

It is important to understand when to use an **ohmmeter**. You must have all power to a circuit turned off before you use an ohmmeter to take measurements since the ohmmeter presents very low impedance (resistance) to the circuit. When impedance is low, it will create a short circuit path if you place the meter leads across the terminals that have voltage present. This will cause the large current to create sparks and possibly cause molten metal to explode into your face; it will certainly damage the meter. In most cases you will use the ohmmeter to make a **continuity test** with power off, after you have used a voltmeter to locate a faulty switch, motor winding, or wire. When you are making the continuity test with an analog motor, you should zero the meter and use the lowest ohm setting. This setting will detect the slightest amount of resistance in the switch or wire. Resistance that is very high is called infinity and indicates that an open circuit is present in the contacts or the wire. If you are testing a wire, this test indicates that the wire is faulty and must be replaced. If you are testing a switch, be sure you are on the correct terminals that make a contact set, and then change the switch to its other positions. If the meter reading remains at infinity, the switch contacts are faulty, and the switch must be replaced.

Another reason to use an ohmmeter is to measure the actual number of ohms in the windings of a motor, transformer, or coil. When you are testing a single-phase motor, you should measure the resistance in each winding accurately so that you can determine that the winding with the highest resistance reading is the **start winding** and the winding with the lowest resistance reading is the **run winding**.

If you are troubleshooting electronic circuits, you will also need to measure the amount of resistance accurately. For example, when you are trying to identify the terminals of a transistor, silicon controlled rectifier (SCR), or triac, you will need to measure the resistance and the polarity of the meter probes. Be sure to switch the meter to the setting that will provide the most accurate resistance reading.

18.8.1 Testing with a Megger

A **megger**® is a specialized ohmmeter that uses approximately 600 to 2500 volts to check cables and motor windings for leaking current through their insulation. The voltage to the circuit must be turned off. The megger puts the voltage into the motor winding or cable for a short time, usually less than several minutes, and tests for leaking current, which indicates a breakdown of the resistance in the winding or cable. The results of the leaking current through the insulation is displayed on the megger in units of resistance (megohms or millions of ohms). A low reading (less than about 100 megohms) indicates possible winding or cable problems. You are looking for very high resistance readings and the higher the better. A high resistance reading on a compressor winding indicates that the insulation is in good working condition and the motor is not likely to short out. Sometimes moisture

will get into the motor winding and cause the resistance of the motor winding to drop lower than it normally would be. The megohm test is usually used on compressors or other expensive motors that you suspect have a shorted winding or a winding that is going bad and starting to draw too much current. You should test the suspected motor windings before you decide to remove it and replace a compressor from the system since a compressor change out requires removing or storing the refrigeration charge as well as changing out the motor. If you are fairly sure a compressor winding is shorted, the megger test will verify your suspicions. A regular ohmmeter can find major shorts and opens in a wire or winding, but it cannot measure the higher values of resistance like a megger. The word *Megger* is a registered trademark of AVO International (formerly the Biddle Co.), but the term is widely used for all similar testers regardless of manufacturer.

18.9 WHEN TO USE A CLAMP-ON AMMETER TO TROUBLESHOOT

A clamp-on ammeter is a valuable troubleshooting instrument that can indicate the presence of current in a circuit. For example, if you suspect that a motor with three-phase supply voltage has one blown fuse, you can quickly determine which leg of the three phases has the blown fuse by measuring the current in each of the three legs with the clamp-on ammeter. The leg that has the blown fuse will indicate 0 amps, and the two legs where the fuses are good will indicate current that is higher than rated for the circuit, because the motor will be drawing locked-rotor current since it will not be able to start. This test is more effective than the voltage drop test, which is subject to feedback voltages.

The clamp-on ammeter is also useful for determining the current load of any circuit wire. This is especially useful in determining how close the circuit wire is to overloading, which will cause a fuse to blow. You can also determine if a motor is beginning to wear out and have bad bearings or lack of lubrication. If the motor current is higher than its rating, you can begin to suspect that it is overloaded. If the load is within tolerance, you can suspect that the motor needs lubrication or the bearing is bad.

The clamp-on ammeter can also be used to adjust belts on belt drive fan systems. You can attach the clamp-on ammeter, adjust the belt tension, and watch the clamp-on ammeter. If the current increases above the motor rating, the belts are too tight. If the current rating is too low, it indicates the belt is not tight enough.

18.10 DIVIDING THE HVAC ELECTRICAL SYSTEM INTO THE LOAD CIRCUIT AND THE CONTROL CIRCUIT TO MAKE TROUBLESHOOTING EASIER

The part of the electrical system that has the motors in it is called the load circuit. Other electrical loads may include lights and electric heaters. The control circuit includes all of the switches and devices that control voltage to the coil of motor starters, relays, or solenoids. When you are troubleshooting a circuit that has pushbutton switches controlling a motor starter coil, you can quickly eliminate the parts of the circuit that are functioning and the parts that are not by seeing if the coil of a motor starter is energized. If the coil of the motor starter is energized, all of the switches and the coil are operating correctly, and you should shift your test to the load circuit where the motor is connected to the motor starter. Another way to understand this concept is to think about troubleshooting an HVAC or refrigeration system the way you would try to locate a specific card in a deck of playing cards. Suppose I pull one card from the deck and then ask you to guess what the card is. As you know, there are 52 cards in a standard deck of cards that has hearts and

diamonds that are red cards, and clubs and spades that are black cards. So if you just start guessing, without a method, it may take you 52 guesses to get the correct one. This section will show you a method that you can use later when you are troubleshooting an HVAC or refrigeration system.

For this example let's say the card I have pulled from the deck is a 9 of hearts. Using this process the first question would be to ask if the card is a black card. The answer is no, so we can deduce that the card must be red. The important thing is that this one question eliminates 50% of the deck. We now know the card must be a heart or diamond, so we can stop thinking about all of the spades and clubs in the deck.

The next question will be whether the card is a heart. By now you can see it does not matter whether I ask if the card is a heart or if I ask if it is a diamond, because the answer will indicate which one it is. In this case the answer is yes, the card is a heart, so I can stop thinking about all of the diamonds. With two questions, we have eliminated 75% of the cards.

Now that I know the card is a heart, I will continue to ask questions that eliminate the remaining cards 50% at a time. Since I know there are 13 hearts ranging from a high of ace to a low of 2, my next question will be whether the card is lower than an 8. The answer is no, so I have eliminated the 2, 3, 4, 5, 6, and 7. The remaining cards are 8, 9, 10, jack, queen, king, and ace. Since the queen or jack is the midpoint, I would next ask whether the card is below a queen. The answer is yes, so now I know the card must be a jack, 10, or 9. The next question is whether the card is lower than a 10. Since the answer is yes, I now know the card is the 9 of hearts.

You may be wondering what this card example has to do with troubleshooting. The four suits of cards—diamonds, spades, hearts, and clubs—represent the systems on most HVAC or refrigeration systems: electrical loads, controls, mechanical parts, and refrigeration parts. When you come upon a broken HVAC or refrigeration system, your first questions should determine which of these systems are working and which are inoperative. You need to develop questions and tests that will eliminate the most possibilities. Let's say that you have found the indoor unit is operating correctly, but the outdoor unit will not energize. The next test would be to see if you can get the electrical system energized at the outdoor unit. You can see that this process cuts down the number of parts of the HVAC system that could possibly be faulty.

When you are testing the electrical system, you should check the control circuits and the loads they control, such as the compressor and fan motors. The control circuit for a packaged heating and air conditioning system is shown in Figure 18–1. It consists of the 24 volts supplied by the transformer secondary winding at terminals R and C. It also has the cooling thermostat, the anti-short cycle timer, the compressor contactor coil, and the fan relay coil. If you use the process of selecting tests that will eliminate large portions of the HVAC or refrigeration system, like the compressor or fan motors, it will logically help you find the problem on any HVAC or refrigeration system every time you troubleshoot.

18.11 WHY IT IS IMPORTANT TO TEST A CIRCUIT WITH POWER APPLIED

You may wonder why it is important to test a circuit with voltage applied to the circuit. You may even think that this is a dangerous practice. However, it is a very safe way to test a circuit and is also the most reliable. The problem with testing a circuit for continuity with the power off is that it is a slow process because all of the wires must be disconnected before they can be reliably tested, and because it is easy to have faulty readings from parallel circuits that you are not aware of. If you expect to become a qualified HVAC or refrigeration technician, you must be able to repeat the voltage-loss test explained in the next sections without making mistakes. You must also be able to make the test quickly and accurately.

18.12 TROUBLESHOOTING THE CONTROL CIRCUIT OF AN AIR-CONDITIONING SYSTEM

The control circuit of an air-conditioning system provides the voltage and current to energize the coil of the compressor contactor. When the coil is energized, its contacts close and allow voltage to flow to the compressor motor windings. Testing the control circuit in an air-conditioning system is a simple procedure. The focus of this test is to identify all the parts in the control circuit that are wired in series with the coil and identify if voltage is provided all the way through the circuit to the coil. You should remember that the control circuit may have a problem with the thermostat, wires, or other components that may have an open, may be faulty, or may not operate correctly. The problem could be any one of several things, such as the loss of the power supply, a loose or broken wire, a faulty switch contact, or a faulty load (compressor contactor coil or relay coil). This means that you must troubleshoot the complete circuit instead of picking on individual components and testing them at random. The following procedure is called the *voltage-loss test*, and you should learn to use this test whenever you are trying to locate a faulty component or wire in an inoperable circuit. We will use the control circuit diagram shown in Figure 18–2 that has all of its test points identified with numbers 1–11.

Start the procedure by testing for voltage across the load terminals (the two terminals of the compressor contactor coil identified as 8 and 9). If voltage is present at terminals 8 and 9 of the compressor contactor coil, the power supply and all the pilot switches and interconnecting wires are working correctly. The remainder of the troubleshooting test should focus on the coil. If there is voltage at terminals 8 and 9, and the compressor contactor relay coil is not energized and pulling in the contacts, you should turn off power and test the coil with an ohmmeter for continuity.

If no voltage is present at the terminals of the coil at test points 8 and 9, the remainder of your troubleshooting tests should focus on the loss of voltage somewhere between the 24-volt power source at test points 1–2 and where the voltage travels

Figure 18–2 Control circuit diagram of an air-conditioning system provided for troubleshooting showing test points 1–11.

through the thermostat, the high-pressure switch, the contacts of the anti short cycle timer (ASCT), and then onto the compressor contactor coil. If voltage is not present at the power supply at test points 1–2, be sure to check for a blown fuse, open disconnect, bad transformer, or other power supply problems. At this point in the test, all that you are concerned with is that the control voltage (24 volts) is available at test points 1–2.

After you have determined that the proper amount of supply voltage is available, the remainder of the tests should focus on the thermostat, high-pressure switch, ASCT contacts, and interconnecting wires. The fastest and most accurate test for locating an open wire, loose terminal, or faulty switch is to test for a voltage drop (loss). The **voltage drop method of troubleshooting** requires that power remain applied to the circuits while you touch the meter probes to each test point. *You must be aware of where you are placing your hands, tools, and meter probes at all times because severe electrical shock could result if you come in contact with the electrical circuit.* Even though leaving power applied to a circuit while you test it presents an electrical safety condition, it is necessary because control circuits tend to be complex and you may not be able to find the problem with the power off.

Figure 18–2 shows the test points where you should place the voltmeter probes to execute the test. Remember that this test can be used to locate problems in any circuit, regardless of the number and types of switches in the circuit. The test points are identified as 1 through 11. The first test should be made with the voltmeter probes touching points 1 and 2, which is across the power supply. Use this test to make sure that your meter is on the right range. From this point on through the remainder of the test, leave one of the voltmeter probes on point 2 or point 9, which is essentially the same point since they both connect to the return terminal of the transformer. The next step is to move the other probe to point 3. If no voltage is available at point 3, and you had voltage at point 1, then the wire between the thermostat and the power supply is open. You can turn off the power and test the wire for continuity to verify that the wire has an open. You should remember that a continuity test uses the resistance range on the VOM meter. If the wire has an open, the meter will indicate infinite resistance; if the wire is good, it is said to have continuity and the meter will indicate low resistance. It is also important to remember that the wire must be isolated (one end removed from its terminals) so that you do not measure resistance from a parallel circuit that may be connected to wire you are testing. If your continuity test indicates that the wire has an open, you should remove and replace the wire to put the circuit back in service.

If you measured voltage when the meter probe was on point 3 and point 9, the circuit is good to point 3. You can move the meter to test point 4. If you do not have voltage at point 4 and you had voltage at point 9, then the thermostat is open in its circuit R–Y. Remember that the thermostat contacts for cooling are R–Y and should be closed when the thermostat is turned to the coolest setting, so you should have voltage at points 4 to 9. You can check to be sure that the thermostat is set to air conditioning and is set to the coolest temperature setting. If the thermostat is set to cooling and is set to its coolest temperature setting, and no power is present at points 4–9, you can suspect the thermostat's R–Y circuit is faulty and the thermostat must be replaced.

If you are testing for the thermostat circuit R–Y that will not de-energize the relay coil, be sure to test point 4 to ensure that voltage is not present there when the thermostat is set to its highest temperature setting. If voltage is still present at points 4 to 9 even when the thermostat is set to its highest temperature setting, you can suspect that its contacts are stuck together and the thermostat must be replaced.

To continue the test for a loss of voltage in this circuit, we will resume from the last point where voltage is present at points 4 to 9, and continue the test by moving the probe to point 5. If the voltmeter indicates that voltage is present at point 4, and not at point 5,

then there is an open in the wire that connects the thermostat and the high-pressure switch. Since the voltage test indicates a loss of voltage at point 5, you can turn off power to the circuit and test this wire for continuity. If this test indicates an open, you should change the wire.

If you have voltage at points 5 to 9, you have determined that the circuit is operating correctly to this point and you can move to point 6. If your test indicates that you do not have voltage at points 6 to 9, you should suspect that the contacts of the high-pressure switch are open. You can try to reset the high-pressure switch and see if you have voltage at points 6 to 9. If you do not have voltage at point 6 when the switch is reset, you should turn off power and test the switch for continuity. If the switch is faulty during the continuity test, and it is not open due to high pressure, you should replace the switch. You should also be aware that the high-pressure switch may open when there is a problem in the refrigeration cycle, such as when the condenser coils become dirty. If you get the system to run after resetting the high-pressure switch, you should be sure to check the refrigeration pressures to ensure that they are within safe limits and clean the coils if the pressure is too high.

If you have voltage at point 6, your next test should be at points 7 to 9. If you do not have voltage at points 7 to 9, then the wire between the high-pressure switch and the ASCT contacts has an open in it. You can turn off power and test this wire for continuity. Replace the wire if it has an open.

If you have voltage at points 7 to 9, you should move the meter lead from point 7 to point 8. If you do not have voltages at points 8 to 9, you should suspect that the ASCT contacts are open. The ASCT is the **anti short cycle timer**, which provides a short 2 to 3 minute time delay anytime the compressor contactor is energized. You can wait 2 to 3 minutes after its coil is energized to see if its contacts close. If its contacts do not close after the time delay, you should suspect that it is bad and needs to be replaced. Many of these devices are solid state and cannot be repaired.

If you have voltage at terminal test points 8 to 9, you have voltage directly across the compressor contactor coil and it should become energized. If the coil does not become magnetic when voltage is applied directly across its terminals, you should suspect that it has an open in its coil. You should turn off power and test the coil for continuity. If the coil is faulty, you will need to remove and replace the contactor. This is perhaps the most important point in troubleshooting. The main function of all of the wires and switches in every control circuit is to provide the proper voltage to the control circuit load, which is usually a coil. In this circuit the load is the compressor contactor coil. If you measure voltage at its terminals and its contacts are not pulled in, you can predict that the circuit wires and switches are operating correctly, and you should suspect that the coil is faulty. You can turn off power to the circuit and test the compressor contactor coil for continuity. If the coil is good, the test should indicate resistance in the range of 21 to 2100 ohms, depending on how much wire is in the coil. If the test indicates infinite resistance, the coil is open and should be replaced. Actually you will need to replace the entire relay most times as the coil is not sold separately in many cases. Also remember that you should not apply voltage to a coil that is not installed in a relay.

One additional point you need to remember if you are using terminal point 2 instead of terminal point 9 as the common terminal: you should be aware that you could have an open in the wire between test point 9 and test point 2. If this is the case, you can determine that the wire is faulty or open by leaving one meter lead on test point 8 and checking for voltage at test point 2. If you have voltage at test points 8 to 2, you should move your meter lead to test point 9. When your meter leads are on test points 8 to 9 and you do not measure voltage, but you did at test points 8 to 2, you can suspect that the wire between test points 9 and 2 is open.

18.13 FAN AUTO MODE AND FAN ON MODE

The evaporator fan in a residential HVAC can operate in the "automatic mode" or in the "on mode," which is always on. A switch on the thermostat allows the switch to be set for the auto mode, which causes the fan to come on only when the air-conditioning compressor is running or when the furnace is in its heating cycle, or the on mode, which causes the fan to run constantly. When you are troubleshooting, a good procedure to follow is to try to operate the fan in manual mode. If the fan operates in manual mode, you will know that the fan is functioning correctly. If the fan does not operate in auto mode, after testing it in manual mode, you can suspect that the components and wiring in the control circuitry for the auto mode is causing the problem.

18.14 TROUBLESHOOTING AN ELECTRIC FURNACE

In this section you'll see a diagram of an electric furnace, shown in Figure 18–3. We will use the troubleshooting procedure again, and you will see how you can use it to troubleshoot any type of equipment and isolate the problem in the system. When you are testing the system, the first thing you want to establish is whether the system has voltage supplied to it. The first point that voltage comes into the system is at test points 1 and 7, which are directly after the **circuit breakers**. If you have 240 volts at test points 1 and 7, you know that voltage is available to the system. The next test that we will make is to determine if the heating contractor is closed or not. If the heating contractors are closed, we can ignore the control circuit. If the contacts for the heating contractor are not closed, we will need to begin our troubleshooting at the control circuit.

Our next test in this diagram is to look for voltage on the load side of the contacts at terminals 3–7 to determine if the contacts are closed. If you place your voltmeter leads on terminals 3 and 7, you should measure 240 V. If you do measure 240 V, the contacts to the heater are closed and you can ignore the control circuit. We will continue our test to determine why the electric heating element is not energized.

If the heater does not heat up, leave one lead of your voltmeter on test point 7 and move the other lead to test point 4. If you had voltage at test point 3 but not at test point 4, you should suspect that the wire between test points 3 and 4 is bad. At this point, turn off power to the furnace, remove the wire between test points 3 and 4, and test it for continuity. Replace it with the proper type of wire and test the system again.

If you have voltage at test points 4 to 7 and the heating element is still not heating up, leave the voltmeter lead on terminal 4 and move the other meter lead to test point 6. If you have voltage at test point 7 but not at test point 6, you can suspect that the wire between those two test points is bad, and you can turn off power and test it for continuity.

If you have voltage at test points 6 to 4, your next test would be to move the meter leads so that you are touching test point 4 and test point 5. If you have voltage at test point 5, you know that the temperature limits switch is closed and now you have voltage directly across the heating elements. If the heating elements do not heat with voltage applied directly to them, you can suspect that the heating element has an open. You can turn off power and test the heating element for continuity. You can also examine the coils to see if there is a burned spot or physical open in the coils. At this point, you will need to install new heating coils.

18.14.1 The Coil of the Contactor Is Not Energized

In this scenario the contactor coil is not energized. When you test for voltage at test point 2 to test point 7, you have voltage, and when you test from test point 3 to test point 7, there is no voltage. Since there is no voltage at test point 3, it indicates that

the contacts are not closed. Now our tests will focus on the control circuit where the coil is.

Figure 18–3 Wiring diagram and schematic (ladder) diagram of an electric furnace provided for troubleshooting. *(Courtesy of Nordyne)*

the contacts are not closed. Now our tests will focus on the control circuit where the coil is.

In this test, you should first determine if there is voltage at the transformer secondary at test points 9 and 10. If you have 24 volts at test points 9 and 10, you have determined that the transformer is supplying the 24 volts to the control circuit. If you do not measure 24 volts at test points 9 and 10, the transformer is not operating correctly and it should be replaced.

If you have voltage at test points 9 and 10, you can leave the meter lead on test point 10 and move the other lead to test point 8. If you do not have voltage at test point 8, the fuse is open. If you have voltage at test point 8, then the fuse is working correctly and you can move to the next test point. Test point 13 is where the "W" terminal of the thermostat provides voltage to the contactor coil. You can check this voltage at test point 13 to test point 10. If you have voltage at test points 13 to 10, you can move on to test points 11 to 12, which is directly across the coil of the contactor. If 24 volts is present at terminal points 12 to 13, then the coil should be energized. If the coil is not energized,

then the coil has an open, and you can turn off power and check for continuity. If you do not have voltage at test points 12 to 13, then you would check for loss of voltage in the wiring. If the coil has an open, you should replace the relay.

18.15 TROUBLESHOOTING A PACKAGED GAS HEATING/ELECTRIC AIR-CONDITIONER COMPRESSOR CIRCUIT

Figure 18–4 shows the schematic (ladder) diagram and the wiring diagram for a packaged gas heating system with electrical air conditioning. This section will show you some typical troubleshooting procedures for a packaged-type heating and cooling system. If you are called to troubleshoot the system and the compressor and condenser fan do not run, you can start the procedure by testing for incoming voltage at L1 and L2. You should have 240 V at L1 to L2. If you do not, you need to go back to the circuit breaker in the electric panel or the fuse disconnect for the system to determine where voltage has been interrupted. If voltage is present, you can leave the meter lead on L1 and move the other meter lead to where voltage comes in the contacts. If you have voltage at this point, you know the wire is good. If you lose voltage, you know the wire between the two points is bad, and you can turn off power and test it for continuity.

The next test should be where voltage comes out of the contacts. If you have voltage at this point, you have determined that the contacts for the compressor contactor are

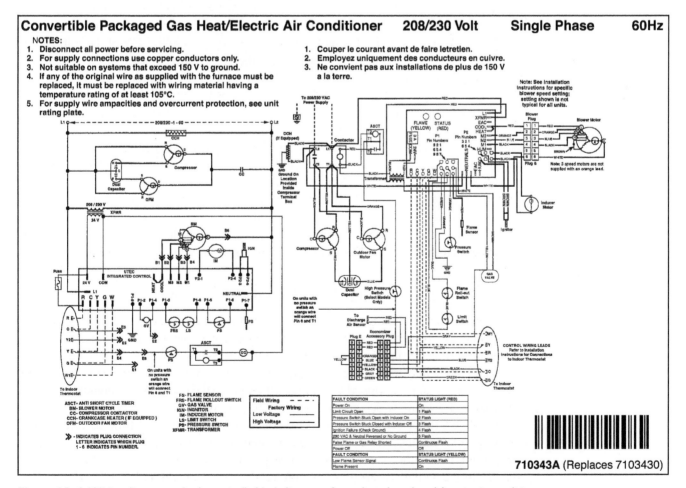

Figure 18–4 Wiring diagram and schematic (ladder) diagram of a packaged gas heat/electric air conditioner. *(Courtesy of Nordyne)*

closed. You can leave the terminal on this point and move your other meter lead to where voltage comes in the capacitor. If you have voltage at this point, you have determined that voltage is provided to the **start capacitor** and to the common terminal of the compressor and condenser motor. If the compressor still does not run and you have voltage at the run (R) terminal and at the start (S) terminal on the left side of the motor and you have voltage at the common (C) terminal on the right side, you can predict that the compressor motor has an internal open that is either caused by the internal overload being open or the run winding or start winding has an open. You can turn off the power to the system and test the compressor windings for continuity. Be sure to provide ample time to allow the compressor motor to cool down in the event the internal overload has tripped.

18.15.1 How to Troubleshoot a Compressor That Tries to Run and Hums Loudly

If the compressor tries to start and hums loudly, you should only leave it run for a few seconds as you make tests and be sure to turn it off immediately so that the compressor does not overheat. When the compressor tries to start and hums, it indicates that voltage is present, and that at least one of the windings is drawing current. You can use a clamp-on ammeter to determine whether the start winding or the run winding is drawing current by placing the meter around the wire connected to the run winding first and turning on power. If it shows a current draw, you have determined that the run winding is pulling current. You can turn off power and move the meter to the wire connected to the start winding. If the start winding is also drawing current, you can suspect that both the start and run winding are good, and the compressor piston is stuck so that the compressor cannot run. You can use the techniques you learned in the chapters about motors to increase the size of the start capacitor or add a hard-start kit to see if you can get the compressor to start. If the compressor does start, you should leave it run for several hours before you turn it off and try it again. If the compressor does not start, you will need to remove and replace the compressor.

If you measure current at the run winding but not the start winding, you should suspect that the start winding or the start capacitor is faulty. You can turn off all power and test the start winding for continuity and check the capacitor. If your tests determine that the start winding has an open in it, you will need to remove and replace the compressor.

If the compressor does not hum when you try to start it, and you do not measure any current at the start winding and the run winding, you should suspect that the open in the compressor is in the common side of the windings, where the internal overload is usually located. Be sure to allow extra time for the internal overload to cool down and close before you make a decision that the compressor is bad and needs to be replaced.

18.15.2 Troubleshooting the Condenser Fan Motor

If the compressor is running but the condenser fan is not running, the system will quickly trip out on high pressure. When you check the system, you may find that both the compressor and the condenser fan are not running. The compressor will not be running because the internal overload may trip due to the high pressure, or both of them may be off because the high-pressure switch has tripped and turned off power to the compressor contactor coil. You will need to determine which condition is keeping the system off and try to set it up so the compressor contactor can be energized for at least a short amount of time so you can proceed with your next tests. Once you get the compressor contactor closed and the compressor running, you will only have a few seconds to determine that the condenser fan is not running. Since the compressor is running, you have determined that the compressor contactor is closed and power is available all the way to the point on the left side of the motors and to

the point on the right side of the motors. At this point you can predict that the problem is either that the condenser fan has an opening in one of its windings, that the start capacitor for the condenser fan is not providing voltage to the start winding, or that the wiring to the motor is bad. You can check this by trying to run the fan motor again and putting a clamp-on ammeter around the wire going to the start winding. If you do not measure any current in the start winding, you will need to make one more test to determine if the problem is the start winding in the motor or the capacitor. If you take a voltage measurement at the start winding, you may find voltage but it could come from the capacitor or it could feed back through the other windings. This means that you will have to turn off power, disconnect the fan motor, and test it for continuity. If both the start winding and the run winding have continuity, then you should suspect that the start capacitor is the problem. If you do not have continuity at the start winding or the run winding, you should replace the fan motor with a new one.

18.15.3 Troubleshooting the System if the Contacts for the Compressor Contactor Do Not Close

If you measured voltage where it comes into the contacts but do not have voltage where it comes through the contacts, it indicates that the compressor contactor has not closed its contacts. If this occurs, you need to move your tests to the control circuit that provides voltage to the compressor contractor coil. The source of voltage for the control circuit is 24 volts that comes from the secondary side of the transformer. You can test for 24 volts at this point. If you have 24 V at this point, you know that the transformer is working correctly. If you do not have voltage at these points, you should check for 240 volts at the primary windings of the transformer. If you have voltage at the primary but not at the secondary windings, you should suspect that the transformer is faulty. Turn off power and test it for continuity. If the transformer windings have an open, you will need to replace the transformer.

If you have 24 V at the transformer you can continue your test through the remainder of the control circuit that provides voltage to the coil of the compressor contactor.

The next test point in this procedure is to ensure that voltage from the transformer secondary through the fuse to terminal R. You can test for 24 V at test at this point. If you have voltage at this point, the fuse is good. The next test is to determine if voltage made it through the cooling circuit of the thermostat. If you have voltage at this point, voltage has made it to the thermostat and is available at terminal Y. If you do not have voltage at this point, but you did at the previous point, then you should check the thermostat to ensure that it is set for cooling and is at the lowest temperature. If it still does not provide voltage through the thermostat, then you can suspect that the thermostat is faulty.

If you have voltage at this point, the thermostat is working correctly, and your next test is to ensure that the high-pressure switch is closed. To determine if the high-pressure switch is closed you need to measure voltage where it comes through the switch. If you do not have voltage at point 12 to point 9, it indicates that the high-pressure switch is opened. At this point it is important to understand that the high-pressure switch is probably not faulty, rather a high-pressure condition has occurred and you should check for things that would cause the high pressure, such as a dirty condenser coil or the condenser fan not running. The high-pressure switch typically has a reset button that you can depress to reset the switch. When you do this you should get voltage through the switch and you can continue checking the system. If the compressor contactor coil now becomes energized and closes the compressor contacts, and the compressor starts to run, you can predict that your problem is with a high-pressure condition in the system. You can then make the necessary repairs by cleaning the condenser coil or ensuring the condenser fan is running.

If the compressor contactor does not energize, you can test for voltage directly across the coil at point 12 to point 14. If you have voltage at point 12 to point 14, your next test will be to ensure that the time-delay circuit is working correctly. The time-delay system is called the anti short cycle timer (ASCT), which provides a short time delay every time the system calls for cooling. You can test the ASCT by waiting 2 to 3 minutes and then testing for voltage at test point 13 to test point 14. If you do not have voltage at point 13 to point 14 after the time delay has occurred, you can predict that the timer is bad and should be replaced.

If the timer is working correctly, you should be able to measure 24 V at point 13 to point 14. This indicates that voltage is provided directly across the coil of the compressor contactor, and it should become energized and pull the compressor contacts in. You should be able to visually observe the compressor contacts changing from their open state to the closed state. You may also be able to hear this condition occur. If the compressor contacts do not change states when the coil is energized, the compressor contactor coil has an open. You can turn on voltage and test it for continuity. If the coil has an open, you will need to replace the entire contactor.

18.16 TROUBLESHOOTING A PACKAGED AIR CONDITIONER WHEN THE COMPRESSOR WILL NOT TURN ON

Figure 18–5 shows the schematic (ladder) diagram and the wiring diagram for a packaged air-conditioning system. This section will show you some typical troubleshooting procedures for a packaged air-conditioning system. If you are called to troubleshoot the system and the compressor does not run, you can start the procedure by testing for incoming voltage at L1 and L2. You should have 240 V at L1 and L2. If you do not, you need to go back to the circuit breaker in the electric panel or the fuse disconnect for the system to determine where voltage has been interrupted. If voltage is present at L1 to L2, you can leave the meter lead on L1 and move the other meter lead to the point where voltage comes into the contacts. If you have voltage at this point, you know the wire is good. If you lose voltage at this point, you know the wire is bad, and you can turn off power and test it for continuity.

The next test should be where voltage comes through the contacts to L1. If you have voltage at this point, the contacts for the compressor contactor are closed, and you can leave the terminal on L1 and move your other meter lead to terminal C on the compressor. If you have voltage at this point, you have determined that the wire is good. If you do not have voltage at this point, you can suspect the wire is open.

If you have voltage at terminal C on the compressor, you can leave your meter lead there and move the other meter lead from L1 to terminal R on the compressor. If you had voltage at the previous point and you do not have voltage at terminal R on the compressor, you should suspect that the wire has an open in it. You can turn off power, test the wire for continuity, and replace it if necessary.

If you have voltage at terminal R to C on the compressor, you will need to make further tests on the compressor. The problems may include an open winding in the compressor, a thermal overload that is opened in the compressor, a start capacitor that is bad, or a stuck piston in the compressor. Remember, if the compressor hums and tries to start, you can use the troubleshooting procedure listed in 18.15.1.

If you suspect that the start capacitor is bad, you can put a clamp-on ammeter around the wire that is connected to the start terminal (S) on the compressor to see if it

Figure 18–5 Wiring diagram and schematic (ladder) diagram of a packaged air conditioner. *(Courtesy of Nordyne)*

pulls any current while you try to start it. If the compressor does not draw current at the S terminal and hums loudly, it has lost the start winding or the start capacitor. You can turn off power and test the start winding for continuity. If the compressor start winding has continuity, you can suspect that the start capacitor is not operating correctly. You can test it and change it if necessary.

If you have determined that voltage is provided to the start capacitor and to the common terminal of the compressor and condenser motor and it does not hum and show any current draw on the start and run windings, you can suspect that the compressor motor has an internal open that is caused by either the internal overload being open or the run winding or start winding having an open. You can turn off the power to the system and test the compressor windings for continuity. Be sure to provide ample time to allow the compressor motor to cool down in the event the internal overload has tripped.

18.17 TROUBLESHOOTING A SPLIT-SYSTEM AIR CONDITIONER WHEN THE COMPRESSOR DOES NOT RUN

Figure 18–6 shows the schematic (ladder) diagram and the wiring diagram for a split-system air-conditioning system. This section will show you some typical troubleshooting procedures for a split-system air-conditioning system. If you are called to troubleshoot the system and the compressor does not run, you can start the procedure by

Figure 18–6 Wiring diagram and schematic (ladder) diagram of split-system air conditioner. *(Courtesy of Nordyne)*

testing for incoming voltage at L1 and L2. You should have 240 V at L1 to L2. If you do not, you need to go back to the circuit breaker in the electric panel or the fused disconnect for the system to determine where voltage has been interrupted. If voltage is present at L1 to L2, you can leave one meter lead on L1 and move the other meter lead to terminal T2 at the output of the contacts. If you have voltage at this point, you have determined that the contacts for the compressor contactor are closed, and you can leave the terminal on L1 and move your other meter lead to terminal C on the compressor. If you have voltage at terminal C on the compressor the wire between the contacts and terminal C on the compressor is good. If you do not have voltage at this point, you can suspect the wire is open. At this point you should be able to determine if the compressor is trying to start and is humming. Refer back to section 18.15.1 to determine how to troubleshoot a compressor that hums, but does not start.

If you have voltage going into the contacts but not at terminal T2 coming out of the contacts, then the compressor contactor contacts are not closed, and you will need to move your tests to the control circuit where the compressor contactor coil is located. We will discuss those tests in section 18.19.

If you have voltage at terminal R and C on the compressor, and the compressor does not try to start and does not hum, you will need to make further tests on the compressor. You are looking for an open in the common circuit of the compressor motor windings or at the internal thermal overload. The problems may include an open winding in the compressor, a thermal overload that is opened in the compressor, a start capacitor that is

bad, or a stuck piston in the compressor. Remember, if the compressor hums and tries to start, you can use the troubleshooting procedure listed in 18.15.1. If the compressor does not hum and try to run, the problem is a loss of voltage at terminal C, an open at the **common point** in the motor, or an open in the internal overload.

If the compressor tries to run, you may also have a faulty start capacitor. If you suspect that the start capacitor is bad, you can put a clamp-on ammeter around the wire that is connected to the start terminal (S) on the compressor to see if it pulls any current while you try to start it. If the compressor does not draw current at the S terminal and hums loudly, the run winding is pulling current and you can suspect that the start winding is open or the start capacitor is faulty. You can turn off power and test the run winding for continuity. If the compressor run winding and the start winding have continuity, you can suspect that the start capacitor is not operating correctly. You can test it and change it if necessary.

If you have determined that voltage is provided to the start capacitor and to the common terminal of the compressor and condenser motor and it does not hum and show any current draw on the start and run windings, you can suspect that the compressor motor has an internal open that is caused by either the internal overload being open or the run winding or start winding having an open. You can turn off the power to the system and test the compressor windings for continuity. Be sure to provide ample time to allow the compressor motor to cool down in the event the internal overload has tripped.

18.18 REVIEW

Now that you have seen all of the features of troubleshooting you should remember several important points. First, always work safely and use lockout and tagout procedures when you are troubleshooting. If you do need to test parts of the circuits with the voltage turned on, you must be very careful to be aware of the points where voltage is live so you do not touch any of them with your hands or any other parts of your body. You should also remember that troubleshooting is the process of determining what systems are working correctly on a HVAC or refrigeration system, so that you can eliminate them from your process and focus on the part of the system that is malfunctioning. You should understand that there is a test that has been presented in this text that will allow you to test any component or control of the electrical system and verify whether the part is good or bad. These tests include in-circuit tests and tests for when a part is removed from a system. If possible, you should test components while they are in circuit. If you have determined that a part is faulty, you should remove it and test it again when it is out of circuit to verify your previous findings.

When you are troubleshooting, you should use electrical ladder diagrams and wiring diagrams whenever possible. If you have a wiring diagram and need a ladder diagram, you can convert the diagram using the procedures shown in Chapter 6. You can also convert a ladder diagram to a wiring diagram when you are installing components or need the wiring diagram for other tests.

The final point to remember about troubleshooting is to use all of your senses—vision, hearing, smell, and touch—to evaluate an HVAC or refrigeration system that is malfunctioning. Also be sure to ask personnel that are around the HVAC or refrigeration system how the HVAC or refrigeration system operates under normal conditions. Check other similar HVAC or refrigeration systems if you have them available to verify operation. Try to operate all of the sections or components of the HVAC or refrigeration system in manual mode before you test them in automatic mode.

You must learn to develop a strategy that strictly follows procedures. You should select tests that will eliminate the largest portions of the HVAC or refrigeration system like the example that showed the best procedure of identifying a card in a deck of cards.

Finally you should keep very accurate and detailed records of your troubleshooting and maintenance functions.

Job Assignment for This Chapter

You receive a work order to troubleshoot a packaged air-conditioning system that has gas heat. It is summer and the unit is in the air-conditioning mode. The complaint is that the unit is not cooling, and your job is to create a process to troubleshoot the problem. Part of your process should include identifying what parts of the system you can ignore and where you should focus your troubleshooting. You need to explain to the home-owner what tests you will make and what troubleshooting processes you will use. You will also need to explain why you can ignore parts of the system.

Solution to the Job Assignment

Since the complaint is that the system will not cool, and it is summer, you can ignore all of the parts of the system that provide heating. You can use the procedures outlined in section 18.16. The first test that you want to make is to determine if the compressor contactor is energized and has closed its contacts. If the compressor contactor contacts are closed, you can exclude all of the control circuit and thermostat, since they have done their job and the coil is energized. This leaves you with only the compressor motor, the start capacitor, and all of the interconnecting wires. If the compressor contactor is not energized and its contacts are not closed, then you would switch your tests to the control circuit, which includes the transformer secondary winding, the thermostat cooling circuit, and the anti short cycle timer (ASCT) relay. You will need to check for voltage across the coil terminals of the compressor contactor. If voltage is present, you can suspect that the contactor coil is bad. If you do not have voltage, you will need to check the circuit with power on to find where voltage is lost. You should be able to explain all of this to the system owner.

Assessment

QUESTIONS

1. Explain why you should use the voltage-loss method to detect an open in a circuit.
2. Explain the difference between a symptom and a problem.
3. Explain when it would be appropriate to swap a part that you thought was faulty.
4. Explain why it is important to use the same procedure every time you troubleshoot an HVAC or refrigeration system.
5. Explain when you should use an ammeter to make a test in a circuit.

TRUE OR FALSE

1. When you test a component with an ohmmeter, you should always ensure the power is turned off.
2. When you test a component with a voltmeter, you should always ensure the power is turned off.

3. When you are called to troubleshoot an HVAC or refrigeration system, you should use all of your senses to determine what parts of the HVAC or refrigeration system are operating correctly.

4. You could use your "sense of feel" to determine if a fan or other part of the system is vibrating excessively.

5. A clamp-on ammeter could be used to determine which line of the three lines that supply voltage to a motor has a blown fuse if the motor is humming and trying to start.

MULTIPLE CHOICE

1. A wiring diagram
 a. shows the sequence of operation for the circuit.
 b. shows the location of each component in the circuit.
 c. shows how each component operates.

2. A ladder diagram
 a. shows the sequence of operation for the circuit.
 b. shows the location of each component in the circuit.
 c. shows how each component operates.

3. Troubleshooting may best be described as
 a. finding the part of the HVAC or refrigeration system that is broken.
 b. identifying the parts of an HVAC or refrigeration system that are working correctly and through a process of elimination determining what is faulty and not working correctly.
 c. swapping parts until the HVAC or refrigeration system starts running again.

4. The control circuit
 a. includes the motor and other loads in a circuit.
 b. includes switches and other controls in a circuit.
 c. all of the above.

5. The load circuit
 a. includes the motor and other loads in a circuit.
 b. includes switches and other controls in a circuit.
 c. all of the above.

Websites

Websites change constantly. Use the key word search to locate sites that have similar content. The products that will be found at each site are indicated with the key word.

Products: Air conditioners
Key word: Nordyne Air Conditioners

Products: Packaged air conditioners
Key word: Nordyne Packaged Air Conditioners

Products: Wiring diagrams for air conditioners and refrigeration systems
Key word: HVAC and Refrigeration Wiring Diagrams

Chapter 19

Electronic Devices for HVAC Systems

Job Assignment for This Chapter

You are working on a new furnace that has an electronic circuit board to control ignition and the gas valve. The furnace will not light off and you suspect that the electronic circuit board is bad. You are suggesting to the homeowner that the board needs to be changed out. You have tested the board and know that it does not have any voltage anywhere on the board after 120 volts AC is supplied to the board. The homeowner asks you what components on the board you suspect are bad and how you might test them. Your job is to explain how the board plugs into the 120 volt AC that supplies power to the furnace and provides the DC voltage for electronic control.

Terms You Will Learn in This Chapter:

Anode
Atom
Base
Cathode
Collector
Conductor
Diac
Diode
Electrons
Emitter
Full-wave bridge rectifier

Half-wave bridge rectifier
Insulator
Lattice structure
Light-emitting diode (LED)
NPN transistor
N-type material
Operational amplifier
PN junction
PNP transistor
Proportional valve
P-type material

Rectifier
Semiconductor
Silicon
Silicon controlled rectifier (SCR)
Solid-state relay (SSR)
Transistor
Triac
Unijunction transistor (UJT)
Valence electron
Variable-frequency drive (VFD)

After reading this chapter, you will be able to:

1. Explain P-type and N-type material.

2. Identify the terminals of a diode and explain its operation.

3. Explain the operation of PNP and NPN transistors.

4. Identify the terminals of a silicon controlled rectifier (SCR) and explain its operation.

5. Explain the operation of a triac.

6. Understand the operation of a variable-frequency drive (VFD).

7. Understand the operation of electronic circuit boards to control furnaces.

19.0 OVERVIEW OF ELECTRONICS USED IN HVAC AND REFRIGERATION SYSTEMS

Electronics are used in all areas of HVAC systems such as electronic thermostats; electronic boards to control gas valves, ignition, and blower speeds; and variable-frequency drives that control motor speeds. Electronics make systems more efficient and more reliable. In this chapter you will find that electronic components provide functions that are similar to electrical components and systems you already understand. The information in this chapter will provide names for all components and explain the operation of electronic devices such as diodes, transistors, and SCRs that have become commonplace in heating, air-conditioning, and refrigeration systems because they provide better control than electromechanical devices and are less expensive to manufacture. Electronic components are also called solid-state components. You will also gain an understanding of P-material and N-material, which are the building blocks of all electronic components. After you get a basic understanding of P-material and N-material, you will be introduced to diodes, transistors, SCRs, and triacs, and you will see application circuits of each of these types of components. The theory of operation and troubleshooting techniques for each type of device will also be presented so that you will have a good understanding of how to determine whether a device or circuit board is working correctly. At first, most HVAC technicians think that it is difficult to troubleshoot electronic devices because you cannot see their operation. After you fully understand the theory of operation of each component and basic circuits, you will find it easy to determine if they are working correctly or if they are faulty and must be replaced. You will also find in this chapter that if you determine that a component or circuit is faulty, you will change out the complete device or board, rather than try to repair an individual component.

19.1 CONDUCTORS, INSULATORS, AND SEMICONDUCTORS

In earlier chapters you learned that **atoms** have protons and neutrons in their nucleus and electrons that move around the nucleus in regions of space called shells. The number of **electrons** in the atom is different for each element. For example, you learned that copper has 29 electrons and three of them are located in the outermost shell. The outermost shell is called the valence shell and the electrons in that shell are called

valence electrons. The atoms of the most stable material have eight valence electrons, which are found as four pairs. This means that an atom may have five, six, or seven atoms, and it will take less energy to add electrons to get a full shell (eight), or it may have one, two, or three electrons, and it will take less energy to give up these electrons to achieve the stable configuration of a full shell with eight electrons.

A **conductor** is a material that allows electrons (electrical current) to flow easily through it, and an **insulator** does not allow current to flow through it. An example of a conductor is copper that is used for electrical wiring. An example of an insulator is rubber or plastic. The atomic structure of a conductor makes it easier for electrons to flow through it, and the atomic structure of an insulator makes it nearly impossible for any electrons to flow through it.

Figure 19–1 shows a simplified atomic structure of a conductor. Atoms of conductors can have one, two, or three valence electrons. The atom in this example has one valence electron. Since all atoms will gain or lose sufficient electrons to achieve a configuration of eight electrons (four pairs) in their valence shell, it takes less energy for conductors to give up these electrons (one, two, or three) so that the valence shell will become empty. At this point the atom becomes stable because the previous shell becomes the new valence shell and it has eight electrons. The electrons that are given up are free to move as current flow.

Figure 19–2 shows the simplified atomic structure of an insulator. Insulators will have five, six, or seven valence electrons. In this example you can see that the atom has seven valence electrons. This structure makes it easy for insulators to take on extra electrons to get eight valence electrons. The electrons that are captured to fill the valence shell are electrons that would normally be free to flow as current.

Semiconductors are materials whose atoms have exactly four valence electrons. Since these atoms have exactly four valence electrons, they can take on four new valence electrons like an insulator to get a full valence shell, or they can give up four

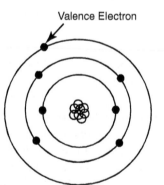

Figure 19–1 Simplified atomic structure of a conductor.

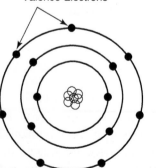

Figure 19–2 Simplified structure of an insulator.

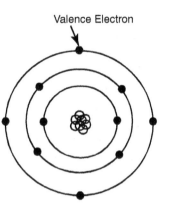

Valence Electron

Figure 19–3 Simplified atomic structure of a semiconductor material.

valence electrons like a conductor to get an empty outer shell. Then the previous shell that has eight electrons becomes the valence shell. Figure 19–3 shows the simplified atomic structure of a semiconductor material.

19.2 COMBINING ATOMS

When solid-state material or other material is manufactured, large numbers of atoms are placed together. The structure that becomes most stable at this point is called a **lattice structure**. Figure 19–4 shows an example of the lattice structure that occurs when atoms are combined. In this diagram you can see that atoms of **silicon** (Si), which is a semiconductor material with four valence electrons, combine so that one valence electron from each of the neighbor atoms is shared, so that all atoms look and act as if they have eight valence electrons each.

19.3 COMBINING ARSENIC AND SILICON TO MAKE N-TYPE MATERIAL

Other types of atoms can be combined with semiconductor atoms to create the special material that is used in solid-state transistors and diodes. Figure 19–5 shows a diagram of four silicon atoms combined with one atom of arsenic. From this figure you can see that arsenic has five valence electrons, and when the silicon atoms are combined with it they create a very strong lattice structure. You can see that each silicon atom donates one of its valence electrons to pair up with each of the valence electrons of the arsenic atom. Since the arsenic atom has five valence electrons, one of the electrons will not be paired up and will become displaced from the atom. This electron is called a *free electron* and it can go into conduction with very little energy. Since this new material has a free electron it is called **N-type material**.

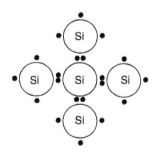

Figure 19–4 Atoms of silicon semiconductor material combine to create a lattice structure.

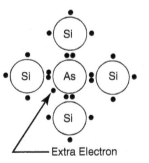

Figure 19–5 N-material formed by combining four silicon atoms with a single arsenic atom.

Extra Electron

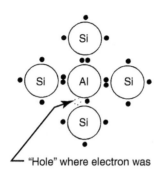

Figure 19–6 P-material formed by combining four silicon atoms with a single aluminum atom.

"Hole" where electron was

19.4 COMBINING ALUMINUM AND SILICON TO MAKE P-TYPE MATERIAL

An atom of aluminum can also be combined with semiconductor atoms to create the special material called P-type material. Figure 19–6 shows a diagram of four silicon atoms combined with one atom of aluminum. From this figure you can see that aluminum has three valence electrons, and when the silicon atoms combine with it they create a very strong lattice structure. You can see that each silicon atom donates one of its valence electrons to pair up with each of the valence electrons of the aluminum atom. Since the aluminum atom has three valence electrons, one of the four aluminum electrons will not be paired up and it will have a space where any free electron can move into it to combine with the single electron. This free space is called a *hole*, and it is considered to have a positive charge since it is not occupied by a negatively charged electron. Since this new material has an excess hole that has a positive charge, it is called **P-type material**.

19.5 THE PN JUNCTION

One piece of P-type material can be combined with one piece of N-type material to make a **PN junction**. Figure 19–7 shows a typical PN junction. The PN junction creates an electronic component called a **diode**. The diode is the simplest electronic device. When DC voltage is applied to the PN junction with the proper polarity it causes the junction to become a very good conductor. Conversely, if the polarity of the voltage is reversed, the PN junction becomes a good insulator.

19.6 FORWARD BIASING THE PN JUNCTION

When DC battery voltage is applied to the PN junction so that positive voltage is connected to the P-material and negative voltage is connected to the N-material, the junction is *forward biased*. Figure 19–8a shows a battery connected to the PN junction so that it is forward biased. In this diagram you can see that the positive battery voltage

Figure 19–7 An example of a piece of P-material joined to a piece of N-material.

P N

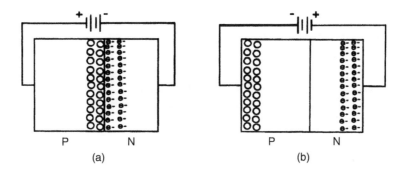

Figure 19–8 a A forward-biased PN junction. **b** A reverse-biased PN junction.

causes the majority of holes in the P-type material to be repelled, so that the free holes move toward the junction, where they will come into contact with the N-type material. At the same time the negative battery voltage also repels the free electrons in the N-type material toward the junction. Since the holes and free electrons come into contact at the junction, the electrons recombine with the holes to cause a low-resistance junction, which allows current to flow freely through it. When a PN junction has low resistance, it will allow current to pass just as if the junction were a closed switch.

It is important to understand that up to this point in this text, all current flow has been described in terms of *conventional current flow*, which is based on a theory that electrical current flows from a positive source to a negative return terminal. At this point you can see that this theory will not support current flow through electronic devices. For this reason *electron current flow theory* must be used when discussing electronic devices. In electron current flow theory, current is the flow of electrons and it flows from the negative terminal to the positive terminal in any electronic circuit.

19.7 REVERSE BIASING THE PN JUNCTION

Figure 19–8b shows a battery connected to the PN junction so that it is reverse biased. In this diagram you can see that the positive battery voltage is connected to the N-type material, and the negative battery voltage is connected to the P-type material. The negative voltage on the P-type material attracts the majority of holes in the P-type material, so they move away from the junction. Thus, they cannot come into contact with the N-type material. At the same time the positive battery voltage that is connected to the N-type material attracts the free electrons away from the junction. Since the holes and free electrons are both attracted away from the junction, no electrons can recombine with any holes. Thus, a high-resistance junction is formed that will not allow any current flow. When the PN junction has high resistance it will not allow any current to pass, just as if the junction were an open switch.

19.8 USING A DIODE FOR RECTIFICATION

In electronic circuits it is difficult to manipulate AC voltage. With the advent of electronics, DC voltage and current can be manipulated (changed) to control a variety of components in an HVAC system such as the speed of a fan motor, the opening and closing of a gas valve, the use of electronic air cleaners, or the creation of ignition system controls. Since most air-conditioning and heating equipment is supplied with AC voltage, electronic components are required to change the voltage into DC voltage so it can easily be changed. One or more diodes are used to change AC voltage to DC voltage.

(a) (b)

Figure 19–9 a PN junction for a diode. **b** Electronic symbol of a diode. The anode is the arrowhead part of the symbol, and the cathode is the other terminal.

Figure 19–10 A single diode used in a circuit to convert AC voltage to DC voltage.

An electronic diode is a simple PN junction. Figure 19–9 shows the symbol for a diode. You can see that the symbol for the diode looks like an arrowhead that is pointing against a line. The part of the symbol that is the arrowhead is called the **anode**, and it is also the P-type material of the PN junction. The other terminal of the diode is called the **cathode**, and it is the N-type material. Since the anode is made of positive P-type material, it is identified with a + sign. The cathode is identified with a − sign, since it is made of N-type material. Figure 19–10 shows a diode connected in a circuit that has an AC power source. The AC power source produces a sine wave that has a positive half-cycle followed by a negative half-cycle. The diode converts the AC sine wave to half-wave DC voltage by allowing current to pass when the AC voltage provides a forward bias to the PN junction, and it blocks current when the AC voltage provides a reverse bias to the PN junction. The forward-bias condition occurs when the AC voltage sine wave provides a positive voltage to the anode and a negative voltage to the cathode. During this part of the AC cycle, the diode is forward biased and has very low resistance, so current can flow. When the other half of the AC sine wave occurs, the diode becomes reverse biased, with negative voltage applied to the anode and positive voltage applied to the cathode. During the time the diode is reverse biased, a high-resistance junction is created, and no current will flow through it.

Rectification is the process of changing AC voltage to DC voltage. One of the main jobs of the diode is to convert AC voltage to DC voltage. Most electronic circuits used in heating, air-conditioning, and refrigeration equipment need some DC voltage to operate. Since the HVAC equipment is connected to AC voltage, a power supply is required to provide regulated DC voltage for the solid-state circuits, and the diode is part of the power supply that rectifies the AC voltage to DC.

19.9 HALF-WAVE AND FULL-WAVE RECTIFIERS

When one diode is used in a circuit to convert AC voltage to DC voltage, it is called a **half-wave rectifier**, since only the positive half of the AC voltage is allowed to pass through the diode, while the negative half is blocked. The **rectifier** shown in Figure 19–10 is a half-wave rectifier. The half-wave rectifier is not very efficient, since half of the AC sine wave is wasted.

If four diodes are used in the circuit, they can convert both the positive half-wave and the negative half-wave of the AC sine wave. Figure 19–11 shows a circuit with four

Figure 19–11 Four-diode, full-wave rectifier. This type of rectifier is often called a full-wave bridge rectifier, since the diodes are connected in a bridge circuit.

diodes used to rectify AC voltage to DC voltage. Since the four diodes can convert both the positive half and the negative half of the AC sine wave, this type of rectifier is called a **full-wave rectifier**. It is important to understand that the full-wave bridge consists of four diodes. Some circuit boards provide four individual diodes on the board, while other boards package the four diodes into an integrated circuit (IC) that has four leads. Two of the leads are marked AC and this is where the AC voltage is supplied to the bridge. The other two leads are identified as DC+ and DC− and these leads will provide DC voltage out of the bridge. If only one of the diodes in a bridge rectifier goes bad, the bridge will provide approximately half its rated voltage. If two or more diodes go bad in the bridge rectifier, the DC output voltage is usually zero.

19.10 THREE-PHASE RECTIFIERS

Larger three-phase AC compressor and fan motors used in air-conditioning, heating, and refrigeration systems can have their speed changed to run more efficiently by changing the frequency of the voltage supplied to them. In these applications, six diodes are used to convert three-phase AC voltage to DC voltage, and then a microprocessor-controlled circuit converts the DC voltage back to three-phase AC voltage. The frequency of this voltage can be adjusted to change the speed of the motors. Figure 19–12 shows a six-diode three-phase rectifier. Notice that the supply

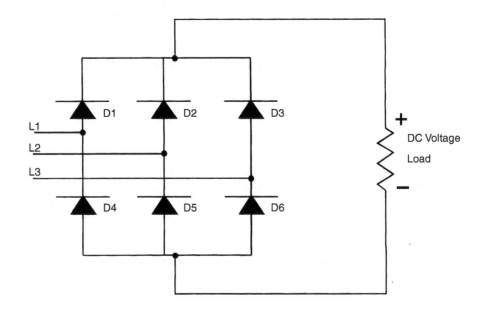

Figure 19–12 A six-diode bridge used to rectify three-phase AC voltage to DC voltage.

Figure 19–13 **a** Using the battery in an ohmmeter to forward bias a diode. The diode should have low resistance during this test. **b** Using the battery in the ohmmeter to reverse bias a diode. The diode should have high resistance during this test.

voltage to the diodes is three-phase AC voltage, and the output voltage from the rectifier consists of six positive half-waves.

19.11 TESTING DIODES

One of the tasks that you must perform as a technician is to test diodes to see if they are operating correctly. One way to do this is to apply AC voltage to the input of the diode circuit and test for DC voltage at the output of the diode circuit. If the amount of DC voltage at the power supply is half of what it is rated for in a four-diode bridge rectifier circuit, you can suspect that one of the diode pairs has one or both diodes opened. If this occurs, you can use an ohmmeter to test each diode to determine which one is faulty.

When you are testing the diodes with an ohmmeter, it is important that all power to the diode circuit be turned off. You should remember from earlier chapters that the ohmmeter uses an internal battery as a DC voltage source. Since you know the diode can be tested for forward bias and reverse bias with a DC voltage source, you can use the ohmmeter as the voltage source and the meter to test for high resistance and low resistance through the diode junction. Figure 19–13a shows an example of putting the positive ohmmeter terminal on the anode of the diode, and the negative ohmmeter terminal on the cathode of the diode, to cause the diode to go into forward bias. During this test the diode is forward biased and the ohmmeter should measure low resistance. When the ohmmeter leads are reversed as in Figure 19–13b so that the negative meter lead is connected to the anode of the diode, and the positive meter lead is connected to the anode of the diode, the diode is reverse biased. When the diode is reverse biased, the ohmmeter should measure infinite (∞) resistance. If the diode indicates high resistance when it is reversed biased and low resistance when it is forward biased, it is good. If the diode indicates low resistance during both the forward and the reverse bias tests, the diode is shorted. If the diode shows high resistance during both tests, it is opened.

Figure 19–14 **a** Typical diode with anode and cathode identified. **b** Power diode with anode and cathode identified.

19.12 IDENTIFYING DIODE TERMINALS WITH AN OHMMETER

Since the ohmmeter can be used to determine if a diode is good or faulty, the same test can be used to determine which lead of a diode is the anode and which lead is the cathode. When you use the ohmmeter to test the diode for forward and reverse bias, you should notice that the ohmmeter indicates high resistance when the diode is reverse biased and low resistance when the diode is forward biased. When the meter indicates low resistance, you know the diode is forward biased, so the positive lead is touching the anode and the negative lead is touching the cathode. This method will work when you are testing any diode. If the diode has markings, you can identify the cathode end of the diode because it has a strip around it. Figure 19–14 shows two types of diodes with the anode and cathode identified in each.

19.13 LIGHT-EMITTING DIODES

A **light-emitting diode (LED)** is a special diode that is used as an indicator because it gives off light when current flows through it. Figure 19–15a shows a typical LED, and Figure 19–15b shows its symbol. From this figure you can see the LED looks like a small indicator lamp. You will likely encounter LEDs on various controls such as thermostats. The major difference between an LED and an incandescent lamp is that the LED does not have a filament, so it can provide thousands of hours of operation without failure.

The LED must be connected in a circuit in forward bias. Since the typical LED requires approximately 20 mA to illuminate, it is usually connected in series with a 600-Ω to 800-Ω resistor that limits the current. If the voltage is higher, the size of the resistance will be larger. Figure 19–16 shows a set of seven LEDs connected to provide a seven-segment display. The seven-segment display has the capability to display all numbers 0–9. Seven-segment displays are used to display numbers on thermostats and other electronic devices.

LEDs are also used in optoisolation circuits where larger field voltages are isolated from smaller computer signals. The LED is encapsulated with a phototransistor. When

Figure 19–15 **a** A typical light-emitting diode (LED). Notice that the LED looks like a small indicator lamp. **b** The symbol for an LED.

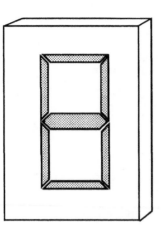

Figure 19–16 LEDs used in seven-segment displays. The seven-segment display can display numbers 0–9.

the input signal is generated, it causes current to flow through the LED, and light from the LED shines on the phototransistor, which goes into conduction and passes the signal on to the computer.

19.14 PNP AND NPN TRANSISTORS

Two pieces of N-type material can be joined with a single piece of P-type material to form an **NPN transistor**. A **PNP transistor** can be formed by joining two pieces of P-type material with a single piece of N-type material. Figure 19–17 shows the electronic symbol and the material for both the PNP and the NPN transistors. The terminals of the transistor are identified as the **emitter, collector,** and **base**. The base is the middle terminal, and the emitter is the terminal identified by the arrowhead.

19.15 OPERATION OF A TRANSISTOR

A **transistor** can be connected in a circuit to perform a wide variety of functions. The simplest function for a transistor to provide is that of an electronic switch. Figure 19–18 shows a transistor as an electronic switch. In this type of application, the base terminal of the transistor acts like the coil of a relay, and the emitter–collector circuit acts like the contacts of a relay. When the proper amount and polarity of DC voltage

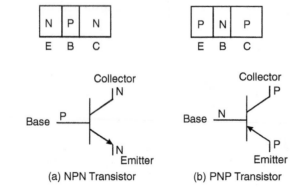

Figure 19–17
a Electronic symbol and diagram of an NPN transistor. **b** Electronic symbol and diagram of a PNP transistor.

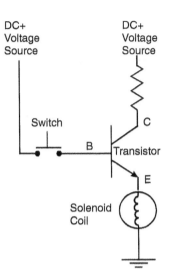

Figure 19–18 A transistor used as an electronic switch.

is applied to the base of the transistor, the resistance between the collector and emitter is relatively low, which allows the maximum amount of circuit current to flow through the emitter–collector circuit. The transistor at this time acts like a relay with its coil energized.

When the polarity of the voltage on the base of the transistor is reversed, the emitter–collector circuit is changed to a high-resistance circuit, which acts like the relay when the coil is de-energized. The major advantage of the transistor is that a very small amount of voltage or current on the base can switch the transistor from high resistance to low resistance. Since the base current is very small and the current flowing through the collector is very large, the transistor is called an *amplifier*. Transistors are used in a variety of applications including thermostats, compressor-protection circuits, and variable-frequency motor drives.

Figure 19–19 shows a PNP transistor and an NPN transistor as two PN diode circuits. The equivalent diode circuits are shown with each transistor to give you an idea of how the two junctions work together inside each transistor. Since each transistor is made from two PN junctions, each junction can be tested just like a single-junction diode for forward bias (low resistance) and reverse bias (high resistance). If you must work on a number of systems that have electronic circuits, you may purchase a commercial-type transistor tester that allows you to test the transistor either in the circuit or out of the circuit.

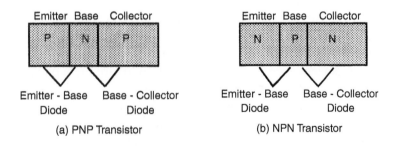

Figure 19–19 **a** A PNP transistor shown as its equivalent PN junctions. Each PN junction can be tested for forward bias and reverse bias. **b** An NPN transistor shown as its equivalent PN junctions.

Figure 19–20 Typical transistors that are used for power control and switching in HVAC equipment.

19.16 TYPICAL TRANSISTORS

You will be able to identify transistors by their shape. Small transistors are used for switching control circuits, and larger transistors are mounted to heat sinks so that they can easily transfer heat. Figure 19–20 shows examples of types of transistors.

19.17 TROUBLESHOOTING TRANSISTORS

Transistors can be tested by checking each P and N junction for front-to-back resistance. Figure 19–21 shows these tests. You can see that each time the battery in the ohmmeter forward biases a PN junction, the resistance is low, and when the battery reverse biases the junction, the meter indicates high resistance. You can test a transistor in this manner if it has been removed from the circuit. You can also test transistors while they are connected in circuit with a commercial-type transistor tester. The transistor tester performs similar front-to-back resistance tests across each junction.

19.18 UNIJUNCTION TRANSISTORS

There are approximately two dozen different types of electronic devices that are made by combining a number of different sections of P- and N-material. Some of these devices are designed specifically for switching larger voltages and currents, and other devices like the **unijunction transistor (UJT)** are designed to produce a small pulse of voltage that can be used to turn on or bias other switching devices. The UJT is used to produce a pulse of voltage that is used to turn on **silicon controlled rectifiers (SCRs)**. SCRs are used to control large DC voltages and currents, and they will be introduced in section 19.21. As an HVAC technician, you may be more familiar with current-switching devices such as SCRs or triacs, but you must remember that they will not operate without the devices such as UJTs that are used to turn them on and off.

The unijunction transistor is made of a single PN junction. Figure 19–22 shows a diagram of the PN-material and the electronic symbol for the UJT. The terminals for the UJT are called base 1, base 2, and emitter. From the diagram you can see that base 2 and base 1 leads are mounted in the large section of N-type material. This means that if you measure the resistance between base 2 and base 1 terminals, you will measure the same amount of resistance regardless of the polarity of the meter leads. When you measure the resistance between base 2 and the emitter, you will find that it is like a diode

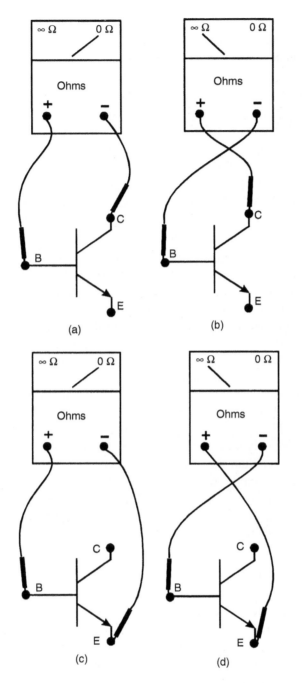

Figure 19–21 a Testing the base-collector junction of an NPN transistor for forward bias. **b** Testing the base-collector junction of an NPN transistor for reverse bias. **c** Testing the base-emitter junction of an NPN transistor for forward bias. **d** Testing the base-emitter junction of an NPN transistor for reverse bias.

Figure 19–22 a Electronic symbol for the unijunction transistor. The terminals for the UJT are base 2, base 1, and emitter. **b** P-material and N-material in the unijunction transistor.

PN junction, and the polarity of the ohmmeter battery will cause the PN junction to be forward biased one way, and reverse biased when you reverse the leads. The base 1 to emitter junction also acts like a diode junction.

19.19 OPERATION OF THE UJT

The simplest way to explain the operation of a UJT is to show it in a typical circuit. Figure 19–23 shows the UJT in a circuit with a resistor and capacitor connected to the emitter terminal, and a resistor connected to each of the base terminals. When voltage is applied to the resistor and capacitor connected to the emitter, the capacitor begins to charge. The size of the resistor controls the time it takes for the capacitor to charge. This time is referred to as the *time constant* for the circuit.

The same circuit voltage is applied to the resistors that are connected to base 2 and base 1. The resistors connected to base 2 and base 1 create a voltage drop with the internal resistance of the UJT. When the charge in the capacitor grows to a value that is larger than the voltage drop across the UJT and the resistor connected to the base 1 lead, a current path is developed through the emitter and through the resistor that is connected to base 1. When current flows through this path, it creates a voltage pulse that increases to its maximum voltage very quickly. The waveform of this pulse is shown in the diagram, and you can see that it turns on and increases to its maximum value immediately. The duration of the pulse depends on the internal resistance of the UJT and the ratio of the sizes of the base 2 and base 1 resistors.

When the UJT is producing the pulse, the capacitor is discharging, and it begins to charge up immediately and repeats the process as its voltage charge increases to a point at which it is larger than the voltage across the UJT. At this point the UJT produces another pulse. The time between pulses is determined by the size of the resistance and capacitance of the resistor and capacitor connected to the emitter. If the amount of resistance is increased, the time for the capacitor to charge will increase and the time between pulses will increase. If the amount of resistance at the emitter is reduced, the

Figure 19–23 A UJT connected to an oscillator circuit to produce sharp output pulses. The size of the resistance and capacitance connected to the emitter controls the frequency of the pulses.

capacitor will charge more quickly and create pulses that are grouped more closely together. This type of circuit is called an *oscillator*, and the number of pulses the oscillator produces in 1 second is called the *oscillator frequency*.

19.20 TESTING THE UJT

The unijunction transistor can be tested like a PN junction. First, you must isolate the UJT from its circuit, then you can test the PN junction between base 2 and emitter, and then test the PN junction between base 1 and emitter. Use an ohmmeter and switch the polarity of its probes so that you forward bias and reverse bias each junction. Remember that when you forward bias a PN junction, it should have low resistance, and when you reverse bias a PN junction, it should have high resistance.

19.21 THE SILICON CONTROLLED RECTIFIER

The silicon controlled rectifier (SCR) is made by combining four PN sections of material. Figure 19–24a shows the PN-material for the SCR, and Figure 19–24b shows its electronic symbol. The terminals on the SCR are identified as anode, cathode, and gate. Since the SCR is basically a diode that is controlled by a gate, its symbol uses the arrow from the rectifier diode that you studied at the beginning of this chapter. When the SCR is turned on, it can conduct large amounts of DC voltage and current (over 1000 V and 1000 A) through its anode and cathode. The major difference between the SCR and the junction diode you learned about in section 19.5 is that the junction diode is always able to pass current in one direction when the diode is forward biased. The SCR is forward biased by applying positive voltage to its anode and negative voltage to its cathode. At this point the SCR still has high resistance at its anode–cathode junction. If a positive voltage pulse is applied to the SCR gate, the SCR's anode–cathode junction will have low resistance, and the SCR will be in conduction. When the pulse is removed from the gate of the SCR, it will remain in conduction because positive current that comes through the anode will replace the voltage the gate provided. The only way to turn the SCR off is to provide reverse bias voltage to the anode–cathode or to reduce the current flowing through the anode–cathode to zero. You should remember that the AC sine wave has zero voltage right before it provides the negative half of its

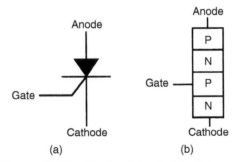

Figure 19–24 **a** Electronic symbol for the silicon controlled rectifier. The terminals of the SCR are the anode, cathode, and gate. **b** P-material and N-material in an SCR. The P-material and N-material are combined to make a PNPN junction.

waveform. This means that if the SCR is powered with AC voltage, the SCR will be turned off when the AC waveform goes through 0 V and then to its negative half-cycle. When the AC voltage waveform goes positive again, a gate pulse can be provided and the SCR can go into conduction again. The gate provides a pulse that is used to cause the SCR to go into conduction.

19.22 OPERATION OF THE SILICON CONTROLLED RECTIFIER

Figure 19–25 shows an SCR connected in a circuit to control voltage to a DC load. The source voltage for this circuit is AC voltage. The main advantage of the SCR is that it will not go into conduction until it receives a pulse of voltage to its gate. The timing of the pulse can be controlled so that it can be delivered anytime during the half-cycle, which controls the amount of time the SCR will be in conduction. The amount of time the SCR is in conduction will control the amount of current that flows through the SCR to its load. If the SCR is turned on immediately during each half-cycle, it will conduct all the half-wave DC voltage just like a normal diode rectifier. If the gate delays the point at which the SCR turns on and goes into conduction at the 45° point of the half-wave, the amount of voltage and current the SCR conducts will be 50% of the fully applied voltage.

The other important feature of the SCR is that it will go into conduction only when its anode and cathode are forward biased. This means that if the supply voltage is AC, the SCR can go into conduction only during the positive half-cycle of the AC voltage. When the negative half-cycle occurs, the anode and cathode will be reverse biased and no current will flow. This means that the SCR will automatically be turned off when the negative half of the AC sine wave occurs. Since the positive half of the AC sine wave occurs for 180°, the SCR can provide control of only 0–180° of the total 360° AC sine wave.

It is also important to understand that since the turn-off point of the SCR is fixed to the point where the sine wave begins to go negative, the SCR can be controlled only by adjusting the point where it is turned on. The point where the SCR is turned on and goes into conduction is called the *firing angle*. If the SCR is turned on at the 10° point in the AC sine wave, its firing angle is 10°. If the SCR is turned on at the 45° point, its firing

Figure 19–25 A silicon controlled rectifier shown in a circuit controlling DC voltage and current to its load.

angle is 45°. The number of degrees the SCR remains in conduction is called the *conduction angle*. If the SCR is turned on at the 10° point, its conduction angle will be 190°, which is the remainder of the 180° of the positive half of the AC sine wave.

19.23 CONTROLLING THE SCR

Figure 19–26 shows an SCR in a circuit with a UJT connected to its gate. The load in this circuit is a DC motor. This type of DC motor is often used as a damper control motor. The circuit is powered by AC voltage, and the variable resistor in the oscillator (capacitor–resistor) circuit sets the timing for the pulse that is used to energize the gate of the SCR. You should notice that a diode rectifier provides pulsing DC voltage for the capacitor, which charges to set the timing for the pulse that comes from the UJT. Since this DC voltage comes from the original AC supply voltage, it will have the same timing relationship of the original sine wave. This means adjusting the pulse from the UJT to turn on the SCR gate at just the right time to control the firing angle of the SCR from 0–180°. In reality the firing angle is usually controlled from 0–90°, which gives sufficient range of control to adjust the output DC voltage that is sent to the DC motor. You should remember that the speed of the DC motor can be controlled by adjusting the voltage sent to the armature and field. This diagram shows the waveform for the voltage at each point in this circuit. The load in this circuit could also be any other DC-powered load.

19.24 TESTING THE SCR

You will need to test the SCR to determine if it is faulty. Since the SCR is made of PN junctions, you can use forward-bias and reverse-bias tests to determine if it is faulty. In this test you should put the positive probe on the anode and the negative probe on the cathode. At this point the ohmmeter will still indicate that the SCR has high resistance.

Figure 19–26 An SCR used to control a DC motor. A UJT is connected to the gate of the SCR to control its firing angle.

If you use a jumper wire and connect positive voltage from the anode to the gate, you should notice that the SCR will go into conduction and have low resistance. The SCR will remain in conduction until the voltage applied to its anode and cathode is reverse biased, or until the voltage applied to the anode and cathode is reduced to zero. This means that you can turn the ohmmeter polarity switch to the opposite setting, or you can remove one of the probes and the SCR will stop conducting. It is important to understand that the amount of current to keep the SCR in conduction is approximately 4 to 6 mA. This means that some high-impedance digital volt/ohmmeters will not have enough current when set to the ohms range to keep the SCR in conduction. If this is the case, you may need to test the SCR with an analog ohmmeter. The analog ohmmeter is a type of ohmmeter with a needle and scale. You should also test the SCR for reverse bias to ensure that it has high resistance. Sometimes an SCR will not go into conduction because it has developed an open in its anode–cathode circuit. Other SCRs may stay in conduction at all times, which means the SCR is shorted.

19.25 TYPICAL SCRs

SCRs are available in a variety of packages and case styles that include three terminals, and larger types that have threads so that they can be mounted directly into heat sink material. The heat sink material is made from metal and in some cases includes a fan that moves air over fins that are molded into the material to help remove the large amounts of heat that build up in the devices. Figure 19–27 shows examples of different types of SCRs.

19.26 THE DIAC

The unijunction transistor provides a positive pulse for the gate of the SCR that allows the SCR to control large DC voltages and currents. The **diac** is an electronic component that provides a positive and a negative pulse used as a trigger signal for a device called a **triac**. The triac is similar to the SCR except that the former can conduct voltage and current in both the positive and negative directions. This means that the triac provides a controlling function for AC voltage and current that is similar to that of the SCR. As an HVAC technician, you may not notice the diac because it is used as a device to produce a firing pulse for the triac. You will learn in a later section how the triac is used in fan speed controls and other AC voltage control circuits. Figure 19–28 shows the electronic symbol for the diac. You should notice that the diac symbol consists of two arrows, which show that voltage pulses can be positive or negative. Since the diac symbol has two arrows, it can be shown in two different styles.

Figure 19–27 Typical SCRs shown in a variety of packages and case styles. *(Copyright of Motorola, used by permission)*

Figure 19–28 Electronic symbols for the diac. Note that the diac has two arrows, and it can be represented by either symbol.

Figure 19–29 a Diac in a circuit. **b** The pulse produced by the diac.

19.27 OPERATION OF THE DIAC

When voltage is applied to a diac in a circuit, its PN junction will remain in a high-resistance state and block the voltage until the voltage level reaches the breakover level. This means that when AC sine wave voltage is applied to the circuit, the sine wave voltage will increase to its peak level, return to zero, and then increase to negative peak voltage. Each time the voltage increases toward its peak and exceeds the breakover level of the diac, the diac provides a sharp output pulse of the same polarity as the voltage that is supplied to it. For example, if the breakover voltage level is 18 V, the diac will block voltage until the AC voltage exceeds the 18-V level. At that point, the diac will produce a pulse that is approximately 18 V. Figure 19–29a shows the diac in its circuit, and Figure 19–29b shows the pulse that it creates.

19.28 THE TRIAC

The *triac* is basically two SCRs that have been connected back to back in parallel so that one of the SCRs will conduct the positive part of an AC signal and the other will conduct the negative part of an AC signal. As you know, the SCR can control voltage and current in only one direction, which means that it is limited to DC circuits when it is used by itself. Since the triac acts like two SCRs that are connected inverse parallel, one section of the triac can control the positive half of the AC voltage, and the other section of the triac can control the negative half of the AC voltage. Figure 19–30a shows the electronic symbol of the triac, and Figure 19–30b shows the arrangement of its P-type and N-type materials. The terminals of the triac are called main terminal 1 (MT1), main terminal 2 (MT2), and gate. Figure 19–30c shows examples of typical triac semiconductor

Figure 19–30
a Electronic symbol for the triac. **b** P-type and N-type materials in the triac. **c** Typical triac semiconductor devices.
(Copyright of Motorola, used by permission)

devices. Since the triac is basically two SCRs that are connected inverse parallel, MT1 and MT2 do not have any particular polarity.

19.29 USING THE TRIAC AS A SWITCH

The triac can be used in an HVAC circuit as a simple on-off switch. In this type of application, the MT1 and MT2 terminals are connected in series with the AC load. When the gate gets a positive pulse signal, the triac turns on for the positive half of the AC cycle. When the AC voltage waveform returns from its positive peak to zero volts, the triac turns off. Next, the negative half-cycle of the AC voltage waveform reaches the triac, and it receives a negative pulse on its gate and goes into conduction again.

This means that the triac looks as though it turns on and stays on when AC voltage is applied. The load connected to the triac will receive the full AC sine wave just as if it were connected to a simple single-pole switch. The major difference is the triac switch can be used for millions of on-off switching cycles. The other advantage of the triac switch is that the gate pulse can be a very small amount of voltage and current. This allows the triac to be used in temperature control where the temperature-sensing part of a thermostat can be a small solid-state sensing element called a *thermistor*. The sensing element can also be a very narrow strip of mercury in a glass bulb that is very accurate but can carry only a small amount of voltage or current.

Another useful switching application for a triac is the solid-state relay. Section 19.39 will explain AC and DC solid-state relays. Figure 19–31 shows a triac in a thermostat connected to a 24-V AC gas valve in a furnace. The temperature-sensing element is connected to the gate terminal of the triac. When the temperature increases, it makes the mercury in the temperature-sensing element expand and make contact between the two metal terminals. The small amount of voltage flowing through the sensing element is sufficient to cause the triac to go into conduction and provide voltage to the gas valve.

Figure 19–31 A triac used as a switch to turn on voltage to a gas valve for a furnace. The triac receives its gate signal from a small amount of voltage that moves through the temperature-sensing element.

19.30 USING THE TRIAC FOR VARIABLE VOLTAGE CONTROL

The triac can also be used in variable voltage control circuits, since it can be turned on anytime during the positive or negative half-cycle, in much the same way that the SCR is controlled for DC circuit applications. In this type of application, a diac is used to provide a positive and a negative pulse that can be delayed from 0 to 180° to control the amount of current flowing through the triac. This type of circuit can be used to control the amount of current and voltage supplied to electric heating elements. This allows the amount of current and voltage to be controlled from zero to maximum by adjusting R2, which in turn allows the temperature to be controlled very accurately. Figure 19–32 shows a diagram of a triac used to control an electric heating element powered by an AC voltage source. Notice that a variable resistor is connected with a capacitor to provide an oscillator pulse to the diac. Since the resistor and capacitor are connected to an AC voltage source, the pulse will be both positive and negative as the AC sine wave changes polarity. The triac is connected to the same AC voltage source so that the timing of the pulse from the diac to its gate will always be synchronized with the polarity of the voltage arriving at the main terminals of the triac.

19.31 TESTING THE TRIAC

Since the triac is made from P-type material and N-type material, it can be tested like other junction devices. The only point to remember is that since the triac is essentially two SCRs mounted inverse parallel to each other, some of the ohmmeter tests will not be

Figure 19–32 A triac used to control variable voltage to an AC electric heating element. Notice that a diac is used to provide a positive and a negative pulse to the triac gate.

affected by the polarity of the ohmmeter leads. In the first test of a triac an ohmmeter should be used to test the continuity between MT1 and MT2. When no gate pulse is present, the resistance between these terminals should be infinite regardless of which ohmmeter probe is placed on each terminal. Figure 19–33 shows how the ohmmeter should be connected to the triac. In Figure 19–33a you can see that the positive ohmmeter probe is connected to MT1 and the negative probe is connected to MT2. Since no voltage is applied to the gate, the resistance should be infinite. When voltage from MT2 is jumped to the gate, the triac will go into conduction and the ohmmeter will indicate low resistance.

In Figure 19–33b you can see that the positive ohmmeter probe is connected to MT1, and the negative probe is connected to MT2. When voltage from MT1 is jumped

Figure 19–33 **a** Testing the triac by placing the ohmmeter positive probe on MT1 and the negative probe on MT2. When gate voltage is applied from the MT2 probe, the triac goes into conduction. **b** Testing the triac by placing the ohmmeter positive probe on MT1 and the negative probe on MT2. When gate voltage is applied from the MT1 probe, the triac goes into conduction. **c** The jumper is placed between MT2 and the gate.

to the gate, the triac will go into conduction and the ohmmeter will indicate that the resistance is low. It is important to remember that the triac will stay in conduction only while the gate signal is applied from the same voltage source as MT1. As soon as the voltage source is removed the triac will turn off. Figure 19–33c shows the gate receiving voltage from MT2, and the triac is in conduction.

19.32 OPERATIONAL AMPLIFIERS

Operational amplifiers have become a common component in heating, air-conditioning, and refrigeration systems because they allow very small signal voltages from sensors to be amplified to a usable voltage to turn compressors and other loads on and off. The **operational amplifier**, also called an *op amp*, is made of dozens of PN-junction devices such as transistors that are manufactured on a single integrated circuit (IC). As a technician, you will find them in most of the newer solid-state controls such as thermostats, voltage monitors, and other protection and control devices.

The op amp provides two basic functions. First, it can take a very small signal and amplify it. For example, a sensor can send a 100-mV signal to an op amp, and the op amp can multiply the signal strength by 10 times ($\times 10$) so that the output signal will be 1 V, or the op amp can multiply the signal by 100 times ($\times 100$) so that the output signal will be 10 V. Second, the op amp can change the polarity of the voltage of the input signal or it can keep it at the same polarity. For example, if a negative sensor voltage is applied to the inverting input terminal of an op amp, the output signal will be positive. If the negative input signal is applied to the noninverting input, the output will remain negative. Each op amp input can also receive a signal that spans -0.5 V to $+0.5$ V and produce an output voltage that spans -12 V to $+12$ V. Figure 19–34 shows the electronic symbol for an operational amplifier with all of its terminals identified.

In Figure 19–34b you can see that the op amp has two input terminals identified as terminals 2 and 3 and an output terminal identified as terminal 6. Terminal 2 is identified by a negative sign ($-$), which indicates that this is the *inverting input*. If an op amp receives a signal on the inverting input, the voltage polarity will be inverted when the signal reaches the output. Any signal the op amp receives on its noninverting input will remain the same polarity when it reaches the output. The op amp also has two terminals where power supply voltage is applied. It is important to understand that when the op amp amplifies a signal, the output signal cannot be any larger than the supply voltage that is connected to the op amp at its V+ (terminal 7) and V− (terminal 4) terminal. Typical power supply voltage

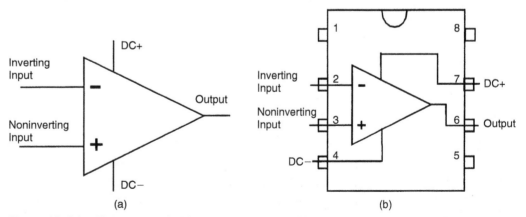

Figure 19–34 a Electronic symbol for an operational amplifier (op amp). **b** Pin outline of integrated circuit (IC) for an op amp.

can be 12, 15, or 18 V DC. You should also notice that the op amp is mounted on an eight-pin integrated circuit, which means that the chip will have eight pins. The chip has an identifier mark at the top of the chip to help identify the number 1 terminal.

19.33 THE OP AMP AS AN AMPLIFIER

Like a transistor, the op amp has the ability to amplify an input signal. The amount of amplification is called *gain*. The major difference between the transistor and the op amp is that the amount of amplification (gain) for the transistor is designed into the transistor when it is manufactured. In contrast, the amount of gain for the op amp is determined by the external resistors connected to it. This means that it is possible to design a circuit where an op amp can have any amount of amplification. The general values of amplification are between 10 to 100.

Figure 19–35 shows a typical op amp circuit. The resistor that is connected to the noninverting input is called the input resistor (R_{in}), and the resistor at the top of the op amp that is connected between the output terminal and the noninverting input is called the feedback resistor (R_{F}). When the input signal is connected to the inverting input, the op amp is called an inverting input. The amount of gain for the inverting op amp is determined by the formula

$$\text{gain} = \frac{R_F}{R_{in}}$$

The formula for the noninverting op amp is

$$\text{gain} = 1 + \frac{R_F}{R_{in}}$$

In this figure you can see that the resistor connected to the noninverting op amp is 10 kΩ and the feedback resistor is 100 kΩ, so the gain is 10 for this circuit. If the input resistor is changed to 1 kΩ, the gain for the circuit will be 100. If a 0.1-V signal is sent to the input, the output will produce a 1-V signal if the gain is 10, and the output signal will be 10 V if the gain is 100.

19.34 CONTROLLING A RELAY WITH AN OP AMP

One circuit you will find in air-conditioning and refrigeration systems is a relay controlled by an op amp. This type of circuit gives a temperature sensor that produces a millivolt signal the ability to control a 10-hp, three-phase fan motor. Figure 19–36

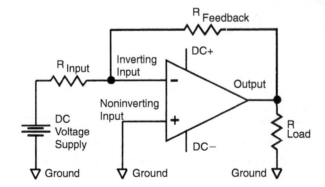

Figure 19–35 An op amp with a resistor connected to the inverting input and a feedback resistor.

Figure 19–36 A millivolt signal from a temperature sensor is used to control a small relay, and the relay controls a motor starter and 10-hp three-phase fan motor.

shows an example of this circuit. In this circuit the op amp receives a small signal from a temperature sensor. The coil of a small relay is connected to the output of the op amp. The contacts of the small relay control voltage to the coil of a motor starter and the three-phase motor are connected to the contacts of the three-phase motor starter. The gain of the op amp is 100.

When the temperature sensor sends a 0.1-V signal to the input of the op amp, the op amp increases the signal to 10 V at the output, which sends voltage to the coil of the small relay. When the coil of the small relay is energized, its contacts provide voltage for the coil of the motor starter, and the contacts of the motor starter close and provide 240 V AC to the 10-hp fan motor. When the temperature drops, the voltage from the temperature sensor drops below 0.1 V, and the small relay coil drops out. When this occurs, the motor starter coil becomes de-energized, and the fan motor turns off.

19.35 A HEATING AND COOLING CONTROL SYSTEM

As you know, the op amp has the ability to receive a signal that spans a range from negative to positive and to produce an output signal that spans a range from negative to positive. Figure 19–37 shows a control circuit for a heating and cooling system. The thermostat for this system is called a resistive temperature device (RTD). The RTD is connected to three other resistors in a circuit called a bridge. The variable resistor in the bridge is used to set the desired temperature that controls when the air conditioning should turn on and when the heating should turn on. The desired temperature is called the setpoint for the system. For this example circuit, the variable resistor is set to turn on the air conditioning when the temperature is above 72° and turn on the heating when the temperature is below 72°. A dead band of + 1° and −1° can be added into this control circuit by adjusting the offset at the op amp.

The bridge circuit allows very small changes in temperature to be converted to changes in voltage. The RTD senses changes in the temperature and produces a different amount of resistance for every temperature. Since this change of resistance is fairly linear, the system's temperature can be controlled very accurately. When the temperature changes, the resistance in the RTD changes and the bridge becomes unbalanced and produces a positive voltage if the temperature is higher than 72°, and a negative voltage if the temperature is lower than 72°. The amount of voltage will increase to a maximum

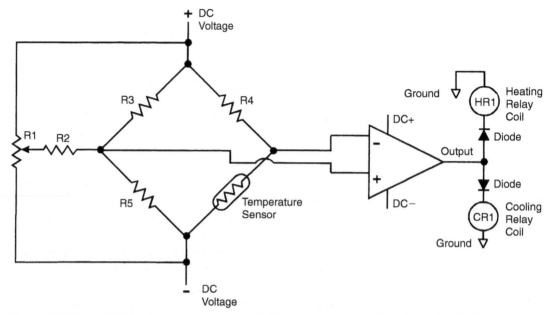

Figure 19–37 An RTD (resistive temperature device) is connected to a bridge circuit. The RTD acts as a temperature-sensing element. The signal from the bridge is sent to an op amp where it is amplified by ×100. The coil of a heating relay and the coil of a cooling relay are connected to the output of the op amp. Diodes are used to control when the heating or cooling relay should be energized.

of +0.15 V when the temperature is 5° above the setpoint and a maximum of −0.15 V when the temperature is below setpoint.

The second part of the circuit is an op amp. You can see that the signal from the bridge circuit is connected to the inverting input of the op amp. The gain of the op amp is set for × 100, so the output signal will be +15 V at its maximum and −15 V at its minimum. You should notice that two diodes are connected to the output terminal of the op amp. The coil of a cooling relay is connected to the top diode, and the coil of a heating relay is connected to the bottom diode.

When the thermostat senses that the temperature is sufficiently above the setpoint, it sends a small positive voltage of 0.1 V to the inverting input of the op amp. The op amp changes the positive voltage signal to a −10-V signal. Since the voltage is negative, it can pass only through the top diode and energize only the cooling relay coil. When the contacts of the cooling relay close, they energize a compressor motor, and the system begins to cool the controlled space. When the temperature in the space cools down, the voltage from the bridge will decrease, and when the input voltage drops below 0.08 V, the output signal will drop below 8 V and the relay coil will drop out.

If the temperature in the space gets below the setpoint, a negative signal is sent by the bridge circuit to the input of the op amp. Since the op amp inverts the signal, the output signal will be positive when the temperature is below the setpoint. When the temperature is below the setpoint enough to produce a −0.1-V signal, the output of the op amp will increase to −10 V, which is sufficient to energize the relay coil. Since the signal is positive this time, the current can pass only through the bottom diode to the coil of the heating relay. The contacts of the heating relay close and energize an electric heating coil.

When the controlled space begins to heat up, the signal from the bridge circuit begins to decrease, and the coil to the heating relay will become de-energized when the input signal drops below 0.08 V. It is important to remember that the amount of positive or negative voltage must be larger than 0.1 V to energize the coils, and when it drops below 0.08 V, the coils will drop out.

| 19.36 | USING PROPORTIONAL VALVES INSTEAD OF RELAYS |

If the temperature control system must be more sensitive, the control devices can be **proportional valves** that control the flow of hot and cold water through a heat exchanger, rather than relays that turn on and off. A proportional valve is like the dimmer switch that controls a light or like the volume control on a radio. When the valve receives a small amount of voltage, it opens slightly, and as the amount of voltage increases, the valve opens wider until it is fully open. The advantage of the proportional valve is that if the thermostat senses that the temperature is slightly below the setpoint, the valve controlling the hot water can be opened slightly to allow a small amount of hot water to enter the heat exchanger. Since a fan blows air over the heat exchanger, the amount of heat (in BTUs) that will be moved will be minimal. If the temperature difference at the thermostat becomes larger, the proportional valve can be opened farther to allow more hot water to pass through the heat exchanger.

Figure 19–37 shows a control circuit that is similar to the previous control circuit that uses an op amp to control relays. In this circuit, the op amps control proportional valves. Figure 19–38 shows the operation of the valve in relation to the voltages it receives. You can see that the valves will be fully open when they receive 10 V, and they will be half open when they receive 5 V. The valve will open 10% for every 1 V it receives.

You should notice in the circuit that again diodes are used to ensure that the correct valve is operating when it is positive and when it is negative. The RTD and the bridge circuit operate the same as they did in the previous circuit. In this circuit when the temperature changes slightly, the amount of signal will be slight, but it will be amplified and sent to the correct proportional valve so that it can open wider or close down depending on whether the change of temperature is more or less than the previous value. The setpoint is used to determine when the signal will change from positive to negative, and this in turn will determine whether the signal is received by the heating valve or the cooling valve. The amount of dead band can be adjusted by changing the offset resistor on the op amp. Typically the dead band is 1–1.5°.

Figure 19–38 An op amp controlled heating and cooling system. This system uses proportional valves instead of relays to control when the system produces heating or cooling.

Figure 19–39 A typical 1-hp variable-frequency drive amplifier. *(Courtesy of Rockwell Automation's Allen-Bradley Business)*

19.37 CONTROLLING THE SPEED OF A THREE-PHASE MOTOR WITH A VARIABLE-FREQUENCY DRIVE

The speed of a single-phase or three-phase AC induction motor can be controlled by adjusting the frequency of the voltage that is sent to it. In modern air-conditioning and refrigeration systems, it is important to be able to vary the speed of motors in the system to provide more efficient use of electricity and to provide a system that is able to control temperatures more precisely.

In the 1980s, manufacturing technology for transistors and other solid-state devices made it possible for transistors to be able to control larger voltages and currents. Prior to that development, triacs and SCRs had been the devices used for controlling large voltages and currents. With the advent of larger transistors, the frequency of larger voltages could be manipulated to a point where the speed of three-phase and single-phase motors could be controlled by changing the frequency of their supply voltage.

Figure 19–39 shows a typical frequency drive. The drive in the picture costs approximately $450 and can control compressors or fans up to 1 hp. Larger drives can control motors up to 100 hp. In some air-conditioning systems the frequency control board is built into the control cabinet with the other air-conditioning controls.

19.38 OPERATION OF THE VARIABLE-FREQUENCY DRIVE CIRCUIT

Figure 19–40 shows the circuit diagram for a typical **variable-frequency drive (VFD)**. You should notice that the circuit consists of three sections. The first section is the *rectifier section*, where a three-phase diode bridge rectifier changes the three-phase AC voltage to pulsing DC voltage. The second section is the *filter section*, where the pulsing DC voltage is smoothed to pure DC. The third section is the *transistor switching section*, which produces the three-phase AC voltage at the desired frequency.

In the rectifier section you can see that six diodes are connected in a bridge circuit to convert the three-phase 60-Hz voltage to DC voltage. The actual rectifier section is

Figure 19–40 Circuit diagram of a variable-frequency drive. *(Courtesy of Rockwell Automation's Allen-Bradley Business)*

mounted on a single module and can be exchanged by removing the three input voltage wires and the two DC bus connections. Each diode in the module can still be tested for front-to-back resistance ratio with an ohmmeter just as if it were an individual diode. The output of the rectifier section is 12 half-wave pulses 60° apart. The output of the bridge rectifier is connected to two large copper conductors that are called the *DC bus*.

The filter section of the circuit consists of several capacitors that are connected in parallel with the DC bus and a large inductor that is connected in series with the DC bus. The capacitors charge and discharge in synchronization with the input voltage. This causes the half-wave signal to be converted to pure DC. The capacitors are used to filter the voltage part of the waveform, and the inductor is used to filter the current part of the waveform.

The transistor section consists of six transistors, two for each output phase. You should notice that one transistor of each phase is connected to the positive DC bus, and the second transistor is connected to the negative DC bus. Each transistor is controlled by a base firing circuit that is controlled by a microprocessor chip (small computer). At the appropriate time each transistor is turned on in six distinct steps. Figure 19–41 shows an example of the waveform that results from turning the transistor on in six steps. You can see that the transistor that is connected to the positive DC bus conducts

Line to Neutral Voltage

Figure 19–41 The six-step waveform of the output transistors. *(Courtesy of Rockwell Automation's Allen-Bradley Business)*

Line Current

and produces the positive part of the output AC waveform, and the transistor that is connected to the negative DC bus produces the negative part of the output AC waveform. The first set of transistors produces the A phase for the three-phase output waveform, and the second set produces the B phase. The third set of transistors produces the C phase of the AC output waveform. You can see that the output frequency can be adjusted by changing the timing of the base firing circuit for each set of transistors. Typical control for a variable-frequency drive is from 0 to 120 Hz, which allows the motor to be controlled between 0 and 200% of its rated rpm. In reality the motor is generally adjusted from 60% to 130% of rated rpm, which provides sufficient control for the system being adjusted. It is important to note that the VFD can only rule the motor for a short period of time at RPMs above 100% because the motor will overheat.

You can see how important it is to be able to adjust the speed of a very large fan to provide minimal air flow when the temperature conditions are near setpoint, and larger air flows when the system conditions warrant. The same is true of controlling the speed of a compressor motor. When the temperature is slightly above the setpoint, the compressor can turn more slowly and pump less refrigerant, and when the temperature is far above setpoint, the frequency of the voltage that is sent to the compressor motor can be increased to cause the motor to rotate more quickly and pump more refrigerant.

19.39 SOLID-STATE RELAYS

Another solid-state device that you will frequently see used in HVAC systems is called a **solid-state relay (SSR)**. The solid-state relay uses AC or DC voltages for the control part of the circuit that acts like the coil of a regular relay, and it can switch AC voltages through the part of the circuit that acts like the contacts. The solid-state relay uses SCRs, transistors, or triacs to control various loads. It operates much like an electromechanical relay in that it is energized by a small voltage that controls a larger voltage and current.

Figure 19–42 shows a picture of a typical solid-state relay and provides a block diagram of the operation of the relay. Figure 19–43 shows the electrical diagram for a solid-state relay that uses a light-emitting diode (LED) and a phototransistor to accept the DC input, and a triac to control the AC voltage and load in the output. The LED and phototransistor are called an *optoisolation device* or an *optocoupler*. When a DC input signal is received at the SSR, it is converted to light through an LED. The LED and phototransistor are manufactured in an IC so that its light shines from the LED directly

Figure 19–42 A solid-state relay rated for DC input and 120 V AC output. *(Courtesy of Johnson Controls, Inc.)*

Figure 19–43 An electronic diagram of a solid-state relay that has a DC input signal and an AC output signal.

on a phototransistor. When the light strikes the phototransistor, its collector–emitter circuit goes to low resistance, and the collector current flows through it to the base of the triac. The AC load is controlled directly by the triac. Notice that the power source for the load must be isolated from the input signal.

19.40 TROUBLESHOOTING SOLID-STATE DEVICES

At times you will suspect that solid-state devices are not operating correctly. You can perform a number of simple tests during troubleshooting. Most solid-state devices consist of one or more junctions between P- and N-material. The basic test for a PN junction is the front-to-back resistance test if the device is not connected in a circuit. Another way to test solid-state components is to identify the function of the component. For example, if you are troubleshooting diodes in a bridge rectifier, you can test for AC voltage on the input side of the circuit and check for the proper amount of DC voltage on the output side. If the DC voltage is lower than specified or if it is 0 V, you have detected that one or more diodes are faulty and the bridge circuit must be replaced.

A transistor functions like a relay in that when the proper polarity of DC voltage is applied to the base of the transistor, the resistance between the emitter and collector should become low just as in the contacts in a relay. In the case of the transistor, the resistance in the emitter and collector circuit will decrease as the voltage to the base is increased. You can prove this by using three voltage measurements. The first measurement should be between the base terminal and the common point in the circuit, the second measurement should be between the emitter and common, and the final measurement should be between the collector and the common. The difference in the voltage between emitter and common, and the collector and common, will indicate how much current is flowing through the transistor. When the transistor is operating correctly and in full conduction, the voltage reading at the emitter will be very similar to that at the collector. If the voltage at the collector is very low, it means that the resistance in the emitter–collector circuit is high, and the transistor is acting like an open switch.

The sections on SCRs and triacs provide troubleshooting tips for you to review. The op amp is made of hundreds of transistors, and the simplest way to troubleshoot it is to test for an input signal and check for an output signal. If you have an input signal and you do not have an output signal, be sure that the op amp has the proper voltage supplied to it and that the polarity of the input signal is correct.

The solid-state relay (SSR) is available for AC circuits and DC circuits. Be sure that the SSR is the correct one for your circuit. You can test the SSR just as you would a regular relay. The coil of the relay must have the proper amount and polarity of voltage for the solid-state circuit that acts like the contacts to close. If you have the proper voltage at the coil and the contact circuit will not close, you should change the SSR. Another problem with the SSR is that its contact circuit will remain closed after voltage to the coil is removed. This problem is similar to having the contacts of a relay become welded closed.

19.41 TYPICAL ELECTRONIC CIRCUITS FOUND IN HVAC SYSTEMS

Some of the electronic circuits that you will find in an HVAC system include the electronic ignition controls on the newer high-efficiency furnaces, electronic control of motor speeds whether it is on a circuit board or on a variable-frequency drive (VFD), electronic circuit boards that control gas valves and oil burners, electronic circuit boards that control sequences on electric furnaces and heat pumps, and electronic air cleaners.

19.42 ELECTRONIC MOTOR SPEED CONTROL

Electronic motor speed control is provided on newer high-efficiency furnaces to control the speed of fan motors and for newer high-efficiency air-conditioning systems to control the speed of compressors. Variable-speed motor control can be provided by a variable-frequency drive (VFD) or from a motor speed control board. The circuits inside both of these electronic systems provide the same function, in that they vary the frequency of the voltage going to the motor to control its speed and vary the voltage and current supplied to the motor to control its torque. The VFD was discussed in detail in section 19.37, so this section will discuss controlling speeds with an electronic board. The variable-speed control board shown in Figure 19–44 provides several functions. One problem that causes motors to wear out prematurely is when they receive full applied voltage when a relay or contactor is first closed to energize the motor. The motor speed control board allows less than full voltage to be supplied to the motor when it is first energized to minimize the wear on the motor windings. Voltage and current are slowly increased to provide the motor the amount of torque needed to turn its load. The control board then varies the frequency sent to the motor, which changes the motor speed.

The variable-speed motor controls also provide maximum savings in power consumption by only running the fan motor or compressor at the speed necessary to provide the correct amount of heating or cooling. This means that a motor may run at a slower speed to save electrical energy, which means the electric bill for running the system will be smaller. Figure 19–45 shows a circuit board that controls the fan circuit for a heat pump. This type of system must be combined with electronic controls to continue to provide comfort to the homeowner. One of the things that customers look for when purchasing a new furnace or air-conditioning system is the efficiency rating for the system. High-efficiency systems utilize variable-speed motors to minimize power consumption.

Variable-speed control is provided on some residential air-conditioning compressors but is more than likely found on larger commercial air-conditioning and refrigeration compressors. Variable-speed control boards also allow the speed of the fan motor to be increased gradually to full speed when the motor starts up, which substantially reduces the amount of noise the fan makes. The motor can also be accelerated and decelerated at

Figure 19–44 Bryant Carrier furnace electronic control board. *(Courtesy of Carrier)*

a variable speed to ensure that duct work does not make noise (pop) when a furnace or air-conditioning fan is turned on and turned off. Figure 19–46 shows an electronic circuit board that controls the time delay for the fan of a standing pilot type furnace.

Figure 19–45 Bryant Carrier heat pump and air handler fan circuit board. *(Courtesy of Carrier)*

Figure 19–46 Protech Rheem Ruud furnace control board. *(Courtesy of Rheem Air Conditioning Division)*

19.43 **ELECTRONIC IGNITION CONTROL**

Another application for electronic control boards is the control of electronic ignition for furnaces. The ignition system may use a spark between the electrodes on an igniter or it may use a hot surface igniter that is a resistive surface component that receives electricity and heats up until it glows hot enough to ignite the gas pilot. Figure 19–47 shows an electronic control board that is designed to work with hot plate igniters. The board is also designed to allow the gas valve to open once the pilot is established and then close when the cycle is complete. The thermostat sends the signal to the electronic control board to turn the furnace on. Figure 19–48 shows a control board for a direct-spark ignition control. This type of electronic control board provides a pulse of electricity at the electrode to cause it to create a high voltage spark in the pilot flame. This spark will cause the pilot gas to ignite and establish a pilot flame. The pilot flame heats the sensor to a point where it sends a signal back to the control to indicate that the pilot flame is established. At this point the control circuit turns off the spark and sends a signal to open

Figure 19–47 Control board for hot plate igniter. *(Courtesy of Rheem Air Conditioning Division)*

Figure 19–48 Direct-spark ignition control board. *(Courtesy of White-Rodgers)*

the gas valve. When the thermostat senses that the room is at temperature, it sends a signal to the control board to turn off the gas valve and complete the cycle.

19.44 ELECTRONIC GAS VALVE CONTROL AND INTEGRATED FURNACE CONTROL

Electronic control systems called integrated furnace controls are available for gas furnaces with standing pilots as well as furnaces with electronic ignition systems. Figure 19–49 shows an integrated furnace control board for a standing pilot furnace that controls the opening and closing of the gas valve for the system. This control board controls the ignition sequence that allows the hot plate igniter to ignite the pilot flame on the gas furnace, and then controls the opening and closing of the gas valve. If the system has a combustion fan to move exhaust gases out of the furnace, it can also be controlled by this type of board. This type of control integrates an indicator lamp that is used to help in troubleshooting.

(a) (b)

Figure 19–49 a Furnace electronic control board. **b** Diagram of an electronic control board. *(Courtesy of White-Rodgers)*

TIP for Changing Out an Electronic Board

If you must change out one of these electronic boards, you can use a digital camera to take several pictures of the wires and the board prior to disconnecting them. This will help you identify the wiring connections when you are reconnecting the wires onto the new board. When you have completed making all the connections to the new board and have tested it to ensure that it operates correctly, you can delete the electronic pictures from your camera. The color pictures from the digital camera will help you identify not only the terminals where the wires are connected but the color of the wires that are used in the system. You can use a variety of zoom levels to show you the necessary details.

19.45 ELECTRONIC AIR CLEANERS

An electronic air cleaner provides a means to remove dirt and dust particles from the air without using paper or media-type filters. One of the problems with the paper or media-type filters is that they remove the most dust and dirt when they are becoming clogged, which means that the airflow through them is reduced or restricted. Restricted airflow creates problems with both air-conditioning and heating cycles of new high-efficiency furnaces and air conditioners. The electronic air cleaner allows dirt and dust to be removed from the air without restricting airflow.

A modern electronic air cleaner is called an electrostatic precipitator and consists of two stages—the ionizing stage and the collector stage. The air cleaner works on a theory of operation of having the air and dirt particles charged positively and an electric collector grid that is charged negatively. When the dirt particles in the air move past the negatively charged grid, they will be attracted to the collector plate and pulled from the air. Wires in front of the collector grid are used to charge the dirt particles with a positive charge. Figure 19–50a shows a picture of an air cleaner connected to a furnace, and Figure 19–50b shows the air cleaner with its coils pulled out. Figure 19–51 shows a close-up picture of the pre-filter, the charging wires, and the collecting grid. Figure 19–52 presents a diagram that shows how air flows through the air cleaner. In the first stage, dirt, dust, pollen, virus spores, bacteria, and mold particles are charged positively in the ionizing section by a number of wires that have a very high positive charge applied to them. The second stage of the air cleaner is the collector section, which has a number of metal plates that are charged with a positive or negative charge. The plates in the collector section are located so that every other plate is positive and the other is negative. The positive plate repels the positively charged dirt particles and causes them to move closer to the negative collector plates where they are strongly attracted. The electronic air cleaner uses an extremely high voltage, up to 6000 volts DC, and it is oscillated at high frequency to create the positive charge on the dirt and the negative charge on the collector. When large pieces of dirt are attracted to the collector, you will typically hear a loud snapping noise as the static electricity from the different charges allows the dirt to be attracted to the grid. It is important that you do not allow anything to touch the ionizing wires or the collector grid when power is applied. It is also important to handle the charging wires and the collector grids carefully during cleaning or installation so that you do not bend them. The distance between the positive ionizing wires and the charged grid is carefully maintained to prevent electrical arcing between these two high-voltage components. If you bend either one and they become closer than designed, an electrical arc will continue to jump between them and eventually burn out the electronic control circuit or damage the wires or the grid. Newer models have an air flow switch that

(a) (b)

Figure 19–50 **a** Electronic air cleaner connected to a furnace and return air duct. **b** Electronic air cleaner with its coils pulled out.

Figure 19–51 Coils section of an electronic air cleaner.

senses when the furnace fan is moving air, and then it turns on the electronic circuit and charges the plates and wires. It is also important to mount the electronic air cleaner on the return air side of the furnace duct work. The air cleaner must be mounted where the condensate from the air-conditioning coil or water from the humidifier does not drip on the coils.

A metal mesh pre-filter is used to collect very large particles of dust or debris that might damage the air cleaner coils. The pre-filter should be in front of the electronic plates and it should be checked and cleaned frequently. If the pre-filter is allowed to become clogged, you will have the same problems with air restriction that you have with paper or media-type filters. It is very important to be aware that the pre-filter is replaced correctly after it has been removed and cleaned. If you are not careful, the

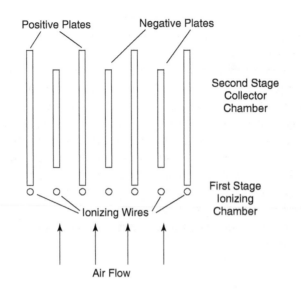

Figure 19–52 Diagram showing positive plates, negative plates, ionizing wires, and direction of air flow.

metal pre-filter can come into contact with the electronic grid and cause serious damage. The only maintenance the air cleaner requires is to clean the pre-filters and periodically clean or wash the collector grid. Be sure to read the manufacturer's recommendation for cleaning and follow the procedure carefully.

If you wash the pre-filter or the collector grid, make sure that both of them are completely dry prior to returning power to the electronic air cleaner. The electronic air cleaner should last a very long time since it does not have any moving parts. Typically electronic air cleaners are damaged by moisture that is left behind when the pre-filter or the electronic collector section are washed. For that reason it is very important to allow air to flow through these parts of the system after they have been cleaned for several hours or up to one or two days before turning power back on to the air cleaner. If you wash the collector system or the pre-filter and water remains on either of them when you return power to the electronic air cleaner, it will typically arc severely and eventually burn out the electronic control system.

19.46 REVIEW

All solid-state components are made of one or more PN junctions. The diode is the simplest PN junction, and it is used to convert AC voltage to DC voltage. The diode must be forward biased (positive voltage on the anode) to allow current to flow through it. When it is reversed biased (negative voltage on the anode), it will not allow current to flow. When one diode is used in a rectifier, it is called a half-wave rectifier. If two or four diodes are used in a rectifier, it is called a full-wave rectifier. When four diodes are used, it is called a bridge rectifier. Rectifiers are used in most solid-state circuits to provide DC voltage from an AC source.

Diodes and other PN junctions can easily be tested with a front-to-back resistance test. A good PN junction should show infinite resistance when reverse biased, and low resistance when it is forward biased.

Transistors are available as PNP or NPN devices. The transistor has a base, emitter, and collector and is a variable resistance device. When voltage is applied to its base, the resistance in its emitter–collector circuit will decrease. The silicon controlled rectifier

(SCR) has an anode, cathode, and gate. DC current will flow through the anode–cathode circuit when positive voltage is applied to the gate and to the anode, and negative voltage is applied to the cathode.

The triac has main terminal 1 (MT1), main terminal 2 (MT2), and a gate. The triac allows AC current to flow between MT1 and MT2 when a voltage is applied to the gate.

The op amp has an inverting and a noninverting input, an output terminal, and power supply terminals. When a positive signal is received at the inverting input, a negative signal is supplied at the output terminal. When a positive signal is received at the noninverting input, a positive output signal is provided to the output terminal.

A variable-frequency drive provides a variable-frequency signal that can control the speed of an AC motor such as a compressor or fan motor. The solid-state relay (SSR) comprises an LED and a phototransistor. When a DC signal is applied to the input (coil) of the SSR, the transistor goes into conduction and allows current to flow through its terminals (contacts).

Job Assignment for This Chapter

You are working on a new furnace that has an electronic circuit board to control ignition and the gas valve. The furnace will not light off and you suspect that the electronic circuit board is bad. You are suggesting to the homeowner that the board needs to be changed out. You have tested the board and know that it does not have any voltage anywhere on the board after 120 volts AC is supplied to the board. The homeowner asks you what components on the board you suspect are bad and how you might test them. Your job is to explain how the board plugs into the 120 volt AC that supplies power to the furnace and provides the DC voltage for electronic control.

Solution to the Job Assignment

The problem stated that you were not able to locate any voltage anywhere on the board after 120 volts AC is supplied to it. You have learned in this chapter that AC voltage is converted to DC voltage through one or more diodes that act as rectifiers. If the board has four diodes, it is called a bridge rectifier that provides full-wave DC. If the board has a fuse in line between the AC and the rectifier, you should check it visually and with an ohmmeter to determine if it is bad. If 120 volts AC is supplied directly to the rectifier, you should check for DC voltage across the output (DC terminals) of the rectifier. If you get no voltage, you should suspect that two or more of the diodes are bad (open). The bridge may have four individual diodes or have the four diodes in the side of a chip with four leads. Two of the leads will be marked AC and the other two leads will be marked DC+ and DC−. You should place your voltmeter across the two leads marked AC to check for AC voltage. If you have the correct amount of AC voltage (120 volts), you can set your voltmeter for DC voltage and check for DC voltage. If you have low or no DC voltage, you can suspect that the rectifier bridge is bad and can change the board out. Typically HVAC technicians do not change out electronic components; rather it is appropriate to change out the complete board. You can explain to the homeowner that if you change components on a board, you cannot guarantee the remainder of the board, and other components may have been stressed and fail soon after you complete this repair. If you replace the board, it will come with a warranty and you can be sure the parts on the board circuit are all working correctly.

Assessment

QUESTIONS

1. Explain P-type and N-type materials.
2. Explain the operation of a diode (PN) junction and show the input AC waveform and the output DC waveform.
3. Identify the terminals of a transistor and explain its operation.
4. Explain the operation of the SCR and the type of circuit where you would find one.
5. Explain the function of a unijunction transistor when it is used with an SCR.

TRUE OR FALSE

1. The triac provides switching similar to that of an SCR, except the triac operates in an AC circuit.
2. The light-emitting diode (LED) is similar to a DC light bulb in that it has a very tiny filament.
3. The main function of diodes and SCRs is to convert AC voltage to DC voltage.
4. The variable-frequency drive has the ability to change the frequency of voltage sent to a motor to control the motor's speed.
5. The UJT and diac are important solid-state devices that provide pulse signals to other components to use as a firing signal.

MULTIPLE CHOICE

1. A PN junction is forward biased when
 a. positive voltage is applied to the N-material and negative voltage is applied to the P-material.
 b. negative voltage is applied to the N-material and negative voltage is applied to the P-material.
 c. its junction has high resistance.
2. The op amp is an important solid-state device because it
 a. can amplify very small signals from thermocouples and RTDs to larger voltages and currents.
 b. can provide a negative firing pulse to the gate of an SCR.
 c. can be used as a seven-segment display for numbers.
3. A circuit that has one SCR in it can
 a. control AC voltage.
 b. control DC voltage.
 c. control both AC and DC voltage.
4. The transistor operates like a relay in that
 a. its emitter is like the coil, and its base and collector are like a set of contacts.
 b. its collector is like the coil, and its base and emitter are like a set of contacts.
 c. its base is like the coil, and its emitter and collector are like a set of contacts.

5. The solid-state relay
 a. uses an LED to receive a signal and a transistor to switch current.
 b. uses a capacitor to receive a signal and an IC to switch current.
 c. uses a transistor to receive a signal and an IC to switch current.

PROBLEMS

1. Draw the symbol for a diode and identify the anode and cathode.
2. Draw the symbols for an NPN and a PNP transistor and identify the emitter, collector, and base.
3. Draw the symbol for a silicon controlled rectifier (SCR) and identify the anode, cathode, and gate.
4. Draw the electrical symbol for a unijunction transistor and identify base 1, base 2, and the emitter.
5. Draw the electrical symbol for a triac and identify MT1, MT2, and the gate.
6. Draw the electrical symbol for an op amp and identify the inverting input, the noninverting input, the output, and the feedback resistor.

Websites

Websites change constantly. Use the key word search to locate sites that have similar content. The products that will be found at each site are indicated with the key word.

Johnson Controls
Products: Solid-state relays and controls
Key word: Johnson Controls

Ranco Controls
Products: Solid-state controls
Key word: Ranco Controls

Honeywell Controls
Products: Thermostats and solid-state HVAC controls
Key word: Honeywell Controls

Robertshaw Controls
Products: Solid-state HVAC controls
Key word: Robertshaw Controls

ABB
Products: Variable-frequency drives for HVAC
Key word: Variable-frequency drive

Chapter 20

Direct Digital Control (DDC) Systems and Programmable Logic Controllers (PLCs)

Job Assignment for This Chapter

You are called to a job where the HVAC system is controlled by a direct digital controller (DDC). The temperature for the heating cycle is controlled by an analog water valve that opens between 0 and 100% to provide hot water to the heating coil. The temperature of the water is 180 degrees, but the room is not staying warm enough. Your job is to check the heating loop to determine why the room is not staying warm enough.

Terms You Will Learn in This Chapter

A/D converter
Analog signals
ASCII
Binary-coded decimal (BCD) numbering system
Binary numbering system
Closed loop
Control loop
Controller
D/A converter
Damper motor actuator

Decimal
Digital
Direct digital controller (DDC)
Discrete
Error
Floating-point number
Hexadecimal numbering system
I/P (current over pressure) converter
Input module
Integer

Limit switch
Microprocessor
Networking
Octal numbering system
Op amp
Open loop
Output module
P/I (pressure over current) converter
Pneumatic sensor
Process variable (PV)
Programmable logic controller (PLC)

Proportional, integral, and derivative (PID)
Proximity switch
Push-button switch
Scan cycle
Selector switch
Setpoint (SP)
Solid-state integrated circuit
Summing junction
Temperature indicator
Thermocouple
Transmitter
Valve actuator

After reading this chapter, you will be able to:

1. Identify the basic parts of a closed-loop system and explain their operations.
2. Explain the operation of a temperature input to a DDC system.
3. Explain how a DDC system controls an output.
4. Explain the difference among discrete, analog, and digital input or output.
5. Describe the four basic parts of any programmable controller.
6. Explain the four things that occur when the PLC processor scans its program.
7. Explain the function of an input module and describe the circuitry used to complete this function.
8. Explain the function of an output module and describe the circuitry used to complete this function.

20.0 OVERVIEW OF DIRECT DIGITAL CONTROL SYSTEMS

In the previous chapters, most HVAC systems that were discussed involved systems for residential or small commercial buildings that are fairly easy to heat and cool. Today, HVAC systems must be designed to control complex structures such as hospitals, schools, and multifloor office buildings. The conditions inside these multiroom structures can vary to the point where one side of the building needs cooling, whereas other rooms need heating. Also, building environments will change as the number of people in the conditioned space changes from just a few people to large crowds. The HVAC control systems for these buildings must be able to meet all these changing conditions. In the 1960s to 1980s, most controllers used pneumatics or op amps to provide the control functions. In the 1980s and 1990s, sequencers and microprocessors were incorporated with op amps to provide digital control. Today, modern **direct digital control (DDC)** systems tend to have digital controllers (**microprocessors**) or hybrid-type controls, with a digital controller and pneumatic input and output devices. DDC systems allow HVAC controls to be integrated into larger networks, where they can become interfaced with colored operator screens or connected to the Internet so they can be monitored from remote cities. In this chapter you learn about **discrete** (on–off), analog (minimum to maximum), and digital devices and controllers. A **digital** device has two states, on and off. All the terminology of DDC systems is explained and examples are provided. This chapter also explains how modern DDC controllers use the functions of *gain, reset*, and *rate*, also called **proportional, integral, and derivative (PID)**, to control closed-loop-system response.

Modern DDC systems are made in a variety of sizes designed to control everything from single-family homes to large commercial buildings. These DDC systems can be set up to control temperature for heating and cooling, air flow, air quality, and fire and smoke alarms. They can be monitored locally or from a remote site, and setpoints can be programmed on a local panel or from a remote monitoring location.

This chapter helps you understand the parts of a typical controller and how they work together as a system. Many kinds of systems are on the market today, so you may need to go to a company school to learn its system in depth and become certified to

work on their controllers. This chapter provides you with a comprehensive overview that will allow you to begin working on these systems or continue for more schooling.

20.1 BASIC PARTS OF ANY CONTROL SYSTEM

It is important to understand that the basic parts of any control system will have the same names and provide the same functions regardless if the controller is pneumatic, **op amp,** or a microprocessor-based system. The names are also the same, regardless if the system is controlling temperature, water flow, air flow, steam pressure, water level, or water temperature. Figure 20–1a shows the basic parts of a control system shown in a block diagram. Figure 20–1b shows a process diagram of the same system. It will be easier to understand these basic parts of a control system if they are applied to a specific application. We will use a simple HVAC direct digital control system with one thermostat that adjusts a variable hot-water valve to cause the conditioned space to get warmer. From the process diagram you can see that the system starts with a **setpoint (SP)** signal. The SP is the desired temperature in the conditioned space. For example, if the desired temperature of the conditioned space is 70°F, the SP must be adjusted to 70°F.

The next part of the control system is the **process variable (PV),** which is the signal that comes from the sensor. For example, in the heating system, the PV is the signal from the room thermostat, which is the sensor for this system. The PV is also called the feedback signal; for this example the room thermostat will be able to provide an analog voltage (0–10 V) that represents the actual temperature in the conditioned space. The process variable is the present value or actual value of the temperature at the instant the sensor reading takes place. Since the sensor reading is continuous, the PV will change

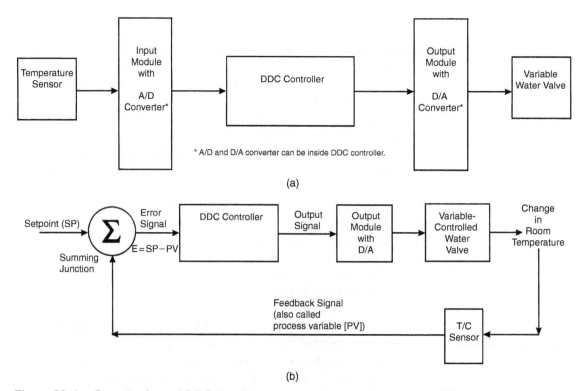

Figure 20–1 **a** Example of typical DDC closed-loop system. **b** Process diagram for a DDC system.

continually to indicate the changing temperature of the conditioned space on a **temperature indicator** that displays the temperature.

The **summing junction** is the place in the control system (inside the DDC controller) where the SP is compared to the PV. This means that if the SP is 70°F and the PV signal indicates the actual room temperature is 66°F, the difference is 4°F. The summing junction is identified by the Greek letter sigma (Σ).

The difference between the SP and the PV is called **error**: The error can be a positive value if the SP is larger than the PV, or it can be a negative value if the SP is smaller than the PV. When the **controller** is in automatic mode, which is also called **closed loop** mode, it uses a control calculation called an *algorithm* that determines how much the output should be opened or closed. The algorithm uses gain, reset, and rate to adjust the output signal in response to the amount of error. For example, a technician can set the gain, rate, and reset to cause the hot-water valve to open 50% if the temperature error is 1°. The technician can also set them to cause the valve to open 100% if the error is 1°. If the valve opens 100% when the temperature error is 1° different, the room temperature may overshoot. If the gain, rate, and reset are set too small, the water valve cannot open wide enough to allow the room to heat up fast enough. Setting up the control depends on the size of the heating and cooling equipment and how fast the system is supposed to change. When a system is in **open loop** mode its control is manual and the output is set manually.

Gain, reset, and rate are also called proportional, integral, and derivative (PID). Later in this chapter you will learn about the effect of the PID values and how they can be adjusted to change the speed of response for the system.

20.2　How the Hot-Water System Works

The hot-water heating system in this application is used to warm a conditioned space, such as an office building. From the control diagram you can see the components for the parts of the basic control system. The controller for this application is a single-point DDC controller, which means it has one input and one output. The SP in this example is entered by someone in the room by adjusting the thermostat to the desired temperature. The sensor for this system is a temperature sensor, also called the process variable. The SP is compared to the PV signal at the summing junction, which is inside the controller. The signal that comes from the summing junction is called the error signal, and it is sent to the controller part of the system.

The controller sends an output signal to an amplifier that controls the position of the hot-water valve. The amplifier adjusts the hot-water valve from 0 to 100%, with a 0–10-V DC signal. The controller sends the 0–10-V signal as an output signal to the amplifier. When the controller sends a signal of 1 V, which is 10%, the hot-water valve is adjusted to 10% open. If the signal is sent for 50%, 5 V is sent to the valve, and if the signal is set for 100%, the controller sends a 10-V signal to the valve.

20.3　Types of Signals Used in DDC Systems

The DDC system tends to look complex when you first examine it because it uses a variety of types of signals: on–off, which is called discrete; analog, which allows the value to be represented by values between a minimum value and a maximum value; and digital (or numerical) values, which allow sensor signals to be represented by numbers. It may help to see an example of each type of numbering system that you use every day.

20.4 NUMBERING SYSTEMS FOR DDCs

Since the direct digital controllers use a microprocessor for control, they use a variety of numbering systems. The most common of these numbering systems are decimal, binary, octal, and binary-coded decimal numbering systems. The DDC uses these different numbering systems to transmit information internally between the input and output modules, memory, and the program that is stored in the processors. These numbering systems are also used externally between all the devices interfaced to it.

20.4.1 Decimal System

The easiest numbering system to understand is the system you learned in grade school. It is called the **decimal** system, or base ten system, because it uses ten values, or digits. *Deci* is also the Greek word for ten. These ten numbers start with zero, which has the smallest value, and 9, which has the highest value. The values for the decimal system are 0, 1, 2, 3, 4, 5, 6, 7, 8, and 9. The decimal system also uses these values in various positions for weighting the numbers. Since nine is the largest value in the decimal system, a column to the left of the first number must be added to count values higher than nine. For example, to get ten, you write 1 to the left of 0. Now you have two columns. The column on the left indicates you have counted beyond nine. The values 1–9 that you put in the left-hand column indicate how many times you have counted beyond nine. It is easier to explain by saying the values in the first column are 0–9. This is called the ones column. Sometimes the ones column is called the units column. The column to the left of the ones column is called the tens column. The first time you count past nine, you put a 1 in tens column and set the value in the ones column back to 0. We think of 10 as having two digits, with 1 in the tens column and 0 in the ones column.

When you count beyond 9 the second time, you actually count beyond 19. This time you place a 2 in the tens column and again put the lowest value, 0, in the ones column. The number is now 20 (or two, zero). The process is repeated every time you count to the highest value in the ones column. For instance, 30 comes after 29, 60 comes after 59, and so on. In each case, when you reach the highest value in the ones column, you add 1 to the tens column and start the ones column at the lowest value.

The highest value you can count with two digits is 99. You use the same theory to count beyond 99 as you did every time the ones column came to 9. In this case, you add one to the next column, which is the hundreds column, and return the other two columns, tens and ones, to their lowest value, which is 0. Now you have the next number larger than 99, which is 100 (or one, zero, zero). You call this number one hundred. When the tens and ones columns become full (199) again, you add one more value to the hundreds column to reach 200. This process is repeated until you reach 999. At this time the same rules are used: add one to a new column to the left of the hundreds column. This new column is called thousands. When the thousands column becomes full, the next column is ten thousands, and the next column is hundred thousands. Figure 20–2a shows the placement and names of the four columns.

You should begin to see that as each column is added in the decimal numbering system, it is exactly ten times as great as the column before it. In other words, the column to the left of the tens column is the hundreds column because $10 \times 10 = 100$. The column next to the hundreds column is the thousands column because $100 \times 10 = 1000$. The column next to the thousands is ten thousands because $1000 \times 10 = 10,000$. No matter how far you count in the decimal system, the next column to the left is ten times as great as the preceding column.

Let's examine several numbers in the decimal system and show their values. For example, 764 has a 7 in the hundreds column, 6 in the tens column, and 4 in the ones

Thousands Hundreds Tens Ones

(a)

Figure 20–2 a Column names for the decimal numbering system. **b** The value 764 shown as a decimal number with column weights.

Hundreds Tens Ones
7 6 4

$$\left\{ \begin{array}{l} 7 \times 100 = 700 \\ 6 \times 10 = 60 \\ 4 \times 1 = + 4 \\ \hline 764 \end{array} \right.$$

(b)

column. You show the number's value by multiplying 7 by 100 to get 700, multiplying 6 by 10 to get 60, and 4 by 1 to get 4. Figure 20–2b shows that when you add these values, you have a total of 764.

Examining the decimal system, the one with which you are most familiar, gives a basis for looking at the other systems. The decimal, binary, octal, and hexadecimal systems all work the same way in that they have weighted columns and values that repeat in each column.

20.4.2 Binary Numbering System

The **binary numbering system** is the common numbering system used for DDCs and computers. The binary system is called the base two system. This means the binary system has just two values, 1 and 0. Since they are the only two values in the system, weighted columns must be used to count beyond these values. The reason the binary system is used in computers and DDCs is that these systems mimic the electrical states of on and off. The value 1 is assigned when a circuit is on and the value 0 is assigned when a circuit is off.

The best way to understand the binary system is to show the base ten values from 0 to 15 and their equivalent binary values. See Figure 20–3. From this example you can see that both numbering systems start with the value 0. The next value for each system is 1. These first two values for both numbering systems look identical. The binary system must use a second column once the values go beyond 1. Actually, the binary system operates just like the decimal numbering system in that each time a number reaches the highest value in a column, the next column to the left is added. Since 1 is the highest value in the binary system, you must add a column to the left of the first column to count beyond 1. This new column is called the twos column.

Decimal	Eights	Fours	Twos	Ones
0	0	0	0	0
1	0	0	0	1
2	0	0	1	0
3	0	0	1	1
4	0	1	0	0
5	0	1	0	1
6	0	1	1	0
7	0	1	1	1
8	1	0	0	0
9	1	0	0	1
10	1	0	1	0
11	1	0	1	1
12	1	1	0	0
13	1	1	0	1
14	1	1	1	0
15	1	1	1	1

(Binary)

Figure 20–3 Counting to 15 (decimal) in the binary numbering system.

Since the binary system is a base two numbering system, the weight of each column is established by multiplying the value of the previous column by 2. Since the ones column is full, you must add 1 to the twos column and put the lowest value in the ones column. This means the binary value for 2 is 10 (one, zero), which is a 1 in the twos column and 0 in the ones column. The binary value for 3 is 11 (one, one), or 1 in the twos column and 1 in the ones column. Now that both columns are full, or have the highest value in them, a column must be added to count beyond 3. This new column is called the fours column. The binary value for 4 is 100 (one, zero, zero), 1 in the fours column, 0 in the twos column, and 0 in the ones column. You can see that 5 is 101 (one, zero, one), which is 1 in the fours column, 0 in the twos column, and 1 in the ones column. Another way to say this is that 5 is made up of one 4 and one 1, or $5 = 4 + 1$.

The next columns in the binary system are eights, sixteens, thirty-twos, and sixty-fours. Again, you can see each new column is found by multiplying the previous column by the number of digits in the numbering system. Since the binary system is a base two system, it has two digits, and each column is multiplied by 2 to get the weighted value of the next column.

Another point to remember is that you must identify the numbering system in which you are working. Since 10 (one zero) can represent 10 in the decimal system and 2 in the binary system, a subscript is used to identify the numbering system being used. The subscript (10_2) is used to show you are using the binary system. If you are using the decimal system, the subscript value is a 10 (10_{10}). The subscript for the decimal numbering system is usually omitted because this system is the most used in daily life. If there is any possibility that a number may be confused, it is very important to use the subscript.

As we said, the binary system is used in digital electronic systems, such as programmable controllers, because the two values in the system, 1 and 0, represent the two possible electrical conditions of a switch. In most electrical systems, on is designated with 1 and off is designated with 0. The 1 and 0 also represent the two states that magnetic memory can use to store numbers and other data. This means the computers and processor chips inside programmable controllers can handle numbers and data because they have virtually thousands of binary circuits. By designating each binary column with a different weighted value, such as ones, twos, fours, eights, or sixteens, the processor can handle very large values.

20.4.3 Binary-Coded Decimal System

The **binary-coded decimal (BCD) numbering system** is a combination of the binary and the decimal numbering systems. The system works basically like this: binary values are the only values used in the system. This means only 1s and 0s are allowed. The weighting of the binary values are traditional binary weights for the first four columns, so the ones, twos, fours, and eights columns of the binary system are used. Each group of four binary columns allows you to count binary from 0 through 9. (Actually, we could count 0–15, but since we will convert these four columns to a decimal number, we need to count only the values 0–9.)

The first group of four binary digits are used together to represent numbers in the ones column of the decimal system. The next four binary columns are placed to the left of the first four columns; this set of columns is used to represent the tens column in the decimal system, enabling you to count values 10 through 99. If you need to count beyond 99, another four binary digits are grouped to give you the hundreds column. An example of BCD-weighted columns is shown in Figure 20–4a, and a table that shows how to count from 0 to 20 in BCD format is shown in Figure 20–4b.

Tens Column				Ones Column				Decimal Column Weights
Eights	Fours	Twos	Ones	Eights	Fours	Twos	Ones	Binary Column Weights

(a)

Decimal Value	BCD Value Tens Column	BCD Value Ones Column
0	0000	0000
1	0000	0001
2	0000	0010
3	0000	0011
4	0000	0100
5	0000	0101
6	0000	0110
7	0000	0111
8	0000	1000
9	0000	1001
10	0001	0000
11	0001	0001
12	0001	0010
13	0001	0011
14	0001	0100
15	0001	0101
16	0001	0110
17	0001	0111
18	0001	1000
19	0001	1001
20	0010	0000

(b)

Figure 20–4 **a** Names of columns for the binary-coded decimal numbering system. **b** Counting to 20 in the binary-coded decimal (BCD) numbering system.

The BCD numbering system is a very usable number system because it can communicate in binary with the computer, but it is also coded into decimal values for humans to read. The BCD system is generally discussed in terms of the number of decimal units the BCD represents. In other words, a one-decimal-digit BCD needs 4 binary digits, or bits. A two-decimal-unit BCD number needs 8 binary digits, or bits, and a three-decimal unit BCD number needs 12 bits. The BCD value needs more bits to operate than the binary system, but it is easier to read. If the computer is communicating with its memory, or modules, it will probably use the binary system. If the DDC is communicating an LED data display or thumbwheels, it may need to convert its binary numbers to BCD so humans can read the values in decimal form. The most common devices that use the BCD numbering system are thumbwheels and 7-segment LED displays. It is essential that you fully understand numbering systems and are capable of converting numbers from binary to BCD and from BCD to decimal.

20.4.4 Octal Numbering System

The **octal numbering system** is another numbering system commonly used in DDCs and computers. The octal system is also called the base eight system, and it has eight values, 0–7. The octal numbering system was very useful in early DDCs because of the abundance of 8-bit memory chips. Since newer DDCs do not use 8-bit memory, the octal numbering system is not as widely used as it once was, but you will still find it in use in many input and output modules and their addressing schemes.

Since the octal numbering system can count only the values 0–7, you will need to add columns to count larger values, just as you did in the decimal and binary system. Figure 20–5a shows how to count to 64 in the octal numbering system. The column

Octal Numbering System

Decimal	Octal		Decimal	Octal		Decimal	Octal
0	0		23	27		46	56
1	1		24	30		47	57
2	2		25	31		48	60
3	3		26	32		49	61
4	4		27	33		50	62
5	5		28	34		51	63
6	6		29	35		52	64
7	7		30	36		53	65
8	10		31	37		54	66
9	11		32	40		55	67
10	12		33	41		56	70
11	13		34	42		57	71
12	14		35	43		58	72
13	15		36	44		59	73
14	16		37	45		60	74
15	17		38	46		61	75
16	20		39	47		62	76
17	21		40	50		63	77
18	22		41	51		64	100
19	23		42	52			
20	24		43	53			
21	25		44	54			
22	26		45	55			

(a)

Decimal Value	Hexadecimal Value	Binary Equivalent Value
0	0	0000
1	1	0001
2	2	0010
3	3	0011
4	4	0100
5	5	0101
6	6	0110
7	7	0111
8	8	1000
9	9	1001
10	A	1010
11	B	1011
12	C	1100
13	D	1101
14	E	1110
15	F	1111

(b)

Figure 20–5 **a** Octal numbering system counting 0 to 64. **b** Hexadecimal numbering system counting 0 to 15.

weights for the octal numbering system are determined just like columns in other numbering systems, by multiplying each column by the base number of the system. The columns in the octal numbering system are ones, eights, sixty-fours, etc. As you will notice, the values 8 and 9 do not appear in the octal numbering system.

The octal numbering system is commonly used in DDC systems, especially for identifying 1/0 modules. It is useful because some DDCs have the ability to utilize older input/output modules with 8-bit 1/0 structure. In this way, the processor can use one 1/0 line for each octal digit. Remember, if you use the octal numbering system, you must use the subscript 8 to identify the system.

The octal numbering system is not especially suited for conversion to other systems. Instead, it is meant to be used on its own. This means that you simply use the octal system and don't worry about converting. If the values 0–7 or 10–17 are used as addresses

for modules, just use them, and don't try to convert them to decimal or binary. You will soon begin to understand how the system works as you use it for module addresses. You will also find it is quite useful because each data memory word is either 8 bits or 16 bits long, and the octal numbering system represents these values well.

20.4.5 Hexadecimal Numbering System

The **hexadecimal numbering system** is another numbering system used in DDCs. The hexadecimal numbering system is unique because it uses the numbers 0–9 and the letters A–F to count from 0 to 15. The important part is that this system only uses a single digit to display values 0–15. This is important when 16-bit systems are used and memory is limited, especially in addressing. The hexadecimal system is used primarily for addressing, so it is not often converted to decimal. Each hexadecimal number has a binary equivalent value. Figure 20–5b shows the hexadecimal and equivalent binary values for 0–15.

20.4.6 Integers

When the decimal numbering system is used in a DDC, it can be used to represent whole numbers (integers) or numbers with decimal points, called floating-point numbers. The integers used in DDCs can be any positive number between 0 and +32,767 or any negative value 0 to –32,768.

Integers are all the positive and negative whole numbers used in the decimal system. This means that only whole numbers and no fractions or decimals are used. You will need negative numbers when you are indicating variables, such as temperature, that are below zero or pressures that are below 0 PSIG, which we know as vacuum.

The maximum values of the positive and negative numbers that can be used in a DDC depend on the type of processor chip. Most DDCs use 16-bit words to store the integers, so the maximum positive and negative values comes from the largest binary value that can be stored in 16 bits. If the number can be either positive or negative, one of the bits is used as a "sign bit," so the largest value is 15 bits. The sign bit is usually stored in the bit location on the far left of the word. For example, 2 to the 15th power is 32,768. Including the value 0 in our system, the exact range is –32,768 to +32,767.

If you want a system that can have only positive numbers, you can use all 16 bits for values, and you do not need to reserve 1 bit as the sign bit. The value of 2 to the 16th power (2^{16}) is 65,636. If you want the system to include zero, you have values that range from 0 to 65,535. In the binary system with 16 bits, each with a 1 as its value, the binary number is 1 1 1 1 1 1 1 1 1 1 1 1 1 1 1 1, and its decimal equivalent is 65,535.

Some smaller, less expensive DDCs are designed to use only integer numbers. You can see that if the DDC provides only integer values, the numbering system is limited to the values –32,768 to +32,767. This can be a problem when you have modern manufacturing facilities that can manufacture more than a million products in one day, such as aluminum can or plastic bottle manufacturing facilities. In these applications a more sophisticated and more expensive DDC is used that uses floating-point numbers, which involve decimal points or exponents. For example, 1.0^{e6} is a number stored with an exponent; its value is 1,000,000 (one million).

20.4.7 Floating-Point Numbers

Floating-point numbers are numbers that have a decimal point. They can be stored as exponential numbers if they are too large or extremely small. In some DDCs, floating-point data are displayed in 7-digit precision and can range from $\pm1.1754944e{-}38$ to $\pm3.4028237e{+}38$.

It is important to understand that floating-point numbers are stored in separate memory from decimals. In most DDCs floating-point numbers are stored in two consecutive 16-bit words. The numerals are stored in one of the two words and the exponent is stored in the other. Some processors may not have a floating-point numbering system available because it takes a special processor to be able to calculate the double word and it takes extra memory to store floating-point values. If the floating-point number is shown with a lowercase *e*, as in 1.234e6, it means the number involves an exponent. The exponent represents the number of places to the right or left of the decimal point the value should have. If the exponent is positive, it represents the number of places to the left of the decimal, and if the exponent is negative, it represents the number of places to the right of the decimal point.

In the number 1.234^{e6}, the actual number has six digits to the left of the decimal point, so the value of the number without the exponent is 1,234,000. The exponents make it easier to store the numbers and allow for numbers that are much larger than the 16-bit integer values allow. If the number has a negative exponent, such as 1.234^{e-6}, you move the decimal point 6 places to the left, which means you add five zeros to the left of the first digit as place holders. This number is 0.000001234, and you can see that the exponent would allow the number to be shortened while maintaining its accuracy. Exponents are extremely valuable in handling very large values and very small decimals where accuracy in important, such as in process control and motion control. For example, if the exponent is positive 20, as in 1.23^{e20}, it indicates a very large number, 123,000,000,000,000,000,000.0, which is very difficult to display and use.

Essentially, a processor operates much faster and uses less memory when only integers or whole numbers are used. If the values are rather large, the decimal point is generally unnecessary. If you are designing a small project that uses only integers, you should select a DDC that provides only integer numbers to save money and increase speed. If you are using the system for process control or motion control, it is important to select a DDC system that provides floating-point numbers so a decimal point or an exponent can be used to maintain the accuracy needed to control the process. With the advent of newer microprocessing chips, newer DDCs will provide both integer and floating-point systems with very little extra cost. The next problem becomes providing sufficient memory to store these larger values because they take more bits to store than a simple integer.

20.4.8 ASCII System

ASCII is a system that has been created to transmit all the letters of the alphabet as well as numbers and other characters you would find on a typing keyboard. ASCII is an acronym that stands for American Standard Code for Information Interchange. You may have seen this system when you sent text between computers and printers. The ASCII system may be used when the DDC is sending information to displays and other text systems, such as printers for reports or data. The system consists of 128 two-digit hexadecimal numbers that represent the 128 letters, numbers, and characters used on a keyboard. The system assigns a separate two-digit number for each lowercase and uppercase letter of the alphabet. The ASCII system transmits the two-digit packets between computer printers and displays.

20.5 PARTS OF THE DDC SYSTEM

Now that you have learned about all the types of numbering systems used in DDCs, you are ready to learn more about the parts of the DDC system. All DDC systems have the same basic parts regardless of their brand names. They include the controller, an input

module for discrete (on–off) sensors, and the output module for discrete outputs. Some systems will have analog input and output modules to control analog sensors and outputs. The next sections explain the features and operation of each of the parts of the system in more detail.

20.5.1 Controller

Modern controllers in DDCs have a microprocessor that continually completes a number of tasks in a sequential order. During each scan cycle, it checks all its inputs, executes its program statements, and sends signals to its outputs. The part of the cycle where the processor checks its inputs is called the *read* part of the cycle. The part of the cycle where it sends signals to its outputs is called the *write* cycle.

If the inputs are discrete (on–off), the processor looks for a 1 or a 0. A 1 indicates an input sensor is on, and a 0 indicates it is off. The processor writes a 1 or a 0 to the output module to indicate if the output should be on or off. If the inputs and outputs are analog, the processor uses an analog value (a number between 0 and its maximum value) to represent the condition of the signal. If the microprocessor and analog modules use an 8-bit structure, they provide 256 values from 0 to 255; if they use a 12-bit structure, they provide 4096 values from 0 to 4095; and if they use a 16-bit structure, they provide 65,536 values from 0 to 65,535. It is important to remember that all data passed between the processor and its modules must be digital (numerical) in nature. Figure 20–6 shows an example of a typical DDC.

| 20.6 | ON–OFF (DISCRETE) CONTROL |

The simplest type of signal to understand is the on–off signal, which is also called a discrete signal. An example of an on–off signal is the wall switch that controls the lights in your room. If you turn the switch to the on position, the lights are full on. If you turn the

Figure 20–6 Example of a typical direct digital controller. *(Courtesy of Siemens Building Technologies, Inc.)*

switch to the off position, the lights are completely off. There are only two conditions for a discrete signal, which makes them very easy to troubleshoot. If the switch is in the on position, you can test it for continuity, and it should show 0 Ω. If the switch is in the open condition, you can test it for continuity, and it should show an infinite amount of resistance.

When a DDC system uses a discrete input, such as a thermostat, it can tell the controller only if the thermostat is open or closed. If it is a heating thermostat, it will be open when the temperature it is sensing is above the setpoint, and it will be closed when the temperature is below the setpoint. If the DDC is using a discrete signal to turn on a circulating pump, the pump will be completely on when the discrete signal is on, and it will be completely off when the discrete signal is off.

20.6.1 Input Modules

The **input module** for a DDC can be located directly on the controller, or in larger systems the modules can be mounted remotely from the controller and connected through cables. The input module may have a status light (indicator) for each input to show the troubleshooting technician if the signal is present at the input terminal. The input module typically has several parts in its circuit to drop the voltage from the sensor to a lower level that the processor can use. If the input sensor uses 110 V AC, the input module will have several resistors and a bridge rectifier, which converts the signal to DC. The DC signal is then sent to the light-emitting diode, which is part of an optocoupler. The light from the light-emitting diode is focused on a phototransistor, which turns on when light is present and passes a low-voltage digital DC signal to the microprocessor. The input module uses light inside the circuit to provide isolation between the 110-V AC voltages or other voltages used with the sensors and the low-voltage digital circuits used in the processor. The circuit also provides a small status light called a status indictor to show when the circuit receives voltage from an input sensor.

20.6.2 Output Modules

The **output module** for the DDC, like the input module, can be located directly on the controller or mounted remotely and connected to the controller through cables. The processor sends a digital 1 to an output module to turn it on and a 0 to turn it off. During each scan, the processor executes its program, which indicates which outputs should be turned on at what time. Immediately after the program is scanned, the processor sends a 1 or a 0 to each output.

The electronic circuit inside the output module acts like a switch that opens or closes to allow voltage and current to flow to the output device connected to the module. For example, if the output device is the 110-V AC coil for a motor starter that controls a pump motor, L1 of a 110-V AC source voltage is brought to one terminal of the output module. When the processor turns that circuit on, it energizes a circuit inside the module that completes the circuit, similar to the way switch contacts close, and allows voltage and current to pass through the module circuit onto the coil of the motor starter. If the processor sends a signal to turn off the output, it causes the circuit in the output module to create an open in the circuit similar to the way switch contacts open, which causes the voltage and current flow to stop.

The actual electronic circuit of the output module is similar to the circuit in the input module in that it provides a change of voltage between the microprocessor and the output device and opto-isolation between the high and low voltage parts of the circuit. The processor sends a small DC voltage signal to the output module when it is to be

energized. In some output modules this small DC voltage goes through the DC section of a four-diode bridge rectifier circuit. When voltage and current flow through the DC part of a bridge rectifier, AC current can flow through the AC part of the rectifier. In this application, the bridge rectifier acts as an electronic switch that controls the flow of AC voltage and current. If the output module is designed to control DC voltage, the small digital voltage signal is sent from the processor to the output module circuit, where it is sent through a light-emitting diode. The light-emitting diode sends light that is focused on an opt transistor, which controls the DC voltage to the output device. Other types of output module circuits can use several other electronic devices, such as light-controlled SCRs or light-controlled triacs. The main point to remember is that in all output module circuits, the circuit is acting like a set of switch contacts that open or close to allow current to flow to the device connected to the output module.

20.6.3 Isolated Output Modules (Relay Output Modules)

One problem with solid-state circuits used in output modules is that they bleed a small amount of current when they are in the off state. Some output devices cannot work correctly with the small amount of bleed current, so output modules are available that use relay contacts to turn the voltage on and off. The relay is integrated into the output module circuit so that the small signal the controller sends to the module is used to control a circuit that provides higher voltage to the coil of the relay. When the relay coil receives the voltage, it energizes and causes its contacts to transition (change state). Most of the relay output modules provide a set of SPDT contacts that provide a set of normally open and a set of normally closed contacts that are controlled by the coil. The voltage that the contacts control must be provided from an external source.

20.7 ANALOG SIGNALS

At first **analog signals** seem very complex. Once you see a common example of an analog signal, the concept is much easier to understand. An example of an analog signal is the volume control on your radio or television. The analog signal can be any value between 0 and 100%, such as 25%, 50%, or 75%. Another example of an analog system is a dimmer switch used to control the lights from 0 to 100%. Some examples of analog signals used in DDC systems include controls for fan speeds, opening and closing water valves, pump speeds, and damper positions that set these devices from 0 to 100%.

If the DDC is controlling a circulating pump with an analog system, it can send a signal to a variable-frequency drive that will set the pump motor speed at any value between 0 and 100%. You will see that the major advantage of analog signals is that they allow control to be more precise.

Later in this chapter you will learn that a DDC system is a digital system, so it requires input and output modules to interface with its sensors. If the sensor uses an analog signal, the input module needs an electronic circuit called an analog-to-digital converter. If the DDC is controlling an analog device, such as the variable-speed drive that controls the pump speed, it will need an output module that has an analog signal.

20.7.1 Typical Analog Signals

Electrical analog signals can be voltage, current, resistive, or frequency signals. Voltage and current are the most common signals, and resistive and frequency signals are usually converted to voltage or current. Analog signals can also be pneumatic (air); air was

Percentage	0–4095	0–10 V	4–20 mA	3–15 psi	Motor rpm
100%	4095	10 V	20 mA	15 psi	3600 rpm
75%	3072	7.5 V	16 mA	12 psi	2700 rpm
50%	2048	5.0 V	12 mA	9 psi	1800 rpm
25%	1024	2.5 V	8 mA	6 psi	900 rpm
0%	0	0 V	4 mA	3 psi	0 rpm

(a)

No. of Bits	Digital Value	Digital Range	0–10 V Resolution	4–20 mA Resolution
8 bits	256	0–255	0.039 v/bit	0.06250 mA/bit
10 bits	1024	0–1023	0.00977 v/bit	0.01563 mA/bit
12 bits	4096	0–4095	0.00244 v/bit	0.00391 mA/bit

(b)

Figure 20–7 **a** A table of analog values for percentage, milliamps, voltage, RPM, and DDC values 0–4095. **b** A table showing D/A and A/D converter resolution values.

commonly used in earlier HVAC control systems. Today you will find a variety of each of these types of signals used in DDC systems. Figure 20–7a shows a table that compares all the analog signals you will find in a DDC system.

In analog systems we can describe the action as a percent, or we can use the actual values of the system, such as revolutions per minute or an amount of light. Since we generally use voltage or current to control these parameters in a DDC, we can begin to talk about the voltage and current systems that are commonly used. From the table you can see that the voltages can be 0–5 V, 0–10 V, –5 to +5, or –10 to +10. Of these, using 0–10 V makes it easy to calculate percentages in your head.

From the table you can also see that current can also be used; the values for current systems are 0–20 mA or 4–20 mA. The 4–20-mA system does not start at zero, but we still refer to it as a minimum to maximum (min-max) analog system. Offset analog systems, such as 2–10-V or 4–20-mA system, are the most common systems used. The system starts at 4 mA instead of zero so it can indicate when a broken wire or open circuit has occurred in the system. If the offset is not used and the analog system value uses 0–20 mA or 0–10 V, the system could have a zero value when the system is at its minimum point or when you have an open in the circuit. This makes it difficult to determine which condition is causing the zero value. For example, if the system is using 4–20 mA, a technician can quickly make a current measurement and determine if the system is indicating 4 mA because it is at its minimum value or if it is indicating 0 mA because there is a broken wire somewhere in the electrical system. Since broken wires and electrical opens are the most common problem in DDC systems, the offset system is the most commonly used analog system because it helps determine which problem has occurred. Pneumatic systems also use an offset, because they are set up to use 3–15 psi. If the air pressure goes to zero, the technician knows there is a leakage problem, and if the signal goes to 3 psi, the technician knows it represents the minimum value.

20.7.2 Analog Output Module with a D/A Converter

Another type of module that most DDC systems provide is an analog output module. Most analog output modules provide variable voltage or variable current in the standard values, such as 0–10 V or 4–20 mA. The main part of the circuit in the analog output module is an analog-to-digital (D/A) converter. The **D/A converter** works on a principle of converting a digital number into changes of voltage. The D/A converter is a

solid-state integrated circuit (IC) chip that is designed for different amounts of resolution. Figure 20–7b shows the digital values for an 8-bit, 10-bit, and 12-bit A/D or D/A system. This figure also shows the amount of resolution for each system. For example, if the chip uses an 8-bit circuit, the resolution for the circuit is 1 part in 256. This means that if the D/A controls 0–10 V, the smallest amount of change of voltage is approximately 0.04 V. If the output is 4–20 mA, the smallest amount of change is approximately 0.0625 mA. Since the processor can send 256 numbers with 8 bits, the range is 0–255 if we include 0 as the starting point. You can determine the resolution for 0–10 V by dividing 10 V by 256. You can determine the resolution for 4–20 mA by dividing the 16-mA difference between 4 and 20 by 256. Some DDC systems use 10- or 12-bit D/A converters because they provide more resolution. A 10-bit system has 1024 values, so we use 0–1023 for the range. A 12-bit system has 4096 values, and when we use 0 for the starting number, the range is 0–4095. The resolution for a 12-bit DA that uses 0–10 V is approximately 0.00244 V, which can be found by dividing 10 V by 4096. The resolution for a 4–20-mA circuit is approximately 0.00391 mA, determined by dividing 12 by 4096. You can see that the larger the number of bits in the D/A converter, the more resolution the module can provide. This means that the analog output voltage or current from the module that is used to control things such as damper actuators, valve actuators, or variable-frequency drives (VFDs) that control fan speeds or pump speeds can process and control the outputs with more precision. It is important to understand at this time that if the resolution in the analog circuits in the damper actuator, valve actuators, and VFDs does not match the resolution in the output module, the smaller resolution will be the best resolution the system can have. This means that if the A/D converter in the analog output module is 12 bits and the damper actuator uses an 8-bit circuit, the best the resolution for the system can be is 8 bits.

The processor controls the value (number) sent to the analog output module, which determines the amount of voltage or current it provides for the output device. The simplest way to think about this type of control is to use percentage. If you are trying to control the speed of a pump motor from 0–3600 rpm, you can see that a 50% value from the processor will cause a 50% voltage signal to be sent from the analog output module. When the VFD receives the 50% signal, it causes the pump motor to run at 50%, or 1800 rpm. The amount of change can be controlled at higher or lower values in this manner.

20.7.3 Analog Input Module with an A/D Converter

The analog input module works like the analog output module, except this module uses an analog-to-digital (**A/D**) **converter** to convert changes in voltage or current into a digital number. The process basically works the same as far as resolution in the voltage or current values and the range of numbers in the processor. The analog input module receives voltages or currents from temperature, pressure, flow, and position sensors to indicate values from 0 to maximum. Some flow sensors use a paddlewheel, which produces a string of pulses that are converted to a frequency. The frequency is converted to voltage through an F/V converter, and that value can be read into the analog input module as a change of voltage. Other analog sensors, such as temperature sensors, may use thermocouples or RTDs. A thermocouple produces a millivoltage signal and an RTD produces a change of resistance. This means that special input modules for thermocouples and RTDs must be used for these sensors. Most DDC systems have thermocouple or RTD modules available, or you can use a thermocouple **transmitter** that converts the millivoltage from the thermocouple to a 0–10-V or 4–20-mA signal. A transmitter converts one form of energy to a usable signal.

20.8 INPUT DEVICES

This section discusses the various types of devices that you will find connected to the DDC as an input. The simplest types of devices are contacts from relays and switches. This section shows the types of devices that control the contacts used for inputs on the DDC.

20.8.1 Switches and Relay Contacts

When you work with contacts in switches and contacts in relays you will notice they have common terminology. The main difference is the symbols that are used to identify each. Figure 20–8a shows symbols for switch contacts, and Figure 20–8b shows symbols for relays contacts. The simplest set of contacts is the single-pole, single-throw (SPST) switch. The word *pole* refers to the number of sets of contacts, and the word *throw* refers to the number of places the switch can make contact. Throw is also used to describe the number of combinations the switch has to make contact with another terminal. You can use an ohmmeter to test the continuity (resistance) of each set of contacts to determine if they are open or closed. A closed set of contacts will have low resistance, and a set of open contacts will have high (infinite) resistance.

In the SPST, the switch has one set of contacts that has two different places where it can make contact. The symbol for the SPST set of relay contacts shows a common point between a set of normally open (NO) and a set of normally closed (NC) contacts. The next configuration is the double-pole, single-throw (DPST) switch. DPST contacts have two separate sets of contacts, each of which has only one contact point to touch when the switch is closed. The two sets of contacts are completely isolated from each other, but they both move open and closed at the same time.

The final configuration for switch contacts is the double-pole, double-throw (DPDT) switch. This configuration is similar to SPDT, except it has two separate sets of contacts that are each SPDT.

When you purchase a relay or switch, you will be asked to specify the configuration of the contacts. If you are replacing an original switch, you should try to replace it with the exact configuration. There may be times when the more expensive DPST or DPDT configurations are in stock and available but the SPST or DPST are not available. In this case you can use any of the four switch contact configurations as SPST switches. You could also use a DPDT switch as a DPST switch or a DPDT as an SPDT switch.

Figure 20–8 a Switch contacts for a single-pole, single-throw (SPST), single-pole, double-throw (SPDT), double-pole, single-throw (DPST), and double-pole, double-throw (DPDT) switch. **b** Contacts for various types of SPST and DPDT relay contacts.

20.8.2 Push-button Switches

Push-button switches are used in DDC systems as start-and-stop switches that are used to start or stop pumps, fan motors, and other devices. A push-button switch can operate a single set of normally open or normally closed contacts. Figure 20–9a shows examples of common symbols for the push button. The push-button switch can also operate multiple sets of contacts, such as single-pole, double-throw, which is a two-position push button, and double-pole, double-throw, which is a double-pole push button. Another special type of push-button switch is called the maintained type. The first four push-button switch symbols in the figure represent momentary push buttons, which return to their open positions after they are depressed and released. The fifth symbol shows a maintained push-button symbol, which indicates this switch will stay in the position that is activated. The fifth symbol showing the maintained contact also shows another symbol convention, where multiple sets of contacts that are operated by the same push button are connected with a dashed line. The dashed line connecting multiple sets of contacts indicates that when the switch is activated, all the sets of contacts will change condition at the same time.

The push-button operator can be added to a contact block that usually has one set of normally closed contacts and one set of normally open contacts. You can add two or three contact blocks on any switch, and all the sets of contacts will change state at the same time. You can use an ohmmeter or a diagram sheet to determine which terminals on the contact block act as a set of normally open or normally closed contacts.

In some diagrams the abbreviation PB is included with the push-button switch symbol. If multiple push buttons are used in a system, they are identified as PB1, PB2, PB3, and so on. This is a standard method of identifying multiple sets of contacts.

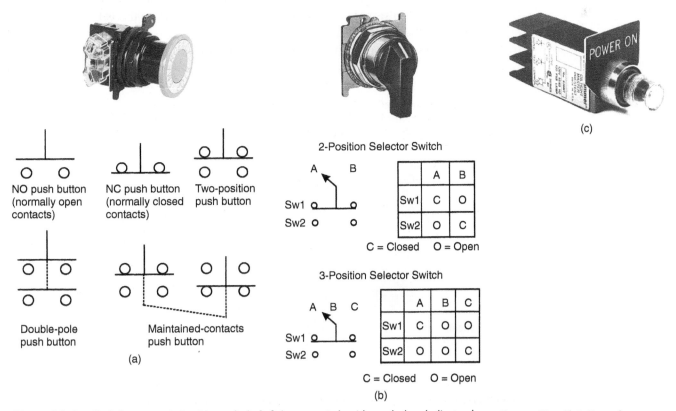

Figure 20–9 a Push-button switch with symbols. **b** Selector switch with symbols. **c** Indicator lamp. *(Courtesy of Eaton/Cutler-Hammer)*

20.8.3 Selector Switches

The operation of a **selector switch** is similar to that of the maintained push-button switch, in that it will stay in the position to which it is changed. This function is quite useful in HVAC circuits because it allows an operator to select an application and leave the switch in that position.

Selector switches are also similar to push-button switches in that they require manual movement to cause the switch contacts to change position. Instead of pressing the switch to cause operation, you turn the selector switch with a handle to cause the contacts to change. Figure 20–9b also shows a picture of a selector switch. The symbols for the two-position and three-position selector switches are provided in this figure. The two-position selector switch is normally used for such applications as on–off, up–down, and other dual conditions. These diagrams are used as the switch symbol in all electrical diagrams. You should also notice that the selector switch abbreviation is SS. If multiple selector switches are used in a circuit, they are identified as SS1, SS2, SS3, and so on. This figure also shows the switches in typical circuit applications.

20.8.4 Indicator Lamps with Push Buttons

Another type of discrete device is the indicator lamp. The indicator lamp is actually a discrete output, but it can also have contacts added to it so that it works like a switch when it is depressed. Figure 20–9c shows an example of an indicator lamp.

20.8.5 Limit Switches

Limit switches are widely used in DDC systems to indicate the position of damper and valve actuators. Figure 20–10a shows a limit switch. The arm of the limit switch is moved when the position of the actuator changes. It provides feedback to the DDC controller when it sends a signal to move the damper or valve. When the valve moves to the end of its travel, the switch is tripped and it sends a discrete (on–off) signal to the controller. Figure 20–10b shows the electrical symbol for various types of limit switches.

Figure 20–10 a A cutaway picture of a typical limit switch. *(Courtesy of Eaton/Cutler-Hammer)* **b** Electrical symbols for limit switches.

There are basically four types of limit switches: normally open held open, normally open held closed, normally closed held open, and normally closed held closed. You should notice that the electrical switch symbol is essentially the same for all types of switches, whereas the type of operator varies slightly. The operator part of the symbol is shown as part of a triangle that is attached under the switch portion. The triangle signifies the limit switch cam or arm, which causes the limit switch contacts to activate. You can see from the electrical symbols that the location of the switch portion will be above the line for a limit switch that is held open and below the line for a limit switch that is held closed. If the switch is shown in the closed position in the symbol, it is considered normally closed. If the switch is shown in the open position, it is considered normally open. Remember that the normal position is the position the switch would be in had no machine motion occurred. The normal position is also the position the contacts would be in had you removed the switch from the machine and bench-tested it. The abbreviation for the limit switch is LS. If multiple limit switches are used in a circuit, they are identified as LS1, LS2, LS3, and so on. The limit switch can also activate multiple contacts. These contacts can be groupings of normally open and normally closed contacts. They can be activated as multiple sets of single-pole contacts or as combinations such as single-pole, double-throw or double-pole, double-throw. The switch contacts are generally mounted in the switch body so that they can be changed without having to unmount the switch. Some switch bodies also have plug-in contacts, which allow for quick replacement.

20.8.6 Proximity Switches

A **proximity switch** is an electronic limit switch. It can determine when the switch is close to a metal or nonmetallic target. A target can be a part of a damper or valve actuator or it can be a small square of metal that is mounted so it can be detected when the actuator moves the assigned distance. The proximity switch puts out a magnetic field, so anytime magnetic metal comes within the detection field (approximately $\frac{1}{8}$ in.), the switch will trigger. The operation of the proximity switch is similar to a limit switch in that it is placed in the DDC system to detect motion of valve actuators or damper actuators.

20.9 TEMPERATURE SENSORS FOR DDCs

Direct digital control systems can use a variety of temperature sensors to provide remote or local temperature sensing. All the temperature sensors must use a thermocouple, RTD, thermistor, or integrated circuit (IC) as the element that converts the temperature value to an electrical signal. The next sections of this chapter explain the operation of and troubleshooting for these devices.

These sensors can be wall mounted like a thermostat or they can be duct mounted. If they are wall mounted, they can be preset from the DDC or they can be designed similar to the thermostat in your home, where you can change the room setting by making a change to the numbers on the face of the thermostat. Some thermostats have the ability to be set locally by someone standing in front of them or remotely by the DDC for conditions such as night and weekend settings, when no one is in the conditioned space.

Duct-type sensors can be a single-point type that measures a temperature and sends it directly back to the DDC as a discrete or analog input. The other type of integrated circuit has the ability to average the temperature and send the average value to the DDC as an input.

Special temperature sensors include freezestats and fixed-over-temperature safety sensors. The freezestat is a specially designed temperature-measuring device that detects conditions that would allow water to drop below the freezing point.

20.9.1 Thermocouples

A **thermocouple** is a temperature sensor that provides a small millivolt signal that ranges between 10 and 80 mV. Thermocouples are made by connecting two dissimilar metals at one end to form a junction. The theory of operation of the thermocouple can be explained by the Seebeck effect, which was discovered by Thomas Seebeck in 1821. The amount of voltage that the thermocouple produces depends on the two types of metal that are used to form the junction. When Seebeck made this discovery in 1821, the thermocouple was not useful because the small amount of voltage produced in the millivolt range could not be amplified at the time.

20.9.2 Seebeck Effect

Thomas Seebeck discovered that if two wires made from dissimilar metals are connected at both ends to make two junctions, when one end is heated, a small amount of current would flow through the circuit. When both ends of the wires are twisted together and one end is heated, the two wires produce an *open-circuit voltage* that is called the Seebeck voltage. The open-circuit voltage is proportional to the amount of heat added to the junction.

If a voltmeter is used to read the millivoltage that the thermocouple produces, a second junction of dissimilar metals is produced at the point where the thermocouple wire is connected to the copper wires of the voltmeter. This junction will produce a small amount of voltage in opposition to the original voltage produced at the thermocouple junction. One way to get around this problem is to extend the copper wire from the meter to a point where a second junction can be produced in such a way that the voltage from this junction can be controlled.

20.9.3 Types of Thermocouples

The most common types of thermocouple are the J-type and the K-type. The J-type is made of iron (Fe) and constantan, which is made from copper and nickel (Cu Ni). Figure 20–11 provides a table for all the thermocouples and shows the wire colors, wire types, temperature range, and millivolt range for each thermocouple. From the table you can see the iron wire produces the positive voltage and the constantan wire produces the

Type	Wire Color	Material	Temperature Range	Millivolt Range
J-Type	+ White − Red	+ Iron (Fe) − Constantan (Constantan is Copper nickel) (Cu Ni)	32 to 1382°F	−8.095 to +69.553 mV
K-Type	+ Yellow − Red	+ Nickel Chromium (Ni-Cr) − Nickel Aluminum (Ni-Al)	−328 to +2282°F	−6.458 to +54.886 mV
T-Type	+ Blue − Red	+ Copper (Cu) − Copper Nickel (Cu Ni)	−328 to +662°F	−6.528 to +20.872 mV
R-Type	+ Purple − Red	+ 13% Platinum / Rhodium − Platinum	32 to 2642°F	−0.226 to +21.101 mV
S-Type	+ Orange − Red	+ 10% Platinum / Rhodium − Platinum	32 to 2642°F	−0.236 to +18.693 mV

Figure 20–11 Table of common thermocouples and their color codes, temperature ranges, and millivolt ranges.

negative voltage. The J-type thermocouple is rated for temperatures from 32 to 1382°F. In some technical literature the highest temperature is rounded off to 1400°F. The millivoltage produced by the J-type thermocouple ranges from –8.095 mV to +69.553 mV.

The K-type thermocouple is made of nickel chromium (Ni Cr) for the positive wire and nickel aluminum (Ni Al) for the negative terminal. The temperature range for the K-type is –328 to 2282°F. The voltage that the K-type thermocouple produces is –6.458 to +54.886 mV.

Three other types of thermocouples that you may find when you are working on DDC systems are the T-type, R-type, and S-type. The positive wire for the T-type thermocouple is made of copper (Cu), and the negative wire is made of copper nickel (CuNi). Recall that copper–nickel wire is also called constantan. The temperature range for the T-type is –328 to +662°F. The amount of millivoltage the T-type produces is –6.528 to +20.872 mV.

The positive wire of the R-type thermocouple is made from 13% platinum and rhodium, and the negative wire is made of platinum. The temperature range for this thermocouple is 32 to 2642°F. The amount of millivoltage the R-type produces is –0.226 to 21.101 mV.

The positive wire for the S-type thermocouple is made of 10% platinum/rhodium and the negative wire is made of platinum. The temperature range for the S-type is 32 to 2642°F. The amount of millivoltage the S-type thermocouple produces is –0.236 to 18.693 mV.

20.9.4 Troubleshooting and Calibrating Thermocouples and Transmitters

Since thermocouples produce small amounts of millivoltage at any given temperature, a millivolts meter can be used to test the output signal from all types of thermocouples. You need to distinguish between an operational test and a calibration-type test. If you want to know if the thermocouple is producing a millivoltage when it is heated, you should put a heat source under the junction and measure the change of millivoltage at the terminals. Remember that some thermocouples produce negative voltage at some low temperatures. If any millivoltage is generated, the junction is operational. You will need to use expensive millivolt meters that are designed for calibration and follow calibration instructions if you want to calibrate the thermocouple. Typically, the thermocouple and its module are sent to someone who can calibrate them professionally if precision calibration is required.

If the procedure allows for field calibration, you can use a thermocouple tester, which you can purchase, that combines the voltmeter and power supply into one tester. The voltmeter is used to measure millivoltage from the thermocouple and the power supply is provided to produce a regulated millivoltage for testing the module or transmitter. This tester has the tables for millivolt signal strength and the cold-junction compensation built into it so the technician needs only to select the type of thermocouple and the temperature range, and the test circuit will provide an accurate reading from this. When the transmitter is being calibrated, the tester can send the voltage signal for the minimum and maximum ranges of each type of thermocouple.

20.9.5 Resistive Temperature Detector (RTD)

In 1821 Sir Humphrey David discovered that the resistance of some metals change when they are heated. This was approximately the same time that Thomas Seebeck discovered that voltage is produced when two different metals are joined to form a junction

and the junction is heated. Since the amount of change in resistance was small and not uniform, Sir Humphrey David's discovery was not evaluated completely until 1871, when Sir William Siemens decided to use platinum as the metal. He found that platinum has a positive coefficient of resistance when heated. This means that the resistance of platinum increases as its temperature increases. He also found that this change was rather linear. In 1932 C. H. Meyers developed platinum into the sensor that we now know as the *resistance temperature detector* (RTD). Early versions of this sensor used a platinum wire that was wrapped around a ceramic stem. Today platinum wire is wound around a glass stem or ceramic bobbin to make the RTD. After the wire is wound onto the element, the entire sensor is sealed in glass in the shape of a bulb so that it is durable and can withstand higher temperatures. The glass bulb can be used as an exposed sensor or it can be shaped so that it will easily fit into a metal sleeve to make a probe.

The latest RTDs use platinum or metal glass slurry to create a thin film that is mounted on a flat ceramic substrate. The *thin-film detector* (TFD) can be manufactured to be used as a small glass bead or it can be sealed in a metal element or in plastic, such as Teflon. The main feature of these RTDs is that they are miniature in size, which makes them useful in temperature-measurement applications where there is not a lot of room for a full-sized sensor. It should be noted that since the material for the RTD is wound on a stem, the entire shaft of the RTD is temperature sensitive, whereas only the tip of a thermocouple is sensitive.

20.9.6 Types of RTDs

RTDs are available in a variety of metals, with platinum, nickel, copper, and tungsten being the most common. Figure 20–12 shows a table that indicates the temperature range and the resistance coefficient *(a)* for these RTDs. The temperature coefficient is the amount of resistance change you can expect for each 1°C of temperature change. For example, the resistance of a platinum RTD is 100 Ω at 0°C. Since its resistance coefficient is 0.0039, it must be multiplied by 100 Ω to determine the amount of resistance change per 1°C. Remember you must convert the Celsius temperature to Fahrenheit to determine the amount of resistance for Fahrenheit.

$$0.0039 \times 100 \ \Omega = 0.39 \ \Omega/°C$$

This table shows the temperature range for platinum as −184°C to 815°C and for nickel as −73°C to 149°C, which is the linear range for these RTDs. The extended range for platinum is −270°C to 982°C, and for nickel the extended range is −101°C to 315°C. These RTDs do not maintain a linear curve when they are measuring temperatures in their extended ranges.

20.9.7 Two-Wire, Three-Wire, and Four-Wire RTDs

RTD circuits are available as two-wire, three-wire, and four-wire RTDs. The symbol for the RTD is a resistor with an arrow through it, indicating it is a variable resistance. The resistor is also identified by R_T to indicate this resistor is temperature sensitive. The

Type	Temperature Range °F	Temperature Range °C	Ohms °C
Platinum	−299.20°F to 1499.00°F	−184.00°C to 815.00°C	0.0039 Ω
Nickel	−99.40°F to 300.20°F	−73.00°C to 149.00°C	0.0067 Ω
Copper	−59.80°F to 300.20°F	−51.00°C to 149.00°C	0.0042 Ω
Tungsten	−99.40°F to 528.80°F	−73.00°C to 276.00°C	0.0045 Ω

Figure 20–12 A table of typical RTDs that shows the type of material, temperature range in degrees Fahrenheit and Celsius, and the number of ohms per degree Celsius.

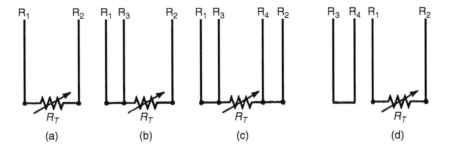

Figure 20–13 a A two-wire RTD. **b** A three-wire RTD. **c** A four-wire RTD. **d** Another diagram for a four-wire RTD.

additional RTDs in a circuit help to prevent temperature error due to offset. Figure 20–13 shows examples of two-wire, three-wire, and four-wire RTD circuits. Since RTDs vary the amount of their resistance when they are heated, they must be connected to a resistive-sensitive circuit such as a Wheatstone bridge to produce a usable signal as a sensor. In some applications the RTD must have additional wires connected to it to provide temperature compensation. Temperature compensation is necessary in some applications because of the difference in temperature between the point where the sensor is mounted and the location where the transmitter or controller is converting the value. Unwanted temperatures can come from the wire heating slightly due to current flowing through it or from temperature changes of the air that is near the sensor and its wires. The compensation occurs because the same amount of current will flow in the compensation lead as in the original lead, and the design of the circuit causes the voltage drop across each set of terminal wires to be the same, which effectively allows them to cancel each other.

20.9.8 Troubleshooting RTDs

The simplest way to troubleshoot the RTD is to use an ohmmeter. Since you know that the RTD must have some amount of original resistance at room temperature, you should be able to disconnect and isolate its leads so you can measure its resistance. If the RTD is platinum, the amount of resistance should be approximately 110. The next step is to heat the RTD and see if the resistance increases. Even though you may not know the actual temperature, this field test will indicate if the RTD can change resistance as its temperature changes. Remember, the resistance change may only be a few ohms. If the resistance does not change or if the resistance is infinity (∞), the RTD should be replaced.

The most common problem with RTDs is that the wire used for the sensor will break and cause an *open circuit*. If the original resistance reading of the RTD is infinity (∞), it indicates the RTD is open and it must be replaced. It is also possible for the RTD to become shorted, which would cause the resistance reading to be near zero and cause it not to change as the temperature changes. Once you have determined that the RTD is operational, its circuit may need to be calibrated. A calibration tester is available that is basically a precision Wheatstone bridge, which allows you to measure the amount of resistance the RTD provides for each temperature throughout its range. You must also be able to accurately control the temperature for each reading if the calibration is to be successful.

20.9.9 Thermistors

The thermistor is a temperature-sensitive resistor much like the RTD. The major difference between the thermistor and the RTD is that the thermistor generally has a negative temperature coefficient and exhibits a greater change of resistance for each degree of

Figure 20–14 Response curves for a thermistor, RTD, and thermocouple.

temperature change. A negative temperature coefficient means that the resistance of the thermistor decreases as its temperature increases. It should be noted that some thermistors are available that have a positive temperature coefficient, but the vast majority used as temperature sensors have a negative temperature coefficient. The main feature that makes the thermistor so useful is that it is far more sensitive than the thermocouple or the RTD. Of importance is the fact that the resistance of the thermistor is not linear.

20.9.10 Linearity of Thermistors

The response graph of the thermistor is different from the thermocouple and the RTD in that it is not linear. Figure 20–14 shows the typical response for thermocouples, RTDs, and thermistors. Since the thermistor response is a curve, it is important to locate linear portions of the curve to provide the best accuracy and repeatability. In some cases you should consult a table to determine the exact amount of resistance the thermistor should provide for any given temperature. You should also notice from the graph of the thermistor that the resistance decreases as the temperature increases, so at some point the temperature will increase sufficiently to make the resistance equal zero, at which point the thermistor is no longer useful as a temperature sensor.

20.9.11 Solid-State Integrated Circuit (IC) Temperature Sensors

When semiconductor material was first manufactured and was used for diodes and transistors, it was found that the material also changed the amount of voltage across the PN junction when forward biased as the temperature changed. The early attempts at using semiconductor material were not too successful because impurities in the material made its response nonlinear. Today, with modern manufacturing technology, the solid-state material used in making ICs provides a linear output response. The semiconductor material also provides a compact sensor when it is manufactured into an IC. The IC sensors are very small and shaped like typical small transistors, which allows them to be placed in printed circuit boards and other electronic circuits.

20.9.12 Thermostats

The thermostats for DDC systems can be discrete (on–off) like a traditional mercury bulb thermometer, or they can be analog. They also can be special-purpose, such as the remote bulb thermostat shown in Figure 20–15. Analog thermostats can have thermocouples or RTDs as their sensors. Figure 20–16 shows examples of several thermostats. Some advanced thermostats are digital thermostats, and they have a microprocessor in them so their temperature can be set locally with someone standing in front of them or be set remotely with the DDC system. Other thermostats have a remote bulb that allows

Figure 20–15 Remote bulb thermostats. *(Courtesy of Honeywell)*

Figure 20–16 Types of thermostats used in DDC systems. *(Courtesy of Honeywell)*

the control to be located inside and the remote bulb located where the temperature is being measured.

Typically, thermostats are used to control the temperature inside a room or office. In DDC systems the duct temperature of mixed air, water temperature, outdoor temperature, and other temperatures determines total enthalpy of the system. Some thermostats are used as a freezestat and send temperature data directly to the DDC controller or to a damper actuator or valve actuator.

Figure 20–17a shows an example of a temperature sensor placed in ductwork to determine average temperature of the air. Figure 20–17b shows a sensor used as an outside temperature sensor, which is protected from rain and direct sunlight. Figure 20–17c shows another type of temperature sensor called a surface-mounted pipe temperature sensor. It is strapped directly on a water or refrigerant line. Figure 20–17d shows a sensor called a flexible temperature sensor. It is designed to be used where a rigid type sensor cannot be mounted to properly measure temperature. Figure 20–17e shows a liquid immersion sensor, which is mounted in a water tank, water-cooled

(a)

(b)

(c)

(d)

(e)

Figure 20–17 a Averaging rigid temperature sensor. **b** Outside air temperature sensor. **c** Surface-mounted pipe sensor. **d** Averaging flexible temperature sensor. **e** Liquid immersion temperature sensor. *(Courtesy of Siemens Building Technologies, Inc.)*

condenser, or directly in a pipe to indicate liquid temperature. This sensor is designed to provide a liquid seal to ensure the liquid will not leak around this sensor.

20.10 HUMIDITY SENSORS

Humidity sensors are an integral part of a DDC system. The humidity sensor may be mounted directly into a duct to measure the humidity of the air that the HVAC system is circulating, or it can be mounted on the wall in a room or office just like a room thermostat. The humidity sensor can be discrete, providing an on–off signal, or it can be analog, providing a voltage or current value that represents the amount of humidity it is measuring. Figure 20–18 shows several examples of humidity sensors.

Figure 20–18 Humidity sensors. *(Courtesy of Honeywell)*

20.11 ## FLOW SWITCHES AND DIFFERENTIAL AND LOW-PRESSURE SENSORS

The DDC system uses flow switches and flow sensors to determine the amount of water or air flow in a system. They can also use differential pressure to measure the amount of air flow. The main difference between a DDC system and a residential HVAC system is that the DDC system is continually monitoring air flow and water flow to determine if changes in pump speeds or fan speeds are needed. It is important to understand that when a fan or pump is turned on, the controller assumes the air or water is flowing. A sensor is needed to measure the amount of flow to verify how much is flowing. If the amount is less than the desired setpoint amount, the controller can increase the analog signal to the pump or fan and increase the amount of flow. If the flow is too large, the controller can decrease the amount. Figure 20–19 shows a discharge air sensor.

Discrete flow switches can be as simple as a lightweight metal paddle that is placed in the flow stream. When the flow is large enough, it moves the paddle enough to cause the electrical switch to operate. Differential, or low-pressure, air switches use sensors that are located in the air flow and measure the pressure drop. These switches can be discrete or analog. Air flow can be calculated by the amount of pressure drop, so some of these analog sensors will have a conversion circuit in them. Figure 20–20 shows an example of a differential pressure switch that can be used to measure pressure drop.

Figure 20–19 Discharge air sensor. *(Courtesy of Honeywell)*

Figure 20–20 Differential pressure sensor. *(Courtesy of Honeywell)*

Another type of flow switch for liquid flow is called a paddle wheel–type flow switch. This type of analog flow switch has a small paddle located in the liquid flow. As liquid moves past the paddle wheel, it causes the wheel to rotate. Each web on the paddle wheel has a magnet on it. The top of the sensor has a circuit that counts each magnet as a pulse as the paddle wheel turns. The pulses create a frequency signal, which is converted to an analog voltage (0–10 V) when the signal is sent through a frequency-to-voltage (F/V) converter.

20.12 PRESSURE SWITCHES

Pressure switches are similar to flow switches in that they can be discrete or analog. Pressure switches can be used to measure steam pressure, high and low refrigerant pressure, or other pressure in the HVAC system. Figure 20–21 shows typical pressure switches. Air pressure or liquid pressure causes the switch to activate, so any leaks in the system will cause the sensors to be inaccurate.

20.13 LEVEL SWITCHES

Another type of sensor used in HVAC systems controlled by DDC systems is the level switch. The level switch is typically a discrete device, whereas the level sensor is typically an analog device. The sensors are used to determine water level in water-cooled condensers or in other parts of the system that store liquid.

Figure 20–21 a Static pressure regulator with one- or two-pipe connection, direct- or reverse-acting, low- or high-capacity controller used with pneumatic actuators to regulate static or differential pressure in central fan installations. **b** Differential pressure sensor with 4–20-mA output signal; two wires with noninteractive zero and span. *(Courtesy of Honeywell)*

(a) (b)

Figure 20–22 a Duct-mounted carbon dioxide (CO_2) sensor. **b** Wall-mounted (CO_2) sensor.
(Courtesy of Honeywell)

(a) (b)

20.14 CO_2 SWITCHES AND SENSORS

Another type of sensor that is used in the DDC system is a CO_2 sensor. These sensors are required by code in some buildings and structures to monitor the amount of CO_2 that may build up. Since CO_2 is colorless and odorless, it can build to dangerous levels and cause occupants in a room to become sick or die. These sensors can be discrete or analog, and they can be wall mounted or mounted directly in the air duct. Figure 20–22 shows examples of these sensors.

20.15 OUTPUT DEVICES USED IN DDC SYSTEMS

The DDC system can control a variety of discrete and analog output. The outputs are energized to control pneumatic systems, motor starters, dampers, and other types of actuators. This section explains the operation of each of these types of output devices.

20.15.1 Pneumatic Outputs

One common type of output device used with DDC is a pneumatic output. A pneumatic output device can be discrete or analog. The discrete-type pneumatic output turns other pneumatic devices on or off as the DDC tells it to. The pneumatic output allows the electronic signals of the DDC to be interfaced directly with existing pneumatic systems. Thus, the DDC can use its remote control features and networking features with an existing pneumatic system. The sensor that interfaces electrical current signal with the pneumatic signal is called an **I/P (I-to-P) converter**. The I in this case stands for current and P stands for pneumatics, or pressure.

If the pneumatic system is sending a signal back to the DDC, a **pneumatic sensor** called a **P/I (P-to-I) converter** would be required. Figure 20–23 shows pneumatic output devices used with DDC systems. If you suspect the pneumatic output or the motor starter is faulty, you can disconnect the signal from the DDC and put in an external signal from another source to test the device. Be sure that it is safe to turn on the output manually.

Figure 20–23 a PP97 pneumatic pressure controller, a one-pipe, pressure-operated device that provides proportional control of pneumatic valves to control steam, air, noncorrosive gas, or noncorrosive liquid pressure. **b** PP903 A differential Pressuretrol, a bleed-type controller providing proportional control of pneumatic actuators and valves. *(Courtesy of Honeywell)*

(a) (b)

20.15.2 A Motor Starter Used as an Output

The DDC system can control both discrete (on–off) and analog output devices. One of the simplest is the motor starter. The DDC sends a signal through its output module directly to the coil of the motor starter. The contacts of the motor starter can be connected directly to a fan motor, pump motor, or compressor. The auxiliary contacts on the motor starter can be used as a discrete input to tell the DDC that the contacts of the motor starter have closed after the signal is sent to the motor starter coil. This signal acts as feedback. Figure 20–24 shows a solid-state motor starter. The solid-state motor starter is similar to an electromagnetic motor starter in that it has a coil and contacts. The biggest difference is that the solid-state motor starter has the ability to provide soft

Figure 20–24 Solid-state reduced voltage starter. *(Courtesy of Eaton/Cutler-Hammer)*

starting so that motors and other large loads do not receive a large surge of current when the contacts are closed. A solid-state starter also allows maximum current levels to be set so that the load connected to it is not damaged by too much current.

20.15.3 Damper Motor Actuators

Another type of output device used with the DDC system is the **damper motor actuator** that controls the opening and closing of air dampers in the system. These controls can be discrete, where they move the dampers from full open to full closed, or they can be analog, where they can position the dampers to any position between full open and full closed. Figures 20–25 and 20–26 shows several examples of dampers and of actuators. One of the pictures shows an actuator attached to the damper. You can troubleshoot the damper motor actuator by sending a test signal to it and setting the valve to 25%, 50%, and 100% to see if the damper changes position. A commercial signal generator is available to send 0–10-V signals or 4–20-mA signals.

20.15.4 Variable-Frequency Drives (VFDs)

The variable-frequency drive (VFD) was introduced and explained in detail in Section 19.37. Figure 20–27 shows a typical VFD used with a DDC. The DDC can send an analog current (4–20 mA) or voltage (0–10 V) signal to the VFD to cause it to change speed. More complex systems can send and receive digital signals with the VFD to change speeds and monitor other conditions that the drive can detect, such as the motor torque or changes in the load that may help detect problems in the system.

(a) (b)

Figure 20–25 a Damper motor actuator connected to dampers. **b** Parallel blade damper. *(Courtesy of Honeywell)*

(a) (b) (c)

Figure 20–26 Modutrol IV motors. **a** Proportional motors operate valves when used with a modulating 2–10-V DC or 4–20-mA input signal, such as an EXCEL 5000 system. **b** Direct-coupled, nonspring return damper actuators used to control dampers in applications such as variable-air-volume (VAV) terminal units and for mounting on ball valves. **c** Spring-returned, foot-mounted, damper motor actuator. *(Courtesy of Honeywell)*

Figure 20–27 Variable-frequency drive (VFD) used to control fan and pump motor speeds.
(Courtesy of Honeywell)

20.15.5 Valve Actuators

A **valve actuator** is another type of output that can be controlled by the DDC. The valve actuator can be discrete and cause a valve to move to the full open or full closed position, or it can be analog and move the valve to any position between full open and full closed. If the valve actuator is the discrete type, it can have a limit switch built into it to send a feedback signal to the DDC to indicate when the valve is in the full open or full closed position. Figure 20–28 shows examples of these valves. You can troubleshoot the

Figure 20–28 Flowrite 599 valve. *(Courtesy of Siemens Building Technologies, Inc.)*

valve actuator by sending a test signal to it and moving the valve to 25%, 50%, and 100% to see if it changes position. A commercial signal generator is available to send a 0–10-V signal or 4–20-mA signal.

20.16 PROPORTIONAL CONTROL (TIME PROPORTIONING) FOR ELECTRIC HEATING

In an electrical heating system, the heating element is controlled by a contactor that has only two states (on and off). The contactor is used as the control device instead of an analog controller because of the expense. Since the output of the process controller is analog (0–100%) and the heating contactor is on–off, the controller must be set up to provide the output signal to the contactor as a ratio of time on versus time off. This type of control is also called *time proportioning*.

Figure 20–29 shows an example of this type of control. From this figure you can see that the coil of the heating contactor is connected to the controller, and the heating element is connected to the contacts of the heating contactor. Anytime the contactor coil is energized, full voltage is sent to the heating element and provides 500 W of heat to the furnace. When the contactor coil is de-energized, no voltage is sent to the heating element and no heat is produced.

The on–off type of control for the heating element can take advantage of the 0–100% analog output of the controller by selecting the ratio control feature of the controller. The ratio control feature uses ratio of on time to off time to get the effects of the analog controller. For example, if the controller determines that the output should be 50%, the ratio of on time to off time is 50%. A time base of 10 s is generally used, which means that for an output of 50%, the heating contactor is energized for 5 s and de-energized for 5 s. If the controller determines the output should be 70%, the contactor energized for 7 s and de-energized for 3 s.

The contactor will be continually turned on and off during each 10-s cycle, as determined by the output percentage. The heating effect from the heating element is similar to that of an element controlled by the more expensive analog amplifier.

Figure 20–29 a Electrical diagram of a 110-V AC heating contactor coil (HC1) controlled by heating controller and with its contacts connected to a 208-V AC heating element. The controller provides proportional control (time proportioning) to the on–off contactor. **b** The timing diagrams for a 50% duty cycle and a 70% duty cycle for the time proportioning control system.

20.17 CALIBRATION

It is very important to ensure that all the DDC input sensors and output devices are calibrated prior to system operation. In most cases you can test the calibration by using an instrument-calibration device that can send and receive a variety of calibrated voltage and current signals. You should know the full-scale signal of every device that you are calibrating, and test it for full operation. It is also important to ensure that the parts of the system in operation during calibration do not pose a safety risk to personnel or other parts of the system.

20.18 DDC SYSTEMS AND SOFTWARE

The controller for the DDC system must be programmed to function properly. The software to program the controller may be proprietary for a specific company system or it may be capable of programming generic systems. Several things are common to all DDC systems regardless of their brand names. For example, every system must be programmed to properly identify its inputs and outputs. The addressing may be built in, or it may depend on the number and types of inputs and outputs that are used. Figure 20–30 shows an example of a Honeywell system and the computer with its software that manages the system. Figure 20–31 shows a typical DDC system. The system that is pictured is the Honeywell Excell 5000 DDC.

The basic elements of a microprocessor controller include the microprocessor, working memory, program memory, clock or timing devices, and a means of getting data in and out of the system. Program tuning or interfacing with a central computer or building-management system is accomplished with a communications port. A battery-backed clock provides timing for microprocessor operation. The clock executes the program instructions in the microsecond range. The basic instruction set for controller operation and application programs are stored in program memory.

Memory size and type vary, depending on the application and whether the controller is considered a dedicated-purpose or general-purpose device. Dedicated-purpose configurable controllers normally have standard programs and are furnished with

Figure 20–30 Computer and software system for programming an Excell 5000 DDC system. *(Courtesy of Honeywell)*

Figure 20–31 The Excell 5000 DDC System. *(Courtesy of Honeywell)*

programmable read-only memory (PROM) or read-only memory (ROM). General-purpose controllers are supplied with field-alterable memories, such as electrically erasable, programmable, read-only memory (EEPROM) or flash memory, to accommodate a variety of individual custom programs. Nonvolatile-type memory is generally used to hold the program for a controller, because it retains the program data during power outages. Nonvolatile ROM or PROM, which is often referred to as firmware, is used to store operating software. The software in the system includes the operating system (OS) and routines for task scheduling. The software must also provide a means for

taking care of I/O, such as I/O scanning, priority interrupt processing, and A/D and D/A conversion. Other variables, such as setpoints, data file information, temperature values, and other system parameters, are also monitored. The program schedules tasks sequentially, and it interlaces routine tasks and other I/O scanning simultaneous. When an external event occurs, such as an alarm or a request to execute an energy-management program, the program requires that normal operations be suspended until the higher-priority task is serviced.

20.19 SYSTEM-LEVEL AND ZONE-LEVEL LOOPS

The DDC can be designed for system-level or zone-level control. Figure 20–32 shows an example of how these two types of controls are handled. Zone-level controllers typically control HVAC terminal units that supply cooling and heating energy to offices and other occupied spaces in the building. They can control heat pumps, fan coil units, VAV terminal units, unit ventilators, space-pressurization equipment, and any other zone-control or unit device.

System-level controllers have more capacity and are more flexible than zone-level controllers. System-level controllers are used in central chiller and boiler plants, equipment rooms, including VAV central supply systems, and built-up air handlers. Control sequences usually contain customized programs written to handle the specific applications. The application of the controller must allow both the number and mix of inputs and outputs to be variable. The number of inputs and outputs required for a system-level controller is usually not predictable. Several different packaging approaches have been used, including fixed I/O configurations, universal I/O configurations, card cages with plug-in function boards, and master/slave I/O modules.

20.20 USING OPERATORS IN PROGRAMMING THE DDC

When the DDC is programmed, it can use a variety of control operators. Figure 20–33 explains the following operators: sequence, reversing, ratio, analog-controlled digital output, digital-controlled analog output, analog-controlled analog output, maximum input, minimum input, delay, and ramp. The table in this figure explains how the DDC uses each of these operators for control.

Figure 20–32 Diagram showing system-level control on the top of the diagram and zone-level control on the bottom of the diagram. *(Courtesy of Honeywell)*

Operator	Description
Sequence	Allows several controller outputs to be sequenced, each one operating over a full output range.
Reversing	Allows the control output to be reversed to accommodate the action of a control value or damper actuator.
Ratio	Translates an analog output on one scale to a proportional analog output on a different scale.
Analog-controlled digital output	Allows a digital output to change when an analog input reaches an assigned value. Also has an assignable dead band feature.
Digital-controlled analog output	Functionally similar to a single switching relay. One state of the digital input selects one analog input as its analog output; the other state selects a second analog input as the analog output.
Analog-controlled analog output	Similar to the digital controlled analog output except that the value and direction of the analog input selects one of the two analog signals for output.
Maximum input	Selects the highest of several analog input values as the analog output.
Minimum input	Selects the lowest of several analog input values as the analog output.
Delay	Provides a programmable time delay between execution of sections of program code.
Ramp	Converts fast analog step value changes into a gradual change.

Figure 20–33 Table showing typical DDC operations. *(Courtesy of Honeywell)*

20.21 DATA FILE INFORMATION FOR A TYPICAL ANALOG INPUT AND OUTPUT

In most DDCs the analog input or output is programmed in a table format. This means a performed table is provided for the programmer to fill in to list information about the analog input or output. Figure 20–34 shows an example table for an analog input. In the table you can see the types of information that the programmer would enter. The left-hand column is filled by the software, and it indicates or prompts what the programmer should fill in. For example, the table prompts the point type, and it gives the programmer choices of "regular or calculation." This means that the input can come directly from a sensor, or it can be a sensor signal that involves a calculation, such as average temperature over the span of several minutes. The next prompt in the table is "Sensor." In this example the sensor is listed as a Platinum RTD (0–100°F), and the next row of the table indicates it will be connected to terminal number 16.

The next prompt is for the programmer to give this sensor a "Use code," which is basically a name that technicians will use to recognize this sensor. In this example the use code name is "Cold deck dry bulb," which tells the technician to look for this sensor in the cold deck area of the system if it needs maintenance. The next prompt is for "Engineering unit." Engineering units are the actual units the sensor is measuring. In this example the units are degrees Fahrenheit. If this sensor were measuring water level, it could show the level in inches, and if the sensor were measuring flow, it could show gallons per minute (gpm). You can see this is an important part of the table, since it adds units to the electrical analog signal that an operator or technician can understand.

The next prompt indicates the number of decimal places for the displayed value. This allows the programmer to set up the display so that it shows the units in a way that aids the operator or technician of this system. The next two prompts allow the programmer to

Point Address	User Address
Point type	Regular or calculation
Sensor	Platinum (0 to 100°F)
Physical terminal assigned	16
Use code	Cold deck dry bulb
Engineering unit	F
Decimal places for display	XXX.X
High limit	70.0
Low limit	40.0
Alarm lockout point	Point address
Point descriptor	Cold deck temperature
Alarm priority	Critical

Figure 20–34 Typical data file for analog input. *(Courtesy of Honeywell)*

indicate the high limit and low limit for this sensor. If the analog value exceeds either the high limit or low limit, an alarm will sound to indicate the system is operating outside the prescribed limits. In this example you can see the low limit is set for 40°F. If the temperature gets much lower than this, there is a chance the coils in the cold deck will freeze and cause damage, so an alarm will sound if this occurs.

The next three prompts indicate the alarm lockout point, the point descriptor, and the alarm priority. All these are used to help identify the alarm and what action you want the controller to take when an alarm condition occurs.

You can see from this table that it is fairly easy to program an input or output sensor in a DDC system. In the main program, the controller will use the analog input as part of a loop. The next section will show how the analog loop is programmed.

20.22 PROGRAMMING A LOOP IN THE DDC

Most DDC systems use tables similar to the one shown in Figure 20–35 to set up a **control loop** for each part of the system. From this diagram you can see the control loop is separated into several sections. The first section is the controller, which takes in a setpoint from an operator or programmer. It also takes in a feedback signal from a sensor and sends it to a control algorithm, which compares it at the summing junction. The error is the difference between the setpoint and the signal from the sensor, which is also called the process variable (PV). The output signal from this area goes through a D/A converter.

The corrective signal is sent from the controller as an analog signal to the final control element, which is generally some type of actuator. For example, this loop could be

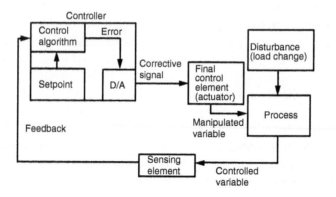

Figure 20–35 A typical loop table for a DDC program.

controlling the temperature of a hot deck by adjusting a hot-water valve. The sensor could be an RTD thermostat, which would send a feedback signal to the controller so the actual temperature in the hot deck can be compared to the setpoint. If the setpoint is 80° and the RTD sensor tells the system that the actual temperature is 78°, the controller would compare the two values and determine that the system needs to be warmer, and it would increase the analog signal that is sent to the hot-water valve actuator. This increase in signal will cause the hot-water valve to open farther, and the hot deck temperature will increase. This is called the process disturbance. The controller continually samples the system by reading the temperature from the RTD sensor. During each scan cycle, the controller compares the two temperatures (setpoint and actual) and determines if the valve's position needs to be changed.

This example helps you understand how the DDC controls each of its loops. The DDC can control a number of loops as well as discrete outputs to make the HVAC system respond in the most economic way.

20.23 TYPICAL CENTRAL FAN SYSTEM LOOPS

Figure 20–36 shows typical HVAC system loops that a DDC can control. From this figure you can see that the following can be controlled: discharge air temperature, mixed air temperature, hot deck temperature, cold deck temperature, humidity or dew point, indoor air quality (IAQ), ventilation, supply fan static pressure, and return fan air flow.

20.24 CONNECTING THE INPUTS AND OUTPUTS TO THE DDC

One of the last steps in the process is to connect the discrete and analog inputs and outputs. Figure 20–37 shows a wiring diagram of a typical DDC. From this diagram you can see the DDC is powered with 24-V AC supply at terminals G and G0. The discrete (on–off) inputs are called binary inputs, and they are connected to terminals E1–E2. Additional inputs can be connected to the universal inputs B1–B5. The discrete outputs are called binary outputs, and they are connected to terminals Q11–Q64. Analog outputs are connected to Y1–Y3. If the system has an analog input, it can be connected to the universal inputs as well. Analog outputs and universal inputs can all use terminal M for the ground side of their signal. The terminal identified as Tool is used to connect the Polycopy controller.

Typical Central Fan System Control Loops

- Discharge Air Temperature
- Mixed Air Temperature
- Hot Deck Temperature
- Cold Deck Temperature
- Humidity or Dew Point Control
- Indoor Air Quality (IAQ) Control
- Ventilation Control
- Supply Fan Static Pressure Control
- Return Fan Air Flow Control

Figure 20–36 Control variables for a simple loop that controls fan systems for the Honeywell controller. *(Courtesy of Honeywell)*

Figure 20–37 Wiring diagram of a DDC system. *(Courtesy of Siemens Building Technologies, Inc.)*

G-GO: AC 24 V Supply
M: Ground (GO) for signals, universal inputs, analog outputs
B: Universal Input
E: Binary input
Y: Analog output
Q: Binary output, various voltages permissible
Tool: Connection for POLYCOPY

20.25 NETWORKING DDC SYSTEMS

One of the main advantages of a DDC system is that it can be connected to a network and controlled remotely. In some cases the control is set up in a special room, where the HVAC technicians for a large building can monitor the entire building from a single control room or console. The **networking** feature also allows several buildings in a complex to be monitored from a single remote site. Figure 20–38 shows examples of a Siemens network system. You can see how the individual controllers are connected to the network to provide an overview of the complete system. Another feature of the network is that its data can be transmitted over dedicated telephone lines or connected through the Internet so that the data are available worldwide. You will need a map or diagram of the network to fully understand the way the DDC system is supposed to operate. Energy management, security, fire control, and other essential functions can be controlled over these networks.

20.26 A FINAL WORD ABOUT PID TUNING

The speed with which a system responds to changes in the system is controlled by changing the PID tuning variable. At this point all you need to know is that PID stands for proportional, integral, and derivative, which are also called gain, reset, and rate control. The PID variables are adjusted after the HVAC system is installed and are controlled by the DDC loops. You will need more information about PID loop tuning before you try to adjust any of these variables.

 The amount of information explaining PID control will be limited in this text to minimize the length of this chapter and the text. I would be happy to send you more information about PID control and how to adjust these variables if you would like to know more about this subject. Please contact me at Tkissell@terra.edu or Tkissell@earthlink.net.

Siemens Building Technologies: System 600 Apogee InSight

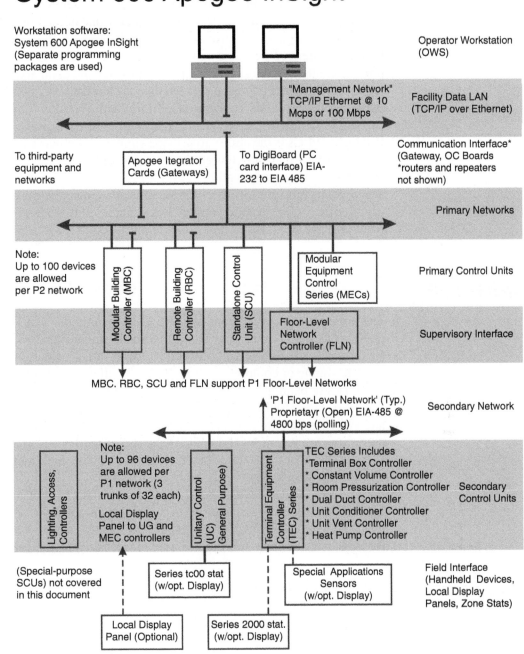

Workstation software: System 600 Apogee InSight (Separate programming packages are used)

Operator Workstation (OWS)

"Management Network" TCP/IP Ethernet @ 10 Mcps or 100 Mbps

Facility Data LAN (TCP/IP over Ethernet)

To third-party equipment and networks

Apogee Itegrator Cards (Gateways)

To DigiBoard (PC card interface) EIA-232 to EIA 485

Communication Interface* (Gateway, OC Boards *routers and repeaters not shown)

Primary Networks

Note: Up to 100 devices are allowed per P2 network

Modular Building Controller (MBC)

Remote Building Controller (RBC)

Standalone Control Unit (SCU)

Modular Equipment Control Series (MECs)

Primary Control Units

Floor-Level Network Controller (FLN)

Supervisory Interface

MBC. RBC, SCU and FLN support P1 Floor-Level Networks

'P1 Floor-Level Network' (Typ.) Proprietayr (Open) EIA-485 @ 4800 bps (polling)

Secondary Network

Lighting, Access, Controllers

Note: Up to 96 devices are allowed per P1 network (3 trunks of 32 each)

Local Display Panel to UG and MEC controllers

Unitary Control (UC) General Purpose)

Terminal Equipment Controller (TEC) Series

TEC Series Includes
*Terminal Box Controller
* Constant Volume Controller
* Room Pressurization Controller
* Dual Duct Controller
* Unit Conditioner Controller
* Unit Vent Controller
* Heat Pump Controller

Secondary Control Units

(Special-purpose SCUs) not covered in this document

Series tc00 stat (w/opt. Display)

Special Applications Sensors (w/opt. Display)

Field Interface (Handheld Devices, Local Display Panels, Zone Stats)

Local Display Panel (Optional)

Series 2000 stat. (w/opt. Display)

Figure 20–38 Block diagram of a Siemens network system. *(Courtesy of Siemens Building Technologies, Inc.)*

20.27 OVERVIEW OF PROGRAMMABLE CONTROLLERS

The programmable controller is also called the **programmable logic controller (PLC)**. As an HVAC technician, you will run into the PLC in a number of places such as controlling large areas in commercial (malls) or industrial HVAC controls. In this part of the chapter you will see how PLCs are used to replace or compliment direct digital controls

(DDCs). The main problem with DDCs is that they are dedicated to specific control of HVAC and do not allow you to add logic or controls for other equipment that may be added to your HVAC control system. The PLC is easy to program; it is an open-type programming system that lets you connect input switches to a variety of output controls such as solenoids, motor starters, and dampers. One of the biggest advantages of using a PLC for control is that it provides status indicators on PLC modules to help you troubleshoot systems. PLCs also make it very easy to control analog inputs and outputs that provide control from 0 to 100% for valve positioners, dampers, pump motor speeds, and fan motor speeds. PLCs also make it easy to integrate to pneumatic controls that have pressure over electrical controls called P/I where P stands for pressure and I stands for current. PLCs also are used for I/P, which have electrical over pressure control. Another feature for PLCs is the ability to provide simple PID process control for a single loop or multiple loops that may be required to control the HVAC system for larger complex buildings. In larger applications, several PLCs can be networked together and can send and receive information to color screens that make setup and control easy for an operator or control technician who would not need to understand PLC controls. These features make PLCs the fastest growing digital control system for HVAC controls.

20.28 THE GENERIC PROGRAMMABLE LOGIC CONTROLLER

The programmable logic controller (PLC) is a computer that is designed to solve logic (AND, OR, NOT) that specifically controls industrial devices such as motors and switches and to allow other control devices of varied voltages to be easily interfaced to provide simple or complex control. Basically all control functions can be broken down into series or parallel functions of switches. Figure 20–39 shows examples of AND, OR, and NOT logic. Switches that are connected in series are called AND logic in a PLC, and switches that are connected in parallel are called OR logic. If closed contacts are used, they are called NOT logic. At this point do not let the term *logic* confuse you. Logic is the word that describes the action switches and contacts will perform when they are wired in series or parallel.

The PLC is designed to have switches such as 110-volt push-button switches and 110-volt limit switches connected to it through an optically isolated interface called an input module. The input push-button switch may be used to start pump motors or compressors, and the limit switches are used to indicate that dampers are at the beginning or

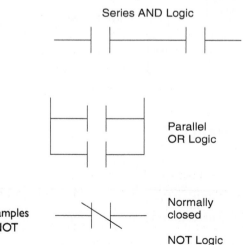

Figure 20–39 Examples of AND, OR, and NOT logic.

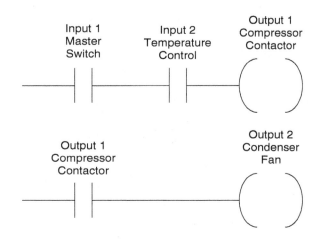

Figure 20–40 An example of a PLC program controlling a compressor and condenser fan. The compressor is controlled by output 1, which energizes the compressor contactor, and the condenser fan is controlled by output 2, which is energized when the contacts from output 1 are closed.

end of their travel or that water level is at a specific level. The PLC can also control 480-volt three-phase motors through 120-volt motor starter coils with a similar interface called an output module. The computer part of the PLC, which is called a central processing unit (CPU), allows a program to be entered into its memory that will represent the logic functions. The program in the PLC uses contact and coil symbols to indicate which switches should control which outputs. These symbols look very similar to a typical relay schematic (ladder) diagram that was shown in earlier chapters. You can view the program through a hand-held terminal or the program can be displayed on a laptop computer screen where the action of switches and outputs are animated as they turn motor starter coils and other outputs on or off. The animation includes highlighting the input and output symbols and changing their color in the program when they are energized.

The PLC was designed to provide the technology of logic control with the simplicity of an electrical diagram so that its inputs and outputs are easy to troubleshoot. Since the logic is programmable, it is easily changed. In a PLC, you will be able to make changes to the operation of an HVAC system in a matter of minutes. The logic can also be easily stored on a disk so that it can be loaded into a PLC anytime, and it can allow for parts of the logic to be used in multiple building controls since their control will be similar.

When PLCs were first introduced, they allowed the control technicians to use them to help troubleshoot large HVAC systems. As PLCs have evolved, they now provide many of the more complex logic functions such as timing, up/down counting, shift registers, and first-in, first-out (FIFO) functions that allow for optimal energy savings and control for a system, as well as allow for creating logic that is specific for providing troubleshooting and diagnostic control for the system.

Figure 20–40 shows an example of a PLC program. You should notice that the program looks exactly like an electrical relay diagram. As the program (diagram) gets larger, additional lines will be added. This gives the program the appearance of a wooden ladder that has multiple rungs. For this reason the program is called a ladder diagram or it may be referred to as ladder logic or a schematic.

20.29 BASIC PARTS OF A SIMPLE PROGRAMMABLE CONTROLLER

All PLCs regardless of brand name have four basic major parts: the power supply, the processor, input modules, and output modules. Figure 20–41 shows a block diagram of the typical PLC. From this diagram you can see that input devices such as push-button switches are wired directly to the input module, and the coil of the motor starter or solenoids are connected directly to the output module. In the PLC the switches and output

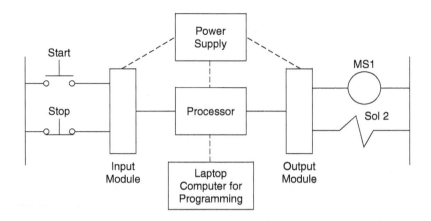

Figure 20–41 A block diagram of the typical PLC.

devices are connected to the modules and the program in the PLC will determine which switch will control which output. In this way, the physical electrical wiring needs only to be connected one time during the installation process, and the control circuitry can be changed unlimited times through simple changes to the ladder logic program.

20.30 THE PROGRAMMING DEVICE

At the bottom of the diagram in Figure 20–41 you can see a programming device. A programming device is necessary to program the PLC, but it is not considered one of the parts of a PLC because it can be disconnected after the program is loaded and the PLC will run by itself. The programming device is used so that humans can make changes to the program, troubleshoot the inputs and outputs by viewing the status of contacts and coils to see if they are energized or de-energized, and save programs to a disk or load programs from a disk. Today it is possible to connect the PLC to an Internet interface and have technicians check on HVAC controls for buildings anywhere in the world from a laptop or a PDA.

In smaller PLCs the programming device is usually a handheld programmer, and in larger systems a laptop computer with PLC programming software loaded on it. The ladder logic program can be displayed on the programming device where it can become animated. This feature is unique to the PLC and it helps the technician troubleshoot very large logic circuits that control complex equipment. When a switch that is connected to the PLC input module is turned on, the program display can show this on the screen by highlighting the switch symbol everywhere it shows up in the program. The display can also highlight each output when it becomes energized by the ladder logic program. When the on–off state of a switch on the display is compared with the electrical state of the switch in the "real world," technicians can quickly decide where the problem exists.

Another feature that makes the PLC so desirable is the fact that it has a wide variety of logic functions. This means that functions such as timing and counting can be executed by the PLC rather than using electromechanical or electronic timers and counters. The advantage to using timer and counter functions in the PLC is that they do not involve any electronic parts that could fail, and their preset times and counts are easily altered through changes in the program. Additional PLC instructions provide a complete set of mathematical functions as well as manipulation of variables in memory files. Modern PLCs such as Allen-Bradley PLC-5 have the computing power of many small mainframe computers as well as the ability to control several hundred inputs and outputs. This means that the control cannot only control the HVAC processes, but also store data on temperatures, pressures, and flows for days on end, which makes it easy to

see utility usage, as well as identify problems for troubleshooting. The exact control of analog outputs means that the system can control multiple zones to within one tenth of a degree and control energy usage to make the system very efficient.

20.31 UNDERSTANDING HOW THE PLC CONTROLS A SIMPLE CIRCUIT WITH A SELECTOR SWITCH AND MOTOR STARTER COIL

The simplest way to understand how the PLC operates is to see how it controls a simple circuit with a selector switch that energizes the motor starter coil. You would find a selector switch in a larger system used to control a chiller water pump through a motor starter. We will compare the PLC-controlled circuit with the hardwired circuit with a selector switch that energizes the motor starter coil. Figure 20–42a shows an example of a traditional circuit where the selector switch controls a motor starter coil. Figure 20–42b shows how this same circuit is controlled by a PLC. The selector switch is connected to the input module of the PLC and the motor starter coil is connected to the output module of the PLC instead of each other. You should notice that each input circuit

(a)

(b)

Figure 20–42 a Ladder diagram of a hardwired selector switch connected to a motor starter coil. **b** Hardware diagram of selector switch wired to a PLC input module and a motor starter coil connected to a PLC output module. **c** Scan cycle for a PLC.

(c)

on the PLC input and output module has an indicator lamp that is called the status indicator. The selector switch is connected to 110-V AC voltage, and when the switch is closed and passing 110 volts to its input circuit on the PLC module, its status light will illuminate. This feature allows you to troubleshoot each switch without using a voltmeter. You can simply look at the status indicator and open and close each switch to determine if they are working correctly. If the status indicator illuminates for a switch, you know that it is working correctly. Each input circuit also has an address number that the PLC recognizes and uses in its program. The selector switch is connected to address I1. In this diagram you can also see that the motor starter coil is connected to the output module at address O1. The output module also has a status indicator for each circuit to indicate when the circuit is energized.

20.32 THE SCAN CYCLE

The next part of the PLC to understand is the *scan cycle*. A diagram of the PLC scan cycle is shown in Figure 20–42c. The **scan cycle** consists of three distinct functions: the *read cycle*, which allows the PLC to check the status of every input; the *solve logic cycle*, where the PLC determines if each line of logic in its program should be energized or de-energized; and the *write cycle*, where the PLC writes the condition, on or off, to every output in its image table. If the condition of the switches in a line of logic in the PLC program indicates that the output should be energized, the PLC writes a "1" in the image register for that output address. If the PLC determines that the output should be de-energized, the PLC writes a "0" in the image register for that output address. This means that the PLC writes a "1" or a "0" in every output address in the output image register during the write cycle part of the PLC scan cycle. When a "1" is placed in the image register address for an output such as O1 where the motor starter coil is connected, the electronic circuit in the output module for that address is energized and 110 V AC is sent to the motor starter coil. The diagram of the scan cycle shows two additional segments of the scan cycle which are the *communications cycle* and the *overhead cycle*. During the communications part of the scan cycle, the PLC sends and receives information to its programming device such as the handheld programmer or laptop computer. A copy of the image register is sent to the handheld programmer or laptop during the communications part of the scan so that you can use the handheld to view the status of each line of the program and the condition of each input or output for troubleshooting purposes. The final part of the scan cycle is the overhead part of the cycle, which the PLC uses for internal purposes.

The PLC executes a scan cycle about every 10 to 20 milliseconds. This means that every 10/1000 of a second the PLC checks the status of every switch to see if it is on or off and then writes the condition of each line of logic (each circuit in the program) to the outputs in the image register to tell each circuit in the output module whether it should turn on or off. The speed of the scan cycle slows down as more lines of logic are added to the program.

If an input is in the "on state" and is providing voltage to the input circuit that it is connected to, two things occur. First, the status indicator for that input is illuminated, and second, the PLC places a "1" in the image register in the address for that input. When the switch changes state to its off state, the status indicator turns off and the PLC places a "0" in the address for the input to tell the PLC that input is now turned off. Since the PLC scan cycle is so quick, it will pick up any change in the condition of any input switch instantaneously. It will also send a signal to the output connected in the program to the switch within the time of the next scan cycle so it looks like the switch is controlling the output instantaneously.

20.33 ## THE OPERATION OF THE PLC IN RUN MODE

When the PLC is set to the run mode, it begins to execute the scan cycle and it checks all of its inputs, solves logic, and writes a "1" or "0" to all of its outputs. Figure 20–43a shows a diagram of the selector switch connected to the input and the motor starter coil connected to the output module. Figure 20–43b shows the program for the circuit with a selector switch that energizes the motor starter coil that is controlling a motor starter for a water pump. From this figure you can see the program. It has a set of normally open contacts that have the PLC address I1 and the name "selector switch." The output is identified as O1 and has the name MS1 Coil.

The PLC program cannot use standard electrical symbols for different types of switches such as limit switches, push buttons, or selector switches. Instead the PLC program uses only the symbols for normally open or normally closed contacts to represent all of the switches in the program. Even though we know the selector switch has a standard electrical symbol that is shown in the hardware diagram in Figure 20–43a,

(a)

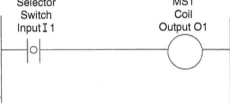

○ Signifies that contacts have not (changed state) "transitioned" and are still open in the program and they are not passing power to the next part of the circuit so the motor starter coil remains in the off state.

(b)

Figure 20–43 a Selector switch in normally open position connected to PLC input module and motor starter coil connected to PLC output module. **b** PLC diagram of selector switch controlling motor starter coil. **c** Image table in PLC for selector switch and motor starter coil.

Image Register

Select SW	I1	0
MS1	O1	0

Selector switch OFF and MS1 coil is OFF.

(c)

when it is used in the PLC program it is always shown as a set of contacts -] [-. The PLC only uses the symbol -] [- for normally open, or -]\[- for normally closed switches. The coil will use the () for the symbol for all of its outputs in its program.

20.34 THE IMAGE REGISTER

The second diagram that is needed to understand how the PLC works is the diagram of the *image register*, which is also called the *image table* in some PLCs. Figure 20–43c shows an example of an image register with just the address for each input and output shown. This example is only showing one input and one output. If the PLC has twenty inputs and ten outputs, the image table will show all of the addresses. The column on the right side of the image table provides a location to place a "1" or "0" to indicate the condition or status of the input or output. The PLC uses the image register to store or "keep track" of the status of each of the inputs and outputs in regards to whether they are on or off. The PLC assigns an "address," which contains the alphabetical letter and a number. For example, all inputs have the alphabetical letter "I" and all outputs have the alphabet letter "O" so the PLC can tell the input addresses from the output addresses. Some PLCs start the first number for its inputs and outputs from zero "0" and continue through the maximum number of inputs and outputs it provides. In our example, we will use input I1 and output O1. The selector switch is connected to the input address I1and the motor starter coil is connected to output O1. You can see in Figure 20–43c that these names have been added to the image table. The PLC will place a 1 or 0 in the address for these inputs at all times to indicate the current state of each component. The PLC does not know the names of the switches and outputs that are connected to the addresses, it only knows the addresses.

Figure 20–43c shows a copy of the image register for the selector switch and motor starter coil in the PLC. You can see the selector switch is assigned address I1, and the output has address O1 for the motor starter coil. You should remember that the selector switch is wired normally opened in the hardware section of the PLC where it is connected to the input module as input I1. The image register for this point in time is shown in Figure 20–43c where input I1 (the selector switch) is opened and the status light on the input module for this input is not illuminated. The PLC has placed a value of 0 in the image register address for address I1. When the PLC solves the logic for this line of the program, it shows the selector switch is off so the output is not energized and the PLC will place a 0 in the O1 address of the image register.

20.35 THE RUN MODE AND THE PROGRAM MODE

When the PLC is in the *run mode*, it is continually executing its scan cycle. This means that it monitors its inputs, solves its logic, and updates its outputs. When the PLC is in the *program mode*, it does not execute its scan cycle. Perhaps a better name for the program mode is the not run mode or the no scan mode. The name *program mode* was originally given to this mode because older PLCs could not have their programming changed while they were executing their scan cycle. Modern PLCs have the ability to have their program edited and changed while the processor is in the run mode. At first glance, changing the PLC program while it is in the run mode looks dangerous, especially when the machine the PLC is controlling is in automatic operation. But you will find in larger control applications, such as pouring glass continually or continuous steel rolling mills, it is not practical to stop the machine process to make minor program

changes such as the changes to a timer. In these types of applications the changes are made while the PLC is in the run mode. As a rule, you should switch the PLC to the program mode if possible when any program changes are being made.

20.36 ON-LINE AND OFF-LINE PROGRAMMING

The programming software for a PLC may allow you to write the ladder logic program on a personal computer, and later download the program from the personal computer to the PLC. If you are connected directly to a PLC and you are writing the ladder logic program in the PLC memory, it is called on-line programming. If you are writing the ladder logic program on a personal computer or programming panel that is not connected to a PLC, it is called off-line programming. Most modern PLC programming software allows the program to be written on a personal computer or laptop computer without being connected to the PLC. This allows new programs or program changes to be written at a location away from the PLC. For example, you may write a program at an office in Detroit, save the changes, and e-mail the file to a company in Chicago where the PLC is connected to the HVAC system.

If you are using a handheld programmer, it must be connected directly to a PLC so that you are writing the program in the PLC memory. Some small PLCs, such as the Allen-Bradley SLC1OO, do not have a microprocessor chip, so you cannot write the PLC program on-line to the processor with a laptop computer. Since the SLC1OO uses a sequencer chip instead of a microprocessor chip, all programming must be completed off-line in the laptop and then downloaded to the PLC memory. An alternative method would be to use a handheld programmer to program the PLC on line.

20.37 STATUS LIGHTS (INDICATORS) ON A PROGRAMMABLE CONTROLLER

One of the advantages of using a programmable controller is that it provides a status light (indicator) for each input and output. The lights are illuminated when an input or output is energized. Figure 20–44 shows a picture of an Allen-Bradley Micro

Figure 20–44 Example of the status indicators on input and output modules for the MicroLogix PLC. The input status lights are across the top, and the output status lights are shown across the bottom.

(Courtesy of Rockwell Automation)

Logix PLC with its status lights on its face. The input status lights are shown in the top left corner of the front of the controller, and the output status lights are shown in the bottom left corner. The status lights are numbered to coincide with the input and output addresses. As a technician, you can make a quick check of the status light for the input or output address when you are trying to determine if it is energized or not. This makes it easier to check the condition of the input or output without using a meter.

20.38 REVIEW

You may need to review several of the previous diagrams to help you get a better understanding of these functions. Back in Figure 20–41 you can see that the PLC has a processor, input modules, output modules, and a power supply. The processor has several sections, including resident memory that contains the instruction set, user memory reserved for programs, and addressable data memory that has room to store timer, counter, and other variables. The variables may be stored as binary data, as an integer (whole numbers), or as a floating point (numbers that have decimal points and can be expressed with exponents).

When the processor is in the run mode, it will continually scan the inputs to see if they are on or off and then update the input image register. Next it will sequentially solve the ladder logic that is stored in its user program area. It completes this part of its cycle by reading the ladder diagram. The processor goes through a fetch cycle to get each address found in the ladder diagram and puts it into the logic solving area. Each address is compared to the value found in its image register address and this value is used to solve the logic. During the next part of the cycle, the processor writes the results of the logic to the output part of the image register. If the output in the ladder logic has contacts assigned to it, like a control relay, the contacts will reflect the status of the output image register address when the processor solves that line of logic. If the output address is connected to the output module, the signal is sent to the module to ensure that the device connected to it is energized.

The final section of each processor scan cycle is called overhead. During the overhead part of the scan, the processor does all of its work outside of solving logic, such as updating the computer screen (monitor) if one is connected, sending messages to its printer, and interacting with the network. Since the main purpose of the PLC is to read inputs, solve logic, and write to its outputs, less than 10% of the scan cycle is devoted to the overhead. The amount of time for the typical scan cycle for a generic PLC is approximately 20 to 60 milliseconds. Newer enhancements to the PLCs try to lower the scan cycle to less than 10 milliseconds.

Job Assignment for This Chapter

You are called to a job where the HVAC system is controlled by a direct digital controller (DDC). The temperature for the heating cycle is controlled by an analog water valve that opens between 0 and 100% to provide hot water to the heating coil. The temperature of the water is 180 degrees, but the room is not staying warm enough. Your job is to check the heating loop to determine why the room is not staying warm enough.

Solution to the Job Assignment

The most likely cause of this problem is that there is not enough hot water going to the fan coil for the room. The first test you would make is to determine what percentage the water valve should be set to. You can determine this by checking the output address for the valve and see what voltage is being sent to the valve. Since the temperature in the room is too cold, the controller should be sending a signal to the valve to open somewhere around 80 to 100%. If the controller is sending a signal larger than 80%, you should check the voltage signal going to the valve; it should be over 8 volts if the valve is using 0–10 volts. If the valve is controlled by 4–20 mA, you should have over 16 mA going to the valve. If the voltage or current signal is sufficient and the valve is not opening, you should suspect that the valve is stuck or not working correctly.

If the signal to the valve is not large enough to cause the valve to open enough to heat the room, you should check to see what temperature value the thermostat is sending the controller. If the temperature signal is not correct, the thermostat may need to be calibrated or changed if it is faulty. You need the correct temperature value sent to the controller and the correct signal to the valve for the temperature to be correct in the room.

Assessment

QUESTIONS

1. Identify the basic parts of a DDC system.
2. Explain the terms *setpoint, error, process variable*, and *output*.
3. Identify the basic parts of a closed-loop system and explain their operations.
4. Explain the operation of a temperature input to a DDC system.
5. Explain the difference between discrete, analog, and digital input.
6. Describe the four basic parts of any programmable controller.
7. Explain the four things that occur when the PLC processor scans its program.
8. The input module has a status indicator for each input circuit. Explain how the status indicator is used in troubleshooting.

TRUE OR FALSE

1. The discrete inputs and outputs have two states, on and off.
2. The octal numbering system uses only digits 0–7.
3. The binary-coded decimal (BCD) system uses binary digits to display decimal numbers.
4. An analog input has only two states, on and off.
5. An isolated output module uses a relay to ensure total isolation.

MULTIPLE CHOICE

1. The range for the most typical analog voltage signals is
 a. 4–20 mA.
 b. 0–10 V.
 c. 2–15 V.
 d. 0–110 V.

2. The range for the most typical analog current signals is
 a. 4–20 mA.
 b. 0–10 V.
 c. 2–15 V.
 d. 0–110 V.

3. A 10-bit A/D converter provides _____ resolution.
 a. one part in 256
 b. one part in 1024
 c. one part in 4096
 d. It is impossible to tell without more information.

4. The main difference between a thermocouple, RTD, and thermistor is that
 a. a thermocouple produces a change in resistance, the RTD produces millivoltage, and the thermistor produces a millivoltage.
 b. a thermistor produces a change in resistance, the RTD produces millivoltage, and the thermocouple produces a millivoltage.
 c. an RTD produces a change in resistance, the thermistor produces millivoltage, and the thermocouple produces a millivoltage.
 d. an RTD produces a change in resistance, the thermistor produces a change in resistance, and the thermocouple produces a millivoltage.

5. A DDC and pneumatic system are
 a. totally incompatible, and parts of the two systems should never be used together.
 b. compatible, and the pneumatic sensors and outputs can be connected directly into the DDC controller.
 c. compatible, and the pneumatic sensors must be connected to the DDC through P/I converters and outputs must be connected through I/P converters.
 d. compatible, and the pneumatic sensors must be connected to the DDC through I/P converters and outputs must be connected through P/I converters.

6. When the PLC processor is in the program mode, it
 a. executes its scan cycle.
 b. does not execute its scan cycle.
 c. is impossible to tell whether the processor is executing its scan cycle because you are not on line.

7. You can use the _____ on the face of a PLC to determine whether an input switch is on or off.
 a. power light
 b. input status indicators
 c. output status indicators

PROBLEMS

1. Count to 15 in the binary, decimal, BCD, octal, and hexadecimal numbering systems.
2. Determine the resolution for a 10-bit system. Use this resolution to determine the analog voltage value a D/A converter would send if the analog voltage range is 0–10 V and the value sent to the D/A converter is 500.
3. Use the diagram in Figure 20–38 and explain how you would network two DDC systems for adjoining doctors' offices that each want an independent DDC system. Draw a sketch of what this system would look like.
4. Use the diagram in Figure 20–37 and sketch a diagram showing how you would connect an analog input to the controller.
5. Use the diagram in Figure 20–32 to explain the difference between system-level and zone-level systems. List two loops you could have in each system.

Websites

Websites change constantly. Use the key word search to locate sites that have similar content. The products that will be found at each site are indicated with the key word.

Products: Direct digital controllers (DDCs)
Key word: Johnson Controls

Products: Environmental controls
Key word: Siemens Automation

Products: Programmable logic controllers (PLCs)
Key word: Siemens Automation

Products: HVAC controllers and control products
Key word: Trane

Products: HVAC automatic controls
Key word: Honeywell

Products: Programmable logic controllers (PLC), Control Logix
Key word: Allen Bradley (Rockwell Automation)

Chapter 21

Getting and Keeping an HVAC and Refrigeration Job

Job Assignment for This Chapter

Your job for this chapter is to create a personal résumé and cover letter and prepare a portfolio of your work that includes some of the diagrams you have worked with and other examples of your work for a future employer. You should also prepare your personal answers for the questions you would get at a typical interview for a job.

Terms You Will Learn in This Chapter

Cover letter

Goals

Job interview

Portfolio

Promotion

Résumé

After reading this chapter, you will be able to:

1. Create a résumé.

2. Create a cover letter.

3. Prepare for a job interview.

4. Identify typical jobs and careers for the HVAC and refrigeration field.

5. Identify five things that you can do on the job to ensure that you will keep your job.

21.0 OVERVIEW OF THE HVAC AND REFRIGERATION INDUSTRY AND EMPLOYMENT OUTLOOK FOR THE FUTURE

U.S. Department of Labor, Bureau of Labor Statistics keeps detailed records of the number of jobs and the number people that have jobs in any field, such as all the fields in HVAC and refrigeration. You can find these statistics by searching on the Internet for the U.S. Department of Labor, Bureau of Labor Statistics. You can also check at this URL: http://stats.bls.gov/oco/ocos192.htm#outlook.

You can then select HVAC job outlook and see the latest information. At the time this book was published, there were over 270,000 persons in jobs related to HVAC. The current statistics show that about 135,000 persons were employed for plumbing, heating, and air-conditioning contractors and about 40,000 were self-employed. The U.S. Department of Labor predicts that the number of jobs in this field for the next 10 years will grow by approximately 18 to 20% per year, which is well above the average for other jobs.

21.1 OCCUPATIONS AND JOBS IN THE HVAC AND REFRIGERATION INDUSTRY

The following is a list of all of the jobs that you could work in using the skills you learned in the classes you are currently taking. Some of the jobs require additional training, certificates, or a degree from an accredited trade school, technical college, community college, or university. Some of the jobs require minimal skills and experience, whereas others require more training or experience. The jobs in this list are shown in alphabetical order and do not indicate an increase of knowledge or skills. You can also use these job names to search for jobs in the newspaper and on the Internet.

- Appliance repair
- Commercial HVAC service technician
- Commercial refrigeration technician
- Construction
- Controls specialist
- Design and construction manager
- Designer/drafter for HVAC
- Ductwork installer

- Electronic controls and digital controls specialist
- Estimator
- Facilities technician for universities with dormitories and labs
- Field service technician
- Foreman
- Furnace/boiler technician
- Helper
- HVAC electrical technician
- HVAC electronics technician
- HVAC installer
- HVAC mechanical technician
- HVAC or refrigeration lab technician
- HVAC senior service technician
- Inside sales
- Installer
- Maintenance technician for manufacturing companies that have HVAC on their process equipment and their buildings
- Manager
- Mechanical engineer (HVAC major)
- Outside sales helper
- Pipe fitter
- Plumbing installer
- Product development engineer
- Project manager
- Refrigeration installer
- Refrigeration service technician
- Refrigeration tech for food processing plants
- Residential HVAC service technician
- Sales (inside over the counter and outside on a route)
- Service manager
- Sheet metal technician
- Small business owner
- Troubleshooter technician

21.2 THE STEPS INVOLVED IN GETTING A JOB

It is important to understand that getting a job is a process rather than an event. This means that you will need to understand the steps involved in this process so that you will understand what employers expect. You should also understand that the process of getting a job takes about 3 to 6 months in some cases, so you will need to start several months before you expect to be on the job. Many people think that they must wait until the end of the school experience to start looking for a job. Instead, you should actually begin the process the day you start taking classes.

When you are looking for a job, you should think like the employer and try to figure out what the employer is trying to do, rather than look at the job as the employee. This will help you understand what skills the employer is looking for. You should also spend time checking out the company you want to work for so you will be familiar with the work or projects the company does and what types of workers it is looking to hire. It is important to understand the kind of skills a company is looking for as you take your classes so you can be more aware of learning that information. You should also understand that the

person who does the hiring is being evaluated on how well they find good employees. If they hire someone who does not work out and has to be fired or who is a problem on the job, their job will be in jeopardy as well, so they will be interested in finding the best employee and making sure the new employee works out on the job.

The steps for getting a job include the following:

- Understand the skills you need to learn to perform the job well.
- Take classes and get good grades in the courses that provide the knowledge to get the skills you need for the job. Be sure to get the best grades you can in your courses. Save course material such as diagrams and problems that you have worked on and solved while you are in classes. This may include pictures of the equipment you have worked on in class.
- Create a list of the skills you have learned. Indicate which skills you are good at and which ones you have mastered, such as learning to use a voltmeter, ohmmeter, and ammeter; learning to read a schematic (ladder) diagram and wiring diagram; and learning to troubleshoot. Keep this list and include it in your résumé.
- Create a résumé to send to potential employers. The résumé is a one- or two-page document that tells the employer everything about you. Typically a résumé will include your name, address and methods of contacting you, the job that you are applying for or your goals, the education that you have, the skills and experience that you have, and any prior jobs that you have had. A sample résumé will be provided in this chapter, and you will see where you can receive help to write your résumé.
- Create a cover letter to accompany your résumé. The cover letter is an introduction of you and your résumé and explains why you are looking for employment, such as if this your first job, or if you are changing jobs to take on more responsibility and earn more money.

21.3 GETTING AN EDUCATION TO LEARN ABOUT THE HVAC AND REFRIGERATION FIELD

The amount of education and training that you will need to get a job depends on the level of job you want and the amount of money you expect to make. For example, you will only need a few courses if you want to be an installer, but the installer's wages are among the lowest in the HVAC field. If you expect higher pay, you will need to be able to troubleshoot and repair equipment. Typically jobs for employers that work only on residential systems will be slightly lower than companies that work on both residential and commercial systems or commercial systems only. You will need to learn about electricity, electronics, and control systems if you expect to be a troubleshooter or installer, and you will also need to understand all the concepts involved with the refrigeration cycle and how different types of heating systems work. Typically the more complex the technology you are learning, the better the pay for the job will be. Remember, your employer can only pay you the amount of money that you are able to earn. This means that you must be able to fix expensive and complex systems so that your employer can charge more to companies that have this equipment. You will also need to work quickly when you are doing service calls so that you can complete as many as possible each day, since your employer makes money by sending bills to people for repairs and installations. If you are repairing equipment, you will find that customer relations and ethical and honest behavior is very important so that your customer will call you back for repeat business. If you do not treat your customers well, they will not call you back and they definitely will

not refer you to others who would become future customers. Remember, without customers, your employer will not have a business and you will not have a job.

You may start your education in a training program and be able to get your first job. You can then continue in an on-the-job training program where you learn more skills at work and take additional courses. You can also start by taking a certificate program and move on to a two-year associate's degree. If you intend to apply for the higher level jobs, you may need a bachelor's degree or a specialized degree such an a mechanical engineering degree for the higher level jobs.

21.4 CREATING A LIST OF THE SKILLS YOU HAVE LEARNED

Many students usually overlook creating a list of the skills they have learned and putting it on their résumé when they are looking for their first job or trying to change jobs. Students looking for their first job are usually put off by job postings or ads that state that the employer is looking for 2 to 5 years of experience when they are trying to hire someone. Since most students coming from schools do not have on-the-job experience, they assume they do not meet the criteria of having experience.

If you list all the skills you have learned and include it on your résumé, you will find that most employers will be willing to consider your skills equal to the years of experience they are looking for. What most students forget is that someone on the job may have 3 years of experience, but those years of experience consist of a few limited skills that are repeated every week. As a student, you can show an employer that you received a wide variety of skills training and that you are proficient in completing these skills. The important point to this is that you must find a way to describe your skills and the level of proficiency you have and put it in your résumé. You also need to take pictures of the equipment that you learned these skills on and show these to the potential employer and discuss how you feel you can do the job. Most employers are more than willing to hire a new person if they have skills and can demonstrate these skills in a job interview.

21.5 CREATING A RÉSUMÉ

One of the most important steps in getting a job is creating a **résumé**. It is important that you understand that a résumé is a piece of paper, usually one page in length, with information about you that you will send to many potential employers. The goal of sending the résumé is getting an invitation for a job interview. The résumé is what each potential employer sees as their "first impression" of you. The employer is looking at your résumé in two ways: first, are you capable of doing the job, and second, will you bring any problems to the job that may create problems with current employees.

Usually an employer will post a job and then start to receive résumés from anyone interested in applying for that job. The employer must look through all the résumés and decide on two or three people to meet in person for a job interview. If you send out résumés to a number of employers and you are not getting any job interviews, your résumé may be weak or may include items that tip off the employer that you would not be a good employee or are not qualified to do the job.

A sample résumé is shown in Figure 21–1. You can see that the résumé shows the person's name, address, phone number, and e-mail. This information is important so that the employer can contact you to discuss the résumé or invite you to a job interview. The next important part of the résumé is the goal that shows what job you are interested in obtaining. Many employers are posting more than one job, such as an installer and a

Jerry Smith
201 East State Street
Anytown, Ohio 45678
555-255-2001
Jsmith@Dnet.net

Career Objective

A position in an HVAC company where I can use my technical skills to install and troubleshoot residential and commercial HVAC equipment.

Education

2007 Associate's Degree in Applied Science, Major in Heating, Ventilating, and Air Conditioning. Terra Community College, Fremont, Ohio
2006 Certificate in HVAC Installation. Terra Community College, Fremont, Ohio

Skills

I have learned the following skills and knowledge during my training:

* Using a volt, ohm, and milliamp meter to troubleshoot.
* Reading and creating electrical schematic (ladder) diagrams and wiring diagrams to install and troubleshoot HVAC equipment.
* Complete understanding of the operation of all electrical controls for HVAC equipment.
* Installation skills for furnaces, air conditioners, and heat pumps.
* Complete knowledge of troubleshooting single-phase and three-phase motors and compressors.
* Excellent customer service skills.

Prior Work Experience

1998–2002 Grill Cook, Joe's Restaurant (perfect attendance and excellent customer service skills)
2003–2006 Electrical Department, City Hardware (Duties included sales knowledge of all electrical components. Helped customers select proper electrical components.)
2006–2007 HVAC Installation Technician, Northwest Heating and Cooling. (Duties included installation of furnaces and air conditioners.)
References are available upon request.

Figure 21–1 A sample résumé for a typical HVAC student.

service technician, so you must indicate on your résumé what job you are applying for, and this information should be directly under your name.

The résumé should contain three additional sections that provide information about you, and they should be put in the order that places your strongest points nearest the top. This additional information should include any education you have received, the skills you have learned, and a list of jobs you have had in the past. If you have just completed a certificate or degree in HVAC and the employer is identifying this as one of the requirements for the job, you should place your education history first. If the employer lists a set of skills the employee should have in the advertisement, then you should place a detailed list of your job skills nearest the top of the résumé and be sure that you emphasize the ones the employer is listing as important for the job.

If the employer is looking for someone with a high level of skills and years of experience, and you have had several jobs and have significant job experience, then you should list your job experience first. Be sure to include details of the things that you did on each job and the skills you have learned. If you do not receive any job interviews when you have sent your résumé to many employers, then you should ask several people to look over your résumé. Sometimes you can get a potential employer's attention by rearranging the

information on the résumé, so that something more important is nearer to the top. You must understand that you have about 5 to 10 seconds (about two sentences) to capture the attention of the reader of your résumé and get them to read your entire résumé. So it is very important to get the most important information about you nearest the top of your résumé.

Another part of the résumé that you should include is your grade-point average, and your attendance record while you were at school or your previous job. It is important that you have good grades and a good attendance record so that you can stress that during this time. If you do not include information about your grades or your attendance record, your employer will ask you about this information if you get to a job interview. This means it is important for you to get good grades while you're at school and have an excellent attendance record whether you're at school or on the job.

A final piece of your résumé is any references that you may have. The reference is a person that knows you and is willing to discuss your strengths and weaknesses with the employer. Many times the employer just wants to talk to someone that knows you and will confirm that what is on your résumé is actually true. Your instructors or previous employers are very valuable as references.

Some people do not see any value in including their first work experiences, which might have been with a restaurant chain or in the service industry, because it does not seem to fit with the HVAC job they are applying for. In fact, this may be a positive part of your résumé because it can show that you are not afraid of hard work and that you have established a history of work attendance. If you have such a work experience, you may be able to get your previous employer to be a reference for you. Remember, the employer reading your résumé usually does not know anything about you or the other persons applying for the job and is trying to narrow the choice down to someone who will be successful on the job. Most employers have been disappointed or "burned" by a previous employee that did not work out and had to be fired, so they are trying to determine that you will not be a problem and will be successful if they hire you. The employer's goal is to hire someone who will add to the company team and help the company grow and make more money. The last thing that an employer wants is to hire someone who is a constant problem and brings down other employees. Your job is to create a résumé that ensures the employer that you will be an excellent employee.

21.5.1 Creating a Time Line of Your Prior Experiences

Another very important part of your résumé is creating the time line of your prior jobs or experiences. This list should be shown in chronological order from the present time, working backwards to the date you finished high school or some other point in time. The main problem that comes up in a résumé is when there are gaps of time left in this list where you cannot account for what you were doing. If you were unemployed, or not working for a period of time, you need to write a simple sentence indicating that you were not working, otherwise the employer will assume something more problematic, such as that you were incarcerated or otherwise detained so that you could not work. Remember, you are trying to convince the employer that you are a safe bet and that you will not bring any problems to the job. If you had an excellent attendance record in your previous employment be sure to stress this on your résumé.

21.5.2 Being Totally Honest on Your Résumé

Another important point when writing your résumé is to be sure that you are honest. Some students think that they should embellish their experiences or what they are able to do and may exaggerate or stretch the truth on their résumé. You should understand

that today an employer can easily verify anything you put on a résumé. An employer also can ask several probing questions to quickly determine your depth of experience, and if you have indicated on your résumé that you can do something or have experience that you cannot back up, you will quickly be eliminated for any consideration of the job. A prospective employer will respect you for being honest. Also remember, if you do get hired because you put something on your résumé that is not true, it is only a matter of time for this problem to be discovered; at that point your employer will lose all respect for you and will dismiss you immediately. At this point your employer will share this experience with any other employers in the HVAC field, and it will be very difficult for you to get another opportunity to work in the field.

21.6 CREATING A COVER LETTER

The **cover letter** is a brief letter that accompanies your résumé. The cover letter is a little more informal than the résumé, and it consists of two or three paragraphs that explain your **goals**, why you are interested in the job, and an explanation or highlights of what is in your résumé. Figure 21–2 shows an example of a cover letter. The first part of the

Jerry Smith
201 East State Street
Anytown, Ohio 45678
555-255-2001
Jsmith@Dnet.net

May 23, 2007

Mr. James White, Owner
Commercial Heating and Cooling
2678 Main Street
Anytown, Ohio 45678

Dear Mr. White,

I am responding to your advertisement in the local newspaper last week for an HVAC technician to install and troubleshoot furnaces and air conditioners. I am enclosing my résumé for this position with your company. I have just completed an Associate's Degree in Applied Science with a major in HVAC. During that time, I learned many skills that have prepared me for this position. My résumé outlines these skills and experiences. I have held several previous jobs and am currently employed in a position where I have gained additional technical skills and customer service skills. I also have a proven track record of working with others as a team member, and I have had an excellent attendance record on these jobs.

I think you will see from the résumé that I can use a volt, ohm, and milliamp meter to troubleshoot a variety of electrical controls and motors. I also have excellent skills in reading and creating electrical schematic (ladder) diagrams and writing diagrams to install and troubleshoot HVAC equipment. I also have a complete understanding of the operation, installation, and troubleshooting of all electrical controls for HVAC equipment. I have two years of experience in installing furnaces, air conditioners, and heat pumps. I have learned the skills used to troubleshoot single-phase and three-phase motors and compressors. I am always eager to continue learning the skills and knowlege to improve myself.

Thank you for your consideration. Please do not hesitate to contact me at the address or phone number provided above if you have any questions.

Sincerely,

Jerry Smith

Figure 21–2 A sample cover letter for a typical HVAC student.

cover letter is used to introduce yourself. The second part of the cover letter explains what job you are applying for, and the third part of the cover letter explains why you feel you are the best candidate for the job. The final part of your cover letter thanks the reader for considering your résumé and offers to answer any questions the reader may have about the résumé. Be sure you provide how to contact you through your mailing address, your e-mail address, and your phone number. Also be sure to sign your cover letter. Remember the function of your cover letter is to introduce yourself and your résumé, and the function of your résumé is to get you invited to a job interview. If you have an excellent attendance record during school or a previous job be sure to include the sentence about this.

21.7 THE JOB INTERVIEW

The next step in the process is an invitation to a **job interview**. If the person reading your cover letter and résumé is convinced that you can do the job and you would become a great employee, you will be extended an invitation to a job interview. The invitation will explain where you are to go for the interview and what time and date you are to meet. You should reply to the invitation as soon as possible. If you cannot attend the job interview for some reason at the time and date indicated, be sure to reply to the employer immediately and indicate when you would be available. Since the employer is extending an invitation, you should make all possible arrangements to make the meeting time and date that is offered. The most important point here is to be on time. You may need to check out where you are going and make a test run the day before so you know exactly what streets or highways you will be using to get there. You should plan to arrive several minutes before your appointment and check your clothing one last time so that you make the best appearance.

After you have accepted the invitation to the job interview and have established a time to meet, you should begin the next step in the process, which is to research as much information about this company as you can. It is important to remember that the employer may have as many as three or four other candidates for the open position, so it will help your position to have as much knowledge as possible about the company. This will come in handy when you are in the interview and you can discuss information about the kinds of jobs their company has and any other details. An employer will consider any knowledge you have about the company a complement from you and make a positive impression of you for taking the initiative to learn about the company without being prompted. If you go into the job interview without any knowledge of the company, you will quickly be embarrassed about your lack of knowledge about your employer.

The other important part of your preparation for the job interview is to gather **portfolio** material that can demonstrate your skills, such as diagrams of equipment you worked on in school. It is also a good practice to take some pictures of the equipment you have worked on and a list of your skills. Another good practice is to take an extra copy of your résumé and cover letter with you to the job interview in case you are asked questions about either. You should also review your cover letter and résumé immediately before your interview to ensure you know everything that is on them.

21.7.1 Dressing for the Interview

Another important point in preparing for the interview is deciding what to wear and how to dress. Remember, you're trying to make a great impression on the employer, and you only have one attempt. This means that the employer is going to assess what you look

like and determine whether you could be sent to a customer's house to represent the company. This issue creates some disagreement between the interviewees and the employer at times, because the interviewees may feel they have the right to wear their hair or facial hair anyway they want, and they can wear body jewelry or piercings because they have done this in their private lives. Remember, you may be asked to wear a uniform or dress in accordance with what the employer wants while you're on the job, because you're representing them. If you have issues about the way you dress, or about jewelry you expect to wear on the job, you may want to bring this up early and allow the interviewer to make a decision immediately.

In recent years, people have elected to wear body piercings and other jewelry, or make a statement with their hair cut, which is totally permissible when you are not at work. Theoretically your employer should not discriminate on the basis of the way you look and present yourself. But in reality their business depends on their customers finding the way you look acceptable. You will find that if the issue of what you wear and how you look is important to you, and you do not want to conform with the standards your employer has set, you may need to pursue a job that does not involve working with other employees or directly with customers. An employer will have reasonable standards and will expect you to follow them, or you should not expect to be hired.

21.7.2 What Will Occur at the Job Interview?

When you are in a job interview, your prospective employer will have a number of questions for you to answer. The answers that you provide to these questions will give the employer a good understanding of your values and what you think is important. Figure 21–3 shows a list of example questions you may be asked. There will be questions

Typical Questions You Will Have in a Job Interview

- Tell me how your previous jobs or experience have prepared you for this position.
- Can you give me an example where you have used teamwork to get a task completed?
- Can you provide an example where you used customer service skills to help your company work out a problem with a customer?
- What motivates you?
- How do you get along with co-workers?
- Describe a difficult situation you have overcome.
- Why are you the best person for this job?
- What interests you about this job?
- What do you know about this company?
- What can you contribute to this company?
- Is there anything about this company you would like to know more about?
- What are your goals in the next 5 years?
- What are your goals in the next 10 years?
- What are your salary expectations now and in 5 years?
- What type of opportunities in this company would excite you?
- Could you tell me why you think job attendance is important for this job?
- Can you give me an example of how you have exhibited exceptional honesty or integrity?

Figure 21–3 Sample questions you may be asked during a job interview.

that will be directly about your level of knowledge and skills, which you can answer verbally or you can provide examples of pictures and diagrams of things that you've worked on. Other questions will try to determine what your values are on such subjects as work ethics, getting along with others, solving problems, and how you would treat customers. You should prepare some example answers to these types of questions.

The person asking these questions will not only be listening to your answers, but they will be looking for other clues that you will provide in your body language. For example, when you first enter the room and meet the person who will interview you, you should listen carefully to hear their name, look them directly in the eye, and shake their hand firmly. At that same time you should tell them your name and that you're happy to have been selected for the interview. When you are answering questions, you should answer each question clearly and concisely. If you need to take a minute to think about the answer, do so as quickly as possible and begin to respond without too much delay. If your answer is delayed for 10 to 20 seconds, the employer will become suspicious that you either are uncomfortable with the subject or you do not understand it at all. If the interviewer asks you a question to which you do not know the answer, be honest and say that you do not know the answer. Many times they will not count this against you; rather they will be impressed with your honesty.

You may want to participate in some practice interviews with your teacher or someone else who would ask you the types of questions you would find in an interview. It is important that you are as relaxed as possible during the interview, so that you represent yourself as best as you can. Finally, it is important to practice looking the interviewer directly in the eyes as much as possible, so that he or she can begin to make a connection with you and feel comfortable in discussing these issues. Remember, the interviewer is trying to find out if you will be a successful employee in just a few minutes of conversation with you, so you must be able to represent yourself in the best light during this conversation. The interview is your chance to sell yourself and convince the interviewer that you are the best person for this job.

Sometime during the interview you will also be able to ask questions about the job and the compensation package, which may include the amount of pay and if there are any benefits. It is important that you ask these questions in a dignified manner and do not dwell on the amount of pay or the types of benefits that you receive. If the employer thinks that this is more important to you than what you will bring to the job and what skills you will be able to perform when you're on the job, the employer will quickly suspect that you will not be a dedicated and a good employee. This may eliminate you from the selection process. For example, if you dwell on how many days off or vacation days that you get, rather than discuss what skills you can bring to the job, the employer is going to suspect that your dedication is going to be minimal. You must be able to convince the interviewer that you're willing to go beyond the minimum expected and produce more for the company than anticipated. You can quickly understand that the employer is going to compare all the individuals interviewing and is going to make a selection based on who will be the best for the company.

At the completion of the interview, you should stand up and shake the interviewer's hand, look him or her in the eye, and express your thanks for the job interview. You should also ask if there was anything else that you might provide that would help him or her make a decision, and remind the interviewer how to contact you. It is permissible to ask the interviewer when a decision will be made. In some small shops you may not be contacted if you're not selected. It's important to understand that you should not call the employer more times to ask if you've gotten the job. If you were not contacted, it is permissible to make one call after the date of the decision to see if a decision has been made yet. If the company indicates that it has made a decision and has

decided on a different person, be sure to thank the interviewer; do not ask why you were not selected. Remember, in most cities all of the employers meet frequently for different functions and will discuss different people that they have interviewed, and if you make a good impression they may refer you to another company even though they did not select you. You can be sure if you made a poor impression they will definitely discuss this with each other.

21.8 AFTER THE INTERVIEW

After the interview you should write a short letter or note thanking the interviewer for taking the time to interview you. You should send this note or letter to the company and address it to the interviewer within two or three days after your interview. Sometimes this will help employers make a difficult decision if they are trying to decide between you and another candidate. By sending a thank-you note, you are showing that you understand their time is valuable and that you value the process they used to make their selection. You should make the thank-you letter very short and to the point and do not include any additional questions. It is acceptable to add a sentence or two to state any information that you think you might have forgotten to provide during the interview, but remember the function of this letter is to thank them for their time.

21.9 IF YOU ARE SELECTED FOR THE JOB

If you're selected for the job, the employer will notify you and ask you when you could come for some final steps in the process. For example, you may need to pass a drug and alcohol test before you can start the job. You may also be asked to pass a physical and provide evidence that you have a valid driver's license. If you have had any problems that may keep you from passing these tests, you need to tell this information to the employer, so the company does not waste money finding it out through the test. It is very easy today to do background checks and determine the status of your driver's license and to determine if you have taken drugs in the past or have an alcohol problem.

The most important part of this section that you need to understand is that employers are willing to pay good wages and benefits to good workers who do not have problems with drugs or alcohol. Your employer must feel comfortable about these issues, as you will be sent on the road in a service vehicle and to customer's homes. The employer may have many years of dedicated service to their customers, and they do not want a new employee to create problems with their customer relationships.

21.10 WHAT AN EMPLOYER EXPECTS FROM THE PERFECT EMPLOYEE

- **Dependability.** Your employer expects you to be dependable and keep your promises. If you tell your boss you will do a specific job or be at work on a specific day, you are expected to be dependable.
- **Ability to follow instructions.** Your employer expects to be able to provide you with a set of instructions to install or troubleshoot, and you will be expected to be able to follow these instructions.
- **Ability to get along with peers.** Your employer expects you to get along with other employees and not cause or create any problems.

- **Ability to work on your own.** Your employer expects you to be able to work on your own. You should be able to receive a job and complete it on your own.
- **Self-confidence.** Your employer expects you to gain self-confidence and become confident enough to work on your own and begin to train others.
- **Pride in your work.** Your employer expects you to have pride in your work and try to improve the quality of your work.
- **Ability to be responsible for yourself and others working with you.** Your employer expects you to take on responsibility and become responsible for yourself and for others who are working with you.
- **Ability to be aware of what is happening around you.** Your employer expects you to learn the routines of work and understand what is going on and what is the next step.
- **Ability to get along with co-workers, supervisors, and your customers.** Your employer expects you to be able to get along with co-workers, supervisors, and customers.
- **Initiative.** Your employer expects you to be a self-starter; when you are given a job, you are expected to get started on it and work on it until you run into a problem.
- **Ethical standards.** Your employer will expect you to be honest in all your dealings with customers and anyone else you work with as an employee.
- **Ability to communicate orally and in writing.** Your employer expects you to be able to write work orders correctly and discuss problems with customers in a way that makes them feel comfortable.

21.11 HOW TO KEEP THE JOB

The next step in the process is to ensure that once you have been given an opportunity to start on a job that everything you do ensures that you are able to keep the job. The following are some of the things that employers will insist on if they intend to keep you long-term.

- **Don't miss work and be on time.** If you know you are going to be late, be sure to call and let the employer know.

 The employer has more than likely made commitments to a customer and expects you to be at work to fulfill those appointments. Customers expect you to be on time when you set an appointment. If you must miss work, be sure to give the employer as much advanced warning as possible so that the work you are scheduled to do can be completed by others or postponed to the customer's convenience.
- **Be interested when you're on the job.** Show that you want to learn as much as possible.
- **Ask questions about the jobs you are doing.** Gain as much information to do the job as well as possible.
- **Be willing to do the jobs no one else wants to do.** You will find there are many small tasks that need to be done, such as cleaning the service vehicle, stocking supplies, and cleaning the shop, that most employees do not like to do. If you step forward and offer to do these jobs, you will make a great impression on your boss. Remember in the end this is the type of person your boss will tend to give more responsibility to and be more willing to promote and give a raise.
- **Demonstrate ethical behavior.** This means be honest, do not try to cheat the customer, do not manipulate your timesheets, and show that you are trustworthy

overall. If you demonstrate any behavior that makes your boss suspect you are not honest and ethical, you will not be considered for a raise or for more responsibility.

- **Create a career plan and follow it.** A career plan may include further training or possibly finishing a degree so that you will be ready when an opportunity arises to be promoted or to take on more responsibility. If you do not have a career plan and have not continued to study and learn new things, you will not be ready when opportunity comes along to be promoted or take on more responsibility.
- **Be observant.**
 - Be able to identify the next step or next tool that is needed in a process.
 - Be able to anticipate what is happening next.
 - Be able to perceive or see the "big picture" in any job.

 One of the biggest complaints most employers have about new employees is that they do not engage and pay attention to what is going on in all parts of the job. For example, if you're not prepared for the job when you go to a customer's home, do not bring all the tools that you need, or need to make several trips and are disorganized, you will begin to make the wrong impression on your boss. Another example of not engaging is to stand around and watch while someone is doing a job or performing a task. At first it is just a small annoyance, but eventually your employer or co-workers will point out these problems to you and expect you to become engaged and catch on as quickly as possible. If they do not have to ask you to do these things, you will make a positive impression on them. Another trick is to ask what the steps to a job are, and where you can do things, even if you are not sure how to do all of the steps. If they have to remind you several times and ask you several times, they will quickly become annoyed and they will look for a reason to dismiss you.
- **Suggest new ways to do things to save time and money.** It is important to understand that there's a fine line with this one. Since you are just starting on the job you do not want to come across as a "know it all," so you need to be careful and phrase your suggestion in the form of a question rather than a declaration telling your boss they're doing it all wrong.
- **Be aware of time and do not bill time poorly.** One of the most important things for a new employee is to determine how the company accounts for time. You may be asked to fill out timesheets that bill your time to customers directly or to the shop. Be sure that you are using acceptable standards that have been established in the shop when filling out timesheets. If you're finding it difficult to account for all your time, this is a good point to ask for help from other employees or your boss to improve your accountability. In other words don't sit around and goof off and expect to be paid by either the customer or the employer. They want to be able to know that they're getting their money's worth when you're turning in timesheets.
- **Be aware of customer needs, clean work site, and be on time.** When you are working on the job at a customer's location, you must thoroughly respect the work site and ensure that it is clean when you leave. Customers also expect you to respect their time. This means that you need to work as quickly as possible on the job and also be on time for any appointments you make.
- **Act professional in everything you do and say.** Another way to impress your boss and the customers that you work for is to learn what are acceptable professional standards and then act in accordance. Your boss and the customers will be watching everything you do and will compare your behavior against what they think is an

acceptable professional standard. They will base their judgment of you on the way that you talk, the way that you dress, and the way that you complete the job.

- **Be an advocate for your boss and company.** Being an advocate for your boss and company means that you do not talk disrespectfully about them to customers or other employees. If you disagree with your boss or company policies, you need to address these privately with the persons concerned. One mistake that new employees make is that they talk to the customer and to other employees when they disagree with policies or decisions that their boss or company makes. Advocates only speak positively about their boss and their company to customers or other employees. Advocates go out of their way to find positive things about their company and their employer and relay these things in their discussions.

If you follow these basic steps and try hard to make improvements in relation to them, your employer will be happy to keep you on as an employee and provide raises and opportunities for you to take on more responsibility. If you violate one or more of these issues, your employer will move you to position where you can do as little damage as possible to the company or its customers and will eventually determine a way to dismiss you. If you are ever dismissed or fired from a job, you will need to have someone go over these issues with you in an honest manner and point out where you may not agree.

21.12 THINGS YOU CAN DO TO LOSE A JOB OR KEEP YOU FROM GETTING A JOB

Another way to think about your job is to be aware of things that your employer would consider serious enough for you not to be hired. For example, if you provide a transcript of your grades during the job interview, and you have a number of low grades, your employer is going to want an explanation for this. Poor grades are hard to explain away, so it is important that you work hard during school to get the best grades possible. If you do get a bad grade in a class, you should consider repeating that course and improving the grade so that the better grade shows up on your transcript. You can easily explain to the employer that you simply had difficulty the first time you took the course, but the second time you took the course you did much better.

Another question your prospective employer will have for you is your attendance record during school. If you're not willing to discuss your attendance record or do not want to show the prospective employer your attendance record because you had excessive absences or were late often, your employer will become suspicious. On the other hand, if you have good attendance, you want to ensure that the employer sees this.

21.13 GETTING THE BEST EDUCATION YOU CAN

Getting the best education that you can may consist of several steps. The first step is to get into classes that introduce you to the HVAC field in electrical and refrigeration technology. Make sure these courses include hands-on experiences that allow you to begin to build the skills that you will need on the job to install and troubleshoot HVAC systems. You can discuss the types of courses that you may need with your

boss or employer as they may be able to help you understand what areas you need to know more about. They may also have a **promotion** in mind if you increase the amount of knowledge you have. Your employer may also be able to suggest some courses that the equipment manufacturers provide to help you learn more about the product line. For example, if you know more about all the furnace and air-conditioning systems a company makes, you would be better at selling new systems to customers. You would also become a better troubleshooter if you know as much as possible about all the products they make, including the units with the newest technology in them. You should also take as many math courses as you can, as well as English reading and writing courses. If you have a chance, take a speech course, so that you become comfortable speaking in public or on the telephone to customers. Some speech courses also show you how to make PowerPoint presentations, which are useful in sales presentations.

21.14 TYPES OF BENEFITS OFFERED BY YOUR EMPLOYER

Some jobs that you might land may have excellent benefits. The better the job, the more types of benefits you will be offered. Benefits may be anything from partially paid health insurance or paid vacation days to dental insurance and paid sick days. Figure 21–4 shows some examples of benefits that the best jobs may offer. Some of the benefits may be offered immediately, whereas other benefits may be delayed for 90 to 120 days after you start to work. The best jobs offer the best benefits, so many more people will be applying for these types of jobs, making the competition for those jobs tough. If you are the best candidate for the job, you will be offered this type of job.

Some benefits, such as a vehicle that you can take home, will save you thousands of dollars on purchasing gasoline and having a second car. You may also be offered some type of retirement plan that will ensure your future. Some employers will also pay tuition for your continued education. These types of benefits are worth thousands of dollars.

- Health insurance
- Dental insurance
- 401K (with partial match)
- Flexible spending account
- Paid vacation
- Paid personal days
- Paid sick days
- Paid holidays
- Short- and Long-term disability
- Tool allowance
- Company truck (take home)
- Company-supplied uniforms
- Bonuses

Figure 21–4 Examples of benefits from a great job.

21.15 REVIEW

The process of getting your first job starts with getting the education or skills needed for the job. Part of this process is to create a résumé and a cover letter. You can then begin to search for jobs. Your job search may take you to the local newspapers, the Internet, and web sites specifically set up for HVAC jobs. If your résumé is outstanding, you may receive an invitation to a job interview where you will be able to represent yourself to the prospective employer. If you get a job offer, you will be expected to exhibit the types of behavior that your employer expects in order to keep the job. You should start your job search approximately 6 months prior to when you expect to be on the job.

Job Assignment for This Chapter

Your job for this chapter is to create a personal résumé and cover letter and prepare a portfolio of your work that includes some of the diagrams you have worked with and other examples of your work for a future employer. You should also prepare your personal answers for the questions you would get at a typical interview for a job.

Solution to the Job Assignment

The solution to the job assignment should include a résumé and cover letter that you have created for yourself. The résumé should include your name, address, phone numbers, and e-mail address if you have one, along with information about your goals, your skills, your educational experience, your previous job experience, and anything else that is significant about you that would make you stand out from other candidates who are applying for this job. Your cover letter should be a brief letter introducing yourself, explaining why you want the job, and showing how you would be the best employee. The questions you should have experience answering for a job interview include why you are interested in this job, what skills you bring to this job, what experience you bring to this job, and why you are the best person that could be hired for this job.

Assessment

QUESTIONS

1. What is a résumé?
2. What is a cover letter?
3. What is a job interview?
4. What type of questions should you be expected to answer in a job interview?
5. List five things that you can do on the job to ensure that you will keep your job.

TRUE OR FALSE

1. You do not necessarily need a résumé if you just go ask somebody for a job.
2. One of the most important things you can stress about a prior job that was not in the air-conditioning field, such as work in a fast-food restaurant, is your attendance record.

3. A self-starter is someone who can be given a minimal amount of information about a job and be able to start the job without too much supervision.

4. In the job interview the first thing you should discuss is the benefits, the pay, and how many days off you're going to get.

5. You should get an invitation to a job interview for every single résumé that you send out.

MULTIPLE CHOICE

1. Your résumé should contain
 a. your name, address, and phone number.
 b. information about your skills.
 c. information about your previous jobs or work experience.
 d. information about your education.
 e. all of the above.
 f. only b, c, and d.

2. Your cover letter should contain
 a. your goals.
 b. an introduction of yourself.
 c. a short statement of why you want this job and why you think you are the best candidate for the job.
 d. a complete list of all the courses that you have taken.
 e. all the above.
 f. only a, b, and c.

3. If you receive an invitation to a job interview, you should
 a. check out the company and gather as much information about it as you can.
 b. prepare a list of likely questions that you'll be asked and practice answering those questions.
 c. determine the location of your job interview so you can tell how long it will take you to get there so that you will not be late to the interview.
 d. prepare a portfolio of information showing your skills or the work that you've done in school.
 e. all the above.
 f. only a and b.

4. Which of the following may keep you from getting a job offer after an interview?
 a. Having poor grades or a low grade-point average.
 b. Making a poor first impression and dressing sloppily at the interview.
 c. Having a poor attendance record at your previous job.
 d. All the above.
 e. Only a and c.

5. Which of the following may get you dismissed or fired from your job?
 a. Being late to the job.
 b. Calling off work too often.
 c. Not being reliable.
 d. Not getting along with other workers or not respecting your boss.
 e. Not taking care of your appearance.
 f. All the above.
 g. Only a, b, and c.

PROBLEMS

1. Create a résumé for yourself and provide it to several people to let them review it and evaluate it.

2. Create a cover letter for yourself and provide it to several people to let them review it and evaluate it.

3. Create a list of typical questions you expect to be asked at a job interview and provide answers for each question. Review these questions and answers with your instructor or others in your class.

4. Create a list of things that you should do when you're on the job that will make you successful. Discuss your list with others in your class or your instructor.

5. Create a list of things that you should not do when you are on the job. Discuss your list with others in your class or your instructor.

Glossary

Across-the-Line Starter A motor starter that allows full voltage to be applied to a motor when the contacts of the starter are closed.

Actuator A device that provides linear or rotary motion such as a pneumatic cylinder that extends to open or close a set of dampers. A rotary actuator rotates a shaft in one complete rotation, which causes a set of dampers to open or close. The motion of an actuator can also be used to control the opening and closing of a valve.

A/D (analog to digital) Converter An electronic circuit that converts analog values such as 0–10 volts DC to digital values such as 0–4095.

Air-Conditioning System An air-conditioning system consists of a condensing section that is located outdoors and has a compressor and condenser coil. The indoor part of the system consists of a furnace or fan unit that moves air over an evaporator coil. The evaporator coil is mounted in duct work, and the air moving over it cools the conditioned space. If the unit is a package system or window unit the evaporator coil and condenser coil are in the same case.

Airflow Switch A switch that is mounted in the duct work in an air-conditioning or heating system. The switch is activated by air flow and usually has a piece of metal (a paddle) that is mounted on the switch actuator and is placed in the flow of air. When air strikes this paddle, it moves the switch actuator causing the switch to change state. When air flow is not present, a spring returns the switch to its normal condition.

Alternating Current (AC) Current that changes from a positive level to a negative level periodically. Its waveform is a sine wave. Example: A 60-Hz alternating current changes from positive to negative 60 times per second and has a frequency of 60 Hz.

Alternator An electro-mechanical device that produces AC voltage when its shaft is turned.

Ambient Temperature The temperature of the medium (air, water, earth) that surrounds an air-conditioning system.

American Wire Gauge (AWG) A standard used for identifying the size of electrical conductors. Gauge numbers have an inverse relationship to size; larger numbers have a smaller cross-sectional area.

Ammeter An instrument that measures electrical current in amperes.

Ampere The unit of measure for current. Current is the flow of electrons in a circuit and in Ohm's law calculations, one amp is equal to one volt divided by one ohm of resistance. One amp is also equal to 6.25×10^{18} (625,000,000,000,000,000,000) electrons flowing past a point in one second.

Analog Meter A meter (usually with a meter movement) that can display the value being measured in a continuous manner by moving a meter needle across a numbered display. The needle is moved across the face of the meter indicating the value being measured. The analog meter is available to measure volts, resistance, current, or power.

Analog Signals Data or values represented by continuously variable in electricity, such as voltage or current. (Example of a variable voltage is a dimmer switch controlling a lamp.)

Anode The positive electrode in an electrolytic cell, toward which negatively charged particles are attracted. The anode has a positive charge because it is connected to the positively charged end of an external power supply. The positive terminal of a diode.

ANSI (American National Standards Institute) An organization in the United States that develops standards for industry.

Anti Short Cycle Timer (ASCT) A timer that is specifically designed to keep a compressor from cycling on and off too quickly. After a compressor turns off, the timer prevents it from coming back on too quickly and allows the system pressure to drop to a level that will not cause the motor to draw excessive current. When a compressor cycles too quickly, the pressure in the system is too high and will cause the compressor motor to draw excessive current and possibly blow a fuse.

Apparent Power Power caused by multiplying voltage times current when the current is caused by inductive reactance, capacitive reactance, and resistance. The amount of apparent power will appear to be larger than true power because the true power has a phase shift caused by capacitance or inductance. True power, by comparison, calculates only current caused by pure resistance. *Note*: Apparent

power is the same value as true power when a circuit contains only resistance.

Armature The moving part of a magnetic circuit, such as the rotating part of a motor or a generator or the movable iron part of a relay.

ASCII (American Standard Code for Information Interchange) A standard for assigning numerical values (0–7) to the set of letters in the Roman alphabet and typographic characters used in computers.

Atom A unit of matter, the smallest unit of an element that consists of a nucleus that has a positively charged proton and neutrally charged neutron. The atom also has one or more negatively charged electrons that moves about the nucleus in an orbit.

Auxiliary Contacts Additional sets of contacts on a relay or motor starter that are not used for switching the main power. The rating of auxiliary contacts is smaller than the contacts used for switching main power. Auxiliary contacts are used for seal-in circuits.

Back EMF (also called counter EMF, or CEMF) Voltage produced by the spinning of a motor's armature. The motor receives voltage in its windings to make the armature (rotor) spin, and when the armature begins to turn, its magnetic field passes the coils of the motor windings and produces a voltage that is 180° out of phase with the applied voltage.

Backup Heat The heating source that is added to a heat pump and is used to provide heat when the heat pump cannot provide enough heat or when it goes into defrost mode.

Ball Bearing A ring-shaped track containing hard metal balls that roll freely, used to reduce friction where a rotating element (such as an axle) is attached to a fixed point.

Base The center terminal of an NPN or PNP transistor, or other electronic device.

Bell-Crank Armature Assembly A relay that has a coil mounted above its armature and a set of movable contacts and stationary contacts that are mounted in the horizontal position. When the coil becomes energized, it pulls the armature upward. The armature is connected to the movable contacts with an arm. When the armature is pulled upward, the motion causes the movable contacts to move forward to make contact with the stationary set of contacts. This motion causes the contacts to move to their closed position.

Bimetal Element The bimetal element is a heat-sensing element that is made of two dissimilar metals used in overloads and thermostats. The two dissimilar metals are designed so that when current flows through them, heat is produced. When the heat increases it causes one end of the metal to move when the other end is mounted in a fixed position. The movement of the element is sufficient to cause a set of contacts to change position.

Bimetal Overload An electrical overload device that is composed of two dissimilar metals. The overload consists of two parts. One part converts electrical current into heat, and the second part of the overload, which is the bimetal strip, senses the increase in temperature and moves. The motion of the bimetal strip activates a set of normally closed contacts and causes them to open. The motor winding is usually connected in series with the bimetal overload, and current to the winding will be interrupted when the motor current becomes excessive.

Binary-Coded Decimal (BCD) Numbering System A numbering system that uses binary numbers "1" and "0" and places them into four weighted locations: 8, 4, 2, and 1. The BCD numbering system represents decimal values 0–9 with the four binary numbers. Additional sets of four binary values can be used to represent additional decimal values in the tens, hundreds, or thousands columns. The BCD numbering system is used because humans can easily convert the binary values to decimal values 0–9.

Binary Numbering System A numbering system consisting of "1" and "0" primarily used in computers and programmable logic controllers (PLCs). Also called a base-2 number system.

Boiler A heating system where water or other fluid is heated to a warm temperature or to a point where the fluid changes to steam. The heated fluid or steam is circulated through piping and radiators that are located in spaces where heat is needed.

Bushing A part of a motor that provides a bearing surface between the moving shaft and the end plates. The bushing is made of a thin bronze tube or sleeve and requires a thin layer of lubrication to complete the low friction surface between the two parts.

CAD Cell A cadmium sulfide cell (CdS) that changes its resistance in response to the change in light that strikes its surface. This device is primarily used in oil furnaces to detect the presence of a flame.

Cam Stat A brand name of a temperature-sensitive switch that is used to turn on the furnace fan when the temperature increases in the plenum; it may also provide a second set of electrical normally closed contacts that are wired in series with the gas valve and will open if the temperature in the furnace gets too high, which indicates a major problem in the furnace. Cam Stat is the name of the company that makes this type of switch.

Capacitive Reactance Opposition in an AC circuit caused by a capacitor. The amount of capacitive reactance increases as the size of the capacitor decreases, or when the frequency of the AC voltage to the circuit decreases. Capacitive reactance is denoted by X_c and is measured in ohms.

Capacitor An electrical component that is made of two conducting plates that are separated by an insulator. When a capacitor is put into an AC circuit, voltage will cause the capacitor to charge. During charging, electrons will flow to one plate until the plate has the same potential as the supply voltage. Since an insulator separates the two plates, current cannot flow directly to the other plate; instead, they must wait until the AC voltage changes polarity during its second half-cycle. The electrons will flow from the first plate that was charged, and they will flow back through the circuit and charge the second plate of the capacitor.

Capacitor-Start, Capacitor-Run (CSCR) Compressor A single-phase compressor motor that has a run winding and a start winding. A run capacitor is permanently connected in series with the start winding. When the motor starts, the run capacitor provides phase shift that creates torque in the start winding to help the motor start. After the motor is running, the run capacitor stays in the circuit and provides additional torque to the motor any time the rpm drops below full speed.

Capacitor-Start, Capacitor-Run (CSCR) Motor A single-phase motor that has a run winding and a start winding. A run capacitor is permanently connected in series with the start winding. When the motor starts, the run capacitor provides phase shift that creates torque in the start winding to help the motor start. After the motor is running, the run capacitor stays in the circuit and provides additional torque to the motor any time the rpm drops below full speed.

Capacitor-Start, Induction-Run (CSIR) Motor A single-phase electric motor that has a start winding and run winding. The start winding is disconnected from the run winding and de-energized after the motor comes up to speed. When the motor stops, the start winding is connected to the run winding again for the next time the motor is started. Most CSIR motors use a centrifugal switch to energize and de-energize the start winding.

Cardiopulmonary Resuscitation (CPR) A first-aid technique used to keep a person whose heart has stopped alive while waiting for more qualified emergency responders to arrive. This technique involves chest compressions to ensure that blood is pumped through the body of a person whose heart has stopped and provides air through the lungs until a qualified emergency medical technician arrives. This first-aid technique should be used by a first-aid responder who has been trained in the most recent techniques.

Cathode The negative terminal of a diode or other electronic device.

Centrifugal Switch A switch that is connected in series with the start winding of a single-phase AC motor and is mounted in its end plate. The purpose of this switch is to open its contacts at approximately 75–80% of full rpm to disconnect the start winding of the motor from the source of voltage. The switch will close when the motor is de-energized and slows to a stop.

Changing Rotation The process of changing the direction of rotation of a motor's shaft to ensure that a fan or pump is rotating in the correct direction. In a single-phase motor, this is accomplished by changing the polarity (switching the leads) of the start winding. In a three-phase motor, this is accomplished by interchanging any two of the three windings with each other.

Circuit Breaker An electrical device specifically designed to protect against overcurrent. The circuit breaker is similar to a fuse in that it protects, but it is different in that it can be reset.

Clamp-on Ammeter An ammeter that has a set of "jaws" that open to allow them to wrap around a wire. The jaws close around the wire and function as a transformer coil to sense the amount of current flowing through the wire. The clamp-on ammeter is a noncontact meter that can quickly clamp around a wire to measure the current flow.

Clapper Armature Assembly The clapper relay uses a yoke that is pulled to the magnetic coil when the coil is energized. The yoke moves a pushrod and this physical movement of the armature of the clapper forces the pushrod and movable contacts upward.

Closed Circuit A circuit that is complete in that it has a power source, a number of conductors, and at least one load.

Closed Loop An HVAC system that consists of a sensor such as a thermostat, a heating device such as a water coil, and a control valve that controls the amount of water flow through the coil, which in turn controls the temperature of the conditioned space (office). A closed loop occurs when the signal from the thermostat is used to control the water valve and cause the temperature in the conditioned space to remain close to the setpoint (as opposed to an open loop where the setting of the valve is manually controlled).

Closed Switch A switch that has its contacts closed and the resistance between the contacts is low, which allows current flow through it.

Coil A part of a relay or solenoid that is made by tightly winding a long piece of wire into loops. When current flows through the wire, it creates a strong magnetic field in the coil that can be used to open and close contacts in a relay, or change the position of a solenoid valve from open to closed or vice versa. Coils of wire are also used in motor windings and transformers.

Collector One of the three terminals of a transistor. The collector attracts the electrons that are given off by the emitter.

Common Point See *Common terminal "C."*

Common Terminal "C" The point in single-phase motor windings where the start winding and run winding are connected.

Compressor An electric motor that directly drives a piston or scroll that is used to pump refrigerant in an HVACR system. In a hermetically sealed compressor, the electric motor is sealed in the same chamber as the piston pump, and refrigerant is used to cool this motor.

Compressor Contactor A relay that has contacts that can carry current in excess of 15 amps. The contactor has a coil that controls a number of normally open contacts that are wired in series with the compressor motor windings.

Compressor Shell The metal container that houses the compressor motor and pump. The shell is sealed so that it does not leak refrigerant. A small amount of oil flows to the bottom of the shell where it is used by the compressor to lubricate its parts.

Compressor Terminals The electrical terminals that are mounted in the side of the compressor shell. The terminals are connected to the run and start winding of the compressor motor for a single-phase motor and are identified as "R" for run, "S" for start, and "C" for common. In a three-phase compressor the terminals are identified as "T1," "T2," and "T3."

Condenser The outdoor unit for an air-conditioning system. The compressor and condenser coil and compressor fan are mounted in the condenser unit. The condenser coil allows the refrigerant vapor to cool down to a point that it turns back to a liquid form.

Condenser Fan The fan on a condenser unit that moves air over the condenser coils.

Condenser Fan Motor The motor that turns the condenser fan.

Conductor A wire that is usually made of copper or aluminum that carries electrical current.

Contactor One or more sets of contacts that are opened or closed by a magnetic coil. The contactor is similar to a relay but normally is larger. By definition, the contactor contacts are rated for more than 15 A.

Continuity When a set of contacts, a fuse, or a wire has low resistance. When switch contacts have continuity, they are considered to be in the closed position. When a wire or a fuse has continuity, it is considered to be good.

Continuity Test A test that measures the resistance of components such as switch contacts, fuses, or wires. A component is said to have continuity when it has low resistance.

Control A device in an electrical circuit that interrupts the flow of current, such as a switch or contacts.

Control Circuit The part of an electrical diagram that contains the switches that control when the loads (motors or electrical resistance coils) turn on or off. In an HVAC system the control circuit usually contains the temperature, pressure, or flow switches.

Control Loop In an HVAC system or a direct digital control (DDC) a control loop consists of a control device (such as a water valve), a sensor (such as a temperature sensor) that provides a feedback signal, and a controller that stores a setpoint (the desired temperature) and calculates the error (the difference between the setpoint and the feedback signal). The controller also produces the amount of output signal to the control valve depending on the amount of error that is present.

Control Transformer A step-down transformer that provides 24 volts AC secondary voltage from 120, 208, or 240 primary voltage AC. A control transformer can also be a step-down transformer that provides 120 volts AC secondary voltage from 208 or 240 primary volts AC.

Controller An electronic system that provides control of an HVAC system. The controller (such as a programmable logic controller, PLC) receives input signals from switches and other controls and sends output signals to relays, motor starters, and solenoids to provide control of air-conditioning systems.

Controls Devices such as switches that control the flow of current in a circuit. When a control is in the closed position, current flows, and when it is in the open position, current does not flow.

Cooling Cycle Referring to a heat pump, the cooling cycle is when the system is running in the air-conditioning mode and liquid refrigerant is pumped to the indoor coil, which provides cool air to the indoor space.

Cooling Thermostat The part of the thermostat (temperature control switch) that controls the cooling cycle of the HVAC system. The cooling thermostat controls current between the R terminal and the Y terminal on the thermostat.

Counter EMF (CEMF) The induced voltage in an AC or DC motor that opposes the applied voltage.

Cover Letter A letter that is written to a prospective employer that accompanies a résumé. The cover letter is used to introduce the person to the prospective employer and explain why he or she is applying for the job and what makes the person someone the employer should invite to a job interview.

Crankcase Heater An electric resistance heater that may be mounted in a cavity under the lowest part of the compressor shell called the crankcase, or it may be strapped around the base of the compressor shell so that its heat causes the oil in the compressor to warm up to a point that it is warmer than the indoor coil. The compressor oil is heated so that it does not migrate to the indoor coil, which is generally warmer than 65 degrees F. If oil migrates out

of the compressor, the compressor pump parts will be damaged from lack of lubrication.

Current The flow of electrons that is measured in amperes.

Current Relay A special relay that has a single set of normally closed contacts and is used to control the start winding of a sealed compressor motor. The coil of the current relay is different from other relays in that it has extremely large wire that can withstand large currents. The coil of the current relay is wired in series with the run winding of the compressor motor, and its contacts are wired in series with the start winding.

Cutaway Diagram A diagram of an electrical or mechanical component that shows the interior parts of the component as if someone used a saw to cut away part of the outside of the part so the interior parts can be viewed to show their location and relationship to each other.

Cut-in Pressure The pressure in a refrigeration or HVAC system that is set to cause the contacts of the pressure switch to close. The cut-in pressure is the lower setting on the pressure switch.

Cut-out Pressure The pressure in a refrigeration or HVAC system that is set to cause the contacts of a pressure switch to open. The cut-out pressure is the higher setting on the pressure switch.

D/A (Digital to Analog) Converter An electronic circuit that is mounted on an IC chip (integrated circuit). The circuit receives digital numbers (in binary format) and converts the value to the equivalent voltage (0–10 volts) or current value (4–20 mA).

Damper Motor Actuator A pneumatic motor or electric motor that is designed specifically for low rpm speed and usually makes only one revolution of its shaft. The motor is connected directly to actuate a set of dampers, and the rotation of the motor shaft turns the dampers from full open to full closed with one revolution of the motor shaft. Some damper motors provide multiple shaft rotations (1–10 revolutions) to cause full movement of the damper.

Data Plate A plate attached to a motor, transformer, or other electrical components that specifies all the electrical information (data) for the component.

Decimal Number system where the values are 0–9.

De-energized The act of turning an electrical component or device off. When no electrical power is applied to a component.

Defrost Control An electrical component that controls one or more sets of contacts by time, temperature, or pressure to ensure that a defrost cycle occurs. This control is used in refrigeration and heat pump systems to ensure that frost buildup on the evaporator coil is kept to a minimum. The contacts may be used to shut down the compressor or energize an electrical heater or a solenoid valve that directs warm refrigerant through the evaporator coil causing the frost to melt. The control may be activated by a timer motor, or when temperature or pressure becomes too low indicating the presence of frost.

Defrost Timers See also *Defrost Control*. A timer specifically designed to energize a defrost cycle and terminate it after a specified time.

Delta-Connected Transformer Transformer windings that are connected in series (in the shape of the Greek letter delta). This connection provides 480 or 240 volts AC.

Delta Connection A three-phase connection where windings are connected in series with the power applied. The windings are connected in the shape of a triangle or the Greek letter delta (Δ).

Delta Voltage A voltage in which the lower voltage is always half of the higher voltage. Three-phase (Δ) voltage has 480 V for the high voltage and exactly half (240) for the lower voltage.

Delta-Wired Motor A three-phase motor connection where windings are connected in series with each other. The windings are connected in the shape of a triangle or the Greek letter delta (Δ) and voltage is applied to the points where each winding is connected to the next.

Diac A two-terminal electronic device that is specifically designed to fire in both the positive and negative directions when its applied voltage reaches a predetermined amount.

Digital An electronic format consisting of two states, on and off.

Digital Meter An electric meter that displays the measured values as numbers. Digital meters are available to measure voltage, current, resistance, power, and frequency.

Diode A two-terminal, solid-state semiconductor that allows current flow in one direction.

Direct Current (DC) Current that flows in only one direction. Batteries are one power source of direct current.

Direct Digital Controller (DDC) A control system or controller that uses digital electronics to control an HVAC system. The DDC can control on and off devices as well as analog devices. Some DDC systems provide loop control.

Direct-Spark Ignition An electronic ignition control that creates a series of electrical pulses (sparks) across an open gap between two metal terminals. The electrical sparks are intense enough to ignite natural gas or LP gas in the pilot light. The timing of the electrical pulses are controlled by an oscillator circuit and occurs several hundred times a

minute to ensure that the gas ignites. When the pilot flame is detected, the electrical circuit that creates the pulses is turned off.

Disconnect (Switch) An electrical switching device that is specifically designed to disconnect electrical power to a circuit or system such as an air-conditioning system or furnace. The disconnect is also designed to be "locked out" to ensure that power is turned off when a technician needs to work on a system with the power off.

Discrete Electronic components that are packaged as individual components rather than together with other components on an integrated circuit. An example would be individual diodes rather than four diodes packaged together on a rectifier chip. A discreate signal is an on/off signal.

Door Heater A special heating strip that is built into the door gasket that is mounted on the door of a refrigerator or freezer. The heat ensures that the door seal stays warm enough that moisture or frost does not form on the door gasket.

Double-Pole, Double-Throw (DPDT) Contacts A switch consisting of two sets of contacts, where each set of contacts has a normally open and normally closed contact connected to a common "C" point. Both sets of contacts are actuated together. The contacts can be in a switch or relay.

Double-Pole, Single-Throw (DPST) Contacts A switch consisting of two sets of normally open contacts. Both sets of contacts are actuated together. The contacts can be in a switch or relay.

Dual-Element Fuse A fuse with two elements. One element is designed to detect short-circuit currents (currents that are five to ten times the normal current). The second element in the fuse is called the overload element. The overload is designed to open when the fuse detects a small amount of overcurrent over a 3- to 5-min period. The dual-element fuse is designed to allow the compressor motor to start and have excessive current during starting for 2–3 s and then continue to provide overload protection in case the motor has an overcurrent that lasts for more than 3–5 min.

Dual Pressure Switch A pressure switch that has two pressure chambers that can sense two separate pressures, such as the oil pressure and refrigerant pressure. The switch also has two separate sets of electrical contacts.

Duty Cycle The time that a motor is energized (running) versus the amount of time the motor is stopped over a given hour. If the duty cycle is listed as continuous, it can be operated continuously without stopping it.

Electric Furnace A furnace that uses electrical resistance coils as its main source of heat. The electrical coils produce large amounts of heat when current flows through them.

Electric Heating Coils Coils of high-resistance wire that produce a large amount of heat when current flows through them. The coils can be placed in a furnace with air blown over them.

Electric Sequencer Relay A special set of relays that have timers built into them and are wired together to create a sequence that ensures one set of heater coils are energized at a time, until the second stage of electrical heating is required.

Electrical Control The switches in a schematic or wiring diagram that provide the sequence to turn loads on or off.

Electrical Load The devices in an electrical system that consume energy and convert it to motion, heat, or light. Typical loads in an HVAC system include indicator lamps, fan motors, compressor motors, electrical heating coils, and solenoid, relay, and motor starter coils.

Electrical Potential Voltage that provides electrical force in a circuit.

Electrical Power The energy in an electrical circuit that is the result of voltage and current. The units of electrical power are watts.

Electrical Shock A condition that occurs when electrical current comes into contact with humans. The effects of electrical shock range from just being noticeable through higher currents that can cause damage to skin tissue and cause your heart rhythm to be interrupted.

Electricity The flow of electrons.

Electromagnetic Induction Occurs when current flows through the primary winding of a transformer and creates a magnetic field. When AC voltage changes polarity naturally during its sine wave, the flux lines from the magnetic field in the primary winding collapse and move (cut across) the wires in the secondary winding. When the flux lines move across the wires in the secondary coil, a current is created that is 180° out of phase with the current supplied to the primary winding. Induced current occurs in the secondary winding, even though it is not physically connected to the primary winding in any way.

Electromotive Force (EMF) Another term for voltage or potential difference.

Electron The negative part of an atom. Electrons are located in orbits (shells) and move around the nucleus (center) of the atom.

Electronic Defrost Control A defrost control that has an electronic circuit that provides the timer function rather than an electrical mechanical timer motor. (See also *Defrost Control* for description of defrost control.)

Electronic Refrigeration Control A refrigeration control that provides the time functions through an electronic type timer instead of an electromechanical timer motor. (See

also *Refrigeration Control* for description of refrigeration control.)

Emitter One of three terminals in a transistor. The emitter provides a source of electrons for the device.

Enclosure The metal box or housing in which equipment is mounted, available in designs for various environmental conditions.

End Plates The ends of a motor that contain the bearings. The end plates are held in place by bolts that go completely through the motor from one end plate to the other.

End Switch See *Centrifugal Switch.*

Energized When a component is connected to a source of electrical potential (voltage).

Error The difference between the setpoint (the desired temperature) and the process variable (the actual temperature). The error in a control loop is used to determine the amount of output signal (0–100%).

Evaporator The indoor coil in a standard air-conditioning system. Liquid refrigerant flows into the evaporator where it changes from a liquid to a vapor as it absorbs heat.

Evaporator Fan The fan that blows air over the evaporator coil. The fan also doubles as the furnace fan when the HVAC system is in the heating mode.

Fan The device driven by an electric motor that moves air across the condenser coils, over evaporator coils, or through the furnace heat exchanger.

Fan and Limit Control A temperature control that has two sets of contacts. The fan switch turns the furnace fan on and off, and the limit control is connected in series with the gas valve and provides overtemperature control for the furnace.

Fan and Limit Switch See *Fan and Limit Control.*

Fan Control A switch that is mounted in a furnace that turns on the furnace fan when the temperature rises to the predetermined setpoint. When the source of heat in the furnace turns off, the furnace plenum cools down below the setpoint and the fan control will turn off the furnace fan motor.

Fan Relay The relay that consists of a coil and one or more sets of contacts in the HVAC system that controls the furnace fan. The coil of the fan relay is connected to the "G" terminal on the thermostat, and the contacts of the fan relay are connected in series with the fan motor.

Fan Switch The switch on the thermostat that has two positions, one (on) is used to turn the furnace fan on constantly, and the other position (auto) connects the fan relay to the temperature activation part of the thermostat so that the fan cycles on and off with the air-conditioning compressor. The fan switch is connected to terminal "G" of the thermostat.

Fixed-Pressure Switch A pressure switch that has a fixed setpoint.

Flame Rod A sensor that detects the presence of a pilot flame in the gas heating system.

Flame Rollout Overtemperature Switch A normally closed, temperature-sensitive switch that is precisely located to sense abnormal flame that occurs outside of the normal position where it is drawn up the chimney. If there is a blockage, the flame will "rollout" near the burners and cause a problem where it can burn wiring or other parts of the furnace where the flame does not belong. The flame rollout switch should not be reset without a technician determining the cause of the flame occurring where it should not be.

Floating-Point Number A number that includes decimal values and/or exponential values. Floating-point numbers allow computers and electronic controllers to use very small decimal numbers or very large exponential numbers.

Flux Lines Magnetic lines that come from the ends (poles) of a permanent magnet or electromagnet. They move from the north pole to the south pole of a magnet.

Four-Lead Motor A single-phase motor with its start winding providing two leads, and its run winding providing two leads.

Frame Size The physical dimension and size of a motor, usually expressed as a NEMA design code number. This allows interchangeability between motors made by different manufacturers.

Frame Type A standardized system for motor sizes that ensures that motor mounting systems are standardized across all motor manufacturers. The frame size standardization ensures that motors with the same frame size number will be a direct replacement.

Freezer Case A display case used for frozen products in grocery stores or other stores. The display case can maintain temperatures below 32 degrees F.

Frequency The number of periodic cycles per unit of time.

Full-Load Amperage (FLA) The amount of current (amperage) a motor draws when it is running at full load. This is also the rating used to determine the proper fuse, circuit breaker, or motor starter for a motor.

Full-Wave Bridge Rectifier A rectifier that changes both the positive and the negative half-cycle of an AC sine wave to pulsing DC. This rectifier provides more power than a half-wave rectifier.

Fuse A device designed as a one-time protection against overcurrent or short-circuit current. The fusible link melts

and causes an open circuit when its current rating is exceeded.

Fuse Disconnect A special switch that is designed to provide a disconnect for an air-conditioning system as well as provide a mounting for fuses. When the disconnect is in the open position, the fuses will be isolated from the source of power, so they can be removed and replaced or tested.

Gas Valve The valve in a gas furnace that is controlled by a solenoid. When the solenoid is energized, the gas valve opens and allows full flow of gas to the burner.

Goals A list of things you would like to achieve in respect to your career.

Good Fuse A fuse that allows current to flow through it and has very low resistance.

Ground The earth, or any point in an electrical circuit that has the same potential as earth. Ground is also a verb meaning that a wire or conductor is connected to a rod that is driven into the earth approximately 8 feet deep.

Ground Fault Interrupter (GFI) A protection device similar to a circuit breaker. It senses small amounts of ground currents and opens the circuit. The GFI must be manually reset.

Grounding Adapter A special plug adapter that allows a three-prong male plug to plug into a two-prong receptacle. You need to be aware of safety when you plug a three-prong male plug into the two-prong receptacle; you may not have a safe ground, and you must ensure that a ground is provided.

Half-Wave Bridge Rectifier A diode rectifier that converts only one-half of an AC sine wave into DC voltage.

Hard-Start Kit A device consisting of a capacitor and solid-state control that is used to help start a single-phase compressor motor. The electronic device controls when the capacitor is switched into and out of the circuit to the compressor run winding. The capacitor provides extra phase shift to the start winding, which helps provide additional torque to the motor for starting. The hard-start kit is used to help compressors that are having trouble starting.

Heat Anticipator A variable resistor found in a heating thermostat. The variable resistor that is used as a heat anticipator is connected in series with the gas valve, and it is set to match the amount of current that is drawn through it. When current flows through the resistor, it provides heat to the thermostatic element and causes it to shut off the furnace before it reaches setpoint so that it does not overheat the room.

Heat Exchanger The part of a furnace generally made of thin metal that has hot combustion gases flowing through it and indoor air flowing over it. The heat exchanger allows the heat from the combustion gas to warm the indoor air and keeps the combustion gas separate from the indoor air, so that carbon monoxide and other harmful gases do not get into the house. Gas furnaces and oil furnaces both have heat exchangers that must be checked periodically for cracks or leaks that would allow harmful gas to get into the home.

Heat Pump An HVAC system that has a special reversing valve that allows the system to provide cooling to the indoor space in the summer and reverse the refrigerant flow to provide heating to the indoor space in the winter. The indoor coil acts as an evaporator in the cooling cycle during the summer, and it acts as a condenser in the heating cycle during the heating season.

Heater A device that can use an electrical resistance coil or a combustion chamber to provide heat to a conditioned space.

Heating Cycle The part of an HVAC where heat is added to the indoor space.

Heating Thermostat A thermostat that controls the heating equipment, which can be electrical resistance heat, gas heat, oil heat, hot water heat, steam heat, heat pump, or other type of heating system. The typical terminals on the heating thermostat are "R" and "W." The heating thermostat can be a stand-alone device or part of a heat/cool thermostat.

Hermetic Compressor A compressor used in an HVAC system that is completely sealed. The electric motor and compressor pump are both sealed in a metal container. The electrical terminal connections for the motor are provided through the metal container by means of a group of electrical terminals that are sealed in the side of the container to provide electrical isolation and a complete seal for the refrigerant part of the system.

Hertz The electrical units for frequency.

Hexadecimal Numbering System A base 16 numbering system that consists of numerals 0–15. In electronic and PLC control systems the numbers are designed to take up only one column. This is accomplished by representing values 0–9 with traditional numbers and two digit values 10–15 with letters of the alphabet where 10 = A, 11 = B, 12 = C, 13 = D, 14 = E, and 15 = F.

High-Efficiency Furnace A gas, oil, or electrical furnace that is designed to have efficiencies at 90% or above. The efficiency rating for furnaces is called Annual Fuel Utilization Efficiency (AFUE).

High-Leg Delta Voltage 208 V provided between L2 and N from a center-tapped delta-wired transformer. The voltage between L1 and N and L3 and N on this transformer is 120 V.

High-Pressure Switch A pressure switch that has normally closed contacts that are opened when pressure increases above a predetermined setting.

Hold-in Current The amount of current flowing through a relay coil that causes a set of relay contacts to remain energized. When voltage is applied to a relay coil, current begins to flow through the coil. The amount of current that causes the contacts to begin to move and change position is called pull-in current. Once the contacts have changed position (from open to closed, or from closed to open) the amount of current required to keep the contacts energized (hold-in current) is approximately 15–20% less than the current that caused the contacts to pull in.

Horizontal-Action Armature A relay that has its stationary contacts mounted horizontally, as are its movable contacts that are mounted on an armature. In this design, the movement of the contacts on the armature is in a sideways direction, rather than upward. The sideways movement needs less energy to move the contacts than if they were pulled upwards.

Horsepower (hp) A unit of power. 1 hp = 33,000 ft-lb/min or 746 Watt.

Hot Surface Igniter A special igniter that is made of ceramic material that glows red hot when current is applied to it. When current is flowing through the igniter, its surface reaches temperatures high enough to cause the gas flowing over it to ignite.

Hot Wire Relay A special motor starting relay that was used extensively to start single-phase hermetic compressors before the advent of electronic controls. The hot wire relay consists of a heating element and contacts that change position from normally closed to open when current flows through the heating element. The heating element is connected so that it senses the amount of current flowing to the motor winding, and the normally closed contacts are connected in series with the start winding. When the motor starts, the heat builds up in the relay within a few seconds and disconnects voltage from the start winding as the motor runs at full speed. When the compressor is turned off, the relay cools down and the contacts return to their normally closed position where they are ready when the motor is started again.

Humidistat A control that senses the amount of humidity in the air. The control is used in residential and commercial HVAC systems to automatically energize and de-energize humidifier equipment at specified setpoints.

HVAC Electrical Symbols Electrical symbols that represent fan motors and compressor motors, electrical heating coils, and a wide variety of controls including pressure switches, flow switches, temperature switches, and level switches.

Ice Maker A system that has a specific job of creating cube or flake ice for residential or commercial consumption. The ice maker has a timed cycle that water is provided to an evaporator plate where it freezes. At the end of the cycle, the ice is harvested into a bin where it is used.

Impedance Total opposition in an AC circuit caused by any combination of resistance, inductive reactance, and capacitive reactance.

Induction Creating current in a coil of wire by passing it through a magnetic field.

Induction Motor An AC motor that operates by passing current through its wire windings (coils). When current flows through the wire coils, it creates magnetic fields. The magnetic fields are near the rotating armature, which causes a magnetic field to be created in it. The relationship to the north and south poles in the motor winding and the armature causes them to attract and repel, which causes the armature to rotate and produces torque at the armature shaft.

Inductive Reactance Opposition in an AC circuit caused by a coil. When inductive reactance increases in a circuit, current decreases.

Infinite (Infinity) The largest value of resistance an ohmmeter can indicate. Infinite resistance is theoretically too large a value to measure. Infinite resistance indicates an open circuit in a motor winding, switch, or fuse.

Input Module A module in a programmable logic controller (PLC) where input switches are connected. Each switch is connected to an individual electronic circuit in the input module, and the input module sends a signal to the PLC to indicate if the switch is energized or de-energized.

Inrush Current Large current caused when voltage is first applied to a motor or other inductive load. The inrush current will return to normal full-load current several seconds after the motor starts and comes up to full rpm.

Insulator A material that does not conduct electrical current easily. These materials have very high resistance. In electronics, an insulator material has 5, 6, or 7 electrons in its valence shell.

Integer A whole number that includes zero and can be positive or negative values.

Intrinsic Standoff Ratio The ratio of the triggering voltage to the applied voltage in a UJT.

I/P (Current Over Pressure) Converter Current over pressure transducer. The input of this sensor is electrical current, usually 4–20 mA. The transducer converts this current proportionally to air pressure, usually 3–15 PSI.

Job Interview A meeting between an employer and a person who is seeking a job. This meeting occurs at the invitation of the employer and is specifically designed for the person seeking the job to sell himself or herself to the employer and explain all of his or her skills and

experiences. The meeting is a time where the employer can ask questions of the person applying for the job to ensure that their skills, abilities, and personality will fit with the position that is open.

Ladder Diagram (also called a schematic diagram) An industry standard for representing relay control logic. The name comes from the fact that the overall form of the diagram looks like a ladder.

Lattice Structure An electronic structure inside an atom where electrons are shared, thereby creating a very strong structure.

Legend A part of a drawing or diagram that explains special symbols and conventions used in the drawing.

Light-Emitting Diode (LED) A two-lead solid-state PN junction device that produces a small amount of light when forward biased.

Limit Switch An electrical switch that is activated by motion. The switch senses linear or rotational motion, which causes its contacts to change state at some specified location.

Line-Voltage Thermostat A thermostat whose contacts are rated for 120 V AC or 240 V AC. The contacts for a low-voltage thermostat are rated for 24 V AC.

Load An electrical device that uses current and converts it to heat, light, or motion, in devices such as a resistance heating element, magnetic coil, motor, solenoid, or indicator lamp.

Load Circuit The part of an air-conditioning or heating circuit that has large loads, such as the compressor, in it. The control circuit has the thermostat and other controls in it.

Locked-Rotor Amperage (LRA) The amount of current a motor winding draws when power is initially applied and the rotor has not begun to rotate. This is the same amount of current the winding would draw if the rotor stopped rotating.

Lockout/Tagout A safety procedure that ensures that the source of supply voltage for a machine is disconnected and locked in the off position so that personnel can safely work on the equipment. The tag identifies who is working on the equipment.

Low-Pressure Switch A pressure switch that has normally closed contacts that are opened when pressure decreases below a predetermined setting.

M (mega) A prefix used with units of measurement to designate 1,000,000 (1 million).

Magnet Any metal that can have its dipoles aligned, or any coil that produces flux lines when current flows through it.

Magnetic Field A field created by flux lines that emanate from the poles of a permanent magnet or from flux lines of an electromagnet. The strength of the field from an electromagnet increases as the amount of current flowing through the coil increases until the magnet reaches saturation.

Megger A megohm meter that produces high voltage and low current. The main purpose of this meter is to measure the resistance of insulation in coils of motors and transformers to determine the breakdown of resistance, which may indicate problems due to the deterioration of insulation or the presence of moisture in windings.

Meter A device that can measure voltage, current, resistance, or wattage. A digital meter presents the value being measured as a digital number, and an analog meter has a meter movement that causes a needle to move as the amount of current being measured increases or decreases.

Meter Movement The internal part of an analog meter that measures current and causes a needle to move across the face of the meter. When the meter is used to measure voltage, the voltage flows through resistance where it is converted to current and the current flows through a coil where it creates a magnetic field. The stronger the current flow, the stronger the magnetic field in the meter movement becomes and causes the needle to deflect more.

Micro A prefix used with units of measure to designate 1 millionth, or 0.000001.

Microprocessor A programmable digital electronic component that is mounted on an integrated circuit (IC). The microprocessor has a central processing unit (CPU) and provides the function of a computer in handheld devices, instruments, and embedded systems such as programmable logic controllers (PLCs).

Milli A prefix used with units of measure to designate 1 thousandth, or 0.001.

Milliammeter An ammeter that measures current down to 1 thousandth or 0.001 of an ampere.

Motor An electromechanical device that converts electrical current to mechanical (rotational) motion at its shaft. The motor has electrical windings that are made from coils of wire, and a movable armature. The windings have current flowing through them to create magnetic fields, and the polarity of the magnetic fields cause the armature to rotate.

Motor Starter A large relay that has a single coil and multiple sets of contacts and a set of overload heaters and contacts that are used to protect the motor against overcurrent. The current that is supplied to the motor flows through the contacts of the motor starter and also flows through the heater element of the overload. When the amount of current becomes too large for the motor to handle safely, the heat that is produced by the heating elements is sufficient to cause the overload contacts to "trip" and open. The contacts

of the overload are connected in series with the motor starter coil, and when the overloads trip and open the contacts, the current supplied to the coil is interrupted so that the motor starter drops out its main contacts. When the main contacts open, the motor stops running.

Multispeed Motor A motor that has more than one speed.

National Electrical Code® (NEC®) A set of regulations governing the construction and installation of electrical wiring and apparatus. The NEC is established by the National Fire Protection Association.

NEMA Design A rating standard that NEMA (National Electrical Manufacturers Association) uses to identify the different amounts of starting current, toque, speeds, and other variables that are available from motors with the same horsepower rating. Letters A, B, C, and D are used to identify the NEMA design values.

NEMA Standards National Electrical Manufacturers Association A group that provides standards for electrical equipment approved in the United States.

Networking A means of connecting two or more computers together so that information and data can be transmitted between them.

Neutral The second line of a single-phase 110-volt electrical system. In some systems the neutral wire is connected to the ground terminal so that they have the same potential.

Neutron The neutral part of the atom. The neutron is located with the proton as part of the nucleus of the atom.

Nine-Lead Motor A three-phase AC motor that has nine leads available for connection that allows the motor to be connected as a delta or wye motor, and its windings can be connected for high (480 V AC) or low voltage (240 V AC).

Nine-Lead Three-Phase Motor The terminals of this motor are numbered 1–9. When the motor is wired for high voltage (480 V AC), T1 is connected to L1, T2 is connected to L2, and T3 is connected to L3. T7 is connected to T4, T8 is connected to T5, and T9 is connected to T6. When the motor is connected for low voltage (240 V AC), T1 is connected to T7, T2 is connected to T8, and T3 is connected to T9. T4, T5, and T6 are connected together.

Normally Closed A condition with switch contacts or relay contacts where they have low resistance. A normal condition is considered when no power is applied to the circuit.

Normally Closed Contacts A set of contacts on a relay or switch that are closed when the switch or relay is in the de-energized state.

Normally Open A condition with switch contacts or relay contacts where they have high resistance. A normal condition is considered when no power is applied to the circuit.

Normally Open Contacts A set of contacts on a relay or switch that are open when the switch or relay is in the de-energized state.

NPN Material Material used to make a transistor emitter, base, and collector where the emitter and collector are made from N-type material and the base is made of P-type material.

N-Type Material Semiconductor material that has a majority of carriers that are electrons and have a negative charge.

Nucleus The center of an atom that consists of protons and neutrons.

Octal Numbering System A base 8 numbering system that consists of values 0–7. The octal numbering system was used with early computer systems that had 8-bit functions.

Ohmmeter A meter that is designed to measure resistance and indicate the value in ohms. A high ohms reading is indicated as infinity.

Ohm's Law A set of mathematical calculations that show the relationship between voltage, resistance, and current. Voltage is equal to the amount of resistance multiplied by the amount of current.

Oil Burner A furnace that burns oil as its primary fuel. The furnace consists of a burner section and a heat exchanger. The burner pumps fuel oil through a small nozzle that causes it to atomize, which makes it easier to burn.

Oil Burner Control A control system that is specifically designed to protect an oil furnace. The control has sensors to determine if the fuel oil has been ignited and it also checks against flame out. If the oil does not ignite during the ignition process, the control cycles the system off so that excessive oil does accumulate in the bottom of the burner.

Oil Pressure Switch A switch that is used in conjunction with an HVAC or refrigeration compressor to ensure that the proper amount of oil remains in the compressor and does not get pumped away from the compressor. The switch checks the amount of oil pressure and gives the compressor 15–20 seconds to build up oil pressure when it is first started.

Op Amp An operational amplifier (op amp) is an electronic circuit that is mounted on an integrated circuit. The main function of this circuit is to provide amplification of an input signal. The op amp is used in a variety of electronic applications where it is combined with a sensor to increase the small signal that the sensor detects and makes it large enough to be usable.

Open Circuit A circuit that has a point in it with extremely high resistance, which stops the flow of electrical current.

The open can be created intentionally with a switch or contacts, or it can be created for safety reasons by a fuse or circuit breaker. Other fault conditions such as broken wires or broken components can also cause the high resistance open.

Open Fuse A fuse that is bad and has extremely high resistance and will not pass current. The open fuse is caused by a fault that causes high current that creates excessive heat through the fuse and causes its link to open.

Open Loop An HVAC control circuit that is running in manual mode. For example, if the loop is controlling water flow to a fan coil, a valve is manually set or adjusted for the proper amount of flow to create the temperature that is desired. If a different temperature is desired, the flow can be adjusted by manually adjusting the flow control valve. In closed-loop mode, the valve is controlled by comparing the feedback signal to the setpoint.

Open Switch A switch that has high resistance between its contacts. When a switch is in the open condition, no current flows through it.

Open Type Motor A motor that is not sealed like a hermetic compressor. An open motor is used for fan, pump, and other general-purpose applications in HVAC systems.

Operational Amplifier See *Op Amp*.

Output Module A module on a programmable logic controller (PLC) or a direct digital control (DDC) system that is designed to take a low voltage signal from the controller and convert it to a full voltage signal such as 12 V DC or 24 V DC or 120 V AC or 240 V AC for use in the system. The full voltage signal can be used to turn on solenoids, motor starter coils, or other small loads.

Overcurrent When current (amps) exceeds the normal rating of a fuse, switch, wire, or load.

Overload A part of a motor starter that consists of a heater element and a set of normally closed contacts. The heater element is connected in series with the motor coils and the contacts are mounted near the heater element. When current flowing to the motor is within acceptable range, the heat that is given off from the heating element is not sufficient to cause the contacts to open. If the current flowing to the motor becomes excessive, it causes the heating element to give off excessive heat, which causes the normally closed contacts to open. The motor starter coil is connected in series with the overload contacts and will become de-energized when the contacts open, which protects the motor against damage from overcurrent.

Packaged Air-Conditioning System An HVAC system that contains the evaporator fan, condenser fan, and compressor in one unit. This unit is specifically designed to be used as a rooftop system or a self-contained system that does not require any refrigerant piping between separate evaporator and condenser.

Parallel Circuit A circuit that has two or more loads connected so that each load receives the same voltage. If two or more switches are connected in parallel, one switch can be opened and current can continue to flow through the other switches if they are closed.

Peak-to-Peak Voltage AC voltage that is measured between the peak of the positive half cycle and the peak of the negative half cycle. This value can also be found by multiplying the peak voltage by 2 if the AC voltage is balanced.

Peak Voltage The amount of AC voltage in one half of a sine wave that is measured from the zero point to the peak in the positive or the negative direction. Peak voltage is also calculated by dividing the peak-to-peak voltage by 2.

Period The time it takes for one cycle of an AC sine wave to occur.

Permanent Split-Capacitor (PSC) Compressor A compressor motor that has a run capacitor connected in series with the start winding. The run capacitor is not disconnected from the start winding, so it stays permanently in the circuit. The run capacitor causes a phase shift to occur in the start winding to help start the compressor.

Permanent Split-Capacitor (PSC) Motor A motor that has a run capacitor connected in series with the start winding. The run capacitor is not disconnected from the start winding, so it stays permanently in the circuit. The run capacitor causes a phase shift to occur in the start winding to help start the motor.

Phase A three-phase electrical power supply provides power on L1, L2, and L3, and a single-phase power supply provides power between L1 and N.

P/I (Pressure Over Current) Converter A device that senses pressure in an HVAC control system and converts it to electrical current. The typical air pressure is 3–15 pounds per square inch and the current is 4–20 mA.

Pilot The small flame for a gas furnace that is used to ignite the main burner flame. A sensor is used to ensure that the pilot flame is burning before the main gas valve allows gas to flow to the main burner.

PN Junction One segment of P-type material that is connected to one segment of N-type material to make a single junction. This is also the basic part of a diode.

Pneumatic Sensor A sensor that actuates from air pressure and can measure level, pressure, temperature, or flow.

PNP Material Material used to make a transistor emitter, base, and collector where the emitter and collector are made from P-type material and the base is made of N-type material.

Poles The ends of magnets or magnetic fields. The poles have polarity, which means they are positive or negative.

The poles can also be parts of the windings of a motor that have a positive or negative magnetic field.

Portfolio A group of examples of your work that shows or demonstrates your knowledge and experience. This is an important part of putting together examples of your lab experience and other school experience that you can show a prospective employer in preparation for your first job.

Potential Relay A special starting relay used to energize and de-energize the start winding on a compressor. The contacts of a potential relay are normally closed and provide a connection for the start capacitor to the start winding. The coil of the potential relay is connected between the start winding terminal and the common terminal on the compressor. When the compressor comes up to speed after it is started, the back EMF from the compressor winding will provide enough voltage (potential) to energize the coil and cause the contacts to open and stop voltage and current flow to the start capacitor and start winding.

Power Work done per unit of time. Power is measured in horsepower or watts: 1 hp = 33,000 ft-lb/min or 746 W.

Power Circuit The part of an HVAC circuit that has the evaporator fan motor, condenser fan motor, and compressor motor connected to it.

Power Factor A measurement of the time-phase difference between the voltage and current in an AC circuit caused when the circuit has a capacitor or inductor connected to it. The power factor is the ratio of real power (in watts) to apparent power (in volt-amperes).

Power Supply A device that converts available power to a form that a system can use; usually converts AC power to DC power.

Pressure Control A sensor that is attached to a switch that can detect the increase or decrease in refrigerant or air pressure. The pressure control can also be used to control the temperature of a refrigeration system since refrigerant pressure can be used to determine temperature at a coil.

Pressure Switch A switch that is activated at a specific pressure. Examples of pressure switches in air-conditioning and refrigeration systems include oil pressure, low refrigerant pressure, and high refrigerant pressure switches.

Primary Control The part of the oil burner system that provides a high voltage spark that is used to ignite the oil during the ignition process. The primary control also provides protection in case the oil does not ignite in the first 30 seconds the burner is turned on, or if the flame burns out anytime during the time the furnace is running.

Primary Winding One of two transformer windings where voltage is taken into the transformer. Supply voltage is connected to the primary transformer winding.

Process Variable The signal that comes from the sensor that provides feedback to the HVAC system. If the control system is controlling temperature, the process variable would be the temperature that is measured by the temperature sensor.

Programmable Logic Controller (PLC) A microprocessor controller that has input modules, output modules, and a controller. The controller has the sequential program saved in its memory and is executed continually as it checks switches that are connected to its input modules, and energizes or de-energizes output devices such as solenoids or motor starter coils that are connected to its output modules.

Programmable Thermostat A thermostat that senses the room temperature and controls the HVAC system according to a schedule established by the homeowner. The thermostat contains an electronic chip that accepts a schedule to turn on or turn off heating and cooling functions.

Promotion A condition on the job where you get an increase in pay and take on additional responsibilities.

Proportional Control An automatic control circuit that uses only gain to calculate the output for the system.

Proportional, Integral, and Derivative (PID) An automatic control system uses gain (proportional), reset (integral), and rate (derivative) to calculate the percentage signal sent to the output. PID adjusts how fast the output of the system is changed when a change is detected at the setpoint or the process variable.

Protectorelay The main part of the primary control for the oil burner. Protectorelay is a brand name control from Honeywell. (See also *Primary Control*.)

Proton The positive part of an atom. The proton is located with the neutron in the nucleus of the atom.

Proximity Switch An electronic switch that can detect position similar to a limit switch. The proximity switch uses an inductive field at the tip of the switch to detect the presence of a metal target. A capacitive proximity switch uses a capacitive field to detect nonferrous targets.

PSC Motor See *Permanent Split-capacitor Motor*.

P-Type Material Semiconductor material that has a majority of carriers that are holes and have a positive charge.

Pull-in Current The amount of current used to pull in a relay coil. After the relay contacts are pulled in, the amount of current reduces since it takes less current to keep the contacts in the closed position after their initial movement.

Push-Button Switch A switch that is activated by a person depressing the switch.

Reach-in Cooler A specialized refrigeration case that allows frozen or refrigerated products to be displayed in a

way that consumers can view and select products while the products stay at the refrigerated temperature. The cooler may have clear glass doors or if the cooler is a case type, the products are placed below the refrigeration line in the cooler.

Reactance The opposition in an AC circuit caused by a capacitor or an inductor (coil). If voltage to a circuit is constant and reactance increases, current will decrease.

Rectifier A device (usually a diode) that conducts current in only one direction, thereby transforming alternating current (AC) to direct current (DC).

Refrigeration Control A control that providre the time function through an electromechanical timer used for defrost and other timing functions.

Relay An electrical device that consists of a single coil and one or more sets of normally open or normally closed contacts.

Relay Coil The part of a relay that is made of a coil of wire through which current flows and creates a magnetic field. The magnetic field causes the contacts to change position.

Relay Contacts The part of the relay that has the electrical contacts (either normally open or normally closed). The contacts change position when the coil is energized.

Resistance The opposition to current flow. The units of resistance are ohms.

Resistor A device that provides opposition in an electrical circuit. The effect of a resistor in a circuit is to cause heat to occur. The heating elements in an electric furnace are resistors.

Résumé A printed document that is created by an individual that outlines the person's education, job history, experience, and skills. The résumé is sent to a number of prospective employers when a person is looking for a job.

Reversing Valve A special solenoid that is used in a heat pump that reverses the refrigeration flow from heating mode to cooling mode. The reversing valve routes the liquid refrigerant to the indoor coil during the cooling cycle, and it routes the liquid refrigerant to the outdoor coil during the heating cycle.

RMS Voltage Root mean square is also called RMS voltage or effective voltage. The RMS voltage can be calculated by multiplying the peak voltage by 0.707.

Root-mean-square (RMS) The effective value of an alternating current corresponding to the DC value that produces the same heating effect. The RMS value is computed as the square root of the average of the squares of the voltage.

Rotor The rotating part of a motor or generator, also called the armature.

Run Capacitor A capacitor that is mounted in a metal container to help dissipate heat. The run capacitor is connected in series with the start winding of a PSC motor and remains in the circuit at all times.

Run Terminal "R" The terminal connected to the run winding or main winding of the motor or compressor.

Run Winding The main winding of a single-phase motor or compressor that has less resistance than the start winding. The run winding remains in the circuit at all times with the motor, while the start winding is de-energized after the motor reaches near full speed.

Scan Cycle The part of a programmable logic controller (PLC) that checks all inputs, executes the logic program, and then writes an "on" or "off" signal to the outputs in the output module. The PLC executes its scan cycle every 30–40 milliseconds.

Schematic (Ladder) Diagram A diagram that is designed to show the sequence of operation of an electrical control system.

Seasonal Energy Efficiency Ratio (SEER) An efficiency rating for air-conditioning equipment. The rating is calculated by taking the Btu rating of the cooling output for a typical cooling season and dividing that number by the total watt-hours consumed by the unit. The higher the SEER, the more efficient the unit is.

Second-Stage Heating The second level of heating for a furnace. In a heat pump system the second stage of heating is an electrical heating coil. In a gas burning furnace, the second stage may allow additional burners or more gas to be used in the ignition process. In an electrical furnace, the second stage of heating is an additional electrical heating coil.

Secondary Winding The transformer has two windings, a primary winding and a secondary winding. Voltage is applied to the primary winding, and voltage comes out the secondary winding. The amount of voltage at the secondary winding will be proportional to the number of turns in the primary and secondary windings.

Selector Switch An electrical switch that has an operator that allows the contacts to be closed or opened when the switch is activated. The contacts in the selector switch will remain in the position they are switched to until the switch is changed to a new position.

Semiconductor A material used in electronic components. Semiconductor material has four valence electrons, which makes it electrically neutral. Extra electrons can be added to make the material N-type material, or extra holes can be added to make the material P-type material.

Series Circuit A circuit that has only one path. Anytime there is an open anywhere in the series circuit, all current flow stops.

Series-Parallel Circuit A circuit that has components connected both in series and in parallel.

Service Factor (SF) When used on a motor nameplate, a number that indicates how much above the nameplate rating a motor can be loaded without causing serious degradation.

Setpoint The desired value of an operational control system.

Shaded-Pole Motor A single-phase AC induction motor that has a single winding. Each pole of the winding has a shading coil that provides the phase shift needed to allow the motor to start. The shaded-pole motor may be stalled without damage and has a very low amount of torque.

Shading Coil A special coil in a shaded-pole motor that causes enough phase shift to create the torque needed to start the motor.

Short Circuit A circuit where resistance in the circuit is near zero. The low resistance causes extremely large, uncontrolled current that can damage the components in the electrical circuit or cause harm to any humans nearby. It is very important to provide fuses or circuit breakers to protect against short circuits.

Short-Cycle Protection A control device for HVAC and refrigeration systems that provides a time delay that is connected in series with the compressor contactor coil. The time delay is set long enough to allow the high-pressure refrigerant at the compressor and condenser to drop sufficiently so that the compressor can start without high head pressure. If the compressor tries to start when the refrigeration pressure is too high, it will draw excessive current and blow its fuses or circuit breakers.

Silicon Material with 4 valence electrons that is used extensively in electronic components such as diodes and transistors.

Silicon Controlled Rectifier (SCR) A solid-state unidirectional latching switch that has three terminals: anode, cathode, and gate. The SCR can convert AC current to DC current and vary the amount that flows through its anode-cathode circuit by controlling the firing angle of the gate.

Sine Wave An outline of the voltage wave form for AC voltage. The sine wave looks like the letter "S" that is placed on its side. The sine wave shows the positive voltage and then the negative voltage of each cycle of the AC voltage.

Single-Phase Compressor A compressor motor that runs on single-phase voltage. The compressor motor has a run winding and start winding that are connected at a point called common "C."

Single-Phase Voltage Voltage that is provided on two lines. For 120 volts AC, the lines are identified as L1 and N. For 240 volts AC, the lines are identified as L1 and L2.

Single-Pole, Double-Throw (SPDT) Contacts A set of contacts that have a common point and can be switched to either of two positions.

Single-Pole, Single-Throw (SPST) Contacts A set of contacts that have one input and one output.

Single-Pole Switch See *Single-pole, Single-throw contacts.*

Six-Lead Motor A three-phase motor that has six leads brought out of its case for connections. The six-lead motor can be connected for low voltage 208/240 V AC or for high voltage 480 V AC.

Slip The difference between rotating magnetic-field speed (synchronous speed) and the rotor (armature) speed of an AC induction motor. Usually expressed as a percentage of synchronous speed.

Soft Start (Motor Starting) An electronic motor-starting circuit where less than full voltage is applied to the motor during starting.

Solenoid A coil of wire that can control a valve or other device that requires linear motion. The coil of a solenoid becomes a very strong magnet when current is supplied to it.

Solid-State Integrated Circuit An electronic circuit that is mounted on a substrate (chip). The integrated circuit has a number of pins that allows it to be mounted in a socket or soldered directly to an electronic printed circuit board.

Solid-State Relay (SSR) A relay that does not have any mechanical parts or contacts. The SSR has a light-emitting diode (LED) that acts like the coil of a relay, and the LED sends a light signal to the base of a phototransistor, which goes into conduction and passes current through its emitter and collector much like the relay contacts. The SSR provides electrical isolation between the input circuit and the output circuit.

Solid-State Starting Relay A relay that is designed to use solid-state components to provide starting current to the start winding of a single-phase motor much like a current relay.

Speed The rate of rotation of the shaft of an electrical motor that is measured in revolutions per minute (rpm).

Split-Phase Compressor A compressor with a single-phase motor that is controlled with a current relay or other starting device. The single-phase motor has a start winding and run winding.

Split-Phase Hermetic Compressor A compressor with a single-phase motor that is controlled with a current relay. A hermetic compressor is one that is sealed in a metal container.

Split-Phase Motor A single-phase motor that has a start and run winding.

Split-System Air Conditioner An HVAC system that has the evaporator in an indoor system, such as a furnace, and the compressor and condenser in the outdoor unit. Refrigeration

tubing connects the condenser coil, compressor, and evaporator coil.

Star Configuration In an arrangement of parallel (bus) connections, a physical configuration such that each device is connected on the bus at the same junction of conductor segments.

Star Connection The arrangement of phase windings in a polyphase circuit in which one end of each phase winding is connected to a common junction. In a three-phase circuit, it is sometimes called a wye connection.

Start Capacitor A capacitor that is connected in the start winding of a compressor or other single-phase motor. The start capacitor is mounted in a plastic case and is only in the circuit for a few seconds.

Start Terminal "S" The lead that is connected to the start winding of a single-phase compressor motor.

Start Winding One of two windings in a split-phase motor. The winding in a split-phase motor that has the highest resistance.

Static Electricity The electrical charge that is caused by the imbalance of positive and negative charges. Static electricity can be created by rubbing two nonconductive materials together and then drawing them apart slightly.

Stator The stationary winding of a motor.

Step-down Transformer A transformer whose voltage in its secondary is lower than the voltage in its primary winding.

Step-up Transformer A transformer whose voltage in its secondary is higher than the voltage in its primary winding.

Summing Junction A point in an automated process loop where the setpoint and process variable signal are compared.

Switch An electrical device with one or more sets of contacts.

Symbols Graphical representation of electrical components.

Synchronous Speed The speed of an AC induction motor's rotating magnetic field. It is determined by the frequency applied to the stator and the number of magnetic poles present in each phase of the stator windings.

Temperature Control A controller that allows you to connect thermocouples or RTDs as sensors and controls heating relays as outputs.

Temperature Indicator A temperature gauge or electronic display that shows the temperature of a system.

Temperature Rise The amount of temperature increase in a motor that is above the ambient temperature. The value is a rating provided for motors so that you can see how much excess temperature a motor can withstand.

Thermal Fuse A specially designed fuse link that is placed in an electrical heating element that senses overtemperature. If the temperature exceeds the setting of the fuse link, it will open and stop current from flowing through it.

Thermal Overload A device used to protect motors that senses current flowing to the motor. When the current flows through the overload, it produces heat. When the current is within acceptable range, the heat is not large enough to cause it to trip. When current increases to an unsafe level, the heat produced by the current causes the overload to trip.

Thermocouple A temperature sensor made of two dissimilar metals. When heat is applied to the tip of the thermocouple, it generates a small amount of DC millivoltage (1/1000 of a volt).

Thermostat A temperature-controlled switch that controls the furnace, air conditioner, and fan motors. The thermostat can be heating only, cooling only, or heat/cool.

Thermostat Subbase A base where the thermostat is mounted.

Three-Phase Open Motor An open motor that has three separate windings. The motor is open in that it has free air that moves over the windings and armature. Three-phase motors are used to drive fan motors and pumps.

Three-Phase Voltage Voltage that is supplied as three separate lines that are identified as L1, L2, and L3. Each phase is 120 degrees out of phase with the other.

Timed Defrost Defrost cycle for a refrigeration system or heat pump that is brought on by a timer control. The defrost cycle turns off with a timer also.

Timer Control A special control that integrates an electronic or motor-driven timer with multiple sets of contacts. The timer allows the contacts to be controlled in an on–off pattern that is programmed for specific defrost or other functions.

Torque A turning force applied to a shaft tending to cause rotation. Torque is equal to the force applied times the radius through which it acts.

Transformer An electrical device with two coils (primary winding and secondary winding) that are located in close proximity. Voltage is applied to the primary winding and voltage is taken off the secondary winding.

Transistor A three-terminal electronic device that has an emitter, collector, and base.

Transmitter An electronic device that allows you to convert the millivoltage signal from the thermocouple to a 0–10-V DC or 4–20-mA signal.

Triac A solid-state bidirectional latching switch that has three terminals: main terminal 1, main terminal 2, and gate. The triac provides full-wave control of AC power.

Troubleshooting A process of finding faults in an electrical or HVAC system. The troubleshooting process works best when you are able to identify everything in the system that is working correctly, and then only check the remaining part of the circuit that is not working correctly.

True Power Power in an AC circuit that is the result of current flowing through a resistive device. True power is measured in watts and does not have any current from capacitive or inductive components.

Turns Ratio The ratio of the number of turns in the primary winding as compared to the number of turns in the secondary winding.

Unijunction Transistor (UJT) A special transistor that has higher impedance than a PNP or NPN transistor. The UJT has three terminals: base 1, base 2, and emitter.

VA Rating for Transformer The power rating of a transformer that is calculated by multiplying the voltage rating of the transformer by its amperage rating.

Valence Electron An electron or electrons in the outermost shell or orbit of an electron.

Valve A plumbing component that can allow or stop the flow of liquid or vapor.

Valve Actuator A manual, electric, pneumatic, or hydraulic device that makes a valve move to a desired position in response to a signal. The motion of the actuator can be linear or rotary.

Variable-Frequency Drive (VFD) An electronic system that connects to a single-phase or three-phase AC motor. The VFD plugs into AC voltage and is converted to pure DC and then back to AC voltage. When the voltage is converted back to AC from the DC voltage, its frequency can be adjusted, which allows the speed to be controlled.

Variable-Speed Drive A controller that changes the amount of field voltage for a DC motor or the frequency in an AC motor to cause a change in its speed.

Vertical-Action Armature Assembly A relay that has its contacts mounted above its coil. When the relay coil is energized, the armature is pulled upward and causes the contacts to change from open to closed.

Voltage The electromotive force (EMF) or pressure in an electrical circuit that causes electrons (current) to flow.

Voltage Drop The voltage that occurs when current flows through a resistor or load in an electrical circuit. In a series circuit, all of the voltage drops across each resistor can be added and their total will equal the supply voltage to the circuit.

Voltage Drop Method of Troubleshooting A troubleshooting procedure that provides means of locating an open circuit in an electrical circuit by applying voltage to the circuit and making a number of voltage measurements. One terminal of the meter must be placed on the return line of the power supply and left there for the duration of the tests. The test procedure then specifies that voltage is checked at each terminal point in the circuit in sequence starting at the power supply. The point where voltage is lost indicates the point where the open circuit exists.

Volt-Ampere (VA) Rating A rating for a transformer found by multiplying the voltage times the current (amperes). This rating does not take into consideration any impedance or reactance losses, so it is an apparent power rating.

Voltmeter A electrical meter that is designed to measure voltage. The voltmeter can be analog (with a needle) or digital with a numerical readout display.

VOM Meter A meter that has the ability to measure AC and DC volts, resistance (ohms), and AC and DC milliamps (current). VOM stands for volts, ohms, and milliamps. This type of meter is also called a multimeter.

Walk-in Cooler A larger refrigeration space that is fitted with a door that allows users to walk in and add or remove products, and then the door is closed making the refrigeration space sealed for better cooling.

Watts The units of electrical power.

Wild Leg Voltage (High-Leg Voltage) 208 V measured between L2 and N in a delta-wired transformer. Typically the voltage between L1 and L3 is 240 V. The voltage between L1 and N or L3 and N is 120 V, or half the line-to-line voltage. If the three-phase delta-wired transformer has a neutral, the voltage between L2 and N will be 208 V and is unusable as a power supply voltage.

Window Air Conditioner A small air-conditioning system that has the evaporator, condenser, and compressor all in one cabinet that fits into a window. The evaporator is positioned so that it provides cool air to the inside, and the condenser and compressor are mounted in the part of the unit that sits outside the window.

Wiring Diagram An electrical diagram that shows the location of all the electrical components in the circuit and the wires that connect to them.

Wye-Connected Transformer A three phase transformer that has one end of each of its windings connected at a common point. The overall shape of the windings look like the letter "Y." The voltage from the three phase wye-connected transformer is 480 line to line, and 277 volts line to neutral.

Wye Voltage In a three-phase transformer, 480 V when it is wired for high voltage and 208 V when it is wired for low voltage. If the three-phase transformer has a neutral, the voltage between each line and neutral for the

208 system will be 120 V. If the line-to-line voltage is 480 V, the line-to-neutral voltage will be 277 V.

Wye-Wired Motor A three-phase motor that has one end of each of its windings connected at a common point. The overall shape of the windings look like the letter "Y." When the three-phase motor is connected in this way, it draws less current during starting than the delta-connected three-phase motor.

Zener Diode A diode that conducts current when it receives reverse voltage that is in excess of its zener voltage value. A zener diode controls, or regulates, the voltage at a constant level for a circuit.

Index